Lockhart and Wiseman's Crop Husbandry Including Grassland

Woodhead Publishing Series in Food Science, Technology and Nutrition

Lockhart and Wiseman's Crop Husbandry Including Grassland

Tenth Edition

Edited by

Alison Samuel

Former Senior Associate Lecturer,
University of Plymouth, Seale Hayne Campus,
Devon, United Kingdom

Louisa Dines

Principal Lecturer in Agronomy,
Harper Adams University, Newport,
Shropshire, United Kingdom

ELSEVIER

WP
WOODHEAD
PUBLISHING
An imprint of Elsevier

Woodhead Publishing is an imprint of Elsevier
50 Hampshire Street, 5th Floor, Cambridge, MA 02139, United States
The Boulevard, Langford Lane, Kidlington, OX5 1GB, United Kingdom

ISBN: 978-0-323-85702-4

For information on all Woodhead Publishing publications visit our website at https://www.elsevier.com/books-and-journals

Publisher: Nikki P. Levy
Acquisitions Editor: Nancy J. Maragioglio
Editorial Project Manager: Emerald Li
Production Project Manager: Sreejith Viswanathan
Cover Designer: Vicky Pearson Esser

Typeset by TNQ Technologies

Contents

Foreword

It is great testament to the expertise and commitment to agricultural education of the original authors that *Lockhart and Wiseman's Crop Husbandry Including Grassland* has now reached its 10th edition. The first edition, published in 1966, reflected the skills of both Jim Lockhart (d. 2012) and Tony Wiseman (d. 2019), in bringing scientific understanding to bare on the practicalities of producing crops. The two authors had overlapping, but also complementary knowledge and approaches. Both had admirable proficiency over many crops, but whilst Jim Lockhart had particular strengths in the science and production of cereals and potatoes, Tony Wiseman had specialist interest in grassland and forages. The *application* of science to the husbandry of crops and grassland was appreciated by generations of students taught at the Royal Agricultural College (now the Royal Agricultural University) at Cirencester. The teaching has had wide and lasting impact as many of those students went on to become farmers, land managers, agronomists, policy makers and a host of other professions connected with the land.

Much has changed in agriculture since 1966. The 1960s witnessed the advent of *the Green Revolution*, a term originally coined to describe yield increases in South Asia, but which became more widely applied to increased agricultural production across the globe. Higher yields were achieved by combining *improved* cultivars with the adoption of new products and technology for cultivation, sowing, crop nutrition, irrigation and protection, to help approach the yield potential of the new cultivars. Throughout this period and in to the 1990s, Lockhart and Wiseman brought out new editions of their book, carefully describing the new agronomic possibilities. The popularity of this book with educators, students, producers and their advisors ensured that the influence of the authors on farming methods reached far beyond those connected with Cirencester and the United Kingdom.

From the 1990s, new editions were produced that increasingly involved coauthors that built on the foundations laid down by Lockhart and Wiseman. This trend has continued in this volume with most of the contributors now from Harper Adams University as well as Aberystwyth and Plymouth Universities. Other developments include an increasing visibility of the importance of the environmental sustainability of production. 'Sustainability' has, of course, been an underlying, if sometimes understated, factor since the first edition. A major aim of all the editions, and particularly the 10th, is to improve the efficiency of the use of resources and to reduce negative effects on the environment. For example, to help ensure that the right products (seeds, nutrients and other agrochemicals) are used at the right time, in the right place, and in the right quantity: the four *rights* of crop husbandry. As described in this volume, new advances in sensing, mapping and variable rate applicators have meant increasing possibilities for precision, both between and within fields. The agri-environment and ecosystems on which production depends are also given due regard in terms of soil health, environmental change, biodiversity and less-intrusive systems of production such as organic farming. The range and number of crops covered remains impressive, but this edition also extends its coverage beyond production in the temperate, oceanic climate of north-western Europe to systems further afield.

I had the great privilege of receiving mentorship from Tony Wiseman during my first lecturing post at Cirencester from 1990 to 1996. Of particular value was observing how he managed to draw on much academic expertise to provide relevant insights and instruction for farming and land management. Tony often held lectures outside in the fields of Gloucestershire, where the students could see first-hand, how the application of science could help meet the requirements of crop production. Since then, I have always tried to remember that the best way to teach agricultural science to budding agriculturalists is to ensure that the students can see the relevance to solving the challenges of farming. This is no less true as society demands more food, yet produced with greater levels of sustainability, in the face of environmental change and political upheaval.

I am delighted that this edition of *Lockhart and Wiseman's Crop Husbandry Including Grassland* has remained true to the applied approach of the original authors and am, therefore, convinced that it will continue to be influential in agricultural education and practice as we approach the middle of the 21st century.

Mike Gooding, B.Sc., PgDip, Ph.D., FRSB
Professor (Honorary), Crop Science
Aberystwyth University

Acknowledgments

The authors are grateful to the following for their contributions to this book

Simon Allen and Matthew Rodenhurst, Harper Adams University, Newport, Shropshire, UK (Chapter 15 *Root crops*).

Matthew Back, Harper Adams University, Newport, Shropshire, UK (Chapter 6 *Diseases of farm crops*).

Andrew Beacham, Harper Adams University, Newport, Shropshire, UK (Chapter 1 *Plants*).

Jonathan Cooper, Harper Adams University, Newport, Shropshire, UK (Chapter 16 *Energy and industrial crops*).

Mitch Crook, Harper Adams University, Newport, Shropshire, UK (Chapter 12 *World cropping systems*).

Edward Dickin, Harper Adams University, Newport, Shropshire, UK (Chapter 11 *Plant breeding and seed production*).

Jonathan Gill, Kit Franklin, Simon Woods, Harper Adams University, Newport, Shropshire, UK (Chapter 9 *Precision crop production*).

Paul Lewis, Harper Adams University, Newport, Shropshire, UK (Chapter 3 *Soil health and management* and Chapter 4 *Fertilisers and manures*).

Paul Lunt, University of Plymouth, Plymouth, Devon, UK (Chapter 2 *Climate and weather*).

Jim Monaghan, Harper Adams University, Newport, Shropshire, UK (Chapter 17 *Fresh produce crops*).

Iwan Owen, IBERS, Aberystwyth University, Aberystwyth, United Kingdom (Chapter 20 *Establishing and improving grassland*, Chapter 21 *Grazing management*, and Chapter 22 *Conservation of grass and forage crops*).

Tom Pope, Harper Adams University, Newport, Shropshire, UK (Chapter 7 *Pests*).

John Reade, Harper Adams University, Newport, Shropshire, UK (Chapter 5 *Weeds of farm crops*).

Joe Roberts, Harper Adams University, Newport, Shropshire, UK (Chapter 7 *Pests*).

Evie West, Totnes, Devon, UK (photographs for Chapter 19, *Introduction to grass production*).

Introduction

Jim Lockhart and Tony Wiseman published their first edition of *Crop Husbandry* in 1966. At the time there were many changes taking place in arable production, and yields were increasing rapidly in the United Kingdom. Concerns in crop production were very different from those found today. Now, over 55 years later, we have produced the 10th edition. It is fitting with the passing of both of the original authors that we look at how crop production changed during their lifetime and what may affect crop production in the future.

Cropping

Until the 1980s, barley was the most commonly grown cereal crop (Fig. 1). Since then, winter wheat has become the major cereal crop in the United Kingdom, while oats continue to be a minor crop. There are many reasons for the shift in popularity of the cereal crops over the years; the main ones being increased yields, changing markets, weather conditions affecting drilling and harvesting and most importantly profitability. In the first edition, spring barley was considered the highest yielding barley crop, and some farmers were growing

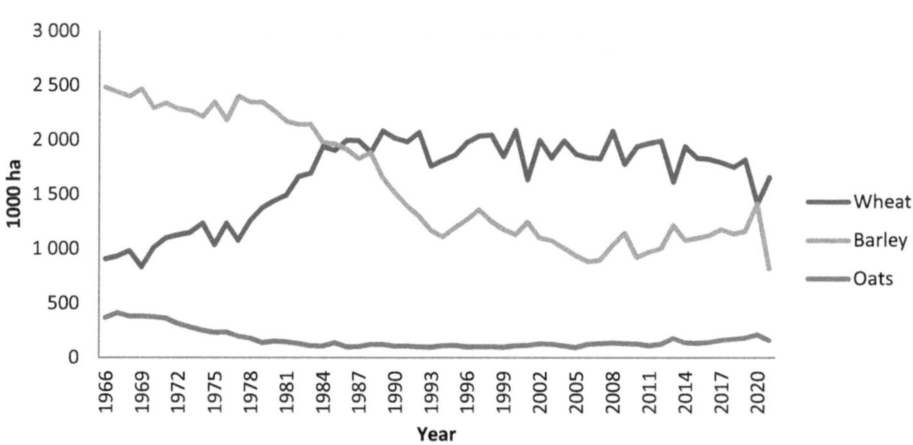

Figure 1 Area of cereals grown in the United Kingdom, 1966–2021 (1000 ha), Defra.

continuous spring barley; this is not the case now. With the increase in understanding of diseases such as barley yellow dwarf virus in winter barley there were many years when winter crops were more popular than spring crops, and in the 1980s many farmers on less fertile soils were growing continuous winter barley. Subsequent problems with barley yellow mosaic virus as well as some poor harvests and the need to diversify rotations has brought this trend to an end and now the popularity has swung back more in favour of spring barley. Rotation and use of green manures were also not considered to be so important with the introduction of more pesticide options and inorganic fertilisers. This has also changed again more recently as we struggle with problems such as pesticide resistance, withdrawal of available pesticides, increases in fertiliser prices, variable weather patterns, diffuse pollution, reduction in soil carbon and climate change.

Of the other arable crops, there have also been some changes (Fig. 2). Most significant is the increase in area of oilseed rape now grown. Once the United Kingdom joined the Common Market in 1973 (now European Union (EU)) there was financial support available for crop production in the Common Agricultural Policy (CAP), either in the form of production support or income support. One of these supports was for oil production from oilseed rape as the EU was not self-sufficient in vegetable oils. This made the crop more viable economically. There were also major developments in breeding making oilseed rape better for oil production for human consumption. Varieties of oilseed rape were also developed for industrial use. Now oilseed rape is the most commonly grown arable break crop which was not envisaged back in the 1960s. Other break crops such as field beans, peas and linseed have fluctuated in areas grown mainly depending on support payments and prices.

Other important changes in arable cropping affected sugar beet and potatoes (Fig. 2). Sugar beet was originally regulated by quotas when the United Kingdom was in the EU. Over the years there have been major reductions in the number of factories and this has restricted areas where the crop is grown. The reduction in area of sugar beet helped fuel the increased popularity of oilseed rape as growers looked for suitable alternative break crops. The area of potatoes grown used to be governed by a UK quota system but after joining the EU this quota system was abolished and there are currently no potato quotas. Potatoes have become a highly mechanised operation, with limited direct farm sales and more being sold on contract to packers for supermarkets and processors; this has led to a reduction in the number of growers.

The area of field vegetables has also declined (Fig. 2); part of this decline is due to improved husbandry and automation meaning that more crop can be harvested from a smaller area; but also the United Kingdom has continued to increase the percentage of imported vegetables especially as some countries are able to produce vegetables at different times of the year from the United Kingdom and sometimes more cheaply.

The importance of some world cropping systems can have a significant impact on UK farming both in terms of commodity prices as well as availability of imported produce which subsequently affects the area of crops grown here; because of this a new chapter has been added in this edition on world cropping systems.

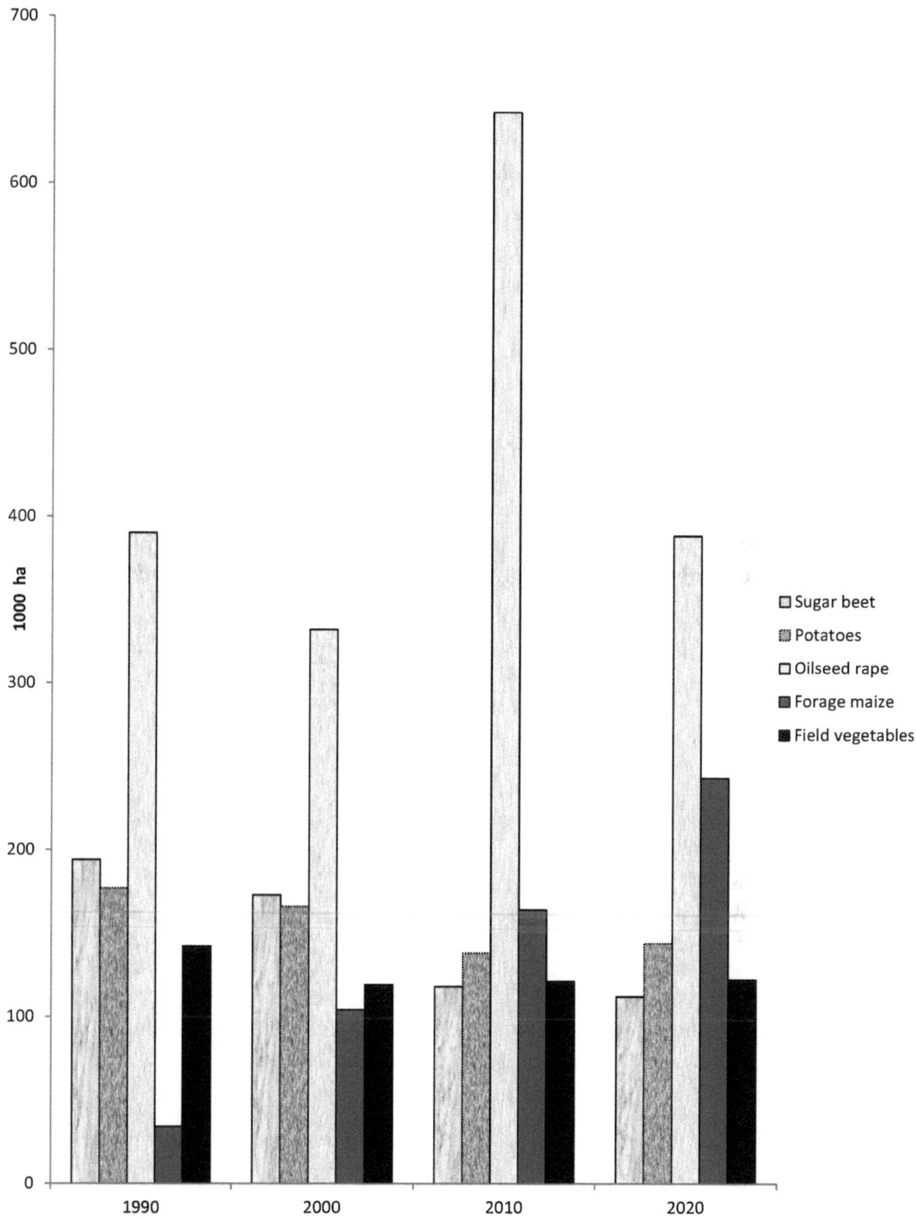

Figure 2 Changes in area grown (1000 ha) of sugar beet, potatoes, oilseed rape, forage maize and field vegetables, in England between 1990 and 2020, Defra.

The other major change in cropping over the years has been the area of forage crops grown, including forage maize (Fig. 2). Again, in the 1960s virtually no maize was grown in the United Kingdom, though there was quite an area of other forage crops such as kale, mangolds, turnips and swedes. The development of new maize varieties with the ability to yield well in the UK climate, as well as the cost of importing livestock feeds, has led to farmers growing more maize for their own livestock feed requirements. There is also increased demand for maize production in arable areas for use in anaerobic digester plants.

Livestock systems in the 1960s relied more on grazed crops as fewer animals were housed during the winter months. With the increase in winter housing and silage rather than hay-based systems, some other forage crops such as kale have declined in importance. Grassland areas have also changed; permanent pasture has remained fairly static but there has been a large decline in the area of temporary grass (under 5 years old). Though the area of temporary grassland has declined over the years many of the ryegrass and clover varieties now grown are higher yielding than the older varieties and have improved feeding quality. In the 1960s ley farming (rotating crops with short-term leys) was important and, according to the first edition of this book, 'mixed farming was found on most lowland farms'. Subsequently, many farms have become much more specialised with fewer mixed farms, especially in eastern England. Now many farmers are looking at more sustainable farming systems especially in relation to increasing soil carbon; subsequently there is increased interest again in ley farming systems. This would have pleased Tony Wiseman as he was a great advocate of the ley farming system.

One other area that has changed since the 1960s is the importance of industrial crops, so a new chapter on these crops was added in the 9th edition.

Crop yields

During the time of the early editions of this book great emphasis was placed on crop yields without any thought of impact on the environment. Today the emphasis has changed due to loss of biodiversity and climate change, so the focus now is to produce safe food using resilient and sustainable farming methods.

Crop yields increased significantly from the 1960s. Wheat and barley yielded similarly in 1966 at around 4 tonnes/hectare (t/ha). Yields increased steadily right up to the late 1990s with most improvement in the yield of wheat. Since the late 1990s yields have plateaued (Fig. 3) despite improvements in varieties and crop husbandry and a small increase in atmospheric carbon dioxide.

In root crops (Fig. 4) there has been a steady increase in potato yields, but it is sugar beet that is the real success story with very good yield improvements, especially over the last 30 years. These yield increases can be attributed to both improvements in crop husbandry as well as in plant breeding which has been aided greatly by British Sugar having their own research farm and dedicated research programme.

Crop inputs

Crop inputs have increased dramatically since 1966 when, for example, in wheat nitrogen rates were only 75 kg/ha with 60 kg/ha phosphorus (P_2O_5) and potassium

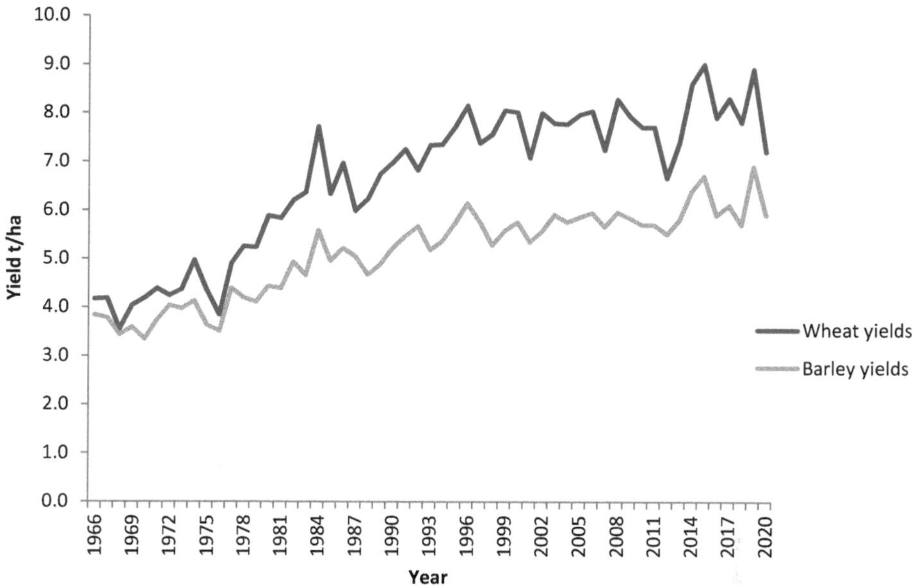

Figure 3 UK wheat and barley yields since 1966 (t/ha), Defra.

(K_2O), a seed treatment was commonly used plus one or two herbicides and occasionally a plant growth regulator was applied. Some inputs were even supported by government grants, such as the liming subsidy.

Fertiliser applications

Application rates of fertiliser nitrogen (N) were one of the main changes to inputs over the years (Fig. 5). Nitrogen rates increased right up to the 1990s at the same time as crop yields were increasing. Since then, nitrogen application rates have remained fairly constant in arable crops. In 1966 ammonium nitrate was only just being developed and used; previously the main nitrogen source other than organic manures was ammonium sulphate. It was not until the 1970s that prilled ammonium nitrate was readily available, which led to increased ease of application. In grassland, Nitrogen inputs peaked in the 1990s but have since declined.

One nutrient that was barely mentioned in the first edition was sulphur. At the time there was sufficient sulphur for crops supplied by atmospheric depositions (though the main bag nitrogen fertiliser then did also contain some sulphur). With the introduction of the Clean Air Act and the subsequent reduction of air pollution, many parts of the United Kingdom, particularly on lighter soils, now suffer from sulphur deficiencies in a range of crops, and applications of fertiliser sulphur have increased.

The first fertiliser manual (RB209) with recommendations for agricultural and horticultural crops was published by MAFF in 1973. This manual has been edited several

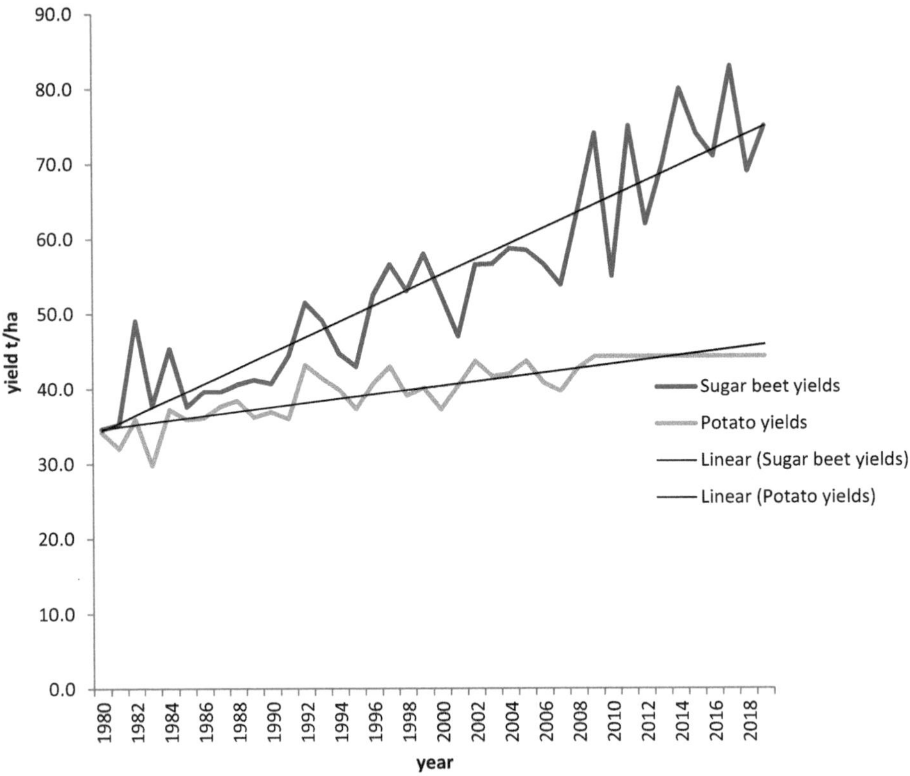

Figure 4 UK sugar beet and potato yields, 1980–2019, Defra.

times and is now the industry standard for fertiliser recommendations. Optimum recommended rates have changed according to trial results as well as commodity and fertiliser price, use of organic manures and legislation.

Crop protection products

The use of crop protection products is an area of crop husbandry that has changed tremendously over the years (Fig. 6). In the first edition, other than seed treatments there were virtually no fungicides available, herbicides were limited as were insecticides. On average in 1974 there were only two treatments applied per crop compared with over 10 now. In some crops, such as sugar beet, hand hoeing within the crop row was commonly undertaken; now a very detailed herbicide programme has been developed.

Many new chemicals, especially fungicides, were developed between the 1970s and 2000. The newer agrochemicals were safer environmentally and were also applied in much lower quantities (g rather than kg of active ingredients/ha). So, in practice, though more treatments are used, lower amounts of active ingredient are applied per

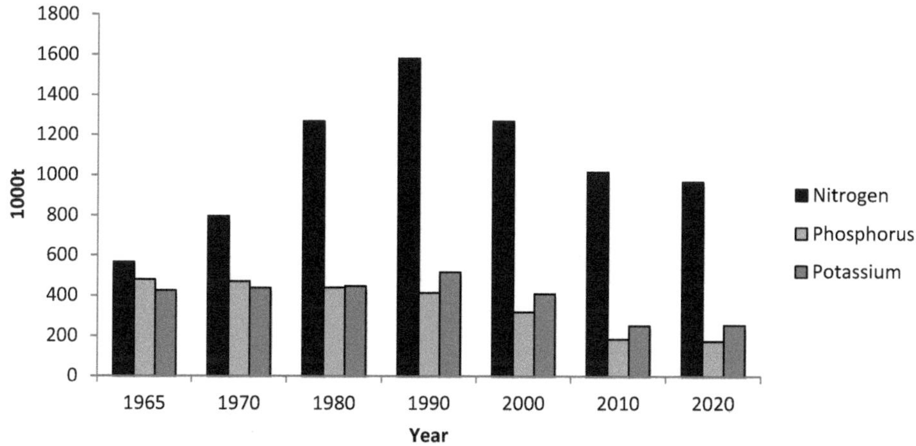

Figure 5 Total fertiliser usage (nitrogen, phosphorus (P_2O_5) and potassium (K_2O)) applied to all crops and grassland in the UK, Defra.

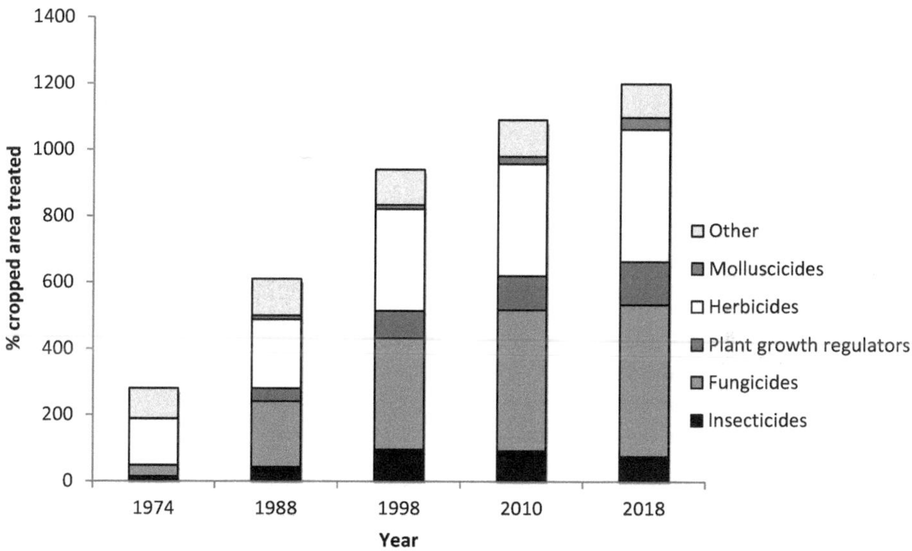

Figure 6 Pesticide-treated area of crops as a percentage of cropped land. Note: Data were for England and Wales in 1974, for Great Britain in 1988 and 1998 and for the United Kingdom in 2010 and 2018.
Adapted from pesticide usage survey results, Defra.

ha (Fig. 7). There has been an increase in the amount of legislation surrounding pesticide use to ensure safety to operators as well as to the environment and food safety. Knowledge of disease epidemiology and pest life cycles has also improved, and decision support programmes have been developed.

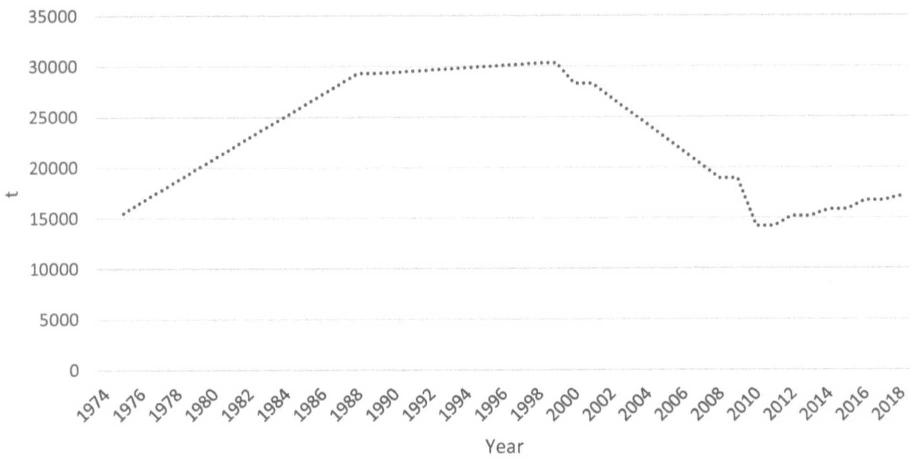

Figure 7 Changes in total amount of pesticides applied (t) (Pesticide Usage Survey, Defra).

In the first edition there was no mention of pesticide resistance. There had been a few cases of resistance to the organochlorine insecticides, but it was not considered an important issue. Now, with 50 plus years of routine use of pesticides, resistance is a common problem to many groups of agrochemicals. Due to the cost of developing new chemicals and the rigours of approval few new crop protection products are being marketed. Many companies are now looking at biological control methods as an alternative to agrochemicals. In the first edition, biological control methods were considered to have little application in this country; how things have changed. It is now standard practice in glasshouse production systems to use biological control methods for control of insect pests and it is even beginning to be looked at for use in the field.

Due partly because of resistance issues and the very high cost of developing new agrochemicals, many farmers now must take a more sustainable approach to weed, pest and disease control. Sustainable control of crop problems is discussed in several of the chapters in this edition; for example, in some vegetable crops there are few approved weed control products and this has led to an increase in development of mechanical hoeing and use of vision guidance systems.

With the increase in complexity of crop protection, a training and regulatory scheme for agrochemical advisers (BASIS) was established in 1978; this meant that the growing number of crop advisers could be trained in weed, pest and disease identification, life cycles and control as well as in safe storage and applications of pesticides. (A similar fertiliser scheme was developed in 1993 (FACTS).)

Varieties and plant breeding

Variety selection has always been very important in crop production. A significant percentage of the yield increases up to the 1990s was attributed to improvements in crop varieties. In the first edition, as now, we recommend readers to consult the recommended variety lists. What is different now is that the majority of plant breeding companies are independent. In the 1960s and 1970s the United Kingdom had its own state-funded plant breeding stations including the plant breeding institute which was very important in producing some of the most commonly grown wheat, barley and potato varieties of the time. In 1987 it was privatised. Since then, private plant breeding companies have taken over and have continued to develop new varieties including hybrids using biotechnological techniques.

Many of the agrochemical companies have now joined or bought out some of the plant breeding companies as they realise how important plant breeding will be in the future in sustainable crop production.

Genetically modified crops are still not allowed to be grown in this country but it will be interesting to see what happens in the near future. The chapter on plant breeding in this edition has been expanded due to all the changes currently taking place.

Crop husbandry

During the 1970s and 1980s there was an increased amount of research work into crop physiology, especially for cereals and potatoes. This led to a better understanding of the interaction between seed rates, sowing dates and varieties and the effect of other inputs, especially nitrogen on crop growth, canopy management and yield. In the first edition, the husbandry of all the different crops was described together; subsequent editions have split many of the crops into different chapters as the amount of crop husbandry knowledge has increased.

Sugar beet is one of the crops whose husbandry has changed most since 1966. Monogerm pelleted (insecticide treated) seed is now the norm and has enabled the crop to be precision drilled, removing the need to hand single. With improvements in varieties, it is now possible to drill the crop earlier, which has also helped to maximise yields. Crop protection programmes are now much more sophisticated. An understanding of soil structure and compaction led to the development of controlled wheeling systems and measures to reduce incidence of soil erosion.

Crop mechanisation and storage

There have been some very major changes in how crops are harvested and stored compared with 1966. At the time of writing the first edition, 30% of the sugar beet crop was still harvested by hand. Potato harvesters were only just being developed and the majority of potato crops were still hand lifted. Potatoes were often stored in clamps outside, whereas now many farmers have temperature controlled, insulated stores. After joining the common market there were some good capital grant schemes available (e.g. FHDS—farm and horticulture development scheme) which encouraged participating farmers to expand, especially in relation to building silage pits, grain and potato stores.

There have been major changes in the size and capacity of machinery over the years. For example, in the 1960s fertiliser spreaders only had a spread of 9 m compared with up to 36 m now. Tramlines were not initially used in cereal crops as there were very few passes across most crops; as inputs increased and farmers required more precision they started drilling tramlines. Now many farmers are using remote sensing techniques and variable rate technology to improve efficiency of application of crop inputs. These precision farming techniques including the use of robots are discussed in the precision crop production chapter added in the last edition and further expanded in this edition.

Seed bed cultivations have varied across the years. In the 1980s many farmers were direct drilling after burning the crop stubble. Once burning was banned many people went back to plough-based systems, but with increasing fuel prices and increased interest in improving soil carbon farmers are again looking at minimum and zero tillage techniques.

In horticulture there have been major developments in crop mechanisation. Many crops are now grown from plants raised in modules; planters can be tractor-less using precision farming techniques; and the use of crop covers has enabled growers to extend the crop season. Several harvesters and field rigs have been developed for harvesting and packing in the fields. With the increasing cost of labour as well as labour shortages more developments are taking place in mechanical harvesting systems. There has also been increased interest and development in vertical growing systems.

Horticultural crops are often cooled as soon as they are harvested and are then refrigerated during storage and transportation in order to maintain quality before arriving on the supermarket shelves (chill chain); this again was not something that was considered in many of the early editions of this book.

Support and legislation

Support to farmers from the CAP was very successful in increasing crop production across the EU. Unfortunately, this led to surpluses known as 'mountains' of some commodities such as wheat which had to be stored in large intervention stores. Subsequently, set-aside was introduced in 1992 to take some arable land out of production. The need for set-aside was removed in 2008 mainly due to a reduction of the grain mountain and also a fall in world supplies. Over the years there have been a number of EU directives that have affected the way the United Kingdom farmed, including the nitrates directive and the introduction of nitrate vulnerable zones and the water framework directive as well as cross compliance rules when claiming the Single Payment Scheme. Some of the directives and schemes were introduced to limit agricultural pollution and soil erosion as well as improving the farmed environment. Now the United Kingdom has left the EU; it is able to develop its own sustainable farming initiatives. In this edition we have expanded the chapter on sustainable crop production techniques to cover these changes.

As well as the various EU directives impacting on arable production, when the United Kingdom joined the EU, it also led to the adoption of metric units. The first edition of crop husbandry was written entirely using imperial units; now of course everything is metricated, though in practice many farmers still use both systems!

Environment and climate issues

During the 1970s and 1980s many people became concerned about loss of habitats and biodiversity, soil erosion and pollution due to the changes in farming practices. Birds are a good indicator of what is happening to farmland biodiversity as they have a wide variety of habitats and food requirements and are fairly easy to monitor. Since 1970 farmland bird populations have fallen so that in 2020 numbers were only 44% of the base numbers in 1970. Some specialist bird species have been more seriously affected. The decline in numbers was greatest during the 1970s and 1980s. The reasons are many, including loss of mixed farms, increased soil drainage plus more autumn cropping and production of silage not hay. As well as changes in land management climate change is also now considered an important threat to biodiversity. The section on climate change in this edition has been expanded due to the potential effect both on UK cropping and farmland biodiversity. UK agriculture could also potentially be very important in helping to mitigate climate change.

In 1985 the first funded Environmentally Sensitive Areas were started in England and this was the beginning of a number of funded farm conservation schemes. Since then there have been various CAP supported agri-environmental schemes introduced. In England about two-thirds of the farmed land are covered by one of the various agri-environment schemes; these current schemes will be replaced, with the new Environmental Land Management Schemes. It is hoped that some of these schemes will help reduce the decline in biodiversity as well as reducing soil erosion and improving soil carbon sequestration. The chapter on sustainable crop production techniques has been expanded to outline some of these new schemes.

In 1966 virtually no one talked about organic farming; it was only after some of these environmental issues were raised, plus financial support, that farmers started converting to this farming system. Jim Lockhart was very much against this type of farming as he had grown up in a time when farmers were trying to reduce the amount of manual work carried out on farms! It was quite a struggle to persuade him to let us put a chapter in this book. We think he would have been quite surprised about the current area of organically grown crops in the United Kingdom and the growing interest in agroecological systems including rewilding and regenerative farming.

Research and advice

In the 1960s and 1970s most agricultural research was provided free by the government (Ministry of Agriculture Fisheries and Food (MAFF) and the Agricultural Development and Advisory Service (ADAS)). The United Kingdom was a net importer of food and the government's target was to promote UK crop production. Up to the 1980s ADAS ran 12 experimental husbandry farms and experimental horticultural stations. These farms were on a wide range of soil types and were used both for applied research but also for highlighting best practice in their area. Field trials were also run by the regional ADAS department staff. Both the experimental farm trials and regional farm trials were used as a way of demonstrating new technologies in the industry. They often trialled innovations coming for the research institutes, and it was the way of encouraging farmers to incorporate new well-researched techniques onto their farms. All this changed after the 1980s when the government started reducing the amount of support for agricultural research. Organisations such as the weed research organisation were closed, ADAS was privatised, and the government pushed for the agricultural industry itself to support applied research through crop levies. Currently, the Agriculture and Horticulture Development Board is the statutory levy board in the United Kingdom, and farmers, growers and others in the supply chain are charged a small levy to fund much needed, independent research and development work. Sugar beet is the research success story where the industry has its own funded sugar beet research programme (British Beet Research Organisation (BBRO)).

In 1979 Jim Lockhart was involved in getting one of the new national cereal events to be held in the Cotswolds—Barley '79' started. These cereal events are now one of the most important annual arable shows in the country. After Barley '79' a group of progressive Cotswold farmers, with support from lecturers and advisers including Jim, started their own private cereal trials organisation; the Cotswold Cereal Centre (CCC). We are sure he would never have imagined how this private organisation would grow! With the loss of government funding for near market research, trials work had to be more farmer funded; exactly what the CCC was doing. Over the years it has grown to have centres in different parts of the country. It joined with Morley (trials and crop advisory group) to form The Arable Group (TAG) and subsequently combined with NIAB to form the national independent, agronomy research and information organisation NIAB TAG.

In the 1960s the government also funded the free advisory service, the national agricultural advisory service (NAAS—later changed to ADAS) employing quite a few staff across England and Wales mainly as general agricultural advisers though there were a limited number of crop and grassland husbandry advisers (Agronomists) as well as plant pathologists, entomologists and soil scientists. When ADAS was privatised, this was at the time when there was growing demand for agronomists especially in relation to using crop protection products, variety choice and fertilisers. Many agronomists are now employed by the agricultural distributors but there is still a place for independent advice from individual agronomists to larger independent organisations. The role of agronomists is currently changing and is less focussed on agrochemicals. Many are taking a more holistic approach to arable farming, from helping with grant applications to advising on sustainable and economically viable crop production across farms.

Education

There has been a decline in the number of universities offering straight agricultural courses over the years; Wye (Imperial) and Seale-Hayne (University of Plymouth) colleges have been the most recent to close. More of the national institutions, now universities (such as the Royal Agricultural University and Harper Adams University), are providing the training, as well as further education colleges. The range of courses available to students has expanded to provide syllabuses that cover the major changes that are currently happening in the industry. Student numbers have fluctuated widely over the years; numbers have often been very dependent on the profitability of farming. With farming becoming more specialised and advanced mechanically and technologically, together with all the legislation that farmers need to follow it is more important than ever that future farmers have a good agricultural education; so the continued requirement for this book.

Sources of further information and advice

Further reading

Anon, 2021. The British Survey of Fertiliser Use on Farm Crops for Crop Year 2020. Defra.

Anon, 2012. Wild Bird Populations in the UK 1970–2011, National Statistics. Defra.

Brassley, P., Harvey, D., Lobley, M., Winter, M., 2021. The Agricultural Revolution. The Transformation of English Farming 1939–1985. Boydell Press.

Garthwaite, D.G., Thomas, M.R., 1999. Pesticide Usage Survey Report 159, Arable Farm Crops in Great Britain. MAFF.

Garthwaite, D.G., Barker, I., Parrish, G., Smith, L., Chippendale, C., Pietravalle, S., 2011. Pesticide Usage Survey Report 235, Arable Crops in the UK 2010. Defra.

Garthwaite, D., Ridley, L., Mace, A., Parrish, G., Barker, I., Rainford, J., MacArthur, R., 2018. Pesticide Usage Survey Report 284 Arable Crops in the United Kingdom. Defra.

Hayhow, D.B., Eaton, M.A., Stanbury, A.J., Burns, F., Kirby, W.B., Bailey, N., et al., 2019. The State of Nature 2019. The State of Nature Partnership.

Knight, S., Knightley, S., Bingham, I., Hoad, S., Lang, B., et al., 2012. HGCA Project Report 502, Desk Study to Evaluate Contributory Causes of the Current 'Yield Plateau' in Wheat and Oilseed Rape. HGCA.

Sly, J.M.A., 1977. Pesticide Usage Survey Report 8: Review of usage of Pesticides in Agriculture and Horticulture in England and Wales 1965–1974. MAFF.

Tree, I., 2018. Wilding. Picador.

Wilkins, R.J., 2000. Grassland in the Twentieth Century. IGER Innovations.

Websites

www.defra.gov.uk.
www.bto.org.
www.statistics.gov.uk.
www.fera.defra.gov.uk.
www.ahdb.org.uk.
www.niab.com.

Plants

1

1.1 Introduction

Plants are living organisms consisting of many specialised individual cells. They differ from animals in many ways and a very important difference is that they can build up valuable organic substances from simple materials such as carbon dioxide and water. The most important part of this building process, called photosynthesis, is the production of carbohydrates such as sugars, starch and cellulose, along with oxygen, using energy provided by light. They have rigid cell walls enclosing a selectively permeable cell membrane which allows the passage of water through it (osmosis). They have specialised organs such as roots, stems and leaves, are mostly immobile and are primary producers of food in most land-based ecosystems.

Most plants consist of roots, stems, leaves and reproductive parts and need a medium in which to grow. These media could be soil, compost, water where plants are grown hydroponically or even air, where the bare roots are sprayed with a fine mist of nutrients and water (aeroponics). In addition to functioning in the uptake of water and nutrients, roots can help to anchor the plant in the growth medium. Stems support the plant, helping to hold the leaves in the optimal positions for light capture and provide routes for transport for substances around the plant. The leaves are the main site of photosynthesis, a process which produces sugars and other molecules for use by the plant.

1.2 Plant groups

There are many ways of classifying plant groups but, from an agricultural and horticultural point of view, a useful way is to divide them into annuals, biennials and perennials according to their total length of life.

1.2.1 Annuals

Annuals are plants that complete their life cycle in one growing season, i.e., starting from a seed, in 1 year they develop roots, stem and leaves and then flower and set seed before dying. Typical examples of annuals grown as crops are wheat, barley and oats.

Lockhart and Wiseman's Crop Husbandry Including Grassland. https://doi.org/10.1016/B978-0-323-85702-4.00012-1

1.2.2 Biennials

These plants grow for 2 years. They spend the first year producing roots, stem and leaves, and the following year producing the flowering stem and seeds, after which they die. Sugar beet, swedes and turnips are typical biennials, although the grower treats these crops as annuals, exploiting their life cycle by harvesting them at the end of the first year when most of the foodstuff is stored up in the root, and before the plant moves on to produce the seed head. Those biennial plants that do behave as annuals and produce a seed head in their first year are called 'bolters' and are effectively treated as weeds in a biennial crop.

1.2.3 Perennials

Perennials live for more than 2 years and, once fully developed, they usually produce seeds each year. Many of the grasses and forage legumes are perennials, as are many of the horticultural fruit crops such as raspberries or apples, and some energy crops such as willow and Miscanthus.

1.2.4 Monocotyledons versus dicotyledons

Plants can also be classified as monocotyledons (monocots) or dicotyledons (dicots). Monocotyledons have one embryonic leaf, or cotyledon and dicotyledons have two. Monocot crops include wheat, maize, rice and other cereals and grasses. Dicot, or broadleaf, crops include potatoes, beans, sugar beet, leafy salads, fruits and vegetables amongst others. The main differences between monocots and dicots are summarised in Table 1.1.

Table 1.1 Summary of the main differences between monocotyledons and dicotyledons.

Monocotyledons (monocots)	Dicotyledons (dicots)
Embryo has one seed leaf (cotyledon). Adventitious (fibrous) root system. Usually narrow-leaved plants, e.g., cereals and other grasses. Vascular bundles are scattered in the stem. Root xylem and phloem form a ring. Leaf veins are parallel.	Embryo has two seed leaves. Primary root system with a tap root. Usually broad-leaved plants, e.g., legumes and sugar beet. Vascular bundles in the stem form a ring. Root phloem forms between arms of xylem. Leaf veins form a net pattern.
These two groups of flowering plants can be further divided for example in the following way:	
Families or orders: e.g., Legumes Genus: Clovers Species: Red clover Cultivar or variety	Leguminosae (Fabacaea) *Trifolium* *Trifolium pratense* Merviot

1.3 Plant structure

The plant can be divided into four parts: the root system, stem, leaf and flower.

1.3.1 The root system

The root system is concerned with the parts of the plant growing in the soil; there are two main types:

1.3.1.1 The tap root or primary system

Dicotyledonous plants possess a primary root called the tap root with lateral secondary roots branching out from it and sometimes additional tertiary roots branching from the secondary roots. This can form, in some cases, a very extensive root system (Fig. 1.1). The root of the bean plant is a good example of a tap root system. If this is split it will be seen that there is a slightly darker central woody core, the skeleton of the root, which helps to anchor the plant and transport foodstuffs. The lateral secondary roots arise from this central core (Fig. 1.2). Carrots and other true root crops such as sugar beet have very well developed tap roots. These biennials store food, usually as carbohydrates, in their roots during the first year of growth to be used in the following year for the production of the flowering shoot and seeds. However, they are normally harvested after one season and the roots are used as food for humans and livestock.

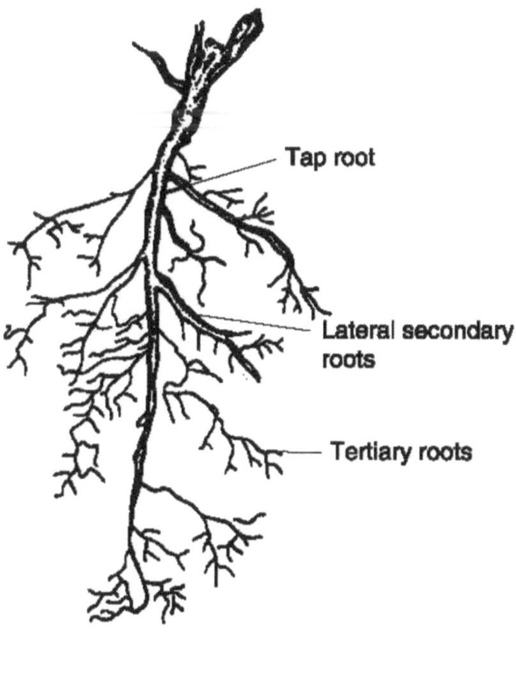

Figure 1.1 Tap root or primary root system.

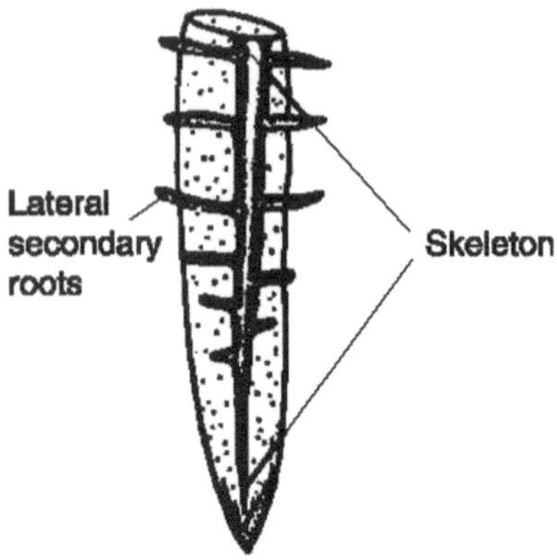

Figure 1.2 Tap root of the bean plant.

1.3.1.2 The adventitious root system

This is found on all grasses and cereals, and it is the main root system of most monocotyledons. The primary root is soon replaced by adventitious roots, which arise from the base of the stem (Fig. 1.3). This forms a large mass of similarly-sized roots. These roots can, in fact, develop from any part of the stem, and they are found on some dicotyledons as well (but not as the main root system), e.g., underground stems of the potato.

1.3.1.3 Root structure

Roots are divided into different sections that perform a range of functions. At the tip of the root is the root cap. This acts to protect the growing root. Its cells can respond to gravity, light and soil particle pressure to guide root growth. It produces a substance known as mucilage which helps the root move through the soil. Behind the root cap is an apical meristem. Cells in the meristem divide and develop into the different

Figure 1.3 Adventitious root system.

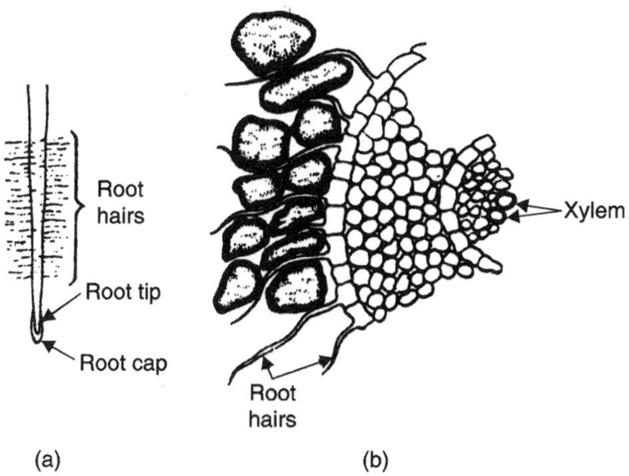

Figure 1.4 (a) Section of root tip and root hair region. (b) Cross-section of root showing the root hairs as tube-like elongations of the surface cells in contact with soil particles.

tissues found within the root. 4—10 mm behind the root cap is found the zone of elongation. The cells in this part of the root can elongate as much as 150 times. This helps to push the root cap and meristem through the soil. Behind this is the zone of differentiation where the different tissues of the root start to form and then the zone of maturation which possesses a large number of root hairs.

Root hairs (Fig. 1.4) are very small white, hair-like structures. They increase the absorptive surface of the root several thousand-fold. As the root grows, the hairs on the older parts die off and others develop on the younger parts of the root. They play a very important part in the nutrition and water uptake of a plant.

1.3.2 The stem

The second part of the flowering plant is the shoot that normally grows upright above the ground. It is made up of a main stem and branches. Stems are either soft (herbaceous) or hard (woody) and in UK agriculture it is usually only the soft and green herbaceous stems which are of importance. These usually die back every year.

Turgor pressure and thickened cell walls in the stem helps to provide a structure that supports the leaves, allowing the best possible access to light, and also the flowers, for presentation to pollinators. Stems provide a route for the transport of water and nutrients up from the roots to the leaves and for sugars and other compounds from the leaves. They also allow the movement of hormone-based signals around the plant to direct responses to the environment and to cause developmental changes. In dicots, the vascular bundles, which carry out this transport, are arranged in a ring close to the edge. This allows growth of the centre of the stem and so sideways (secondary) expansion. This allows the growth of very large stems such as the trunks of trees. In monocots, the vascular bundles are scattered across the stem cross-section, meaning that monocots lack secondary stem growth. Stems are also able to store large amounts of starch and water in parenchyma cells.

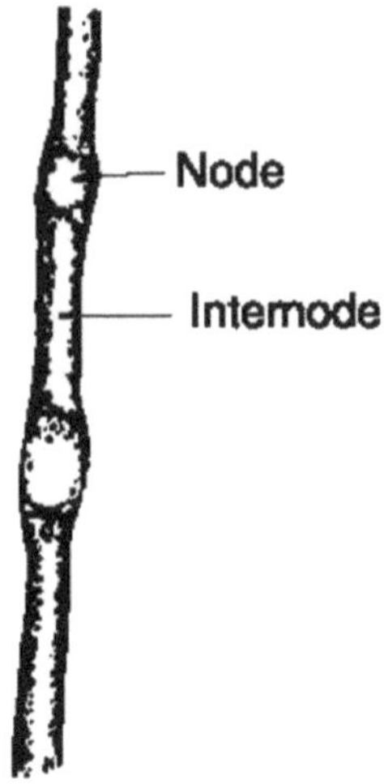

Figure 1.5 Jointed stem.

Stems are usually jointed with each joint forming a node. The part between two nodes is called the internode. At the nodes, the stem is usually solid and thicker, and this swelling is caused by the storing up of material at the base of the leaf (Fig. 1.5).

All stems start life as buds and the increase in length takes place at the tip of the shoot called the terminal bud. If, for example, a Brussels sprout is cut longitudinally and examined it will be seen that the young leaves arise from the bud axis, where an apical meristem is present. This meristem is continually making new cells for the growth of the stem (Fig. 1.6). Axillary buds are formed in the angle between the stem and leaf stalk. These buds, which are similar to the terminal bud, develop to form lateral branches, leaves and flowers.

1.3.2.1 Modified stems

1. A stolon is a stem that grows along the ground surface.
   Adventitious roots are produced at the nodes, and buds on the runner can develop into upright shoots, and separate plants can be formed, e.g., strawberry plants.
2. A rhizome is similar to a stolon but grows under the surface of the ground, e.g., common couch (Fig. 1.7). If a rhizome is detached from the parent plant or cut into sections, the resulting parts can generate a new plant. This is a form of vegetative reproduction and used in the propagation of certain plants.
3. A tuber is really a modified stolon. The ends of the stolons swell to form tubers. The tuber is therefore a swollen stem. The potato is a well-known example with 'eyes' (buds) which develop shoots when the potato tuber is planted. Potato tubers will also turn green by producing chlorophyll when exposed to light; this is another characteristic of stems.
4. A tendril is found on certain legumes, such as the pea. The terminal leaflet is modified as in the diagram (Fig. 1.8). This is useful for climbing purposes to support the plant.

Corms and suckers are other examples of modified stems.

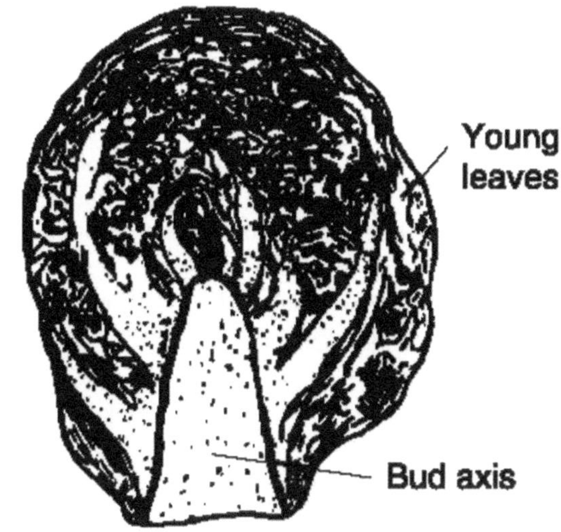

Figure 1.6 Longitudinal section of a Brussels sprout.

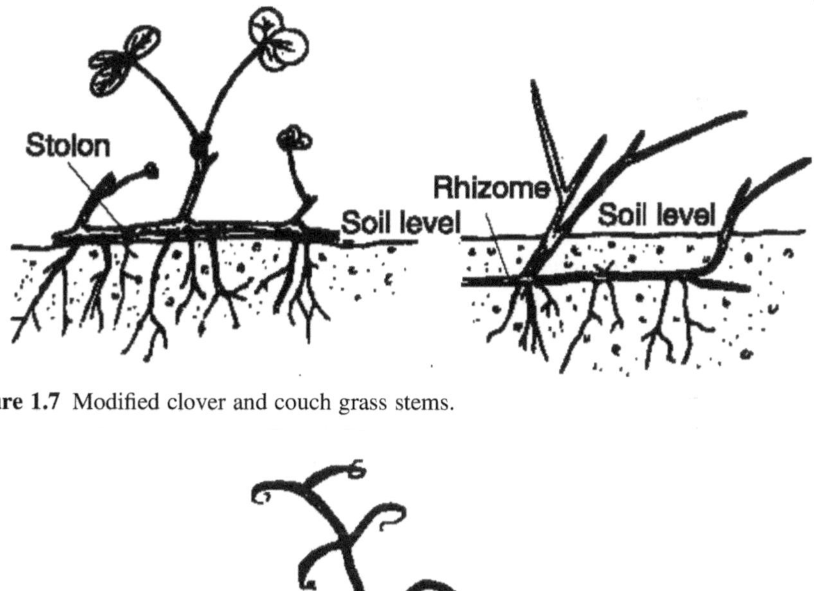

Figure 1.7 Modified clover and couch grass stems.

Figure 1.8 Modified pea stems.

1.3.3 The leaf

Leaves in all cases arise from buds. They are extremely important organs, being not only responsible for the manufacture of sugar and starch via photosynthesis for the growing parts of the plant, but they are also the organs through which transpiration of water takes place.

A typical leaf of a dicotyledon consists of three main parts:

1. The blade.
2. The stalk or petiole.
3. The basal sheath connecting the leaf to the stem. This may be modified (as with legumes) into a pair of wing-like stipules (Fig. 1.9a).

The blade is the most obvious part of the leaf and it is made up of a network of veins formed from vascular bundles, with leaf tissue in between.

There are two main types of dicotyledonous leaves:

1. Those with a prominent central midrib, from which lateral veins branch off on either side. These side veins branch into smaller and smaller ones (Fig. 1.9a).
2. Those with no single midrib, but several main ribs spread out from the top of the leaf stalk; between these, the finer veins spread out as before, e.g., horse-chestnut leaf (Fig. 1.9b).

The veins are the essential supply lines for the process of photosynthesis.

Their vascular bundles consist of two main parts: the xylem, for bringing the required raw materials (water and nutrients) up to the leaf from the roots and the phloem, which carries the finished product (sugars and other carbohydrates) away from the leaf, to growing and/or storage regions.

Leaves can show great variation in shape and type of margin, as in Fig. 1.9. They can also be divided into two broad classes as follows:

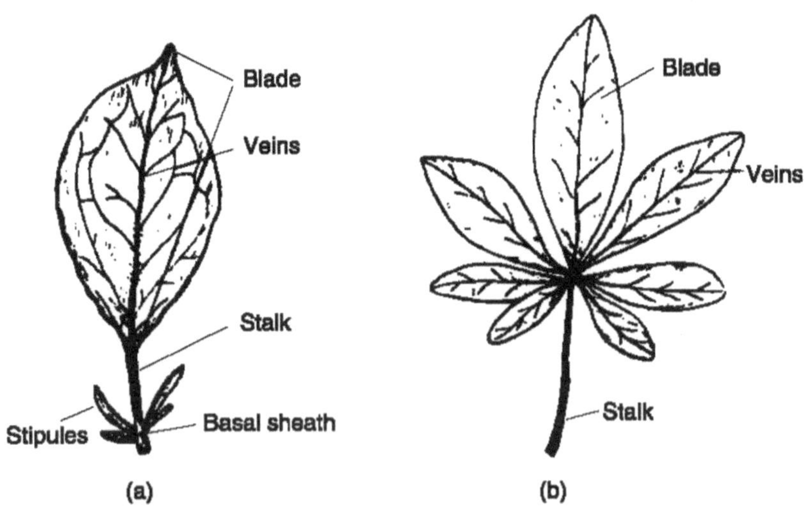

Figure 1.9 (a) Simple leaf; (b) compound leaf.

1. Simple leaves. The blade consists of one continuous piece (Fig. 1.9a).
2. Compound leaves. Simple leaves may become deeply lobed and when the division between the lobes reaches the midrib it becomes a compound leaf, and the separate parts of the blade are called the leaflets (Fig. 1.9b).

Monocotyledonous leaves are generally simpler but still contain a network of (usually parallel) veins for transport.

The leaves are the main parts of the plant where photosynthesis occurs (Fig. 1.10). The leaf surface is often covered with a waxy cuticle produced by the epidermal cells near the surface. The cuticle helps to reduce water loss and protect against pathogens. The blade surface may be smooth (glabrous) or hairy, according to variety. This is important in legumes because it can affect their palatability. Below the epidermis are two types of mesophyll cells. The palisade mesophyll cells contain many chloroplasts and are the main site of photosynthesis, while the spongy mesophyll tissue contains many air spaces between the cells which as important for the exchange of gases into and out of the cells. Below this is the lower epidermis.

A very important feature of the leaf structure is the presence of large numbers of tiny pores (stomata) on the surface of the leaf, in the epidermis layers (Fig. 1.11). There are usually thousands of stomata per square centimetre of leaf surface. Each pore (stoma) is oval-shaped and surrounded by two guard cells. Carbon dioxide to be used in photosynthesis diffuses into the leaf through the stomata. Most of the water vapour leaving the plant, as well as the oxygen produced from photosynthesis, diffuses out through the stomata.

Monocotyledonous leaves are dealt with in Part IV, Chapter 19 (Grassland).

1.3.3.1 Modified leaves

1. Scales are normally rather thin, yellowish to brown membranous leaf structures, very variable in size and form. On woody stems they are present as bud scales which protect the bud; they are also found on rhizomes such as common couch.
2. Leaf tendrils. The terminal leaflet on the stem can be modified into thin thread-like structures, e.g., the pea plant (see Fig. 1.8).

Other examples of modified leaves are leafspines and bracts.

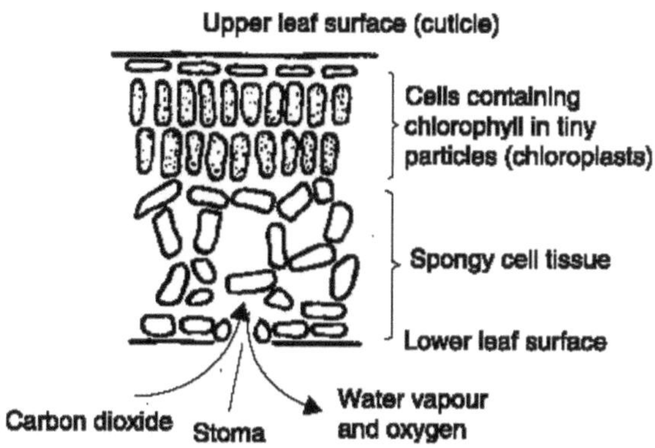

Figure 1.10 Cross-section of green leaf showing gaseous movements during daylight.

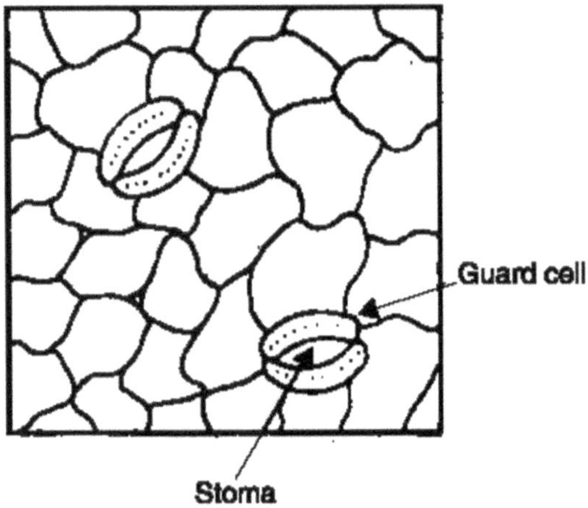

Figure 1.11 Stomata on leaf surface.

1.3.4 The inflorescence

Special branches of the plant are modified to bear the flowers, forming what is called the inflorescence. There are two main types of inflorescence:

1. Indeterminate fluorescence. Here the branches bearing the flowers continue to grow, so that the youngest flowers are nearest the apex and the oldest farthest away (Fig. 1.12a). A well-

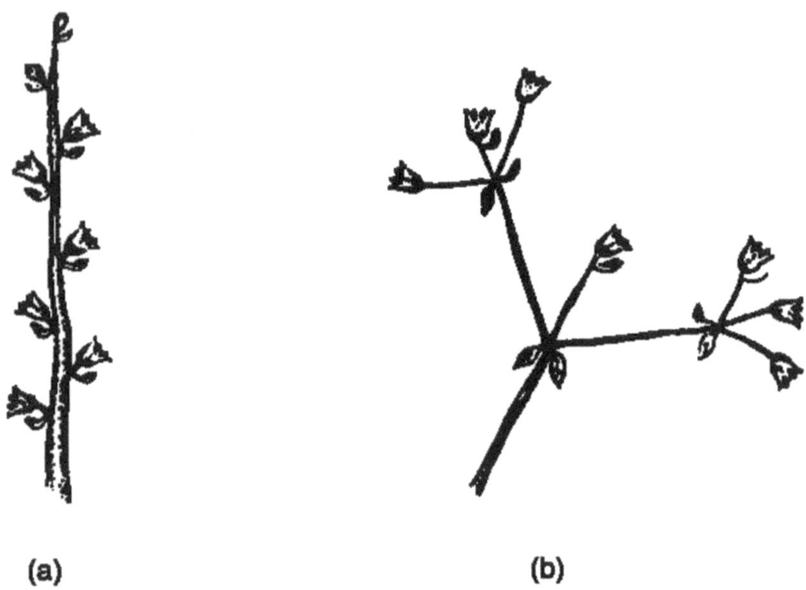

Figure 1.12 (a) Indeterminate inflorescence; (b) determinate inflorescence.

known example of this type of inflorescence is the spike found in many species of grasses such as wheat.
2. Determinate fluorescence. Here the main stem is terminated by a single flower and ceases to grow in length; any further growth takes place via lateral branches, and they themselves eventually terminate in a single flower and growth is stopped (Fig. 1.12b). An example of a species with this type of inflorescence is linseed.

There are many variations of these two main types of inflorescence. For example, some plants, such as sunflowers, have a head called a capitulum, while others, such as carrots and parsley have inflorescences called umbels. The inflorescences of oats and many grass species, meanwhile, are referred to as panicles.

1.3.5 The flower

Plants use a number of different approaches to reproducing via the use of flowers. The majority of plants possess hermaphrodite flowers that contain both male and female parts. However, a small number of plants possess separate male and female flowers on the same plant (referred to as monoecious species) e.g., maize, hazel and squashes, while around 6% of flowering plants species carry male and female flowers on separate plants (referred to as dioecious species) e.g., holly.

The remainder of this section refers to the structure and reproduction of a typical hermaphrodite flower.

In the centre of the flower is the axis that is simply the continuation of the flower stalk. It is known as the receptacle and on it are arranged four kinds of organ. Whilst many flowers are more complicated in appearance than the one described here, they still consist of these four main parts:

1. The lowermost is a ring of green leaves called the calyx, made up of individual sepals. The sepals form the outside of the closed flower bud and so protect the remainder of the flower at this stage.
2. Immediately above the calyx is a ring of petals, known as the corolla. The number, shape and colour of petals on a flower can differ widely between species. Petals are usually brightly coloured and their function is to attract pollinators (usually insects). Flowers can also use scent to attract pollinators. At the base of the petals are often found modified structures called nectaries. These, as their name suggests, produce the sweet nectar that acts as a reward for a visiting pollinator.
3. Above the corolla are the stamens, again arranged in a ring. These are the male part of the hermaphrodite flower. They are similar in appearance to an ordinary match, the swollen tip called the anther sitting on top of the filament, which holds the anther in place. When ripe, the anther bursts open to release the pollen grains. The filaments can be of varying sizes, either keeping the anther inside the flower or allowing it to hang outside (useful in plants which are pollinated by wind rather than e.g., insects).
4. The highest position on the receptacle is occupied by the pistil which consists of one or more small green bottle-shaped bodies called carpels. These are comprised of three parts: the stigma, style and the ovary. The style supports the stigma, which is the site of pollen binding and germination. The ovary contains ovules that, when fertilised, will form the future seeds (Figs. 1.13 and 1.14).

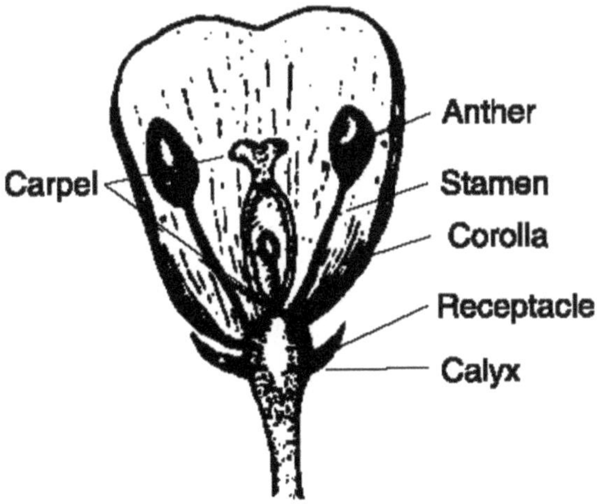

Figure 1.13 Longitudinal section of a simple flower.

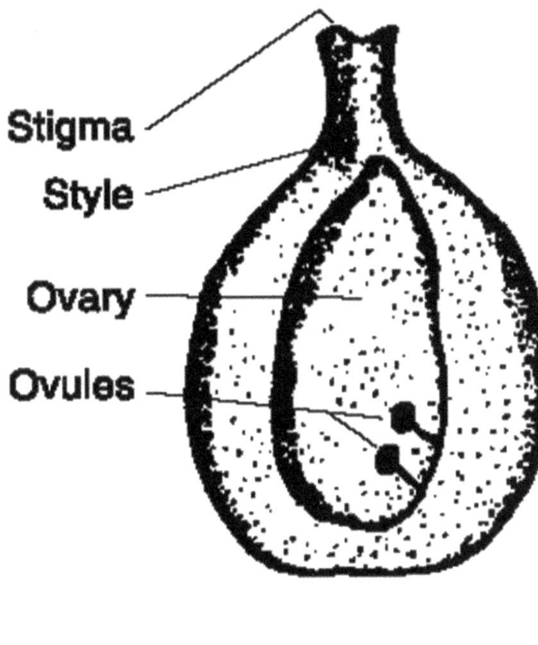

Figure 1.14 Carpel detail.

1.4 Plant physiology

1.4.1 Transpiration

The evaporation of water from plants is called transpiration. It mainly occurs through the stomata and has a cooling effect on the leaf cells. Water in the cells of the leaf can pass into the pore spaces in the leaf, such as in the spongy mesophyll tissue, and then out through the stomata as water vapour (Fig. 1.10).

The rate of transpiration varies considerably depending upon the environmental conditions and status of the plant. It is greatest when the plant is well supplied with water and the air outside the leaf is warm and dry, creating a large water vapour gradient. When the guard cells are turgid (full of water) the stomata are open. When the plant is under drought stress, the guard cells lose water and the stomata close, slowing down the loss of water vapour (transpiration) from the plant. However, this also slows down the rate of photosynthesis. The stomata also close in very cold weather, e.g., 0°C. Transpiration is also retarded if the humidity of the atmosphere is high because there is only a very small water vapour gradient between the inside of the leaf and the outside atmosphere. The stomatal guard cells also close (and so transpiration ceases) during darkness. This is to minimise water loss when photosynthesis ceases and the benefits of CO_2 uptake are lost.

1.4.2 Conduction

The conductive flow of water through the plant takes place in the xylem tissue that runs in bundles along the length of the root and stem and into the organs of the plant. The xylem vessels, which carry the water and mineral salts from the roots to the leaves, are tubes made from dead cells called tracheids and vessel elements. The cross walls of these cells are no longer present and the longitudinal walls can be thickened with lignum to form wood. These tubes help to strengthen the stem.

1.4.3 Translocation

The movement of food materials through the plant is known as translocation. The phloem forms a branched system throughout the plant and transports assimilates from sites of production (sometimes called sources) to sites of demand or consumption (sinks). The phloem tubes therefore carry organic material through the plant, e.g., sugars and amino acids, from the leaves to storage parts or growing points. In contrast to the xylem, the phloem vessels are chains of living cells, are not lignified, and possess cross walls which are perforated. They are sometimes referred to as sieve elements which, along with companion cells that provide energy and proteins, form the sieve tube members.

1.4.4 Osmosis

Much of the water movement into and from cell to cell in plants is due to osmosis. This is a process in which a solvent, such as water, will flow through a selectively permeable membrane (e.g., a cell membrane) from a weak solution to a more concentrated one. The cell membrane only allows the water to pass through, as the molecules in solution are too big. The force exerted by such a flow is called the osmotic pressure. In plants the normal movement of the water is from the soil solution into the cells. However, if the concentration of a solution outside the cell is greater than that inside, there is a loss of water from the cell, and it contracts; this is called plasmolysis.

1.4.5 Uptake of water

Water is taken into the plant from the soil. This occurs mainly through the root hairs near the root tip. Their function is to increase the surface area available for absorption of water. There are thousands of root tips on a single healthy crop plant (Fig. 1.4). As the root grows the hairs are constantly replaced.

A number of processes are involved in the uptake of water into and their subsequent flow through the plant. As the cell contents and intercellular spaces of the roots are usually more concentrated than that of the external rootzone, water moves into roots via osmosis. This creates pressure that can help drive water flow up through the xylem. In addition, transpiration of water from leaves creates a negative pressure higher up the plant, providing an additional driver for xylem sap flow. Finally, water movement along the narrow vascular tubes is aided by capillary action arising from interactions between the water molecules and the walls of the tubes.

The rate of absorption of water into roots is slowed down by:

1. A shortage of water in the soil.
2. A lack of oxygen for root respiration (e.g., in waterlogged soils).
3. A high concentration of salts in the soil water near the roots.

Normally, the concentration of the soil solution does not interfere with water absorption. High soil water concentration can occur in salty soils and near bands of fertiliser. Too much fertiliser near developing seedlings may damage germination and subsequent emergence by restricting the uptake of water.

1.4.6 Uptake of nutrients

The absorption of chemical substances (nutrients) into the root cells is partly due to a diffusion process but it is mainly due to the ability of the cells near the root tips to accumulate such nutrients. Nutrients are taken into the root in the form of charged ions through the root hairs, along with water. The water and solutes move through the cells into the inner ring of xylem. They are prevented from leaking back into the soil between the cell walls by a waxy layer of cells called the Casparian Strip. In this way an electrochemical gradient is produced which allows the flow of nutrient ions into the plant from the soil solution.

1.4.7 Photosynthesis

In photosynthesis, specialised pigments such as chlorophyll A and B use light energy (normally sunlight but sometimes artificial) to change carbon dioxide and water into sugars (carbohydrates) and oxygen in the green parts of the plant. Indeed, it is these pigments which give these parts of the plant their green appearance, because chlorophylls primarily absorb red and blue light but reflect green light. The cells that contain chlorophyll can also possess orange/yellow pigments such as xanthophyll and carotene, and brown pigments called phaeophytins, which absorb different wavelengths of light to the chlorophylls. Crop plants can only build up chlorophyll A and B in the light, and so any leaves that develop in the dark are yellow and cannot efficiently produce carbohydrates. The yellowing of leaves (chlorosis) can also be caused by disease attack, nutrient deficiency or natural senescence (dying off).

The process of photosynthesis may be set out as follows:

(a) The light reactions (light-dependent)

This takes place in the thylakoid membranes inside the chloroplast, an organelle found inside the cells of green plant tissue. The thylakoid membranes, which are the site of proteins containing chlorophyll, are stacked into structures called grana. Light provides energy for the chlorophyll molecules which causes the release of electrons. By passing through a series of proteins found in the thylakoid membranes, these electrons provide energy to form the energy storage molecule adenosine triphosphate (ATP), with water as a by-product. The chlorophyll molecules then regain electrons from the splitting of water into oxygen and hydrogen.

The hydrogen and ATP are then used in the next stage:

(b) The dark reactions (light-independent)

This takes place in the watery stroma of the chloroplast. Here, via a series of chemical reactions known as the Calvin Cycle, and using energy released by ATP molecules, the hydrogen is combined with carbon dioxide to form carbohydrates.

The overall reaction for the two stages of photosynthesis can be written:

$$CO_2 + 2H_2O \rightarrow (CH_2O) + O_2 + H_2O$$

Photosynthesis therefore not only provides the basis for all food production but it also supplies the oxygen which animals and plants need for respiration The carbohydrates (CH_2O) are simple sugars, which can be moved through the vascular system of the plant in solution to wherever they are needed. The simpler carbohydrates, such as glucose, may be built up to form starch for storage purposes or as cellulose for building cell walls. Fats and oils (lipids) are formed from carbohydrates by a process of esterification which produces mostly triglycerides. These are usually found in seeds and are a form of concentrated energy. Protein material, which is an essential part of all living cells, is made from carbohydrates and nitrogen compounds and also frequently contains sulphur. These form from amino acids which are held together in proteins by peptide bonds. Proteins can carry out structural roles, but many act as enzymes which catalyse chemical reactions in the plant.

The amount of photosynthesis that takes place each day is limited by the duration and intensity of sunlight, and the ability of the green parts of a plant to capture it. It will therefore depend upon weather conditions (such as cloud cover) and factors affecting day length, such as latitude and time of year, and shading by other plants, an important consideration in crop canopy and weed management. The amount of carbon dioxide available can also be a limiting factor. Shortage of water, low temperatures and leaf disease or damage can also reduce photosynthesis (Fig. 1.15).

The light reactions and the Calvin Cycle are common to all photosynthetic plants. However, some plants have evolved variations which aid photosynthesis under different conditions. The majority of plants carry out what is known as C3 photosynthesis, where all steps take place in the chloroplasts of mesophyll and other cells. This is so-called because the first product of carbon fixation is a compound containing three carbon atoms. While this method of photosynthesis works well in most environments,

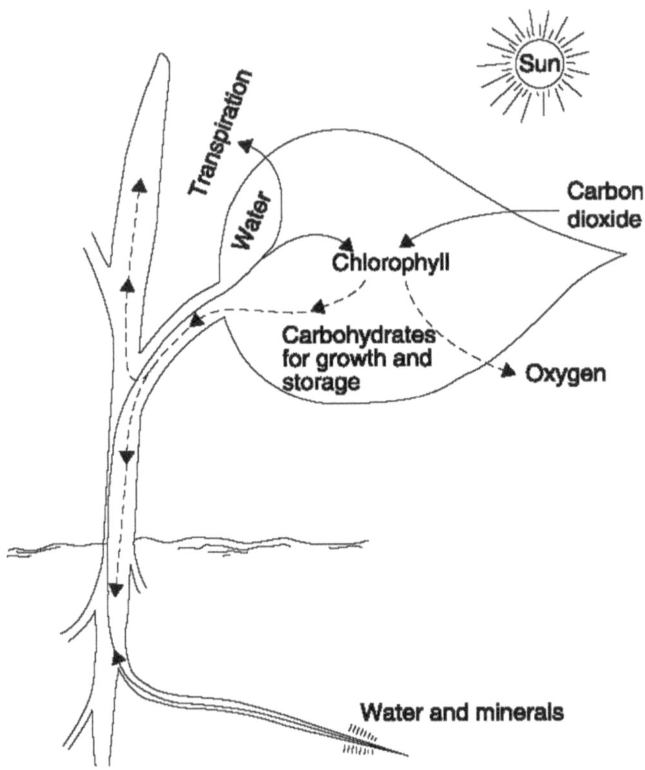

Figure 1.15 Overview of transport and photosynthesis in a plant.

in hot and dry conditions, when stomata close to reduce water loss, it performs poorly as oxygen builds up inside the leaf and photosynthetic enzymes bind this oxygen instead of carbon dioxide.

In order to deal with this, some plants have evolved variations, known as C4 (where the first product of carbon fixation is a compound containing four carbon atoms) and crassulacean acid metabolism (or CAM) photosynthesis. In C4 plants, photosynthetic processes are split between mesophyll and bundle sheath cells which surround the vascular bundles. This helps to reduce oxygen binding to the photosynthetic enzymes. A number of important crops are C4 plants, such as maize. In CAM plants, the stomata are closed in the day to reduce water loss and open at night to allow carbon fixation. Both of these approaches help to increase photosynthesis and reduce water loss in hot and dry environments.

1.4.8 Respiration

Plants, like animals, carry out respiration i.e., they break down, or metabolise, sugars such as glucose in order to release energy, stored again as ATP. To do so, they also need to take in oxygen to act as an acceptor for some of the electrons released during respiratory processes, which results in the production of water. Respiration also results in the release of carbon dioxide. Respiration therefore appears, superficially, to be the reverse of photosynthesis, with carbohydrates and oxygen as the inputs and carbon dioxide and water (plus energy) as the resulting products. However, it is actually more complicated than this and involves different metabolic processes taking place.

There are three main processes. The first is glycolysis, performed in the cell cytoplasm, where simple sugars such as glucose are split to release energy (as ATP) and to form pyruvate, water and an electron carrier molecule, called nicotinamide adenine dinucleotide (NADH). The second stage is carried out in the mitochondria and is called the 'Citric Acid' or 'Krebs' cycle, where the pyruvate is converted to citric acid, which cycles within the system through intermediate molecules, releasing carbon dioxide and generating more NADH and some further ATP. In the third and final stage, known as oxidative phosphorylation, the NADH molecules donate electrons to proteins found in the inner membrane of the mitochondria. Like in photosynthesis, the electrons pass through a series of these proteins to release energy for the formation of more ATP. Oxygen then acts as an electron acceptor and water is produced.

The ATP molecules can then be moved around the plant to provide it with the energy it requires for metabolism, growth and development.

Oxygen availability is important for the growth of plants. Indeed, their growth may be limited by the availability of oxygen in the rootzone. If the air spaces in the soil fill with water and it becomes waterlogged, this lowers oxygen availability and the roots may not be able to access sufficient oxygen for growth. This is termed hypoxia. If the situation persists and the soil has a complete lack of oxygen, caused by long term flooding, this is called anoxia.

1.4.9 Reproduction

1.4.9.1 Pollination

The first stage in reproduction is pollination. Pollen from the anther is transferred to the stigma via pollinators or by the wind. This may take the form of self-pollination, where the pollen is transferred from the anther to the stigma of the same flower, or cross-pollination when it is carried to a different flower, most often on a different plant of the same species (Fig. 1.16). The vectors of pollen transfer can also vary. Some flowers are insect pollinated (entomophilous) and are usually scented and brightly coloured. The pollen itself is usually sticky or oily. Other flowers are wind pollinated (anemophilous) and do not need to be brightly coloured. They produce huge amounts of pollen (as most of it is lost) and the pollen grains are smooth, light and small, aiding their transfer by the wind. The flowers of wind pollinated plants are often unisexual with a predominance of male flowers. Stamens and stigmas often hang outside the flowers, and the stigmas are often feathery to give them a better chance of trapping a pollen grain as it blows past.

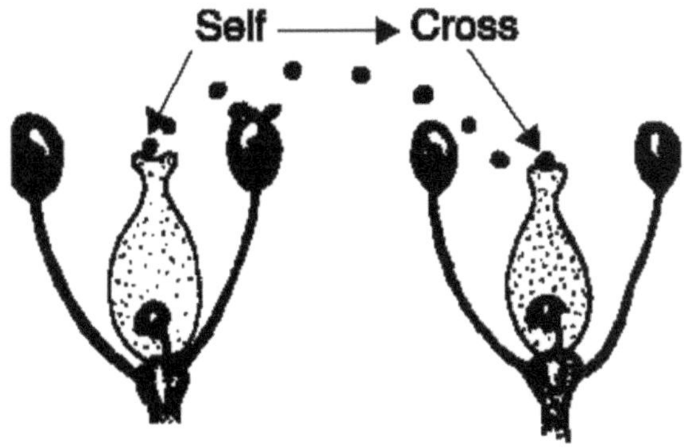

Figure 1.16 Self- and cross-pollination.

1.4.9.2 Fertilisation

After a pollen grain attaches to the stigma, it soon germinates and a pollen tube containing three nuclei (one tube nucleus and two male nuclei) begins to grow. The pollen tube grows down the style of the carpel, towards the ovary sac. When the ovary is reached, the pollen tube bursts. This releases the nuclei. One male nucleus fuses with the egg cell in the ovule and the other joins with a second nucleus to form the primary endosperm nucleus. This double fertilisation is unique to flowering plants.

The ovule itself goes on to form the seed. The ovary also changes after fertilisation to form the fruit which encloses the seed.

With grasses, and therefore cereals, there is only one seed formed in the fruit and, being so closely united with the inside wall of the ovary, it cannot easily be separated from it. This one-seeded fruit is called a grain.

1.4.9.3 Asexual reproduction

Some plants can also reproduce without the processes of pollination and fertilisation. Rhizomes and stolons can generate new plants by forming roots and shoots from their nodes. Plants can also be propagated by the use of cuttings made from stems or leaves. This is sometimes used in horticulture to obtain many genetically identical plants from a single parent. One drawback of these asexual forms of reproduction is a lack of the genetic diversity that is introduced via pollination and fertilisation and which can cause problems with susceptibility to issues such as pests and diseases.

1.5 Seeds

1.5.1 Dicotyledon

A good example of a dicotyledon seed is the field bean. If its pod is opened when nearly ripe it will be seen that each seed is attached to the inside of the pod by a short stalk called the funicle. All the nourishment that the developing seed requires passes through the funicle from the bean plant. When the seed is ripe and has separated from the pod, a black scar, known as the hilum, can be seen where the funicle was attached. Near one end of the hilum is a minute hole called the micropyle (Fig. 1.17).

If a bean is soaked in water the seed coat can be removed easily and all that is left is largely made up of the embryo (germ). This consists of two seed leaves (cotyledons) which contain the food for the young seedling. Lying between the two cotyledons is the radicle (which eventually forms the primary root) and a continuation of the radicle at the other end, the plumule (Fig. 1.18). This develops into the young shoot and is the first bud of the plant that will hold true leaves.

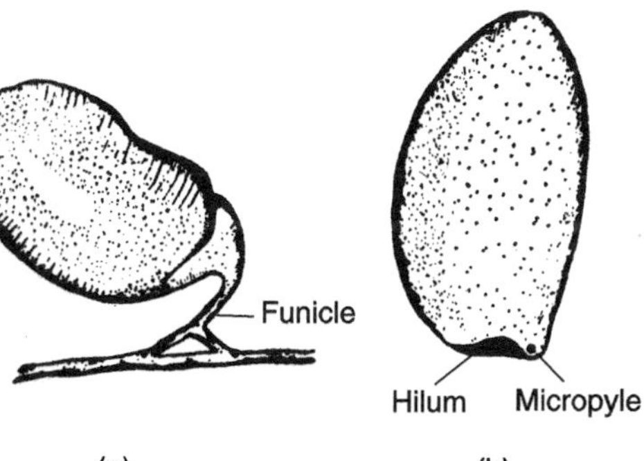

Figure 1.17 (a) Bean seed attached to the inside of the pod by the funicle. (b) Bean seed showing the hilum and micropyle.

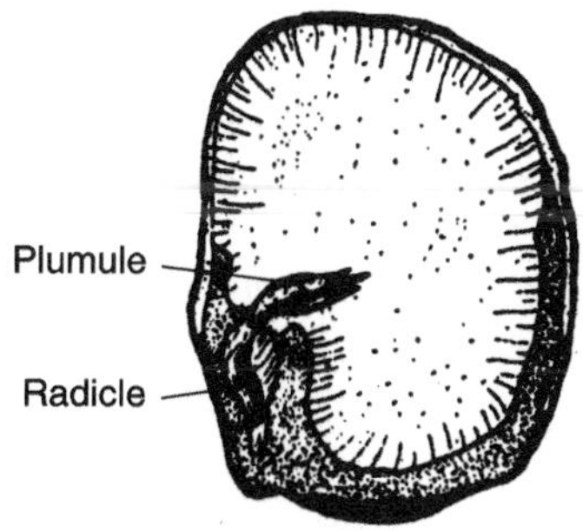

Figure 1.18 Bean seed with one cotyledon removed.

1.5.2 Monocotyledon

The wheat grain is a typical example of a monocotyledon seed. It is not a true seed (it should be called a single-seeded fruit). The seed completely fills the whole grain, being almost united with the inside wall of the grain or fruit. This fruit wall is made up of many different layers which are separated on milling into varying degrees of fineness, e.g., bran and pollards which are valuable livestock feed.

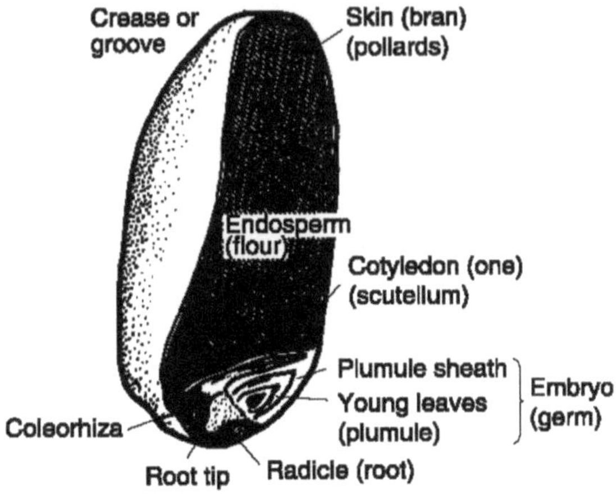

Figure 1.19 Wheat seed cut in half at the crease.

Most of the interior of the grain is taken up by the floury endosperm. The embryo occupies the small raised area at the base. The scutellum, a shield-like structure, separates the embryo from the endosperm. Attached to the base of the scutellum is the radicle, enclosed by a sheath called the coleorhiza. Meanwhile, the cotyledon is enclosed by the coleoptile. The position of the radicle and the plumule (which will form the shoot carrying the first true leaves) can be seen in Fig. 1.19.

1.5.3 Germination

Germination of the seed requires suitable environmental conditions. This ensures that the seed does not germinate when conditions are unsuitable for subsequent growth of the plant e.g., too dry or too cold. Different forms of seed dormancy control when a seed will germinate and under which conditions. Seeds require water, heat and air for successful germination.

Uptake of water into the seed is called imbibition. Imbibition causes the seed coat to swell and break and also causes the production of enzymes within the seed that will be needed for its metabolism. The moisture requirement for the germination of the seed varies between species. For example, cereal seeds and oilseed crops require 18% and 40% moisture content, respectively for germination. This has implications for storage of harvested crops and also for the sowing of seeds by farmers as dry seedbeds will delay the germination and establishment of the crop.

Seeds also require oxygen from the air for germination. This is used via respiration to provide energy for the growing seedling. Air is found in pores in the soil but these can become filled with water after heavy rainfall and reduce oxygen availability, negatively impacting germination.

Seeds of different plant species also require different temperatures for germination. Spring sown crops such as maize and sugar beet require higher soil temperatures for seed germination than e.g., winter wheat. Seeds of the same crop will also therefore take longer to germinate for example, in November versus September, in part because lower temperatures slow the rate of respiration in the seed.

Together, these factors therefore have implications for the selection of both crops and cultivation methods, including the planting depth of seeds. Seeds which are sown at too shallow a depth can be at risk of drying out and consumption by pests, however, those which are sown too deeply may deplete their energy reserves before the shoot can reach the surface and photosynthesis begin. Small seeds such as brassicas, clovers and many of the grasses must be, as far as possible, shallow sown. Their food reserves will be exhausted before the shoot reaches the surface if sown too deep. Larger seeds such as beans and peas can and should be sown deeper.

1.5.4 Germination of the bean—the dicotyledon

Given suitable conditions for germination, the seed coat of the dormant, but living, seed splits near the micropyle and the radicle begins to grow downwards through this split to form the main, or primary, root from which lateral branches will soon develop (Fig. 1.20). When the root is firmly held in the soil, the plumule starts to grow by pushing its way out of the same opening in the seed coat. As it grows upwards its tip is bent to protect it from injury in passing through the soil, but it straightens out on reaching the surface, and leaves develop very quickly from the plumular shoot.

In the field bean the cotyledons remain underground, gradually giving up their stored food materials to the developing plant. This is called hypogeal germination (Fig. 1.21a). However, in the French bean and in many other dicotyledon seeds, the cotyledons are brought above ground with the plumule. This is epigeal germination (Fig. 1.21b).

1.5.5 Germination of wheat—the monocotyledon

When the grain germinates the coleorhiza expands and splits open the seed coat and, at the same time, the radicle breaks through the coleorhiza and enters the soil (Fig. 1.22).

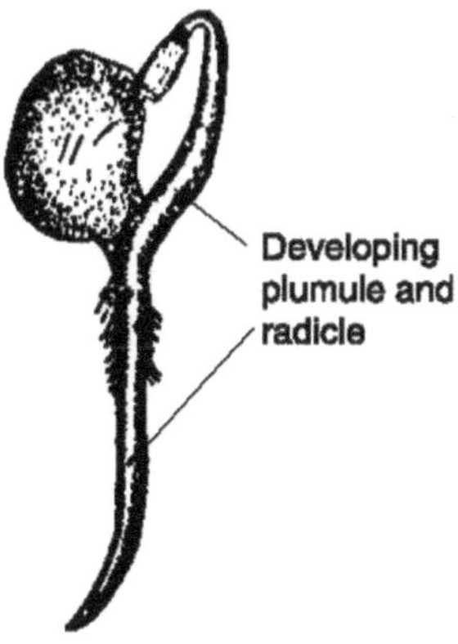

Figure 1.20 Germination of bean seed.

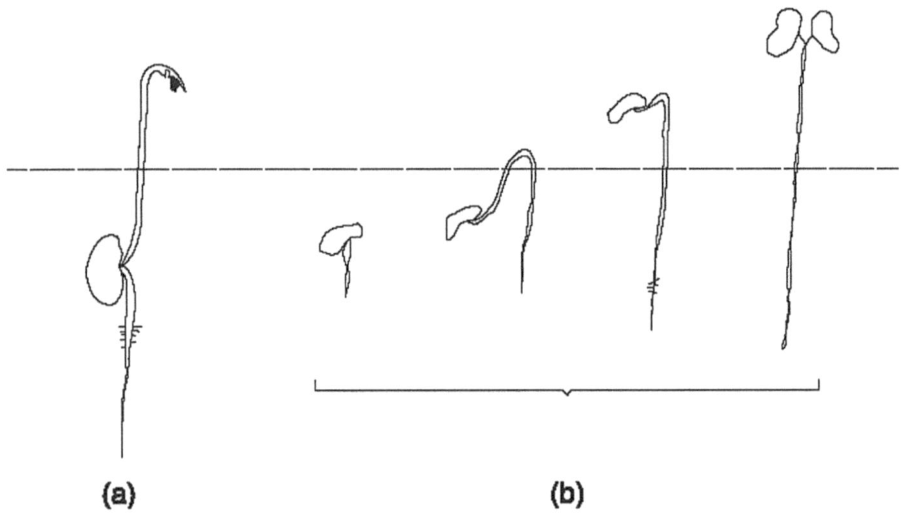

Figure 1.21 (a) Hypogaeal germination; (b) epigaeal germination.

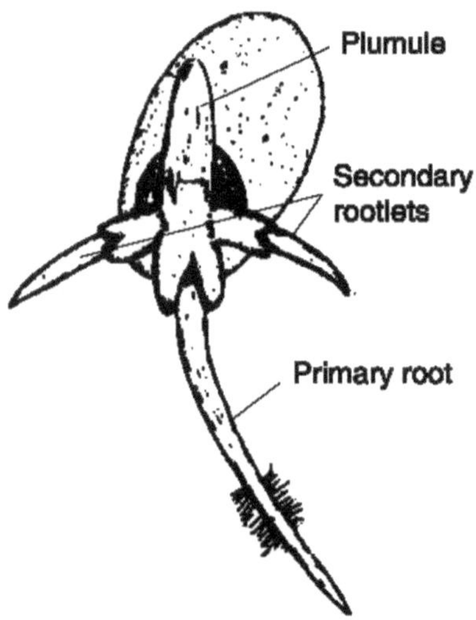

Figure 1.22 Germination of the wheat grain.

The primary root is soon formed, supported by the two pairs of secondary rootlets, but this root system (the seminal roots) is only temporary and is soon replaced by adventitious roots (Figs. 1.3 and 1.23). As the first root system is being formed at the base of the stem, so the coleoptile and plumule start to grow upwards and appear above the ground as a pale tube-like structure. From a slit in the top of the coleoptile there appears the first true leaf followed by others, the younger leaves growing from the older leaves (Fig. 1.24).

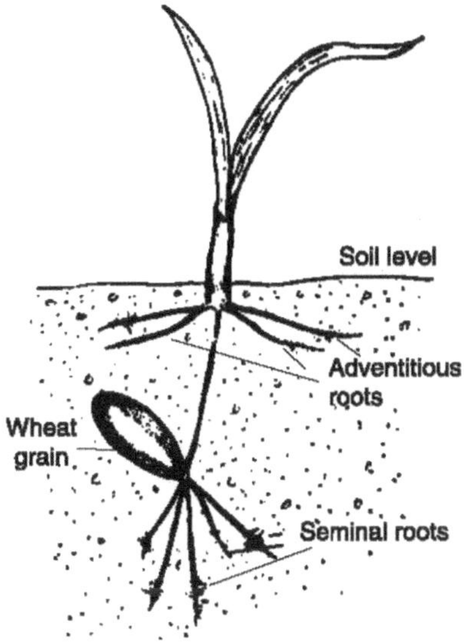

Figure 1.23 Developing wheat plant.

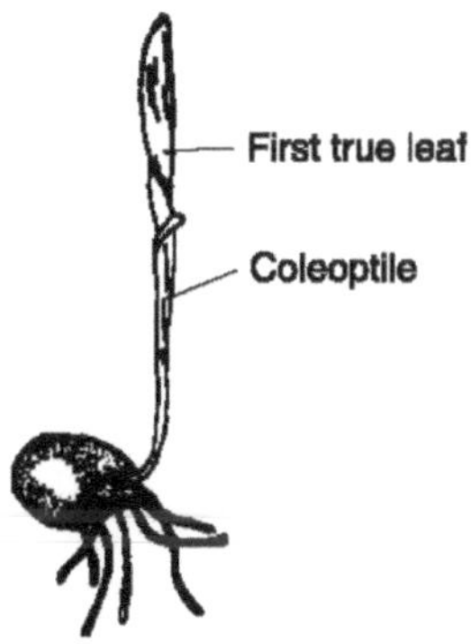

Figure 1.24 Seedling wheat plant.

As the wheat embryo grows the starch reserves of the floury endosperm are used up by the developing roots and plumule. The scutellum has the important function of changing the endosperm into digestible food for the growing parts, converting starch into simple, mobile sugars. It does this by producing the plant hormone gibberellic acid (GA). After production, gibberellic acid moves to the aleurone layer of the endosperm. Here it causes the production of enzymes including alpha amylase, which break down starch stored in the endosperm.

Unlike, in the field bean, where the cotyledons provide the food for the early nutrition of the plant, the wheat grain is dependent upon the endosperm.

When the leaves of the plant begin to manufacture food by photosynthesis, and when the primary root has established itself sufficiently well to absorb nutrients from the soil, the plant can develop independently, provided there is sufficient moisture and air present in the soil, and conditions above ground are suitable for growth.

The main differences between the two groups of plants can be summarised in Table 1.1.

1.6 Plant requirements

To grow satisfactorily a plant needs warmth, light, water, carbon dioxide, oxygen and other chemical elements which it can obtain from the soil.

1.6.1 Warmth

Most crop plants in this country start growing when the average daily temperature reaches 6°C. Growth is best between 16°C and 27°C. These temperatures apply to thermometer readings taken in the shade about 1.5 m above ground. Crops grown in hotter countries usually have higher temperature requirements. Several plants require specific soil temperatures before they can successfully germinate, e.g., maize. Plants require a certain number of 'heat units' before they can successfully complete their development. This explains the difficulty of trying to grow subtropical or Mediterranean plants in the UK, e.g., durum wheat, grain maize or soya. Climate change may make it easier and more successful to grow these crops but the change may not necessarily result in 'global warming' effects and rises in average temperatures in many temperate countries. Instead, increased variation in seasonal and short-term weather patterns may be the result and actually increase the difficulty of growing crops with a higher incidence of extreme weather events such as flooding.

Cold, frosty conditions may seriously damage plant growth. Crop plants differ in their ability to withstand extreme cold. For example, winter rye and wheat can stand colder conditions better than winter oats. Potato plants and stored tubers are easily damaged by frost. Sugar beet plants may go to seed (bolt) if there are frosts after germination producing a vernalisation effect (see below). Frost in December and January may destroy crops left in the ground. Cold may, however, be beneficial. A period of cold temperature is necessary to allow some autumn-sown crops to change from vegetative to sexual development. This is called 'vernalisation'. Some seeds may have to undergo cold conditions before they will germinate. This is known as 'stratification'. Cold temperatures can also help to reduce the incidence of weeds, pests and disease in crops.

1.6.2 Light

Without light, flowering plants cannot produce carbohydrates and will soon die. The amount of photosynthesis that takes place daily in a plant is partly due to the length of daylight, and partly to the intensity of the sunlight. Cloud cover can dramatically reduce the amount of light reaching the ground, and so the amount of photosynthesis that is possible.

Bright sunlight is very important where there is dense plant growth. However too much light can damage the photosynthetic proteins of the plant via the production of reactive oxygen species. The periods of daylight and darkness will vary according to the distance from the equator and also from season to season. This affects the flowering and seeding of crop plants and is another of the limiting factors in introducing new crops into a country.

Light also controls stem growth. The stems of seedlings growing in the dark elongate much faster than those growing in light (a process known as etiolation). This can be a problem as it can lead to long weak stems in some crops such as culinary herbs.

1.6.3 Water

Water is an essential part of all plant cells and it is also required in large amounts to account for losses via the process of transpiration. Water carries nutrients from the soil into and through the plant; it also carries the products of photosynthesis from the leaves to wherever they are needed. It is also used to give the plant rigidity (turgor), is used in many of the biochemical pathways inside the plant, and is also used to regulate the temperature of the plant. On average, plants take up about 200 tonnes of water for every tonne of dry matter produced.

1.6.4 Carbon dioxide

Plants need carbon dioxide for photosynthesis. This is taken into the leaves through the stomata, and the amount which can go in is affected by the degree of stomatal opening. Another limiting factor is the small amount (0.03%) of carbon dioxide in the atmosphere. The percentage of carbon dioxide can increase just above the surface of soils rich in organic matter, where soil bacteria are active and releasing carbon dioxide. This is possibly one of the reasons why crops grow better on such soils. In protected crops, in greenhouses and polytunnels, the levels of carbon dioxide can be artificially raised by the grower to improve production. In some instances this is provided as a by-product from other industrial or agricultural processes such as sugar refinery. As discussed earlier, some plants have a different carbon utilisation pathway which makes them more efficient at using CO_2 and therefore can survive amongst thick tropical vegetation where the CO_2 supply is limited or in hot conditions where stomata frequently have to close.

1.6.5 Oxygen

Contrary to popular belief, plants need oxygen all the time, not just through the night. Oxygen plays a key role in the process of respiration. Seedlings, before they emerge and begin photosynthesis rely entirely on respiration of the seed reserves for energy and therefore require oxygen in the seedbed. Additionally, oxygen is crucial for root growth and survival with oxygen supply to roots being impaired by periods of waterlogging.

1.6.6 Chemical elements required by plants

Many different chemical elements are needed by the plant in order that it may live and flourish. Different elements are required by the crop in different amounts. Elements required in larger quantities are referred to as macronutrients (or major nutrients, see Table 1.2). Those required only at a small level are called micronutrients (also known as minor or trace elements). The exact amount of each nutrient needed by the crop

Table 1.2 Sources and uses of major nutrients.

Macronutrients	Use	Source
Carbon (C) Hydrogen (H) Oxygen (O)	Used in making carbohydrates	The air and water
Nitrogen (N)	Building amino acids and proteins	Organic matter (including farmyard manure, FYM) Nitrogen-fixing soil micro-organisms Nitrogen fertilisers such as ammonium and nitrate compounds and urea
Phosphorus (P) (phosphate)	Essential for DNA replication, energy provision and many chemical reactions	Small amounts from the mineral and organic matter in the soil Mainly from phosphate fertilisers, e.g., superphosphate, ground rock phosphate, basic slag and compounds
Potassium (K) (potash)	Helps with formation of carbohydrates and proteins Regulates water in and through the plant	Small amounts from mineral and organic matter in the soil Potash fertilisers, e.g., muriate sulphate of potash
Calcium (Ca)	Essential for development of growth tissues, e.g., root tips	Usually enough in the soil. Applied as chalk or limestone to neutralise acidity
Magnesium (Mg)	A necessary part of chlorophyll	If soil is deficient, may be added as magnesium limestone of magnesium sulphate, also FYM
Sulphur (S)	Part of many proteins and some oils	Usually sufficient in the soil Atmospheric sulphur absorbed by the soil and plant Added in some fertilisers (e.g., sulphate of ammonia and superphosphate)

varies between crop types and is influenced by factors such as soil type and climate. Lighter soils, such as those containing a high proportion of sand, tend to have lower nutrient levels compared to heavier, clay-rich soils.

Despite this, most soils are usually able to supply the majority of the required nutrients and in farming practice it is predominantly with regard to nitrogen (N), phosphorus (P), potassium (K) and to a lesser extent magnesium (Mg) that there is any widespread necessity to supplement the natural supplies from the soil. However, other macronutrients may be required in certain situations. Calcium is an essential plant nutrient but, as lime, it is regarded more as a soil conditioner for changing pH. Sodium is highly desirable for crops such as sugar beet and fodder beet (which have maritime ancestors), when it can replace some or all of the potassium requirements. Sulphur, in areas away from industry, could be needed for grass crops when cut more than once in the year, and for certain brassicas such as oilseed rape.

Despite being required only in small amounts by the plant, micronutrients are, nevertheless, essential and a shortage or lock-up in the soil, especially of boron and manganese (often as a result of liming or high organic matter), will cause deficiency symptoms in particular crops. Other trace elements include manganese, chlorine, iron, molybdenum and zinc, but these rarely cause trouble on most farm soils (see Chapter 4).

Agricultural sources and applications of these macronutrients are discussed more fully in Chapter 4 (fertilisers and manures) and see also Table 1.2.

1.7 Legumes and nitrogen fixation

Legumes are plants that have a number of interesting features, such as:

- A special type of fruit called a legume, which splits along both sides to release its seeds, e.g., pea pod.
- Nodules on the roots containing special types of bacteria (*Rhizobium*) which can 'fix' (convert) nitrogen from the air into nitrogen compounds. These bacteria enter the plant through the root hairs from the surrounding soil and lead to the formation of root nodules. The bacteria are contained within the nodules. The plant provides carbohydrates for the bacteria and the bacteria provide nitrogen-containing compounds for the plant.

This fixation of nitrogen is of considerable agricultural importance. Many farm crops are legumes, for example peas, beans, vetches, lupins, clovers, lucerne (alfalfa), sainfoin and trefoil. The nitrogen compounds produced by the bacteria in the root nodules are released into the soil or directly into the host plant in the form of soluble nitrates. In the soil, the ammonium part of the compound is changed to nitrate and taken up by neighbouring plants, e.g., by grasses in a grass and clover sward, or by the following crop, e.g., wheat after clover or beans. The amount of nitrogen that can be fixed by legume bacteria varies widely—estimates of 50–450 kg/ha of nitrogen have been made. Some of the reasons for variations are:

- *The type of plant.* Some crop plants are able to fix more nitrogen than others, e.g., lucerne and clovers (especially if grazed) are usually better than peas and beans.

- *The conditions in the soil.* The bacteria usually work best in soils that favour the growth of the plant on which they live. A good supply of calcium and phosphate in the soil is usually beneficial, although lupins grow well on acid soils.
- *The strains of bacteria present.* The majority of soils in any particular country contain the strains of bacteria required for most of the leguminous crops which are grown. However, if a legume is introduced, e.g., lucerne or alfalfa into the UK, and it has not been grown in the field within the previous 3 years, it is necessary to inoculate the seed, i.e., treat it with an inoculum containing *Rhizobium meliloti,* before sowing to encourage effective nodulation.

1.8 The control of plant growth and development

1.8.1 Plant hormones

In order for plants to grow, develop and respond to their environment in an orderly manner, there needs to be some kind of regulation of underlying biological processes such as cell division, elongation and differentiation and the activation of defences against stress. As the plant has no nervous system, this regulation is entirely chemical in nature. The chemicals involved are called plant hormones, or phytohormones and can be divided into a number of different classes:

1. Abscisic acid (ABA)—a growth inhibitor associated with dormant buds and some seeds, and the closure of stomata.
2. Auxins—these substances mostly control cell enlargement and differentiation. Most auxins are made at the apex of stems and in young leaves.
3. Brassinosteroids—involved in cell elongation and division, responses to gravity and stress.
4. Cytokinins (CKs)—associated with cell division in the presence of auxins. Found in large amounts in fruits and seeds where they are associated with embryo growth.
5. Ethylene (ethene)—a gas which controls fruit ripening and senescence. Controls fruit and leaf drop from some trees. Antagonistic to auxins.
6. Gibberellins (GAs)—also involved in cell enlargement, but also play a key role in the breaking of dormancy in seeds and the mobilisation of seed reserves by the production of the enzyme alpha amylase.
7. Jasmonates (JAs)—involved in the response to pests and pathogens.
8. Salicylic acid (SA)—involved in response to pathogens and environmental (abiotic) stress, such as drought, extreme temperature or salinity.
9. Strigolactones (SLs)—involved in a range of processes including shoot branching, environment responses and interactions with soil fungi.

There is a large degree of interaction between the different hormone signalling systems within a plant. Some hormones are able to cause the up- or down-regulation of other hormones and can be involved in the regulation of a number of different processes in a tissue and dose-dependent manner.

One of the ways that plants react to external stimuli is by movement. Roots, shoots, leaves and flowers can all move either towards or away from these stimuli. This movement is called 'tropism'.

1.8.2 Phototropism

This is the movement in response to light. Shoots will be positively phototropic and move towards light. Auxins play a major part in this movement by moving to the dark side of the shoot, encouraging cell growth and thus bending the shoot towards the light source. This is a vital response if maximum sunlight interception is to take place by the plant so that this can be converted to yield.

1.8.3 Geotropism

This is the response to gravity. Shoots are negatively geotropic and therefore grow upwards while roots are positively geotropic and grow downwards. This is particularly important in the germinating and emerging seed and means that the grower does not have to ensure that the seed is planted the right way up. Other organs of the plant have an intermediate geotropism, e.g., leaves and side branches.

1.8.4 Hydrotropism

Roots are positively hydrotropic and will grow towards water.

1.8.5 Thigmotropism

This is a type of tropism related to contact. Tendrils of peas are positively thigmotropic allowing them to wrap around each other and any supporting objects them come into contact with. Similarly, hop stems will climb up strings in the hop garden by wrapping themselves around for support.

1.8.6 The effect of light

As well as movement caused by differentiation in cell enlargement and division, the plant's development is also controlled in part by external stimuli. By far the most important factor is light, although temperature does have a part to play. As discussed, light can affect plants in several ways. Without sufficient light, plants become 'etiolated'—they grow tall and thin, making them fragile and can lack chlorophyll, resulting in a yellow appearance. In addition, the intensity and spectrum of the light has an effect on developmental processes such as breaking dormancy in some seeds and in the regulation of flowering.

Probably the most important aspect of light is its duration. The response of a plant to day length is called 'photoperiodism'. In fact, it is the length of the dark period that is critical. The biggest effect is on flowering, although fruit and seed production, dormancy and leaf fall are also affected. Plants can be categorised into three groups when considering the differences in photoperiod requirements for flowering:

1.8.6.1 Short day (long night) plants

Flowering is induced by dark periods *longer* than a critical length. The length can vary according to species.

1.8.6.2 Long day (short night) plants

Flowering is induced by dark periods *shorter* than a critical length. Many spring crops, e.g., spring wheat and spring barley, are long day plants with the move into summer, and the subsequent shortening of the nights, stimulating them to move from vegetative growth to sexual development.

1.8.6.3 Day neutral plants

Flowering is independent of photoperiod. There are a large number of plants in this category.

Photoperiodism is more important for plants located further from the equator, that is, in areas with significant seasonal variation in day length.

The knowledge of how each species is triggered into flowering is particularly important in the horticultural cut flower and house plant sector. Growers need to synchronise flowering with important dates in the calendar, such as Mother's Day or Christmas. This ensures that their stock will be in perfect condition for the many bouquets that are bought at these times. They can manipulate flowering in the glasshouse or polytunnel by artificially adjusting lengths of daylight using high intensity lamps and exploiting the plant's photoperiodic response.

1.9 Key points

- Plants can be classified according to their growth patterns and structure.
- Plant structure consists of roots, stems, leaves and flowers.
- Plant physiology includes a number of key processes such as photosynthesis, transpiration, uptake and conduction of water and nutrients and translocation of food.
- Plant reproduction involves pollination and fertilisation, leading to the production of seeds. Seed germination varies between plants and is controlled by a range of environmental factors.
- Plants have a number of requirements for growth including warmth, light, water, carbon dioxide, oxygen and soil nutrients.
- Plant growth and development is controlled by a range of different hormones.

Sources of further information and advice

Barnes, C., Poole, N., Poore, N., 1994. Plant Science in Action. Hodder Arnold.
Bidlack, J.E., Jansky, S., 2010. Stern's Introductory Plant Biology. McGraw-Hill, New York.
Mauseth, J.D., 2003. Botany: An Introduction to Plant Biology. Jones and Bartlett, London.
Rost, N.Y., et al., 1984. Botany: A Brief Introduction to Plant Biology. John Wiley.
Sablowski, R., et al., 2010. Plant Biology. Garland Science, Abingdon.
Taylor, D.J., Green, N.P.O., Stout, G.W., Soper, R. (Eds.), 1997. Biological Science 1 and 2, third ed. Cambridge.

Climate and weather

2

2.1 Introduction

Climate has an important influence on the types of crops which can be grown at a given geographic location. It may be defined as a seasonal average of weather conditions in a given area and time of the year, typically averaged over a period of 30 years. Weather is the atmospheric conditions at any one time. Weather patterns are created by the state of the atmosphere and movement of air currents. The Earth is surrounded by an envelope of gas (atmosphere) and oceans cover more than 70% of the surface. These fluids (atmosphere and oceans) move when heated, creating convection currents which transfer heat from equatorial to polar areas.

Weather is the combined effect of such conditions as heat or cold, wetness or dryness, wind or calm, clearness or cloudiness, pressure and the electric state of the air. It has a major influence on the type, yield and quality of the crops grown.

The climate of the United Kingdom is mainly influenced by:

- The amount of solar energy received, which is mainly a product of distance from the equator (50–60° latitude).
- The warm Gulf Stream which flows along the western coasts.
- The direction of the prevailing winds e.g., south-westerly.
- The numerous 'lows' or 'depressions' which cross from west to east and bring most of the rainfall.
- The elevation and distribution of highlands and lowlands; most of the uplands occur on the west coast of the UK or run from north to south through the middle of the country.
- Its nearness to the continent of Europe, from where warm southerly winds blow in summer and cold easterly winds affect the weather in the southern and eastern areas.
- Positioning and strength of the jet stream, which consists of very strong winds 5–10 miles above the earth, determine the strength of prevailing westerly winds across the Atlantic.

The normal positioning of the jet stream occurs to the north and west of the British Isles. When the pressure difference in the North Atlantic between the Azores and Iceland is large, south westerly winds dominate with stronger and more frequent storms. When the pressure difference is small, there is an increased tendency for blocked weather patterns, reducing the influence of Atlantic weather systems.

Local variations are caused by elevation, aspect and slope.

Elevation: Height above sea level can affect climate in many ways. The temperature drops about 0.6°C for every 100 m rise above sea level, reducing the rate of plant growth. Every 15 m increase in height can shorten the growing season by 2 days (one in spring and one in autumn). High land is more likely to be buffeted by strong winds and will receive more rain from the moisture-laden prevailing winds, which are cooled as they rise upwards.

Lockhart and Wiseman's Crop Husbandry Including Grassland. https://doi.org/10.1016/B978-0-323-85702-4.00022-4

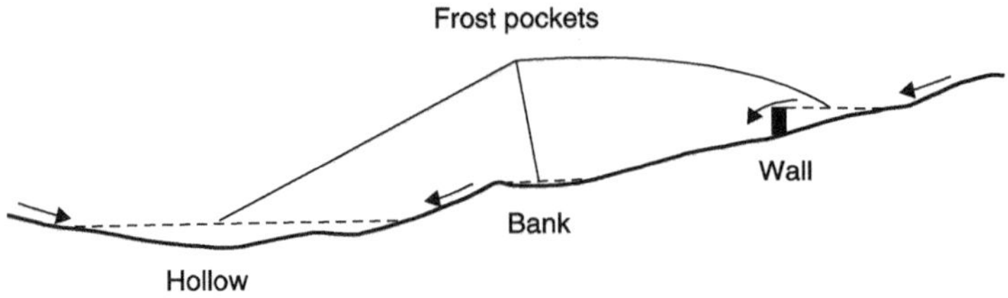

Figure 2.1 Frost pockets formed as cold air flows down a slope.

Aspect: The direction in which land faces can affect the amount of sunshine absorbed by the soil and thus the soil temperature. In the UK the soil temperature of north-facing slopes may be 1°C lower than on similar slopes facing south.

Slope: When air cools it becomes heavier and will move down a slope and force warmer air upwards. This is why frost often occurs on the lowest ground on clear still nights whereas the upper slopes may remain free of frost. 'Frost pockets' occur where cold air collects in hollows or alongside obstructing banks, walls, hedges, etc. (Fig. 2.1). Frost-susceptible crops such as early potatoes, maize and fruit should not be grown in such places.

Crop production can be regarded as essentially the conversion of solar radiation, water and soil nutrients into useful end-products. The influences of various aspects of weather and climate on potential crop production can be numerous and varied. For instance, sunshine and rainfall could be considered as primary controls of production because of their immediate influence on the rate of crop metabolism, nutrient uptake, turgidity, biochemical processes within the plant and crop structure, as well as their effect on soil and air temperature. In addition, there will be an effect from topography, geographical position and proximity to bodies of water. Other climatic elements such as frosts, wind and humidity will also have an effect on production because of their influence on such things as crop disease, pest distribution and direct physical damage.

2.2 Solar radiation

Solar radiation during the growing season is a very significant determinant of crop production because of its role in photosynthesis. In the absence of any other restrictions, crop production or harvestable yield would be determined by the interception of sunlight. The intensity of solar radiation changes with season, latitude, aspect, slope and the amount of cloud cover and levels of atmospheric pollution. More light falls onto the south of England than in Scotland and the difference between the UK and, say, the south of France is even more marked. Yields of sugar beet in the Paris basin are often 25%–30% higher than in East Anglia, whilst the yield of apples in Provence can be 100% higher than in the West Midlands. Other factors may be involved but a large part of the difference is due to the intensity and the duration of sunlight. The efficiency of use of the sun's energy by crops is not good, with a maximum potential of about 8%, but, in commercial field crops, it is nearer 1%. The *interception* of light will vary according to ground cover: early in the season much of the sunlight will fall onto bare ground because of the small size of the plants; and late in the season, with senescence and subsequent leaf loss, the interception of energy will again fall. During the crop's 'grand growth stage' interception of incident light will be near 100%.

2.3 Rainfall

Water supply is vital for crop growth. Crops require a regular and predictable supply of rain during the growing season. Too little rainfall and soil water deficits can reduce plant cell division and cell growth. Eventually photosynthesis will slow down due to the closing of the stomata. Leaves will wilt and their ability to intercept sunlight will be adversely affected and it will be more difficult for the plant to take up nutrients in solution from the soil. The soil water deficit of a crop is determined by the difference between rainfall and rates of plant evapotranspiration from leaf surfaces. Seasonal water deficits can be compensated for by crop irrigation. Too much rainfall during the growing can be almost as problematic, requiring land drainage to prevent soil water logging and allow machinery access for crop management and harvesting operations. Intense rainfall events, such as those observed in recent decades can cause flooding, soil erosion and severe crop damage or loss. In the UK rain comes mainly from the moist south-westerly winds and from the many 'depressions' which cross from west to east. Western areas receive much more rain than eastern areas; this is partly due to the west to east movement of the rain-bearing air and also because most of the high land is along the western side of the country. As the moisture-laden air rises over this high ground it cools, the moisture condenses, becomes heavy and falls as rain. On the leeward (sheltered) side of the high ground the air is drier and warmer. This is sometimes known as a 'rain shadow' effect.

The average annual rainfall on lowland UK areas in the west is about 900 mm and in the east is about 600 mm. Larger differences can be seen by comparing eastern lowlands with western highlands—average annual rainfall for parts of Cambridgeshire is 500 mm whilst in the highest, western facing districts of the Lake District it is 5000 mm!

2.4 Air and soil temperature

The optimum temperature for the metabolic processes within a temperate plant, especially photosynthesis, is about 25°C. Therefore, in most seasons and especially during autumn and early spring, plants will not be metabolising at their optimum rate in the UK. Temperature has a large effect on the growing season, with growth of most crops beginning when night time temperatures rise above 5–6°C and ceasing in the autumn when temperatures drop below this level. This can give growing seasons of as little as 180 days in the highlands of Scotland and more than 350 days in the south west of England.

Soil temperatures are particularly relevant to plant growth during the early part of the growing season. Plant germination is at its fastest at soil temperatures between 15 and 25°C; these temperatures are unlikely to be reached for autumn- and spring-sown crops in the UK. Winter wheat will germinate at temperatures as low as 3°C as long as the soil is not waterlogged and it is not uncommon to find soil temperatures of 10–15°C at 5 cm depth when most drilling occurs in September and early October.

Temperature changes are mainly due to:

- The seasonal changes in length of day and intensity of sunlight.
- The source direction of the wind, e.g., whether it is a mild south-westerly, or whether it is cold polar air from the north or from the Continent in winter, and hot, dry wind from the south in summer. Wind direction depends mainly on the position of areas of high and low pressure in the immediate vicinity of the British Isles, although local winds, e.g., offshore breezes, Fohn effect winds (warm) and katabatic winds (cold) from high ground, can occur independently of the cyclonic system.
- Local variations in elevation and aspect.
- Cloud cover during the night. Night temperatures are usually higher when there is cloud cover which prevents too much heat escaping into the upper atmosphere.

The soil temperature may also be affected by colour—dark soils absorb more heat than light-coloured soils. Also, damp soils can absorb more heat than dry soils. The presence of stones can also increase mean soil temperatures by acting as heat reserves in a similar way to storage radiators in buildings. The average January air temperature in lowland areas along the western side of the country is about 6°C and about 4°C along the eastern side. The average mid-day July temperature in lowland areas in the southern counties is 18°C and 14°C in the north of Scotland. Temperature can also have an effect on the agronomy of the crop. For example, it can affect the rate of breakdown of residual herbicides applied to the soil, it can affect the efficacy of plant growth regulators and it can affect the volatilisation of ammonia from applied urea. High temperatures can also mean that plants with 'soft' new growth are more susceptible to phytotoxic effects from agrochemicals applied onto them.

In near-coastal land areas of the UK, temperatures are closely linked to sea-surface temperatures. This is particularly noticeable along the western coasts of the British Isles where winter warming and summer cooling occurs via energy transferred from The Gulf Stream by the prevailing south-westerly winds sweeping up over the Atlantic. Air temperatures over the sea to the south-west of the UK in the winter months can be 10–15°C higher than those on land. The effects of this are noticeable over most of the UK but are at their most intense over west-facing coastal areas. These are often known as early cropping areas, where potatoes and vegetable crops can be sown during winter and harvested in late spring, e.g., Scilly Isles, Pembrokeshire and Cornwall. Often more than one crop per year can be achieved as there is virtually year-round plant growth and little or no chance of frosts.

The likely incident of frosts in an area will have a large effect on the types of crops grown. Most areas of England and Wales will be subject to periods of night frosts during the winter months. Autumn—sown crops are bred to exhibit a certain level of frost hardiness and will survive most winters unscathed. In areas of Scotland where temperatures are lower and during winter days may never exceed 0°C, the crops are mostly spring-sown and this is certainly the case in the huge wheat growing areas of the US and Canada. High value crops such as soft fruit are most at risk from night frosts occurring in the spring when the flower buds are emerging. This can be partially overcome by the use of shelter, artificial protection with mulches or the use of spray irrigation.

2.5 Effect of weather on soils and crops

Plants growing in a given climate have a potential production which is modified (often limited) by the soil in which they are grown and by individual weather conditions experienced during the growing season. A very important effect of climate on soil is that of the soil condition. The soil processes which can be affected by climate include:

- Littering and humification—the accumulation of raw organic matter and its subsequent conversion to humus. The breakdown of organic matter is quicker in higher temperatures because of the increased biological activity in the soil.
- Decalcification—the removal of calcium carbonate from soil profiles by, among other things, leaching. This will be worse in light soils combined with high rainfall.
- Gleying—the reduction of iron under anaerobic conditions with mottling of ferric concretions. Found in heavy soils with poor drainage characteristics in areas where rainfall amounts mean long periods of waterlogging of the soil.
- Salinisation and alkalinisation—the build-up of salt in the soil and the accumulation of sodium (Na^+) ions on the exchange sites. Usually associated with the use of saline water for irrigation but can occur naturally where high temperatures give high evaporation from the soil surface. Salts are moved to the surface by capillary action and there is not enough rain to wash them away from the rooting zone. The overuse of water for irrigation in some areas has also led to the ingress of sea water into aquifers used for irrigation.

Perhaps the most important interaction between climate and soil is that of soil type and rainfall. At the one extreme this interaction can allow a crop to reach its full potential production and at the other extreme it can severely limit this potential, to the point where near crop failure can occur. Two identical soil types in terms of structure and texture can have very different characteristics depending on where they are located. Take, for example, a sandy loam situated in South Devon and a similar sandy loam in Suffolk. The total rainfall in Devon could be as high as 1000 mm whereas in Suffolk this would be nearer 500 mm. The soil in Devon could well be under permanent grassland because of the difficulty in finding available work days for cultivations in autumn and spring. If it was cultivated, leaching of nitrogen and calcium would be a major concern but water supply to crops would be adequate through the summer and justification for an irrigation system would be difficult. The Suffolk soil would probably be under arable or horticultural production. There would be plenty of available work days but irrigation would be vital for high value crops so that yield potential and quality could be maintained.

The epidemiology and life-cycles of many crop pests and diseases are heavily influenced by weather. Many of the most serious fungal diseases on mainstream crops, such as potato blight are favoured by wet, windy and humid conditions. Crop diseases and the philosophy of integrated crop management will be covered in Chapter 6.

2.6 Human caused climate change

Human cause or influenced climate change is referred to as anthropogenic climate change, which is distinct from natural climate cycles that have occurred over long-time periods throughout Earth's history. Through the burning of fossil fuels and land-use change humans have increased the concentrations of greenhouse gases (GHG), which include carbon dioxide (CO_2), methane (CH_4), nitrous oxides (NO_2), ozone, water vapour, halogens (bromine, chlorine, and fluorine) and halocarbons in the Earth's atmosphere. These GHG act like a blanket around the Earth's atmosphere, trapping in heat energy which would otherwise be radiated back into space. Two key ways in which GHG differ from each other are their ability to absorb energy (their 'radiative efficiency'), and how long they stay in the atmosphere (also known as their 'lifetime'). Some GHG, such as CH_4 and NO_2, have higher radiative efficiency than CO_2 but have a much shorter lifetime in the atmosphere. To make it easier to compare and add up emissions estimates of different GHG, are expressed as their CO_2 equivalent (CO_2e) values, calculated over a time period of 100 years. Making CO_2e the standard unit of measurement for all GHG.

The most extreme emissions scenario, associated with a doubling of atmospheric CO_2 concentration above those of pre-industrial concentrations, could result in a 5°C increase in average global surface temperatures to 20°C. Since the industrial revolution average global surface temperatures have risen from ~15°C to ~16°C, a rise of 1°C. In some parts of the world, such as the Arctic Circle, temperature rises have been higher, up to 2°C. The Paris Agreement signed by 170 nations in 2015 set a target of achieving net zero carbon emissions globally between 2050 and 2100. Net zero carbon places a limit on the amount of greenhouse gases emitted by human activity to the same levels that trees, soil and oceans can absorb naturally. This means that inputs to the atmosphere and uptake by the biosphere would be equal. If reached, this ambitious target would keep average global surface temperatures rises below 1.5°C on pre-industrial levels, avoiding the worst consequences of climate change.

Globally, 2010–20 was the warmest decade on record. Warmer air masses have the potential to hold higher % water contents. On average for typical ambient summer air temperatures a 1°C temperature increase can result in up to a 7% increase in atmospheric water content and therefore an increase in the intensity of rainfall. In the UK, recent decades have been warmer, wetter and sunnier than those in the 20th century. The last decade, 2011–20, was on average 4% wetter than 1981–2010 and 9% wetter than 1961–90. UK summers during 2011–20, were on average 15% wetter than 1981–2010 and 17% wetter than those of 1961–90. Since 2009, the UK has had its wettest February, April, June, November and December on record in a monthly series going back to 1862, as well as the wettest winter on record.

In the UK, the most recent decade (2011–20) has been on average 0.5°C warmer than the 1981–2010 average and 1.1°C warmer than 1961–90 (Fig. 2.2), with warming across all months and slightly less warming in the northern and western half of the UK. All the top 10 warmest years for the UK in the series from 1884 have occurred since 2002. Light intensity during the growing season is a very significant determinant of crop production. In the most recent decade (2011–20) the UK has had on average 4% more sunshine hours than the 1981–2010 average and 8% more than the 1961–90 average, with springs recording the greatest increase.

Climate change is unequivocal but much like weather, predicting climate is complex and relies on probability rather than certainty. Climate change models predict how likely the climate is to change in a geographical area within a given timeframe. With warming there is still a probability in any one year that there will be a cold spell but with climate change the probability of a cold spell becomes lower.

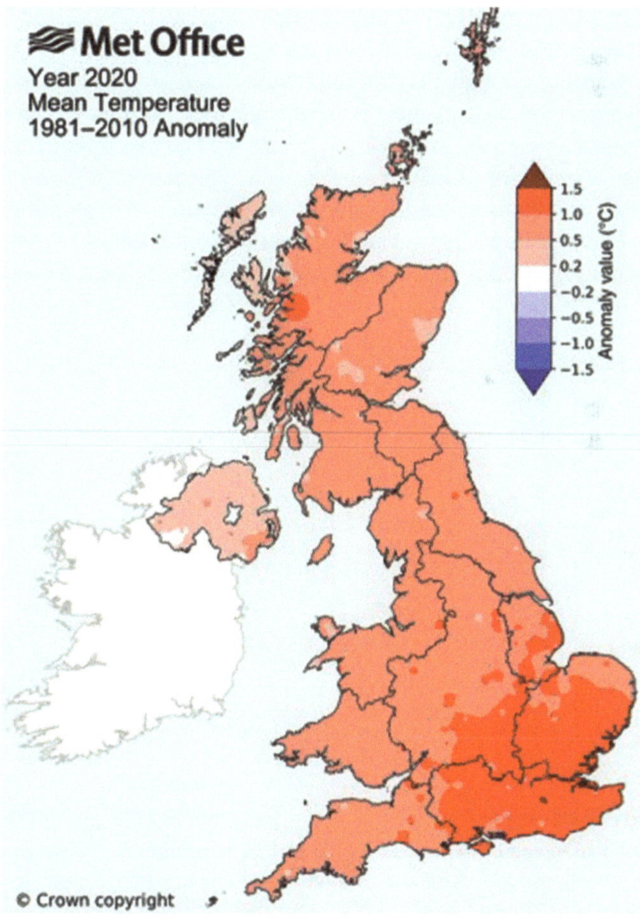

Figure 2.2 UK average temperatures 1981–2010 in comparison to 2020.

2.7 Future climate change predictions

The International Panel for Climate Change (IPCC) and UK Met Office Hadley Centre produce climate change models based on possible future global GHG emissions scenarios. Fig. 2.3 (adapted from IPCC, 2014) shows predicted increases in average global surface temperatures relative to preindustrial levels (pre-1880's) for four possible future emissions scenarios. The four scenarios describe different climate futures, all of which are considered possible depending on the levels of GHG emitted in the years to 2100. The best possible outcome for continued crop production as we know it is that global GHG emissions start declining by 2020 and reach Net Zero if not by 2050 then soon after, limiting increases in pre-industrial average surface temperatures to less than 2°C and possibly lower (Fig. 2.3). The next best scenario and one of concern for agriculture is that GHG emissions continue to rise until 2060 and only then are gradually reduced to Net Zero emissions by the end of the century. Scenario three is referred to as the 'business as usual scenario' where GHG emissions continue to increase at a relatively steady state past the end of the century with a likely 3°C increase on pre-industrial global average surface temperature (Fig. 2.3). If this were not fair warning enough the last and 'worst-case scenario', is that emissions continue to rise year on year at their present trajectory throughout the 21st century and climate feedbacks such as disappearing Arctic sea ice, loss of the Amazon rainforest and release of methane from the circumpolar permafrost region increase GHG emissions further and increase the severity of the situation. We are presently on track for scenario's two and three. However, following recent pledges by nations to reach Net Zero Carbon by 2050, these frightening scenario's seem much less likely.

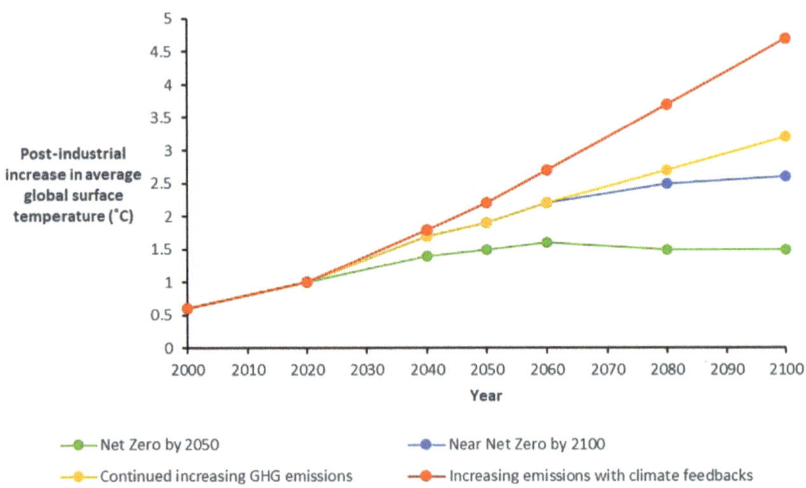

Figure 2.3 Predicted increases in average global surface temperatures (°C) associated with four possible future greenhouse gas emissions scenarios. Figures are approximations from The International Panel for Climate Change (IPCC) four representative concentration pathways (RCPs) (IPCC, 2014).

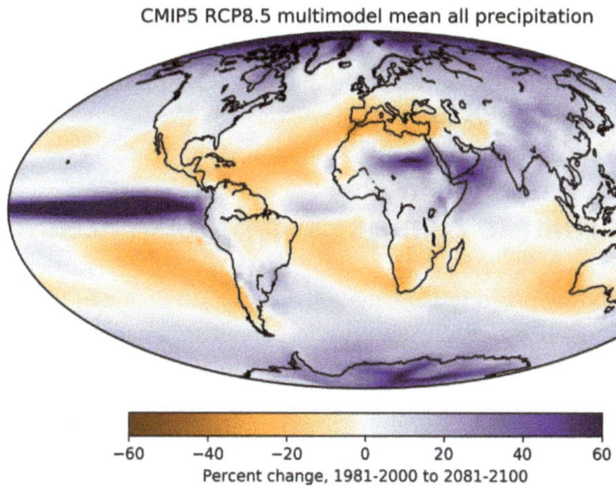

Figure 2.4 Multimodel average percent change in total precipitation (rain and snow) between 1981−2000 and 2081−2100 for the 'worst-case scenario', where emissions continue to rise year on year at their present trajectory throughout the 21st century with possible positive climate change feedbacks (IPCC 2014).

Increases in average global surface temperatures by the end of the 21st century are more likely to range from a very optimistic 1.5°C to a possible upper end value of 3°C.

On average, warming is expected to result in dry areas becoming drier and wet areas becoming wetter, especially at mid- and high-latitudes (Fig. 2.4). Globally it is predicted that there will be an increase in annual rainfall totals due to increases in water vapour in a warmed atmosphere. The average of the models shows large increases in precipitation near the equator. They also show increased precipitation in the Arctic and Antarctic, where cold temperatures currently limit how much water vapour the air can hold. Models for the worst case emission scenario predict a 20% decrease in annual precipitation by 2100 in Mediterranean regions, with similar reductions also in southern Africa. Under the highest emission scenario, California, Mexico, Central America, Chile and Western Australia would become 10% drier; with increases in the frequency and intensity of droughts in these areas predicted under all emissions scenarios. In sub-Saharan Africa and parts of Asia droughts linked with human caused climate change are already having significant impacts on food production. In the UK, increases in the intensity of rainfall could result in higher crop losses in flood vulnerable catchments and reductions in summer rainfalls would increase the need for crop irrigation during the summer months.

Relatively small changes in rainfall or precipitation (including snow) can have a significant effect on agricultural yields for the majority of crop species. Despite the challenges in predicting precipitation, the overall agreement among climate scientists is that annual precipitation will increase in Northern Europe and decrease in Southern Europe, causing increased aridity (Fig. 2.4). Seasonal changes in rainfall patterns are more difficult to predict with certainty. The consensus from climate models is that summer rainfall will decrease in North Western Europe and the UK with increases in autumn and winter rainfall. There is less certainty in climate models when predicting changes in summer rainfall, which typically operates at a more local scale than winter rainfall. Changes in the frequency and intensity of summer rainfall is particularly problematic for crop production and soils.

2.8 Impact of climate change on crop production

The impacts of anthropogenic climate change on agriculture is one of the greatest challenges facing humanity. As we reach the middle part of this century the world could face a 'perfect storm' of problems as food, energy and water shortages interact with climate change. The challenge globally is to increase food production by 10% per decade to feed a growing human population in a changing and increasingly unpredictable climate. Food access will be the principal challenge in low income countries with (1) Increases in global food prices highly likely as a result of increased production and energy costs. (2) More areas of land become too arid for production. (3) Increases in extreme weather events associated with climate change, causing sudden reductions in agricultural productivity.

In recent decades we have observed a systematic geographic range shift, in the spatial pattern of crop planting in response to changing climate conditions. This shift shows a gradual northward movement of southern crops and locally an increase in the elevation of planting sites. All crop species have an optimum climate which meets the plants requirement for germination, growth and reproduction. In a crops optimum climate, losses will be minimised and yield maximised. A plants climate requirements can be defined in terms of mean monthly temperatures, precipitation, % humidity, light levels and the plants ability to cope with high wind speeds, hail and frost.

Scientists studying long-term data sets of biological response to seasonal changes in climate have shown that on average autumn is being delayed and spring events occurring earlier. Table 2.1 shows decadal % changes in key crop related climatic variables in the UK. There has been a significant and sustained increase in growing degree days per year in the most recent decade (2011–20). In the last 50 years, this extension in the growing season equates to a 3 week increase, ~2 weeks in spring and one in autumn. Climate change is leading, on average, to fewer extremely cold days and nights. Data

Table 2.1 Percentage change in average UK climatic variables in the most recent decade (2011–20).

	Average percentage change in comparison 2011–20	
Climate variable	**1981–2010**	**1961–90**
Growing degree days	↑ 7%	↑17%
Number of days of air frost	↓ 16%	↓25%
Number of days of ground frost	↓14%	↓20%

Data source Kenton et al. (2021).

presented in Table 2.1 shows reductions in the annual numbers of air frost and ground frost days in the most recent decade (2011–20). In the UK, widespread and substantial snow events occurred in 2018, 2013, 2010 and 2009, but their number and severity have generally declined since the 1960s. Reduction in the length and severity of winters with fewer and less intense winter frost days in the UK has had an impact on the spread, survival and overwintering of some crop pests and diseases (covered in Chapters 6 and 7).

Conversely, high summer temperatures can significantly impact agricultural productivity, reducing yields within previously high production areas. Temperature impacts on yield depend on the stage of crop development. In seed crops, flowering and grain filling running up to harvest appear to be the most sensitive periods, advancing senescence and reducing harvestable yield. In low latitudes where temperatures are already close to the physiological maxima for crops, predicted average temperature rises of 1–2°C have been shown to be detrimental to rice yield. Increases in night time minimum temperature in the cropping season has been shown to result in a 10% decline in rice yield for every 1°C increase.

In mid and high latitudes, such as the UK, suitability and productivity of crops are projected to increase and extend northwards, especially for cereals and cool-season vegetables, such as asparagus, beets and Brussels sprouts. A 2°C local warming in the mid-latitudes could increase wheat production by nearly 10% whereas at low latitudes the same amount of warming may decrease yields by nearly the same amount. Crops prevalent in southern Europe such as sunflower and soya beans could also become more viable in the UK. Here, yields could increase by as much as 30% by the 2050s, owing to longer planting windows and the warmer growing conditions predicted by climate change. Scientists have also simulated the potential for increases in the land area suitable for crop production in large parts of Canada and the Russian Federation with predicted climate change. Conversely under future climate predictions, the incidence of coastal flooding is set to rise. Predicted sea-level rise of up to 70 cm by the end of the century could result in widespread farmland inundation and crop losses along the east coast of England and low-lying coastal regions around the North Sea through salt water inundation of agricultural soils.

Meteorological records suggest that heatwaves have become more frequent over the 21st century, with increased probability of a heatwave attributed to climate change. Europe experienced a particularly extreme climate event during summer 2018, with average temperatures 6 °C above normal and precipitation deficits of up to 300 mm, which in Northern and Eastern Europe resulted in multiple and simultaneous crop failures. An evaluation of the EU-28 countries over the last five decades to 2018, showed a statistically significant increase in event frequency for droughts (annual increase 1%), heatwaves (6%), floods (29%), and cold waves (10%). In temperate oceanic climates, such as the UK, average cereal yield reductions due to both droughts and heatwaves combined was 6.6%, with the largest losses for barley and no reduction for maize yield. Non-cereal crops such as vegetables, sugar beet, soft fruits, roots and tubers showed the largest yield and production losses in the temperate oceanic climate and warm-summer humid continental climate zones.

Increased rainfall can be beneficial in parts of the crop cycle where water is required for plant growth. However intense rainfall can be problematic causing crop damage, soil erosion, soil waterlogging and flooding. Researchers found from 1964 to 2015 that floods had a significant effect on cereal yield at a regional level with barley showing a 3.4% decline in yield in temperate oceanic countries. However, at a European wide level this reduction in cereal grain yields was more than offset by a 5.3% increase in maize (widely grown in southern Europe), due to wetter growing seasons, increasing maize yield in southern semi-arid climates. Observed increases in maize yields are unlikely to carry on indefinitely as the impacts of climate change on water deficits become more pronounced in semi-arid regions of Europe.

Elevated levels of atmospheric CO_2 generally increase the rate of photosynthesis, known as the CO_2 fertilisation effect. Concentrations of CO_2 are growth rate limiting below 550–600 ppm for C3 crop plants (e.g., rice, wheat, soybeans and barley). Plants exposed to elevated CO_2 show a reduction in stomatal conductance, which can reduce rates of water loss due to evapotranspiration. Most UK crops are C3 plants and are less well adapted for storage of CO_2 compared with C4 plants (e.g., maize, sorghum, sugarcane and millet) that have photosynthetic adaptations to cope with higher temperatures and water shortages (Chapter 1). C4 plants have a unique leaf anatomy which allows CO_2 to be stored in 'bundle sheath' cells and photosynthetic adaptations which enable higher growth rates than C3 plants at higher temperatures. C3 crop plants have to balance CO_2 uptake for photosynthesis via open stomatal cavities in leaves with the potential for water loss through stomatal opening. In mid and high latitudes, with optimal nutrients, above ground productivity of C3 dominated grasslands has been shown to increase by an average of 16% with a doubling of atmospheric CO_2 concentrations. The combined effect of elevated CO_2, higher temperatures, and nitrogen deposition is to increase the rate of plant growth and production. This positive effect of elevated CO_2 levels can easily be neutralised by reductions in water availability.

Research can develop new crops and varieties via traditional plant breeding or by genetic modification (Chapter 11) to maintain yields in drier, warmer conditions. However, ultimately climate change will place greater pressure on wetter, cooler regions like the UK and northern Europe where in coming decades crop production is likely to continue to increase with temperature rises and longer growing seasons.

2.9 Sources of agriculture emissions

Reducing the carbon footprint of our food is vital in reducing climate change. Annually agriculture is responsible for 8%–10% of global GHG emissions. The decarbonisation of agriculture by 2050 is key to reaching net zero carbon and avoiding the worst effects of climate change. Emissions occur from five main sources: (1) Burning of fossil fuels in machinery during crop production; (2) Transport and on-farm processing of agricultural goods; (3) Production and use of fertilisers and pesticides; (4) GHG emissions from livestock production; and (5) During the drainage and cultivation of soils. A significant proportion of current GHG emissions originate from the release of CO_2, CH_4 and N_2O from livestock and arable farming.

Globally, soils hold twice the carbon reserves compared with that held in the atmosphere as CO_2 and plant biomass combined. Climate mitigation measures should consider GHG emissions from soils. The main causes of GHG emissions from soils in the UK and Europe result from farming techniques involving drainage, methods of tillage and the addition of inorganic fertiliser. Healthy functioning soils can make crops more resilient to the impacts of climate change, with increased water holding capacities, water storage and flood mitigation potential (see Chapter 3).

2.10 Potential for cropping systems to achieve net zero carbon emissions

In 2017, land use—including agriculture, forestry and peatland—accounted for 12% of total UK greenhouse gas emissions. Achieving net zero carbon emissions in agriculture will require a mix of Government legislation and incentives to encourage technological change such as:

1. Decarbonisation of fuel for farm vehicles using renewable electricity or hydrogen fuel cells powered by green hydrogen.
2. Precision application of fertilisers, agrochemicals and slurry acidification to reduce the release of ammonia and methane.
3. On farm micro generation of renewable electricity via solar, wind and hydro.
4. Anaerobic digesters to generate methane for electricity generation or farm vehicles.
5. Ground and or air source heat pumps to reduce the energy requirements for space heating.

Also, wide scale uptake of climate smart land management practices such as:

1. Carbon sequestration in tree planting, hedgerow planting, bioenergy crops and the restoration of peatlands.
2. Carbon sensitive soil health and management practices.
3. Alternative land management for flood risk alleviation.

Faced with the challenge of reducing carbon emission, governments are encouraging farmers to recognise the importance of trees, hedges and soils in sequestering atmospheric CO_2 and in reducing the impacts of climate change. In recognition of the ecosystem services provided by trees, such as carbon sequestration, biodiversity, natural flood management, timber and recreation the UK Government has set a target of planting 30,000 ha of trees annually from 2025; raising the existing UK woodland coverage from 13% to 17% by the year 2050. Agroforestry is a land-use systems that combines trees and shrubs with crops on the same piece of land. This management approach has received significant recent public interest for the increases it provides in carbon sequestration and biodiversity enhancement.

Climate smart land management practices which increase carbon storage in soils can contribute to multiple provisioning, regulating and supporting services. Soil management practices such as reduced tillage, mixed planting and the use of cover crops have been shown to reduce carbon emissions from soils and reduce the need for applications of agrochemicals; the resultant reduction in the use of machinery reducing diesel usage by 10%–15%.

Cover crops and retention of crop residuals also shows good potential in increasing soil organic carbon in arable systems. As shown in Fig. 2.5 plants take CO_2 from the

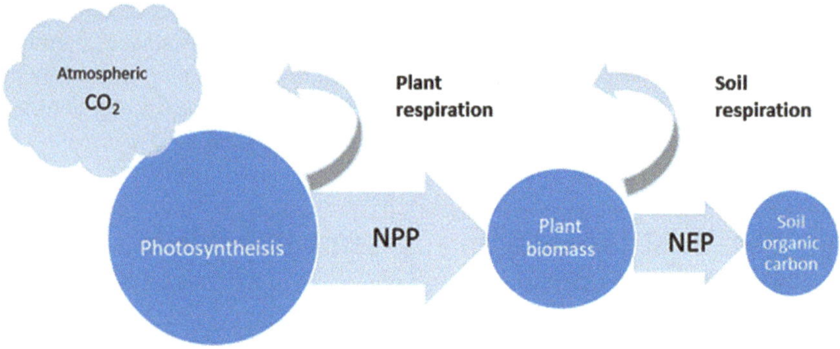

Figure 2.5 Schematic representation of soil carbon storage. *NEP*, net ecosystem production; *NPP*, net primary production.

atmosphere and in the process of photosynthesis assimilate carbon in plant growth. Carbon is held in plant leaves, stems, shoots, seed and below ground in roots (NPP). A proportion of this carbon will be returned to the atmosphere as CO_2 by microbial respiration. The remainder will be stored in the soil as a medium or long term component of soil carbon.

Greater retention of crop residuals can increase soil organic carbon and soil nutrients. The lower the proportion of the plant biomass the farmer removes as a crop the higher the residual carbon returned to the soil. Winter green cover crops can also increase the period of active photosynthesis and therefore the amount of CO_2 removed from the atmosphere. Plant cover crops can help protect the soil from surface erosion during intense rainfall events, reducing surface run-off.

To meet European and UK climate action targets of Net Zero carbon by 2050 and ideally net positive uptake of CO_2 will require payments to incentivise climate smart land management. UK Farmers could receive payment from the Nature for Climate fund for tree planting, peatland restoration and natural flood management. The new Environmental Land Management Schemes also aim to address some of these issues (see Chapter 8).

2.11 Summary

The impact of climate change on agriculture is one of the greatest existential threats to humanity. Arable and grassland systems require predictable weather and climate. It is critical to be able to predict when, what and where to plant and to be confident of viable economic returns. In the UK and Europe, the amount and frequency of summer rainfall is the most significant weather factor effecting arable and grassland production. The climate has begun to change due to human influence and will continue to change in more dramatic ways in the decades to come. Climate models predictions for the UK suggest warmer, generally wetter conditions with potential reductions in summer rainfall and more extreme weather patterns. Agriculture will need to adapt to a changing climate, which in the UK and northern Europe will provide opportunities for new crops and varieties as temperatures rise and growing seasons become longer.

2.12 Key points

- The climate for crop growth in the UK and Northern Europe has changed significantly over the last 70 years with significant increases in: average monthly temperatures; total annual rainfall; intensity of rainfall; duration and intensity of summer droughts and the length of the growing season.
- This rate of climate change will continue to accelerate over the coming decades.
- Farming will need to adapt to the changing climate by: planting heat and drought tolerant crop species; increase use of drainage and flood control measures; increased use of irrigation for drought susceptible crops and the development of new climate resilient crop varieties.
- Climate change will present a challenge and opportunities for increasing yields in some crop species.
- Agriculture will be challenged to reduce the carbon footprint of food production and land-use carbon emissions.
- Climate mitigation measures will include: reducing use of fossil carbon based fuels and agrochemicals; increasing carbon in below and above ground standing biomass; increasing the generation of on farm renewable energy and increasing yields, whilst minimising crop losses.
- Public funding for climate mitigation measures such as carbon sequestration and natural flood management will make an increasing contribution to farm income in the coming decades.

Sources of further information and advice

IPCC, 2014. In: Pachauri, R.K., Meyer, L.A. (Eds.), Climate Change 2014: Synthesis Report. Contribution of Working Groups I, II and III to the Fifth Assessment Report of the Intergovernmental Panel on Climate Change [Core Writing Team. IPCC, Geneva, Switzerland, 151 pp. IPCC_AR5_SYR_Final.pdf. Accessed 07.08.21.

Kenton, M., McCarthy, M., Jevrejeva, S., Matthews, A., Sparks, T., Garforth, J., 2021. State of the UK climate 2020. International Journal of Climatology 41 (Suppl. 2), 1–76. https://doi.org/10.1002/joc.7285.

Soil health and management

3

3.1 Introduction

Sustainable agricultural production is inherently linked to soil characteristics and management. When considering crop husbandry, the soil needs to provide suitable conditions from germination to harvest that ultimately aid both yield and quality. In this context the soil matrix will mainly be linked to plant rooting, support for anchorage and the supply of water, nutrients and oxygen for above and below ground growth and development. Soil temperature must also be suitable. The farmer has a critical role in managing soil for viable crop production, and livestock rearing where appropriate.

Soil covers most of the Earth's land surface and is a multi-functional matrix of living and non-living material, chiefly being composed of solid mineral and organic matter, voids mainly occupied by water and air and a rich diversity of biological species. Most soil has developed over a very long time, with the interactions of geology, climate and living organisms and through the processes of physical, chemical and biological forces. Soils are formed in a mineral state initially by the weathering of parent rock, often accompanied by deposition of material by ice, water and/or wind. Organic matter is added to the soil from the growth and decay of living material. All of this has resulted in much soil complexity and diversity across the Earth. Soil should be considered as a dynamic, rather than inert, entity with ongoing inputs, transformations and losses through natural processes and human intervention.

3.1.1 Soil health

Alongside food production soils have a critical part to play in the context of a sustainable, wider environment. They should be able to support and undertake a diverse functionality successfully. In recent years increasing attention has been placed on linking soil roles to a statement of soil *health* and this term is now being extensively used by farmers, growers and advisers in the agricultural sector. A healthy soil is one that is said to be able to function long-term as a living system supporting and improving plant, animal and human health. In more detail this can be considered as providing a range of services and resources within natural or managed ecosystems, key ones including:

- food and fibre production.
- water filtration and storage, thus helping to regulate water quality and quantity and reduce flooding risk.
- carbon storage aiding climate regulation.
- supplying a diverse habitat and biodiversity within it and above ground.
- absorbing and deactivating contaminants and pollutants.

Lockhart and Wiseman's Crop Husbandry Including Grassland. https://doi.org/10.1016/B978-0-323-85702-4.00023-6

- supply of raw materials e.g., minerals.
- platform for buildings and structures.
- protecting cultural heritage.

It is a range of physical, biological and chemical factors that integrate together to affect soil health and functions. Describing and defining soil health is easier than establishing criteria that decide on whether a soil is healthy or not as currently there are no set widely accepted indicators and levels related to its measurement and recording. Much research and collaboration is taking place in this area of work investigating the value of key soil physical, biological and chemical properties and use of field and laboratory soil tests and techniques. Using these allows ways of monitoring soil health that can be translated into guides or toolkits for farmers, growers and other soil managers e.g., use of score cards or a traffic light system that records whether a soil health indicator is in a good status or otherwise. Examples of soil properties that are being used in this research include:

- physical - soil structure, aggregate stability, bulk density, levels of compaction.
- biological - quantity of soil organic matter and its speciation (total, active, stable), earthworm populations, microarthropod populations, microbial biomass respiration and carbon fluxes, microbial DNA profiling.
- chemical - pH, macronutrients, micronutrients.

Soil quality is another term which is also being increasingly used and is essentially synonymous to soil health as a measure and aim. Optimal soil health is associated with soil functionality that is broad-ranging and that balances key outcomes from soil processes. Plant, animal and human health, environmental quality and protection and productivity are all soil assets that should not be considered individually in isolation. All of these functions are influenced by the farmer and affected by land use and management. Good soil management aims to improve soil health or quality using a multi-functional approach and the farmer needs to consider agricultural productivity alongside other important soil benefits and outcomes. It is important to note that as soils are highly variable in make-up that soil health indicators must be able to be reliably interpreted in relation to such diversity.

3.1.2 Soil characteristics

Soil is finite, considered as non-renewable and is able to be both improved and degraded. Good management is essential to maximise its benefits and functions and reduce the amount that is lost or damaged, for example through erosion of valuable soil material or compaction affecting drainage, aeration, rooting and biological activity. Soil can become much more difficult to manage if the amount of organic matter within it is reduced.

The arrangement of soil layers (*horizons*) that differ in visual and measured characteristics is the soil *profile*. In some cases this may only consist of a shallow surface soil of 10–15 cm on top of rock such as chalk or limestone. In deeper well-developed soils there are usually three or more definite horizons which vary in colour, texture and structure (Fig. 3.1).

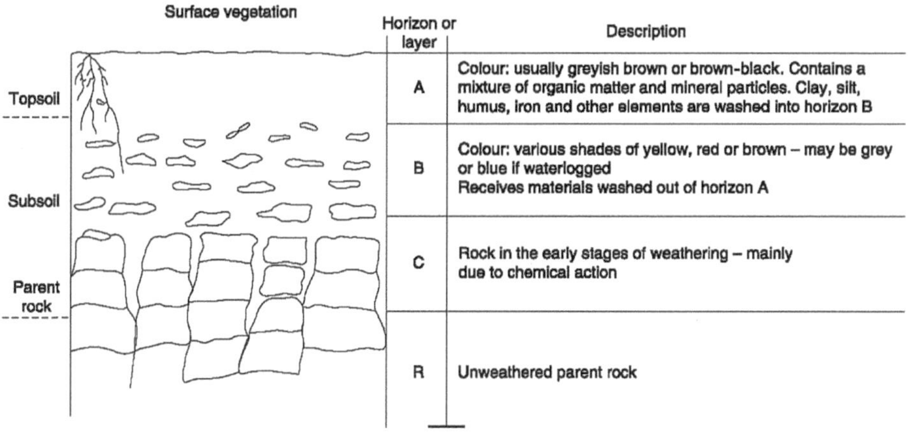

Figure 3.1 Soil profile showing the breakdown of rock to form various soil layers (horizons).

The soil profile can be examined by digging a pit mechanically or with a spade or by taking out cores of soil from various depths using a soil auger. A careful examination of the layers can be useful for assessing soil texture, structure and compaction as well as the soil cropping potential. The colour of the soil in the various horizons is an indication of organic matter content, water content and the presence (and oxidation status) of oxides mainly iron. Generally organic rich soil and moist soil is darker and well-aerated soil is brighter in colour. The colours of iron-containing minerals are important in indicating the drainage status of soil, and hence water and air content.

The topsoil is the uppermost layer, typically up to 30 cm deep, likely to be the greatest depth which a farmer can plough or cultivate and in which most of the plant roots are found. It is the soil zone with most biological activity and the layer most at threat of degradation. The subsoil, which lies underneath, is an intermediate stage in the formation of soil from the rock below. Some deep-rooting plants such as cereals and oilseed rape can grow in the subsoil down to depths of 1.5–2 m.

Soil diversity has led to a range of different classification methods being used for identification and in some cases matching soils to use and management, including crop production. UK soil taxonomy is based on a hierarchical system which uses four levels (*major group, group, subgroup, series*) for increasing the individual soil detail. More general characteristics separate soils at the higher levels and more distinct ones lower down that give more precise subdivisions in parent material mineralogy, texture, depth and colour etc. Most UK agricultural soils, apart from in some upland situations, would be examples of a *brown soil* at the major group level. Where soil series have developed on similar parent materials and occupy specific positions in the environment then they are grouped into a soil *association*. Each association consists of a number of soil series each of which has distinct characteristics, both of parent material and soil profile. The soil series is usually named after the place where the soil was first described. The same soil series can occur in different regions. Practically on farm there may be a number of distinct soil series matching soil variability and giving rise to differences in a range of practical crop and livestock management operations e.g., machinery working days, grazing periods, slurry spreading and irrigation need. Soil associations and soil series have been mapped in the past, partly by sampling and in other areas by extrapolation. Hard copy mapping from previous soil surveys is increasingly being digitised and is available for public access e.g., www.landis.org.uk or www.ukso.org. Soil characteristics (such as depth, stoniness and drainage), together with relief, climate and cropping potential, have also been used to record land for farming potential giving rise to agricultural land use capability classification (Appendix 4).

3.2 Soil formation

There are very many different types of topsoil and subsoil. The differences are partly due to the kind of parent material from which they have been formed. However, other factors such as climate, topography, plant and animal life, the age of the developing soil material and farming operations can all affect the type of soil which develops.

3.2.1 Parent rock materials

- *Igneous rocks* were formed when the Earth's very hot, molten magma cooled, solidified and crystallised. They are either *plutonic* where they solidified beneath the Earth's surface, e.g., granite (coarse crystals), or *volcanic* from eruption and solidification on the surface e.g., basalt (fine crystals). Igneous rocks are very hard and usually weather very slowly.
- *Sedimentary rocks* formed on or near the Earth's surface through transportation and sedimentation of weathered material (e.g., silt and sand), typically carried and deposited by water and wind, or deposition of biological material. Many formed as ocean sediments. Erosion, weathering and precipitation are important processes in sedimentary rock formation. The sediments later became compressed by more material on top and cemented to form rocks such as sandstone and shale. *Chalk* and *limestone soils* were formed from the shells and skeletons of sea animals of various sizes. These rocks are mainly calcium carbonate but in some cases also contain magnesium carbonate. The calcareous soils are formed from them (see Section 3.4.5).
- *Metamorphic rocks*, e.g., marble (from limestone) and slate (from shale), formed when original igneous and sedimentary rocks were changed in various ways, such as by heat or pressure.

3.2.2 Some other soil forming deposits

- *Organic material.* Deep deposits of raw organic matter or peat are found in places where waterlogged soil conditions did not allow the full breakdown of dead plant material by microorganisms and oxidation (due to the anaerobic conditions). There are different peats depending on the type of vegetation and area where they developed. Peat and peaty soils have been formed in this way (see Section 3.4.6).
- *Glacial drift.* Many soils in northern Europe are derived from material deposited by glaciers and are often known as boulder clays. Study of such soils can be very complicated due to the variable nature of the glacial deposits.

- *Alluvium.* This is material which has been deposited recently; for example, by river flooding. It has a very variable composition. The texture depends on the speed of river flow (e.g., fast rivers—stones and sand, slow rivers—silt and clay).

3.2.3 Weathering of rocks

The breakdown of rocks and minerals into smaller particles is called *weathering*. It is mainly caused by physical and chemical processes, the former often being the precursor for the latter.

Physical weathering can happen in a number of different ways. One important method is by frost action. Water can cause pieces of rock surfaces to split off when it freezes and expands in cracks and crevices. The pieces of rock broken off are usually sharp-edged, but if they are carried and buffeted by glaciers, rivers or wind, they can become more rounded in shape, e.g., sand and stones in a river bed. Expansion and contraction of parent material when exposed to varying temperatures will also cause some rocks to fracture. Plants can also cause weathering by growing through rock cracks and exerting enough pressure to cause some shattering.

Chemical weathering is the breakdown of the mineral matter in a developing soil by dissolution, hydrolysis and oxidation and is brought about by the action of water, oxygen, carbon dioxide and nitric acid from the atmosphere, and by carbonic and organic acids from the biological activity in the soil. The soil water, which is a weak acid, dissolves some minerals and allows chemical reactions to take place. Clays are produced by chemical weathering of some primary minerals which then re-crystallise to form secondary minerals. In the later stages of chemical weathering soil minerals can be broken down to release plant nutrients e.g., potassium and magnesium.

In poorly drained soils, which become waterlogged from time to time, various complex chemical reactions (including a reduction process) occur, referred to as *gleying*. This process, which is very important in the formation of some soils, results in ferrous iron, manganese and some other micronutrients moving around more freely and producing colour changes in the soil. Gleyed soils are generally greyish in colour (but may also be greenish or blueish). Rusty-coloured deposits of ferric iron (oxidised iron) also occur in root and other channels and along the boundaries between waterlogged and aerated soil, so producing a mottled appearance.

3.2.4 Other factors in soil formation

- *Climate.* The rate of weathering partly depends on the climate. For example, the wide variations of temperature and the high rainfall of the tropics make for much faster soil development than would be possible in the other climatic regions that are colder and drier.
- *Topography.* The slope of the ground can considerably affect the depth of soil. Weathered soil tends to erode from steep slopes and build up on the flatter land as slopes bottom out. Level land is more likely to produce uniform weathering.
- *Biological activity.* Plants, animals and other organisms, during their life cycles, leave many organic substances in the soil. Some of these substances may dissolve some components of the mineral material. Vegetation such as mosses and lichens can attack and break down the

surface of rocks. Holes made in the soil by burrowing biota such as earthworms, moles and rabbits help to break down soft and partly weathered rocks. Biological activity usually increases with higher temperatures and decreases under waterlogged and/or acid conditions.
- *Time and human activity.* Soil formation is a dynamic process. The length of time necessary is typically very long, but can be variable and is also dependent on the other soil forming factors. Human activity within that time period can have major effects; for example, when specifically considering farming activity, operations such as deep ploughing and cultivation, artificial drainage and liming can speed up the soil formation processes.

3.3 Soil composition and properties

Soil is made up of:

- *Mineral matter* (stones, gravel, sand, silt and clay)
- *Organic matter* (chiefly residues from plants and animals)
- *Water* (a weak acid)
- *Air* (mainly oxygen, carbon dioxide and nitrogen)

Technically not part of the soil, but usually classed within its make-up would also be:

- *Living biota.* Microorganisms (bacteria, fungi etc.) and soil fauna and flora including earthworms.

Soil physically exists as a combination of solid, liquid and gaseous matter. On a typical basis approximately 50% of soil volume is made up of the solid matter (mineral and organic) and the remaining 50% is the *porosity* or pore space, this being air and water filled. The relative amount of air and water and organic matter depends on the soil type as well as climate, cropping and cultivation systems.

3.3.1 Mineral matter and soil texture

Soil mineral matter is classified by particle size distribution which is dependent on soil parent material and degree of weathering. Particles >2 mm diameter make up the soil *coarse fraction* whereas particles ≤2 mm comprise the soil's *fine earth* fraction and are subdivided into sand, silt and clay. A soil's coarse fraction is made up of gravel, stones and boulders. The description of a soil in relation to its proportions of sand, silt and clay is called the *soil texture*. There are large variations in the texture of soils found on farms. Soil texture can be determined accurately in a laboratory (mechanical analysis to establish a particle size distribution), but can also be practically assessed in the field by hand.

If mechanically analysing a soil, the organic matter and mineral coarse fraction are first removed. The fine earth fraction is then divided using sedimentation and sieving techniques. This gives an accurate measurement of the amount and proportion of clay, silt and sand particles present. The particle size classification and equivalent surface area for soil mineral matter are shown in Table 3.1. Because these mineral components

Table 3.1 Soil mineral matter particle size and surface area.

Material	Diameter of particles (mm)	Surface area
Clay	Less than 0.002	100000 × a
Silt	0.002–0.06	1000 × a
Fine sand	0.06–0.02	100 × a
Medium sand	0.2–0.6	10 × a
Coarse sand	0.6–2.0	a
Gravel, stones and boulders	More than 2.0	

have very different properties it is important to know the relative amount of each, hence the importance of knowing the overall soil texture. In the UK soil texture is graded into 11 textural classes (Fig. 3.2).

Hand texturing in the field is a very quick and simple technique; with practice it can be a reasonably accurate method of analysis (see Appendix 1). It uses the difference in feel and binding properties of moist soil to differentiate between the textural classes. Using this the main characteristics of soils dominated by the three mineral fine earth fractions are as follows.

- *Sands.* Feel gritty and rough. Do not bind and are moulded with difficulty.
- *Silts.* Feel smooth and silky. Bind and can be moulded into a soil ball which is easily deformed.
- *Clays.* Feel sticky. Bind and can be moulded to form a strong soil ball which can take a polish.

The texture of a soil has a major influence on many wider factors including:

- *Soil structure.* This is the arrangement of individual soil particles into larger units or aggregates (*peds* and *clods*). It is mainly the clay particles that aid aggregation (*flocculation*) as

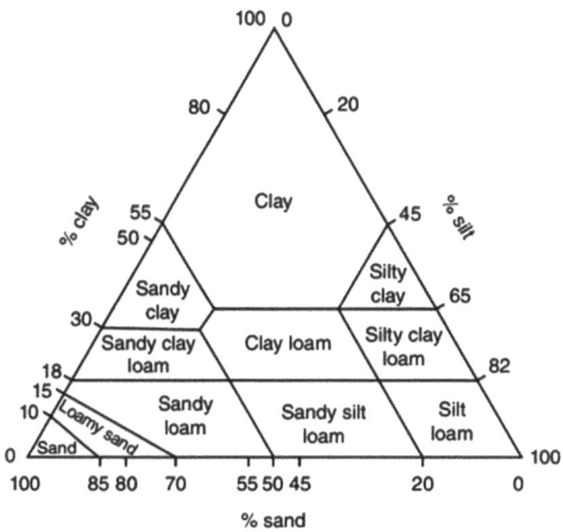

Figure 3.2 Soil texture triangle.

they are charged and bonded to each other. Charged colloidal organic matter particles as well as living organisms such as fungi, actinomycetes and bacteria are also very important in the process of soil aggregation.

- *Air and water supply.* Clay soils have the largest number of small pores which tend to be filled with water rather than air. Some of this water is unavailable for root uptake due to being too tightly soil held. Conversely sandy soils have the greatest number of large pores that tend to be filled with air rather than water. Because of these characteristics clay soils usually need draining for arable cropping and sandy soils are drought prone with irrigation needs for certain crops. It is the medium-textured silty soils that hold the most available water for crop growth. Air and water content will be linked to soil temperature with sands warming up more quickly in spring than clay soils.
- *Cultivation system.* The texture of the soil can affect the cultivation system used to establish crops (Chapter 8). Sandy soils are comparatively easy to work and require little power to cultivate. Clays, however, can be very difficult with higher cultivation power requirements and limited periods when the land can be worked without the risk of damaging soil structure.
- *Nutrient and lime requirements.* Soil textural components have differing abilities both to hold and release nutrients. Most clay minerals (and organic matter components) have negative charges which attract positively charged ions (*cations*) such as potassium, magnesium and calcium. The soil's capacity to hold these is termed the *cation exchange capacity*. The presence of these charged sites affects the availability of nutrients for root uptake through differing attraction, holding and release characteristics. Sand and silt are relatively inert hence little cation exchange capacity. Different soil textures with the same level of acidity have different recommendations for the application of lime. This is due to differences in the soil's cation exchange capacity with respect to the acidic cations of hydrogen and aluminium.
- *Erosion and runoff.* Sandier based soils with low particle cohesion tend to suffer from both wind and water erosion. Highly organic soils can also be affected by wind erosion. Clay based soils are more prone to surface water runoff due to lower water infiltration.
- *Herbicide use.* Sandy soil textures are more prone to herbicide leaching, particularly of some soil applied residuals. Recommended application rates of these chemicals in order to avoid risk of crop damage and loss to the wider environment are affected by the soil texture. On clay and organic soils rates may be higher as some chemicals are adsorbed onto the soil particles and are then unavailable for weed uptake.
- *Cropping.* All of these characteristics of different soil textures influence which crops can be grown successfully and their yield potential.

3.3.2 Soil structure

Soil structure is defined as the size, shape and stability of the aggregates formed from mineral particles and organic colloids (Fig. 3.3).

Unlike texture, structure can be altered naturally by weathering (e.g., lumps changed to crumbs by frost action or by alternate wetting and drying), by the penetration of plant roots and very importantly, by cultivations. Soil peds are naturally formed soil aggregates whereas clods are formed through human intervention. Soils with a high clay or organic matter content tend to have a more stable, stronger soil structure than those containing mostly sand and/or silt. Soils containing free calcium carbonate and, to a lesser extent iron oxides, also tend to have a more stable soil structure.

Figure 3.3 Soil structural units.

Soils described as being well structured through the soil profile have plenty of pore spaces, voids and fissures between the aggregates with evidence of good root growth and biological activity such as earthworms. This will aid soil water behaviour, nutrient transformations and plant residue breakdown, all of which contributes to a healthy soil. Good structure in the topsoil is recognised by a dominance of small, rounded aggregates whereas in the subsoil units will be larger being separated by vertical fissures. Soils with few pores and fissures in both the topsoil and/or subsoil are poorer structured and are often compacted. Compacted soils can affect crop root growth and the soil water-holding capacity, reduce nutrient uptake and make the growing crops more prone to pests and diseases, all of which will ultimately lower crop yield. Compacted soil is much more prone to soil erosion and runoff and the soil is generally not in a healthy state. Soil *bulk density* can be used as an indicator of compaction. This is calculated as the dry weight of soil divided by its volume which will be made up of the volume of both soil particles and the pore spaces. High bulk densities will be linked to impaired soil function due to limited aeration, water movement and structural support.

Soil structure can be assessed in the field by visual examination of excavated material such as soil spits dug with a spade at appropriate locations and depths. Techniques and guides are available to interpret results as to whether structure would be classed as good or poor. Examples include Visual Evaluation of Soil Structure (VESS) and Visual Soil Assessment (VSA). Soil rod-and-cone penetrometers can be used to record levels of soil compaction.

The structure of the soil can be easily damaged by inappropriate operations, e.g., heavy traffic in wet conditions or ploughing when the soil is too plastic and is then capable of being deformed permanently. Overworking poorly structured silty and sandy soils, often with low organic matter, can lead to surface puddling and capping after a period of intense rain followed by surface drying. Surface capping can lead to crop failure if it occurs after drilling a small seeded crop and also increases the risk of erosion and runoff through poor water infiltration.

Tilth is a term used to describe the condition of the soil in a seedbed. For example, the soil may be crumb-like in a finely-divided state or it may be rough and cloddy. Whether a tilth is suitable or not partly depends on the crop to be grown. In general, a small seeded crop requires a finer tilth than one with larger seeds in order to achieve good seed-soil contact for germination.

3.3.3 Organic matter

Soil organic matter (SOM) is mainly made up of plant and animal residues in various stages of decay and stable residual *humus* transformed from original structures. Overall total content varies with soil texture and cropping. The most active fraction of SOM is associated with the most recent additions of plant residues and organic manures. SOM usually increases with clay content, e.g., typical arable cropped, heavy, clay based soils have about 4%–6% organic matter compared with 1%–3% in lighter, sandy soils. Mineral soils where arable crops are continuously grown usually have an organic matter content below 5% compared with up to 10% under permanent grassland with high returns of organic matter from root turnover. There are much higher amounts of organic matter in the peat based soils, these generally being higher than 20% content. True peat contains more than 50% organic matter. When draining, liming and cropping peaty soils they can suffer rapid organic matter breakdown or *wastage*.

Organic matter when added to soil may remain for a short time in the undecayed state. However, it is soon processed by the soil biota including bacteria, fungi and earthworms. Sugars, starch and crude protein are broken down first and the more complex cellulose, lignin and waxes last. Humus produced has no fixed chemical composition and is characterised as being persistent and amorphous. The amount of humus formed is greatest from plants which have higher quantities of strengthening (lignified) tissue, e.g., straw. Organic matter is broken down most rapidly in warm, moist, aerated soils with a high biological activity. Decomposition is slowest in waterlogged acidic conditions, but which can lead to peat formation over longer periods of time.

SOM has many benefits for enhancing soil physical, chemical and biological properties. Humus is very beneficial in restoring, improving and stabilising soil structure aiding soil workability, trafficability and health. It helps to reduce the risk of soil compaction and surface capping. Like clay particles, humus exists in colloidal form. It is a gluey substance which behaves like a sponge; it absorbs water and swells up when wetted and shrinks on drying. Humus colloids are not as gummy and plastic as the clay colloids but they can improve light (sandy) soils by binding groups of particles together. This reduces the size of the pore spaces between the particles and increases the water-holding capacity. Humus can also improve clay soils by making them less plastic and by assisting in the formation of a crumb structure provided that liming material is also present. Increasing the organic matter content of a soil will increase its ability to hold water and the amount of water available for crop growth (available water capacity). It will also increase the soil's cation exchange capacity.

SOM has an important role in the supply of plant nutrients, through mineralisation of nitrogen and phosphorus as organic matter is decomposed and through cation exchange. It also has an important role in the sequestration of carbon, especially under permanent grassland or forestry. Many soils in entirely arable rotations lose more carbon than is returned. Generally SOM will be maintained or increased by growing grass, working-in straw and other crop residues, using cover crops and green manures and by adding bulky organic amendments such as farmyard manure and composts. In areas where erosion by wind and water is common, mineral soils are less likely to suffer damage if they are well supplied with humus. Mineral soils generally of low organic matter will reach a natural equilibrium in SOM content particularly after a number of initial years of aiming to increase it. These maximum levels will vary according to soil type and typical cropping. In a similar way if SOM levels are reducing through soil degradation, poor management and/or lack of organic material input, a low residual level is generally maintained as loss plateaus off and soil activity will therefore continue to function albeit it at a much reduced condition. Claims of permanent soil degradation and finite agricultural productivity do not have scientific proof to back up statements made such as limited harvests remaining from our farmed soils, but poor soil health is an important factor linked to reduced agricultural productivity and environmental condition.

3.3.4 Water in the soil

A healthy, productive soil will usually have a very good supply of water for root uptake and be well drained. Soil is inherently a mass of irregular-shaped particles forming a network of pore spaces or channels in-between and these will be filled with air or water or both. If the pore space is completely filled with water, the soil is waterlogged and is at *saturation*. It is then unsuitable for most plant growth because the roots need oxygen for respiration. Ideally, there should be about equal volumes of soil air and water. Connectivity of the pores within the soil profile will aid drainage.

The size of the pores is also very important in determining the amount of water held in a soil. *Macropores* (>0.5 mm) are the largest pores and are important for drainage and aeration. *Mesopores* are medium sized and are the pores containing water available for crop growth. *Micropores* (typically <0.2 μm) are the smallest pores and these contain water that is not available for root uptake. A sandy soil has a large proportion of macropores so normally more of the total pore space is occupied by air rather than water; it is the opposite for clay soils where there are a larger amount of small pores.

Pore space may be altered by a change in the:

- Soil structure, e.g., by cultivations.
- Amount of organic matter present. The larger the quantity, the greater the water-holding capacity.
- Compaction of the soil. Compaction reduces the number and size of the pores. It can also affect rooting depth and hence the amount of plant available water.

Another important factor is the surface area of the particles. Water is held as a thin layer or film around the soil particles. The smaller the particles, the stronger the attractive forces holding the water. Also, the smaller the particles, the greater the surface area per unit volume. A comparison of the surface area of the various soil mineral particles is shown in Table 3.1. The surface area of the particles in a cubic metre of clay may be over 1000 ha.

When water is received by a dry soil it does not become evenly distributed through the soil profile. The topmost layer is gradually filled first and, as more water is added, the depth of the layer increases. In this layer most of the pore space is filled with water. However, a well-drained soil cannot hold all of this water for very long; after a day or two some of it will soak into the lower layers through gravity and soil permeability through the connected pores and fissures. The movement of water through the soil is a measure of its *hydraulic conductivity*. The soil will eventually reach a stage when the amount of any additional water added by rain or irrigation is matched by the loss of water from the profile through natural or artificial drainage. This is known as *field capacity* and represents maximum practical water retention, typically after around two days post-initial saturation. The amount of water which can be held in this way varies according to the texture and structure of the soil. The weight of water held by a clay soil may be half the weight of the soil particles, whereas a sandy soil may hold less than one-tenth of the weight of the particles.

The ways in which water is retained in the soil can be summarised as being:

- Adsorbed as a film around the soil particles.
- In the organic matter colloids.
- Filling some of the smaller pore spaces (*capillary water*).

Plant roots can easily extract and take up most of this retained water, but as the soil dries out the remaining water is more firmly held and eventually a stage is reached when the plant cannot extract any more. This is the *permanent wilting point* because plants remain wilted and then soon die. At this position there is still a very small quantity of water present in the soil, but it is too tightly held for root extraction. Permanent wilting should not be confused with temporary wilting which sometimes occurs on very hot days. In this situation the rate of transpiration is greater than the rate of water absorption through the roots. Plants recover however when it gets cooler on the same day and transpiration reduces.

The difference between the water amounts held at field capacity and permanent wilting point represents the water available to plants for uptake; this is the soil's *available water capacity*. In clay soils only about 50%–60% of water at field capacity is available; in sandy soils up to 90% or more may be available. It is the silt loams and peaty soils that have the highest available water capacity.

Water in the soil helps to hold soil particles and aggregates together. When a clay soil is at, or above, half field capacity, it is possible to form it into a ball which will not fall apart when handled. As a clay soil becomes wetter it becomes more like plasticine. When there is enough soil moisture to roll the soil into a 3 mm diameter 'worm' without breaking it is at the *lower plastic limit*. Cultivating the soil at this moisture content will damage the soil structure. Some of the water in soils with very small pores moves through by capillary forces, i.e., surface tension between the water and the walls of the fine tubes or capillaries. This is a very slow movement and may not be fast enough to supply plant roots in a soil which is drying out.

Water is lost from the soil by evaporation from the surface and by transpiration through plants. It moves very slowly from the body of the soil to the surface; after the top 20–50 mm have dried out, the loss of water by evaporation is very small.

Cultivations increase evaporation losses. When more water is removed from the soil than is added then a state of *soil moisture deficit* develops and the soil starts to dry. This is effectively soil moisture reduced from its content at field capacity and is a measurement used for crop irrigation scheduling. Plants take up most of the available water in a soil during the growing season and air moves in to take its place. Air moves easily where the soil has large pore spaces, but the movement into the very small pore channels in clay soils is slow until the soil shrinks and cracks, vertically and horizontally, as water is removed by plants.

The water which enters soil soon becomes a dilute solution of the soluble soil chemicals. It dissolves some of the carbon dioxide present and so becomes a weak acid.

3.3.5 Soil aeration

Plant roots and much of the soil biota require oxygen for respiration releasing carbon dioxide in the process. The air found in the soil is atmospheric air which has been changed through respiration and also through various other chemical reactions; as a result it contains less oxygen and more carbon dioxide than air above the ground. After a time this reduction in oxygen and increase in carbon dioxide can become harmful to the plant and other organisms. *Aeration* is the replacement of this stagnant soil air with fresh air. The movement of water into and out of the soil mainly brings about the process, e.g., rainwater soaks into the soil filling many of the pore spaces and driving out the air. Then, as the surplus water moves away by gravity, soaks down to the drains or is taken up by plants, fresh air is drawn into the soil to refill the pore spaces. Additionally, oxygen moves into the soil and carbon dioxide moves out by a diffusion process similar to that which happens through the stomata in plant leaves (Chapter 1). The aeration process is also assisted by:

- Changes in temperature.
- Changes in barometric pressure.
- Good drainage.
- Cultivations.
- Good soil structure.

Size and number of macropores affect the aeration of the soil. Sandy soils are usually well aerated because of their open structure. Clay soils are usually poorly aerated, especially when the very small pores in such soils become filled with water. Good aeration is especially important for germinating seeds and seedling plants.

3.3.6 Soil microorganisms (microflora and microfauna)

There are thousands of millions of organisms in every gramme of fertile soil. They have many key roles in both soil processes and soil functionality. These include the breakdown of organic residues and formation of humus, nutrient transformations and increasing available nutrients for plant uptake, helping to create soil structure and contributing to soil and crop health by breaking down chemical contaminants and pathogens. Many different types are found and are described as follows:

1. Bacteria are the most numerous group, are prokaryotic and are the smallest type of single celled organisms. Most of them feed on, and break down, organic matter. They need nitrogen to build cell proteins, but if they cannot get this nitrogen from the organic matter they may use other sources such as the nitrogen applied as fertilisers. When this happens (e.g., where straw with high carbon to nitrogen ratio is ploughed in) the following crop may suffer from a temporary shortage of nitrogen. Some types of bacteria can convert (*fix*) the nitrogen from the air into nitrogen compounds which can be used by leguminous plants (Chapter 1). Numbers of bacteria vary in the soil; they increase near a source of organic matter. In the area next to plant roots (*rhizosphere*) there is an increase in the number of bacteria compared with the rest of the soil. The rhizosphere bacteria may be important in crop nutrient uptake. Bacteria are found in all soils although fewer are present in acid conditions. Certain soil borne bacteria are associated with crop disease e.g., blackleg in potatoes.
2. Fungi are eukaryotic organisms which feed on, and break down, organic matter. They are the most important organisms responsible for breaking down complex compounds such as lignified (woody) tissue and waxes. They are plant like, but have neither chlorophyll nor proper flowers and are less biochemically versatile than bacteria. The species and number of fungi in the soil are constantly changing, depending on soil conditions and the type of organic material present. Fungi can live in more acid and drier conditions than bacteria; they are the main organisms for decomposing organic material in acid soils. Some fungi form symbiotic associations with plant roots (*mycorrhiza*); these associations can improve nutrient and water uptake and can protect root systems from some pathogens. Most cropping families other than *Brassicaceae* have mycorrhizal interactions. As well as beneficial fungi there can be some disease-producing fungi present in soil, e.g., those causing take-all in cereals, blight in potatoes or club root in brassicas.
3. Actinomycetes are a particular type of bacteria which produce mycelium like fungi. They need oxygen for growth and are more common in the drier, warmer soils. They produce the chemical geosmin which gives freshly wetted soil its characteristic earthy smell. Some types can cause plant disease, e.g., common scab in potatoes (worst in light, dry, alkaline soils).
4. Algae on, or near, the soil surface are very small simple organisms which contain chlorophyll and so can build up their bodies by using carbon dioxide from the air and nitrogen from the soil. Algae growing in swampy (waterlogged) soils can use dissolved carbon dioxide from the water and release oxygen. This process is an important source of oxygen for crops such as rice. Algae are important in colonising bare soils in the early stages of weathering. Blue-green algae (cyanobacteria) can fix atmospheric nitrogen.
5. Protozoa. These are very small, single-celled organisms. Most of them feed on bacteria, fungi and other microbes. A few types contain chlorophyll and can produce carbohydrates like plants. The biomass of protozoa in an arable soil is comparable to the biomass of earthworms.
6. Nematodes are a diverse group of non-segmented roundworms that can be grouped according to their feeding, whether being bacterial or fungal feeders, plant parasites or predators feeding on protozoa or other nematodes. They themselves can be prey to certain fungi. Most nematode species are beneficial contributing to nutrient cycling and disease control, but there can also be detrimental effects of certain species with respect to crop production e.g., root attacking potato cyst nematodes.

3.3.7 Soil meso- and macrofauna

In addition to earthworms, there are many species of small animals present in most soils. They feed on living and decaying plant material and other components of the soil food web. Some of the common ones are slugs, millipedes, centipedes, ants, spiders, mites, springtails, beetles and larvae of various insects such as cutworms, leatherjackets and wireworms. The farmer is mainly concerned with those which damage crops or livestock and those which are beneficial predators of crop pests such as the carabid beetles. Many though have key roles in healthy soils providing organic matter turnover, feeding on fungi and nematodes and contributing to nutrient cycling e.g., microarthropods such as springtails and mites.

Earthworms have a very beneficial effect on the fertility of soils, particularly those under grass. Several million earthworms per hectare have been recorded under grassland. There are many different kinds found in our soils, but most of their activities are very similar. Different species have a range of colours and sizes, together with other characteristics such as feeding and burrowing behaviours. *Epigeic* earthworms live in, and feed on, leaf litter on the soil surface. They tend not to make burrows. *Endogeic* earthworms live in, and feed on, the soil making horizontal branching burrows to move and feed. These are typically shallow burrowers and topsoil dwellers, but some can burrow much deeper. *Anecic* earthworms are generally the largest in size living in deep, vertical unbranching burrows and feeding on soil and organic matter litter—either living plants or, more often, dead and decaying matter. They drag down fallen leaves and other material into their burrows. Their casts, which are usually left on the surface, consist of a useful mixture of organic and mineral matter and lime. Earthworms do not thrive in acid soils because they need plenty of calcium (lime) to digest the organic matter they eat. The greatest numbers are found in loam soils, under grass, where there is usually a good supply of air, moisture, organic matter and lime. Soil disturbance can have a negative effect on earthworm populations. Numbers are often higher under reduced cultivation systems compared with ploughing.

Earthworms act as ecosystem engineers improving soil aggregate stability, aggregate formation, water infiltration into soil and the incorporation of organic matter. The positive effect of earthworms can be seen in both arable and grassland soils with increases in above-ground and below-ground biomass. They are seen as useful indicators in the assessment of the health of a soil.

3.4 Farming and managing different soil types

Soils can be classified and grouped using a variety of means, but texture/geology and organic matter in particular have major effects on the practicalities of farming operations and management. Key different soil groups based on these considerations are therefore to be described in relation to their characteristics, management and cropping suitabilities.

3.4.1 Clay soils

Of all soil types clay soils have the highest proportion of clay particles, at least 30% of the textural make-up. They are described as heavy soils and in detail are composed of clay, silty clay and sandy clay textures. It is the clay material which is mainly responsible for their specific characteristics. Clay particles can group together (*flocculate*) into small clusters or become scattered (*deflocculated*). They usually have a high cation exchange capacity, and can combine through bonding with various chemical elements such as calcium, sodium, potassium and magnesium (*base exchange*); in this way plant nutrients can be held and clay soils generally maintain higher nutrient content indices.

Grouping or flocculation of the clay particles is very important in helping to make clay soils easier to work. Those particles combined with calcium (added as lime) will flocculate easily, whereas those combined with sodium if present will not. The bonding properties of clay are very beneficial to soil structure when the groups of particles are small (like crumbs). Deflocculation can occur when clay soils are flooded with sea water or worked when wet. If the latter occurs they become puddled and if the weather then becomes dry the clay dries into hard lumps that resemble clods. Frost action, and alternating periods of wetting and drying, will help to restore large lumps or clods to more of a crumb-like condition.

Many clay particles have the capacity to expand and contract leading to swell and shrink characteristics. When natural soil cracks occur as a result of shrinkage some air can be drawn in. These cracks remain when soil clods are wetted again and so lines of weakness are formed which eventually help to break the clods down. In prolonged dry weather these cracks may become very wide and deep which later may be very beneficial for drainage, but can sometimes also provide preferential loss pathways for nutrients or contaminants dissolved in water or when applied in slurry.

3.4.1.1 Characteristics

- Clay soils feel very sticky and mould and deform like plasticine when wet.
- They can hold more total water than most other soil types and, although only about half of this is available to plants, crops seldom suffer from drought.
- They swell when wetted and shrink when dried, so a certain amount of natural restructuring can take place in these soils depending on weather conditions.
- They lie wet in winter and so stock should be taken off the land to avoid poaching (the compaction and puddling of soils by animals' hooves).
- These soils are very late in warming up in the spring because water heats up more slowly than mineral matter.
- They normally have good reserves of potassium and naturally higher levels of organic matter than other mineral soils.
- They usually need large, infrequent dressings of lime for pH rectification.

3.4.1.2 Management

Clay dominated soils are often called heavy because for ploughing and subsequent cultivations, compared with light (sandy) soils, two to four times the amount of tractor power may be required. All cultivations must be very carefully timed (usually restricted to a shorter period than on other soil types) so that soil structure is not damaged. Minimum cultivations are often used for crop establishment and many farms only plough rotationally (Chapter 8). Autumn ploughing, to allow for a frost tilth, will aid considerably in the setting up of good seedbeds in the spring.

Waterlogging is often a major problem and good drainage is very necessary for arable cropping. Originally drainage on clay soils was done using 'ridge and furrows', which were set up by ploughing, making the 'openings' and 'finishes' in the same respective places until a distinct ridge and furrow pattern was formed. The direction of the furrows was the same as the fall on the field so that water could easily run off into ditches. This early type of drainage was replaced by using underground mole and pipe-based drainage systems (Chapter 8).

In many clay-land areas, especially where rainfall is high, the fields are often small and irregular in shape because the boundaries were originally ditches which followed the fall of the land. The hedges and deciduous trees, which were planted later, grow very well on these fertile, wet soils. Increasing the soil organic matter, such as by applying straw-rich farmyard manure, working-in cereal straw, compost or grassland residues, makes these soils easier to work.

3.4.1.3 Cropping

Because clay soils can be difficult to manage, only a limited range of arable crops are grown, though yield potential is high. In high rainfall areas the ground is usually left in permanent grass and only grazed during the growing season. Autumn sown crops such as winter wheat, winter oilseed rape, winter barley and winter beans are the most popular arable crops for clay soils. Higher nitrogen contents make them suitable for milling wheats. Sugar beet and potatoes are occasionally grown in some regions. However, there can be difficulties in field access for seedbed preparation and harvesting these crops, especially in a wet season. Establishment costs will be higher on these soils. The best sequence of cropping with such root crops is following a period under grass; then the soil structure is more stable and the soil easier to work.

3.4.2 Sandy soils

These soils are mainly composed of sand (greater than 70%) with very little clay (less than 15%) and/or silt. They generally hold low quantities of organic matter. The soil texture is either a sand or loamy sand. Because these soils contain very little clay or organic matter they tend to be very weakly structured. They are comparatively easy to work. No natural restructuring takes place on these soils so any compaction has to be removed using mechanical cultivations. As sand is inert with little surface bonding on particles the cation exchange capacity of these soils is very low. Due to the large particle size of sand they contain a large number of macropores which tend to contain air rather than water.

3.4.2.1 Characteristics

- Sandy soils can be worked relatively easily at any time, even in wet weather.
- They are normally free-draining, but some drains may be required where there is clay or other impervious material underneath, or a high water table.

- These soils are droughty as they have a very low available water capacity. For some crops irrigation is required.
- They warm up early in spring.
- These soils are prone to leaching and are very responsive to application of nutrients.
- They are weakly structured and can be easily eroded by water (especially on slopes) and by wind.
- They have little natural structure of their own and often need subsoiling at regular intervals to loosen buried compaction (pans).
- Sandy soils can affect intensity of some crop pests and diseases, e.g., slugs are not usually a problem, whereas take-all disease can be very serious in second cereal crops.

3.4.2.2 Management

In many ways sandy soils are the opposite of clays and are often called light soils because, when working them, comparatively little power is required to draw cultivation implements. They are less suitable for minimum and reduced cultivation systems and topsoil loosening is required, usually by ploughing. After inversion few cultivations are required to produce a seedbed, often a one-pass system being used. Sandy soils can suffer from slumping and capping, particularly in low organic matter situations. Due to low nutrient retention, higher amounts of fertiliser are applied to maintain optimum nutrient status. Liming is necessary but must be used carefully—'little and often'. Organic matter, especially as humus, is very beneficial because it helps to hold water and plant nutrients in the soil and improve soil structure and biological activity. On fully limed fields though organic matter break down is very rapid because the soil microorganisms are very active in these open-textured soils that have a good air supply.

Sandy soils are very prone to both wind erosion in exposed situations and water erosion and run-off on slopes; the sand-blasting effect of wind erosion in the spring can destroy a young crop. Various methods are available to reduce the risk of erosion including planting nurse crops, cover crops, applying bulky organic manures and cultivation management to avoid fine seedbeds.

3.4.2.3 Cropping

A wide range of crops can be grown, but yields are very dependent on a good supply of water and adequate nutrient supply. Cereal yields are usually low.

On the lighter sands in low rainfall areas, and where irrigation is not possible, less drought-susceptible crops are grown, e.g., rye, carrots, sugar beet, forage maize and lucerne; lupins are grown in a few areas where the soil is very poor and acidic. To reduce the risk of drought stress, crops are established early to try and encourage deep rooting; better results are produced from winter- rather than spring-sown cereals. Winter barley is often preferred to wheat as it ripens earlier, is deeper rooted and may be less affected by a summer drought. It is important to ensure that there is no soil compaction which would also restrict rooting. Sandy soils are suited to malting barley where it is easier to achieve the low grain nitrogen required for malting.

On the deeper sandy soils, and particularly where the water supply (from rain or irrigation) is good, the main arable farm crops grown are cereals, peas, sugar beet, potatoes and carrots. For root crop production these soils are better as there is easier field access for harvesting in the autumn and winter and, for sugar beet, lower dirt tare. Where there are hedges they are often left tall to act as windbreaks. The trees are usually drought-resistant coniferous types. Livestock can be out-wintered on forage crops on sandy soils with less risk of damage by poaching, even in wet weather. Sandy soils are suited to outdoor pig production, ideally taking place effectively as part of a well-managed cropping rotation.

3.4.3 Loams

Loams are intermediate in texture between the clays and sandy soils and, in general, have most of the advantages and few of the disadvantages of these two extreme types. These soils have developed over a wide range of parent material; they are dominated by what are often called medium-textured soils. The amount of sand, silt and clay present in loam soils varies considerably. Texture classes include the clay loams (resembling clays in many respects), sandy loams (resembling the better sandy soils) and sandy clay loams and silty clay loams. Sandy silt loams and silt loams are also part of the loam family, but will be considered separately in a silty soils category. A large proportion of the sand in this group is coarse and/or medium sized which often makes the soil feel gritty. Climate and topography are the main limitations to crop production.

3.4.3.1 Characteristics

- Loams have average water-holding capacity and so are fairly resistant to drought.
- They warm up reasonably early in the spring.
- They are moderately easy to work.
- Depending how they were formed, and their parent material, some of the loams can contain a large stone content which can affect the sowing and harvesting of some crops.
- Loams are potentially fertile soils in many situations.

3.4.3.2 Management

Loams are moderately easy to work, but should not be worked when wet, especially clay loams. Minimum cultivation systems can be successful on these soils.

3.4.3.3 Cropping

Loams are generally regarded as the best all-round soils because they are often naturally fertile and are relatively easy to manage well. They can be used for growing most crops provided the depth of soil is sufficient. These soils can be used for most types of arable or grassland farming but, in general, are well-suited to mixed farming. Cereals, oilseed rape, potatoes and sugar beet are the main arable crops grown, and leys and forage maize provide grazing and fodder for dairy cows, beef cattle or sheep.

3.4.4 Silty soils

The silty soils are dominated by silt particles, these being at least one third proportionally of the total mineral content. Clay is less than 18% of the make-up. Silty soils comprise of sandy silt loam and silt loam textures. The sand fraction is mainly of very fine sand particles. The soils have a very silky, soft, buttery feel, occasionally with some grittiness. They have generally been formed from glacial, river, marine and wind-blown deposits and usually have deep stone-free subsoils.

3.4.4.1 Characteristics

- Many silt soils need some sort of drainage. In certain coastal areas there has to be a pumped system in order for the water table to be lowered and for arable crops to be grown.
- They are potentially very fertile mainly due to their great depth, very high available water capacity and nutrient retention.
- Good working properties exist provided that organic matter is maintained above 3%.
- Water erosion and run-off can be a problem in some areas.

3.4.4.2 Management

Although very fertile, silts can be difficult to manage. The two main problems are capping and compaction. Capping occurs when heavy rain falls on a very fine seedbed. Silt and clay particles go into suspension in a surface slurry and, as this dries out, it forms an impenetrable harder layer on the surface of the soil. It is particularly damaging if seeds have been sown, but have not yet germinated and emerged. Crops with small seeds will sometimes need to be redrilled if capping occurs. Leaving a rougher seedbed and increasing surface organic matter can decrease the risk of capping.

The lighter silts, as with the sands, can be damaged by compaction if worked in unsuitable conditions. Correction by deep cultivations may be necessary.

3.4.4.3 Cropping

On the heavier silts, mainly grass, cereals or fruit are grown; yield potential is high. However, on the drained lighter silts there are no limitations to crop growth. A wide range of crops can be grown in these situations including wheat, potatoes, sugar beet, vining peas, bulbs and field vegetables. Deep, fertile, silty soils are often associated with high value fresh produce cropping. River and marine silts in high rainfall areas, that are liable to flood, are best left in permanent grass.

3.4.5 Chalk and limestone soils

These are soils derived from calcareous or dolomitic parent material and contain various amounts of calcium carbonate or calcium magnesium carbonate, between 5% and 50%. The depth of top and subsoil may vary from 8 cm to over a metre, but being typically less than 30 cm depth. In general, the deeper soils are more fertile than the shallow ones. The ease of working and stickiness of these soils depend on the amount of clay and chalk or limestone present; they usually have a loamy medium texture. Sharp-edged flints of various sizes, found in soils overlying some of the chalk formations, are very wearing on cultivation implements and tyres, as well as being destructive when picked up by harvesting machinery. In some places the flints are found mixed with clay, e.g., in the clay-with-flint soils.

The soils overlying chalk are generally more productive than are those over limestone because plant roots can penetrate the softer chalk and explore for water. The limestones are mainly harder and more impenetrable. Brashy limestone rock pieces, loosened by cultivations, are a more severe problem to the farmer than the pieces of chalk that work their way to the surface on the downland arable farms.

Dry valleys are characteristic of the limestone and chalk downland. The few rivers rise from underground streams and the deliberate flooding of water meadows in the river valleys used to be a common practice. Watercress beds flourish along some chalk streams. The farms and fields on this type of land are usually large, especially on the thinner soils. There are very few hedges and the trees are mainly beech and conifers. Walls of local stone form the field boundaries in some limestone areas.

3.4.5.1 Characteristics

- Chalk and limestone soils are free draining except in a few small areas where there is a deep clay subsoil.
- They are often shallow and can be prone to drought (limestone soils).
- They are usually found at fairly high altitudes, above 120 m.
- These soils have strongly developed structure with stable aggregates.
- Stones and rock fragments within these soils can affect sowing and harvesting of some crops.
- They naturally contain low levels of some macronutrients. The alkalinity of the soil can cause some micronutrients to be unavailable. Only the deepest and/or sandy outcrops ever need liming.

3.4.5.2 Management

All calcareous soils, except those with a high clay content, are easy to cultivate. Minimum cultivation systems have been very successful on these well-structured soils. As the soils are free draining there is a large window when it is possible to cultivate. Drilling of spring cereals can often take place very early in the spring. Organic matter can be beneficial, but in the alkaline conditions it breaks down fairly rapidly.

3.4.5.3 Cropping

Cereals are good crops for these soils and malting barley has been very successful. Continuous barley and/or wheat production has been common practice, but this has changed partly due to the difficulty of controlling grass weeds and better returns from other crops. Roots such as sugar beet, fodder beet and potatoes (some for seed) are grown on the deeper soils on some farms. Other crops such as oilseed rape, peas, beans, linseed, and leys for grazing, conservation or seed production provide a break from cereals. On the thin limestone soils it is important to establish the winter cereals early so that they are less affected by any summer drought.

As chalk and limestone soils can be found at fairly high altitudes and the fields are often exposed, great care should be taken when growing crops which shed their seed easily, e.g., oilseed rape. Harvesting must be carried out carefully to minimise yield loss (Chapter 14).

Areas of black puffy soil (18%–25% organic matter) are found on some chalkland farms and they require special treatment; cultivations can be difficult and often micronutrient deficiencies, e.g., copper, need correcting.

3.4.6 Peat and peaty soils

These soils are dominated by high quantities of organic matter that produce key characteristics. Peaty soils (peaty loam and peaty sand) contain about 20%–35% of organic matter, whereas there is 35%–50% in loamy peat and sandy peat and over 50% in true peat. All of these organic dominated soils are usually very black or dark brown in colour and feel silky and smooth.

Peat and peaty soils can be found in both upland and lowland situations, being more widespread in the former. Blanket bog peats have formed in waterlogged upland areas overlying impermeable parent material where there is a high rainfall. Plants such as mosses, cotton grass, heather, mat grass, purple moor-grass and rushes grow. The dead material from these plants is only partly broken down by the types of microorganisms that can survive under these acidic waterlogged conditions. This humus material builds up slowly, about 30 cm every century. These peats are very acidic and are low in available nutrients. A certain amount of improvement can be undertaken by drainage, fertilising and liming, but because of their situation they are only suitable for rough grazing or forestry.

There are two important types of lowland peat, the sedge or fen peats and the moss peats. The fen peats developed in marshy land depressions below sea level, where the water coming into the area was rich in nutrients. Successive layers of decaying plant material built up. The main vegetation consisted of reeds, sedges and woody plants. Old bog oaks are sometimes brought up by cultivations. When drained these peats break down to produce very fine particles. These soils are slightly acidic and are low in phosphorus, potassium and some micronutrients. The moss peats developed on boggy land with a high rainfall. The main vegetation was moss which grew and decayed very slowly to produce a raised area (raised moss). When drained these peats have a low pH and are very low in nutrients. Moss peat is more fibrous than fen peat.

3.4.6.1 Characteristics

- These soils are rich in nitrogen, released for plant use by microbial mineralisation, but are poorer in phosphorus, potassium and also micronutrients such as manganese and copper.
- The lowland peats are very fertile soils if drained and fertilised. As well as ditches for drainage those areas below sea level need a pumped drainage system if arable crops are to be grown.

- They are naturally acidic and initially need heavy liming to be made more productive. However, over-liming can lead to micronutrient deficiencies.
- When drained, peat initially shrinks. Afterwards the organic matter breaks down and oxidises away very rapidly and the soil level can fall by as much as 2.5 cm per year; this is called 'wastage'. Eventually any mineral subsoil will come close to the rooting area and have a major effect on the characteristics and potential cropping of the soil.
- Wind erosion or 'blowing' in spring is a serious problem due to the light density of organic matter when dry. Several plantings of crop seedlings, together with the top 5−8 cm of soil and fertilisers, may be blown into the ditches. Deep ploughing and/or cultivation to mix the underlying clay/mineral subsoil with the organic topsoil can prevent this. Husbandry techniques such as straw planting, nurse crops or strip cultivation, together with shelter belts, are also used to prevent wind erosion.
- Weed control can be very difficult. Weeds grow very vigorously with the high nitrogen levels. Residual or soil-applied herbicides are not normally effective as the chemicals become adsorbed onto the colloidal particles of organic matter. Foliar-acting herbicides or mechanical methods are commonly used for weed control on these soils.
- Peat soils have a very high available water capacity for crop growth. Crops rarely suffer from drought on these soils.
- Upland waterlogged soils are at high risk of runoff and erosion and are very susceptible to livestock poaching and damage from farm traffic.

3.4.6.2 Management

Peat and peaty soils are substantial reserves of carbon that has been sequestered over considerable time. More recent intervention with these soils in the uplands for agricultural use through drainage has accelerated organic matter wastage and carbon release as atmospheric carbon dioxide. Attention is now focused on restricting and stopping further losses by rewetting and blocking drainage grips. Peat extraction for fuel or horticultural use is reducing so that these considerable organic matter reserves are maintained as they have key roles in carbon storage and management, regulating water quality, water storage and habitat management.

Past drainage of the lowland peats has produced high quality, fertile farmland for a variety of cropping and food production. Management of these soils to maintain agricultural output and soil conservation and protect from further soil wastage is an important focus for the future. Use of cover crops and less intensive operations are now receiving increased attention.

Where arable crops are being grown heavy applications of up to 25 t/ha of ground limestone may be required to reduce the soil acidity. Sometimes a very acid reddish brown ('drummy') layer builds up between the organic topsoil and any underlying clay subsoil. Peat soils are easy to cultivate and do not suffer from soil compaction. Under-consolidation of the seedbed is common though. The peaty soils with a lower organic matter content suffer less from under-consolidation. Cultivation systems for organic mineral soils that contain about 10%−20% organic matter are similar to those for the non-organic mineral soils.

3.4.6.3 *Cropping*

Drained fen or light peat soils are among the most fertile arable soils. Crops such as potatoes, sugar beet, celery, onions, carrots, lettuce and other salad crops are commonly grown. Cereals produce low yields. On light undrained peats, or where the water table remains high, the main crop is grass. On some of these areas willows are grown. Peaty loams are suitable for growing root crops and cereals. Where the depth of peat is very shallow (due to wastage) cropping is similar to that for the underlying mineral soil type.

3.5 Soil fertility and productivity

Soil fertility is a rather loose term used to indicate the potential capacity of a soil to grow a crop (or sequence of crops). The productivity of a soil is the combined result of fertility and management and is typically characterised as a measured output of crop yield, and where relevant can be extended to include animal output. This can be developed further to consider soil outputs in relation to measured soil inputs e.g., nutrients and water. The fertility of a soil at any one time is partly due to its natural make-up (inherent or natural fertility) and partly due to its condition and management (variable fertility) at that time.

Natural fertility can have an influence on the rental and sale value of land. It is the result of factors which are normally beyond the control of the farmer, such as:

- Texture and chemical composition of the soil mineral matter.
- Topography (natural slope of the land) which can affect drainage, temperature and workability of the soil.
- Climate and local weather, particularly the effects of temperature, and rainfall (quantity and distribution).

Soil management can affect variable fertility and soil condition by changing the:

- Soil structure.
- Amount of soil organic matter and biological activity.
- Amount of soil water through drainage and irrigation (Chapter 8).
- Level of soil nutrients (Chapter 4).
- Soil loss by erosion and runoff (removal by water, wind, livestock and/or cultivations).
- Soil pH.

Careful attention of the above factors should help maintain or increase soil fertility and productivity.

3.5.1 Soil erosion and runoff

As with soil formation, soil loss through erosion is a natural process. It can be defined as soil moved through the action of a vector such as wind and water, but it can be accelerated by extreme weather and poor or inappropriate land use or management. It leads to a loss of the soil resource with detrimental effects on soil functions and therefore damaging soil health. Soil erosion normally occurs very slowly on undisturbed soils with natural vegetation. When soils are cultivated for cropping or grazed by livestock soil erosion and runoff can occur at an elevated level. On cropped land erosion can cause losses of nutrient rich topsoil and deposition of soil sediment on roads and into water courses. Soil runoff of nutrients, organic matter and pesticides can also affect water quality and habitats. Total crop loss can occur in severe cases of wind erosion. Although soil erosion in the UK is not as severe as found in other parts of the world, it is an increasing problem due to the incidence of more extreme weather events and reduced soil organic matter levels. Farmers need to take responsibility to introduce appropriate soil management practices to try to reduce the impact of erosion and runoff on the wider environment.

3.5.1.1 Wind erosion and control

Cultivated, dry, sandy and peaty soils are most susceptible to wind erosion, and particularly so in the spring before crops are fully established. Keeping field sizes small and planting tall shelter belts can help reduce wind speeds, but other management measures are also required. Growing nurse crops or planting straw between rows of susceptible crops have been very successful at reducing wind erosion (Chapter 15). Leaving rough seedbeds and minimising cultivations can also help stabilise the soil, as can spreading mulches or even applying some synthetic polymers.

3.5.1.2 Water erosion, runoff and control

Silty and sandy soils are most prone to water erosion, particularly in areas with a high rainfall and where the ground slopes. Fields very susceptible to erosion commonly have shallow channels or rills, or sometimes even larger deeper gullies, after heavy rain. Surface water runoff happens when the intensity of rainfall is so great that all the water cannot soak into the soil. Livestock management can also affect the amount of soil erosion and runoff on grassland. Grazing with large numbers of stock in wet conditions can cause serious poaching and compaction in the topsoil which can lead to an increased risk of runoff.

Some cropping systems are much more prone to water erosion than others. The most susceptible land use is when crops are harvested in the autumn under adverse soil conditions which can happen with forage maize, root crops and field vegetables. Long term leys and permanent pasture usually pose much less risk. Soils with a good crop cover are much less susceptible than where there are fine seedbeds and/or bare ground.

Reducing compaction, minimising cultivations and using appropriate slope management such as contour working, increasing soil organic matter, using green manures and cover crops, avoiding late planted autumn sown or harvested crops, planting buffer strips and establishing grass leys can all help to reduce the risk of soil erosion on susceptible fields.

3.5.2 Soil acidity and liming

Most farm crops will not grow satisfactorily if the soil is very acidic. In acid conditions aluminium, iron and manganese become more readily available for root uptake. Excessive uptake of aluminium in acid conditions can severely affect crop growth. Some crops are more affected than others. This can be remedied by applying one of the commonly used liming materials.

3.5.2.1 pH, acidity and cropping

All substances in the presence of water are either acid, alkaline or neutral. Acidity and alkalinity are expressed by a pH scale on which pH 7 is neutral, numbers below 7 indicate acidity and those above 7 alkalinity. Soil pH is a measure of the negative logarithm of the hydrogen ion activity (concentration) in solution. Most cultivated soils have a pH range between 4.5 and 8.5. Soil pH can be assessed in the field by using a soil testing kit, or more accurately in the laboratory. Recommendations for soil sampling in order to analyse soil pH are given in Chapter 4 as pH is typically measured alongside levels of available soil phosphorus, potassium and magnesium.

To give crops the best opportunity to grow well, the minimum soil pH on mineral soils should be as shown in Table 3.2. The optimum pH for mineral soils where arable crops are grown is 6.5 and for continuous grass and clover swards the figure is 6.0.

Indications of soil acidity (i.e., the need for liming) are as follows:

- Crops failing in patches, particularly the acid-sensitive ones such as barley and sugar beet. The plants look yellow and stunted with a stumpy root system.
- On grassland, there are poorer, less productive types of grasses present e.g., bents. Often a mat of undecayed vegetation builds up because the acidity reduces the activities of earthworms and bacteria that break down such material.
- On arable land, weeds such as sheep's sorrel, corn marigold and spurrey are common where a soil is, or has a history of, being acidic. Also, in acid conditions residual herbicide activity may be reduced.
- Club root disease in brassicas is aggravated in acid soils. Liming is used as a method of control.

3.5.2.2 Lime requirement and its benefits

The need for lime is based on the *target* pH for the crop to be grown (which is 0.2 pH above *optimum*), the soil texture, soil organic matter and current pH. Grassland has a

Table 3.2 Minimum soil pH requirements.

Cropping	Soil pH
Barley, sugar beet, beans, peas and lucerne	6.5
Maize, oilseed rape, oats, wheat, cabbage and carrots	6.0
Potatoes, rye and apples	5.5
Ryegrass and fescues	5.0

lower requirement than arable crops. Heavy (clay) soils and soils rich in organic matter require more lime to raise the pH than other types of soil. For example, to raise the pH from 5.5 to 6.5 on a loamy sand may require about 7 t/ha of ground limestone, but on a clay soil 10 t/ha of ground limestone may be necessary. The sandy soil, however, will need to be limed more frequently than the clay soil. Lime should be applied before growing acid susceptible crops such as beet or barley, but after growing such acid tolerant crops as potatoes. Lime requirements are shown in Table 3.3 and are based on a liming effect for 20 cm depth of cultivated soil for arable cropping and 15 cm depth of grassland soil.

The main benefits of applying lime are:

1. It neutralises the reserve acidity of the soil by exchanging the hydrogen and aluminium ions on the soil charged sites with calcium or magnesium and releasing them into solution.
2. It neutralises the active acidity of the soil by removing the hydrogen ions in the soil pore space solution through chemical reaction.
3. It supplies calcium (and sometimes magnesium depending on its source) for plant nutrition.
4. It improves soil structure and makes the structure more stable. In well limed soils, plants usually produce more roots and grow better. Bacteria are more active in breaking down organic matter and this also usually results in an improved soil structure so the soil can be cultivated more easily.
5. It affects the availability of plant nutrients. Nitrogen, phosphorus and potassium are freely available on soils maintained at optimum pH. Too much lime is likely to make some micronutrients unavailable to plants, e.g., manganese, boron, copper and zinc. This is least likely to happen in clay soils and most likely to happen on organic soils.

3.5.2.3 Materials commonly used for liming soils

The supplier of lime must give a statement of the neutralising value (NV) of the liming material. This is a measure of its relative effectiveness compared with that of pure calcium oxide (CaO). In addition to NV other measures that contribute to liming material effectiveness are particle size fineness (degree of grinding/sieving) and the hardness of

Table 3.3 Lime requirement of different soils (t/ha).

Cropping	Soil type	pH 6.0	pH 5.5	pH 5.0
Arable cropping	Light	4	7	10
	Medium	5	8	12
	Heavy	6	10	14
	Organic	6	10	14
	Peat	0	8	16
Grassland	Light	0	3	5
	Medium	0	4	6
	Heavy	0	4	7
	Organic	0	4	7
	Peat	0	0	6

the parent material. In general, the higher the NV, the finer the particle size and the softer the material, then the more effective it is. NV and fineness determine a liming material's Reactivity Value which practically relates to speed of action in neutralising acidity and soil efficacy of use.

Ground or screened lime - limestone or chalk (also called carbonate of lime or calcium carbonate $CaCO_3$) is the commonest liming material used. It is generally obtained by quarrying limestone or chalk rock and grinding and/or sieving it to a fine powder.

Magnesian or dolomitic limestone consists of calcium magnesium carbonate $CaMg(CO_3)_2$. It is commonly used as a liming material in areas where it is found. Additionally, the magnesium present may prevent magnesium deficiency symptoms in crops.

Calcareous sand or shell sand is a liming material collected from some beaches where there is a high shell content. The neutralising value is lower than that for ground limestone. It is a useful material in areas far removed from the limestone or chalk quarries.

Some other liming materials are obtained from industrial processes where lime is used as a purifying agent, such as in sugar beet factories; these are called *waste limes*. Sugar beet waste lime is also a valuable source of plant nutrients. Also industrially, *granular limes* have been manufactured by combining very finely ground calcium or magnesium carbonate with an organic binding agent. Granules are typically 2–5 mm in size and the material is spread using traditional fertiliser spreaders.

3.5.2.4 Lime application, costs and losses

For most effective use liming materials should be well worked into cultivated soil to provide good incorporation and to increase efficiency of pH rectification. If there is a need for rapid pH change then a fast-acting liming material should be considered e.g., hydrated lime. But typically most farmers use ground limestone or chalk and arrange for it to be spread mechanically by the suppliers. Where large amounts are required (over 7 t/ha) it is generally preferable to apply it in two dressings, e.g., half before and half after ploughing, to ensure more even distribution through the soil.

The cost of liming is largely dependent on the transport costs from the lime works to the farm. By dividing the cost per tonne of the liming material by the figure for the neutralising value, the unit cost is obtained. In this way it is possible to compare the costs of the various liming materials.

Lime is removed from soil in many ways. Between 125 and 2000 kg/ha of calcium carbonate may be lost annually in drainage water. The rate of loss is greatest in areas of high rainfall, well-drained soils and soils rich in lime. Fertilisers containing ammonium, nitrate, sulphate or chloride ions have the greatest soil acidifying effect. Every 1 kg of sulphate of ammonia removes about 2 kg of calcium carbonate from the soil. Other nitrogen fertilisers have a smaller acidifying effect. Some organic manures can slightly reduce soil acidity. Crops and stock also remove some calcium carbonate, for example a 7 t/ha cereal crop can remove 50 kg and a 500 kg animal sold off the farm removes about 16 kg of calcium carbonate in its bones.

3.6 Evaluating soil health on the farm

The purpose of this chapter has been to consider soils in the context of crop husbandry. Viable and sustainable crop production is inherently linked to many aspects of soil characterisation and farming plants and animals can have both positive and negative effects on soil properties and functionality. By considering these elements in terms of soil health and management, and by aiming to assess and improve soil health, the farmer is playing a significant role in the maintenance and respect of a crucial asset and resource of the Earth. The soil's biological, physical and chemical processes and properties work and interact to produce soil functionality and health and an appreciation and understanding of these when considering crop husbandry and other practices is of significant value. Table 3.4 provides some key indicators of soil health, shows how they can be practically assessed on farm and are matched with principles and practices that aim for maintenance and improvement.

Table 3.4 Soil health on the farm—properties, indicators, assessment and positive principles and practices.

Property	Indicator	Assessment	Positive principles and practices
Biological	Organic matter	Sample for laboratory analysis.	Maintain or improve levels as appropriate. Consider inputs from crop residues (cash and cover crops), bulky manure amendments, grass leys.
	Earthworms	Sample, count and identify.	Leave crop residues as appropriate. Increase organic amendments. Reduce cultivations. Maintain suitable drainage and pH.
Physical	Texture	Assess by hand and use of key.	Match with suitable cropping and cultivations. Drainage as appropriate. Reduce soil erosion and runoff.
	Structure	Assess by spade and visual guide (e.g., VESS).	Minimise compaction and surface capping. Reduced and timely cultivations. Maintain/increase soil organic matter. Consider appropriate cropping.
Chemical	pH	Sample for laboratory analysis.	Maintain optimum pH for cropping and rotational considerations. Use an appropriate liming policy.
	Macro- and micronutrients	Sample for laboratory analysis. Assess soil nitrogen supply.	Maintain macronutrient indices at target levels. Rectify any micronutrient deficiencies. Use a nutrient management planning policy.

3.7 Key points

- Soil is a variable matrix of living and non-living material composed of solid mineral and organic matter, voids mainly occupied by water and air and a rich diversity of biological species.
- In a healthy state soils perform many functions as a living system improving environmental quality and protection, and plant, animal and human health. In a cropping context the soil matrix will mainly be linked to plant rooting, support for anchorage and the supply of water, nutrients and oxygen for above and below ground growth and development.
- Good soil management aims to improve soil health using a multi-functional approach and the farmer needs to consider agricultural productivity alongside other important soil benefits and outcomes.
- A range of physical, biological and chemical properties integrate together to affect soil health and functions. Examples include (physical) texture and structure; (biological) organic matter, fauna and flora; (chemical) pH and nutrients.
- Soils can be classified and grouped in many ways, but texture/geology and organic matter in particular have major effects on the practicalities of farming operations and management. Key different soil groups separated in this way include those described as sandy, silty, loam, clay, chalk/limestone and peaty.
- Soil fertility is the potential capacity of a soil to grow a crop (or sequence of crops) and is inherently linked to soil productivity. Fertility is partly due to a soil's natural make-up as well as its condition and management.
- Key soil health indicators can be practically assessed on the farm and should be matched with good farming principles and practices that aim for soil health maintenance and improvement.

Sources of further information and advice

AHDB, 2022. Nutrient Management Guide (RB209) Section 1: Principles of Nutrient Management and Fertiliser Use. Agriculture and Horticulture Development Board.

Al-Kaisi, M.M., Lowery, B., 2009. Soil Health and Intensity of Agroecosystems. Elsevier-Academic Press, London.

Ashman, M.R., Puri, G., 2009. Essential Soil Science. Wiley-Blackwell, New Jersey.

Avery, B.W., 1990. Soils of the British Isles. CAB International, Wallingford.

Bardgett, R.D., 2005. The Biology of Soil. Oxford University Press, Oxford.

Blum, W.E.H., Schad, P., Nortcliff, S., 2018. Essentials of Soil Science. Borntraeger Science Publishers, Stuttgart.

Bridges, E., 1997. World Soils. Cambridge University Press, Cambridge.

Coleman, D.C., Callaham, M.A., Crossley, D.A., 2018. Fundamentals of Soil Ecology. Elsevier-Academic Press, London.

Davies, B.D., Eagle, D.J., Finney, J.B., 2001. Resource Management: Soil. Farming Press, Tonbridge.

Defra, 2005. Controlling Soil Erosion. Defra Publications, London.

Defra, 2009. Protecting Our Water, Soil and Air. The Stationery Office, Norwich.

Eash, N.S., Sauer, T.J., O'Dell, D., Odoi, E., 2015. Soil Science Simplified. Wiley-Blackwell, New Jersey.

Environment Agency, 2008. Think Soils. Environment Agency.
Gerrard, G., 2000. Fundamentals of Soils. Routledge, London.
Gregory, P.J., Nortcliff, S., 2013. Soil Conditions and Plant Growth. Wiley-Blackwell, New Jersey.
Morgan, R.P.C., 2005. Soil Erosion and Conservation. Blackwell Publishing, Oxford.
Rowell, D.L., 2014. Soil Science: Methods and Applications. Routledge, London.
Soil Survey of England and Wales, 1985. Soils and Their Use in Eastern England; Midlands and Western England; Northern England; South East England; South West England; Wales. Soil Survey of England and Wales, Harpenden.
Weil, R.R., Brady, N.C., 2017. The Nature and Properties of Soils, fifteenth ed. Pearson Education, Harlow.
White, R.E., 2006. Principles and Practice of Soil Science, fourth ed. Blackwell Publishing, Oxford.

Websites

www.ahdb.org.uk/greatsoils.
www.landis.org.uk.
www.wossac.com.
www.sustainablesoils.org.
www.soils.org.uk.
www.ukso.org.

Fertilisers and manures

4

4.1 Introduction

Good nutrient management is a key aspect of viable and sustainable agricultural production. There is the challenge of increasing food production for the growing global population whilst ensuring that environmental impacts are lessened. Supplying optimal levels of nutrients for crop and livestock production as part of profitable farming businesses involves extensive decision making associated with the importance of integrated farming systems and their connection with the wider environment including the abiotic considerations of soil, water and air linked to the biotic aspects of habitats, biodiversity and species.

For viable field crop production, in terms of economic yield and good crop quality, nutrients should be readily available so that uptake is optimised and both nutrient deficits limiting uptake and nutrient excesses that may cause losses to the environment are reduced. Nutrient removals off the field as a result of crop harvest should be considered in nutrient budgeting calculations and decisions so that nutrient inputs help to reduce any nutrient gaps in the system and soil available nutrient levels are maintained at an optimum level that suit both effective cropping and a better environment. An assessment of nutrient budgeting considers at a basic level both crop nutrient supply and harvest offtake, but in more detailed assessments can include all nutrient additions to the soil supplying the crop and all nutrient outputs including to the wider environment away from the crop. It is important that any nutrient supply system minimises the loss of nutrients to the environment as well as avoiding either a deficiency or an excess of individual, required nutrient elements. There are additional factors associated with local soil characteristics, nutrient behaviour and effects of weather and climate that should also be part of any nutrient decision-making process.

Crops will obtain nutrients through a number of supply sources. Crop rooting will source plant food supplied by the soil. This supply is linked to soil processes such as *mineralisation* of soil organic matter and the *weathering* of soil minerals to create pools of readily available nutrients which can be countered by soil processes such as nitrogen *immobilisation* and phosphorus *fixation* that can reduce these pools. Soil nutrient levels can be amended by appropriate use of chemical fertilisers (manufactured inorganic sources of nutrients) or natural inorganic fertiliser materials, used where possible in conjunction with organic materials such as farmyard manure and slurry or non-farm produced amendments. Manufactured chemical fertilisers may also be used for direct crop uptake through foliar feeding applications. Legume crops are used to obtain their nitrogen through biological fixation of atmospheric nitrogen. Nutrients can also be received by soils and crops through deposition from atmospheric sources e.g., nitrogen and sulphur linked to precipitation and dry deposition. Crop

Lockhart and Wiseman's Crop Husbandry Including Grassland. https://doi.org/10.1016/B978-0-323-85702-4.00014-5

nutrient supply from all of these sources should be considered in order to match, but not exceed, crop nutrient demand. Inputs from fertilisers and manuring materials made as part of farming decisions should be based on the premise that crop nutrient supply from other sources does not match crop need. Applications would then make up differences needed or possibly aim to boost soil nutrients to more optimum levels as well as match next crop demand post-application.

4.2 Crop nutrient requirements

There is some debate as to the exact list of essential nutrients needed for effective plant growth as certain nutrients may aid plant performance but may not be considered as exactly essential. Carbon, hydrogen and oxygen are three elements that assist in the chemical behaviour and reactions of nutrients in agricultural systems. Otherwise the main nutrient elements required by crops can be divided up into major (macro) nutrients and micronutrients (trace elements).

- Major nutrients are nitrogen, phosphorus, potassium, sulphur, magnesium and calcium and are taken up by crops in relatively large quantities.
- Micronutrients are manganese, copper, boron, zinc, molybdenum, iron and chlorine and are needed in smaller quantities than the major nutrients.

There are a number of additional elements that may have a role in growth of certain crop plants, but not all, and would therefore not be classed as ubiquitously essential. These include cobalt, selenium, nickel and sodium. Some of these have important roles in animal nutrition and so adequate amounts are important in crops used in grazing systems.

Soil nutrient content is not necessarily matched with plant nutrient availability and can be considered as a nutrient bank with credits and debits, readily available and not so available accounts. Usually $<1\%$ of nitrogen, phosphorus and potassium in soil will be in plant available form. The soil organic nitrogen pool is 50–100 times larger than plant available nitrogen whilst total soil phosphorus or potassium may be 50–500 times larger. Nutrient availability is greatly affected by other soil properties e.g., pH, aeration, and in general plant nutrient uptake from soil is usually as soluble, inorganic forms. Details of nutrients, the chemical forms in which they are taken up by plant roots from soil solution sources and typical plant tissue concentrations are given in Table 4.1.

A deficiency of any one of these nutrients can restrict plant growth and development resulting in decreased yield. Optimum supply of all is therefore needed for full crop potential. Excess levels of nutrients can cause certain toxicity issues in plants, compromise crop quality and lead to agronomic issues such as lodging of cereals. There can also be *antagonistic* effects where excessively large quantities of one nutrient may decrease the availability or uptake of another e.g., relationship between potassium and magnesium. Table 4.2 indicates typical values for nutrient offtakes in various crops.

Table 4.1 Essential nutrient elements required for crop growth.

Element	Chemical symbol	Form absorbed by plants	Typical plant tissue concentration (mg/kg dry matter)
Major (macro) nutrients			
Nitrogen	N	NH_4^+ & NO_3^-	35,000
Phosphorus	P	HPO_4^{2-}	3000
Potassium	K	K^+	30,000
Sulphur	S	SO_4^{2-}	2000
Magnesium	Mg	Mg^{2+}	2000
Calcium	Ca	Ca^{2+}	15,000
Micronutrients			
Manganese	Mn	Mn^{2+}	50
Boron	B	$H_3(BO)_3$	25
Copper	Cu	Cu^{2+}	10
Zinc	Zn	Zn^{2+}	20
Molybdenum	Mo	MoO_4^{2-}	0.5
Iron	Fe	Fe^{2+}	70
Chlorine	Cl	Cl^-	100

Table 4.2 Typical values of crop nutrient removal (fresh material) (kg/ha).

Crop	Nitrogen (N)	Phosphate (P_2O_5)	Potash (K_2O)
Wheat			
Grain 7 t/ha	130	55	40
Straw 5 t/ha	17	4	31
Total	147	59	71
Barley			
Grain 6 t/ha	100	47	34
Straw 4 t/ha	25	3	25
Total	125	50	59
Oats			
Grain 6 t/ha	100	47	34
Straw 5 t/ha	15	5	59
Total	115	52	93
Beans			
Grain 4 t/ha	176	44	48
Potatoes			
Tubers 50 t/ha	150	50	290
Sugar beet			
Roots 45 t/ha	80	36	76
Fresh tops 35 t/ha[a]	120	39	202
Total	200	75	278
Kale			
Fresh crop 50 t/ha[a]	224	60	250

[a]If sugar beet tops or kale are eaten by stock on the field where grown, some of the nutrients will be returned to the soil.

4.3 Major nutrients

4.3.1 Nitrogen

Nitrogen (N) is the nutrient required in the largest quantity by most crops and it generally has a greater effect on crop growth, yield and quality than any other nutrient. It is however complex in its behaviour occurring in wide ranging reactive organic and inorganic forms in plants, animals, soil, air and water. Nitrogen has a key role in controlling the rate of crop growth through leaf growth, photosynthesis, carbohydrate production and protein synthesis. A deficiency or excess can drastically affect crop yield.

4.3.1.1 Soil nitrogen

The principal forms of N in soil are organic compounds and inorganic (mineral) ammonium (NH_4^+) and nitrate (NO_3^-) ions. Most soil N is contained in the organic matter, both fresh and humified, and is slowly converted by soil bacteria and other micro-organisms to inorganic ammonium N which is then rapidly converted by other microbes to inorganic nitrate N. Both these forms are soluble and available for plant root uptake unlike the insoluble organic compounds. The microbial conversion of organic to inorganic N is termed *mineralisation* through the processes of *ammonification* and *nitrification*. Soil contains large reserves of N in its organic matter. In UK topsoil this can amount to some 3000–7000 kg N/ha for arable land and 12,000–20,000 kg N/ha for permanent grassland. Ideally, mineralisation is roughly balanced by the addition of organic nitrogen in decaying crop residues, stubble and added organic amendments such as manures. However, small differences in levels of mineralisation and addition can accumulate over many years resulting in the gradual accumulation or loss of soil organic matter. During the process of mineralisation carbon compounds within the soil organic matter are degraded and used as a source of energy by the mineralising micro-organisms. Nitrogen in excess of the microbial need is liberated into the soil inorganic N pool. However, if the particular organic matter source is very low in N content, the organisms will need extra N to grow. In this situation inorganic N already present in the soil will be used by the microbes and this process is deemed nitrogen *immobilisation*. The ratio between carbon and nitrogen (C:N ratio) in organic matter determines whether immobilisation or mineralisation of N takes place. Incorporating certain forms of organic matter into the soil with a high C:N ratio e.g., cereal straw (80:1) will result in higher levels of immobilisation. Conversely, incorporating organic matter with a low C:N ratio e.g., legume residues (20:1) will lead to mineralisation with the potential for higher quantities of inorganic N then being available to growing plants.

The rate of soil microbial activity will depend on several factors. Soil water content and oxygen supply must be adequate for mineralisation and immobilisation to occur. Temperature is also important with nitrification starting at approximately 5°C. In Europe the rate is usually highest in the spring and autumn when soils are relatively warm and moist and when the soil is cultivated increasing aeration. The high soil temperatures that occur in hot climates can result in high rates of mineralisation and rapid loss of soil organic matter.

The quantities of soil inorganic N formed by microbial activity during plant growing seasons will depend upon the quantity of soil organic matter and the proportion of N it contains. Both of these depend upon previous cropping and soil type. Soil nitrogen supply (SNS) is a useful measure of determining the soil N value for crop uptake. SNS is defined as:

$$\text{Soil mineral nitrogen(SMN)} + \text{estimate of nitrogen already in the crop}$$
$$+ \text{estimate of mineralisable soil nitrogen}$$

where:

SMN (as kg N/ha) is the nitrate N plus ammonium N content of the soil within the normal maximum rooting depth of the crop;
Nitrogen already in the crop (as kg N/ha) is the total crop content of N when sampling for SMN is carried out;
Mineralisable soil nitrogen (as kg N/ha) (additionally available N) is the estimated amount of N which becomes available for crop uptake from mineralisation of soil organic matter and crop debris during the remainder of the growing season after sampling for SMN.

Additional soil nitrogen inputs occur through *fixation* of atmospheric N as soil inorganic N by certain bacteria that are free-living within the soil and those living in symbiosis with plants. The former accounts for little fixed N in the UK, but, agriculturally, symbiotic fixation is more important. Legumes (e.g., clovers, peas, beans, lucerne) have a key role in an ability to fix atmospheric N when in symbiosis with the soil bacterium *Rhizobium*. Fixation is inhibited below pH6 and by high N fertiliser use. Well aerated soil conditions are also important.

4.3.1.2 Loss of nitrogen

Soil inorganic mineral nitrogen, as ammonium and nitrate N, is directly available for crop uptake. Within the soil, ammonium N can be held on the negatively charged sites of clay minerals and organic humic compounds. This will reduce its mobility in the soil compared to the more mobile nitrate ion. Generally, soils have a higher binding capacity for cations than anions and so nitrate in excess of plant uptake is therefore at risk to loss by leaching when excess rainfall or irrigation water drains through the soil. Soils most prone to nitrate leaching losses will be freely draining, light sandy and/or shallow soils, often described as 'leaky'. Husbandry and agricultural practices which encourage high levels of nitrate in the soil include:

- amount of fertiliser N applied exceeding crop requirements
- organic manures applied in autumn/early winter
- following the cultivation of soil in autumn and particularly the ploughing up of grassland
- when soil is left uncropped in winter e.g., spring sown crops.

Nitrate losses through leaching have implications in the eutrophication of surface waters and human health concerns through increased nitrate concentrations in drinking waters.

In addition to nitrate leaching there are other methods of inorganic N loss from soil, these being gaseous losses through the processes of *denitrification* and ammonia *volatilisation*. Denitrification transforms nitrate N to nitrogen gas and returns it to the atmospheric pool. Denitrification also releases quantities of the potent greenhouse gas nitrous oxide (N_2O). The process is anaerobic and often occurs in very wet, warm soils where there has been a recent N application. Denitrifying microbes need readily degradable organic matter as an energy source.

Ammonia volatilisation can provide significant N losses particularly when organic manures and certain inorganic fertiliser sources are surface applied to soils and are not incorporated within a short time period. Elevated ammonia concentrations in the environment are linked to eutrophication and loss of biodiversity in surface waters and sensitive habitats and acidification of soils and surface waters.

4.3.2 Phosphorus

Phosphorus (P) is essential for plant development as it is a constituent of nucleic acids. Compounds of P are also concerned with the vital processes of respiration, photosynthesis and energy trapping and with the efficient functioning and utilisation of nitrogen. Phosphorus is of special importance in the processes concerned in root development and the ripening of seeds and fruits.

4.3.2.1 Soil phosphorus

Phosphorus is present in the soil in three main forms:

- soil solution
- mineral or inorganic phosphates
- organically bound phosphorus within organic matter

Most soils have large reserves of sparingly soluble or insoluble phosphates, but only small amounts of phosphate are usually present in solution. The main source of plant available P is generally termed the labile pool. This provides fairly rapid exchange with the soil solution, thus maintaining the solution concentration which is directly available for root uptake. The remaining fraction of P is the non-labile pool containing larger quantities of insoluble, precipitated, iron, aluminium and calcium phosphates which are only very slowly released into the labile pool. The combination of both organic and inorganic phosphates constitutes the labile and non-labile pools. However, there is no clear distinction by which particular forms can be designated as in one or the other pool. Generally, the labile pool is dominated by inorganic orthophosphate adsorbed on the surfaces of soil inorganic and organic materials. The relationship between the quantity of phosphorus in the labile pool and the soil solution concentration depends particularly on soil texture and pH. Adsorption of orthophosphate is stronger, and therefore phosphate is less available, at low pH. In calcareous soils, precipitation of calcium phosphates is a major factor limiting solubility. Phosphate availability is generally greatest between pH 6.0 and 7.0.

Movement of phosphorus from soluble, adsorbed, labile sources to the non-labile pool (and hence reduction in plant availability) occurs mainly by the formation of slightly soluble compounds. Precipitation as iron and aluminium phosphates will also reduce phosphorus availability. Fixation as insoluble compounds is most likely in calcareous soils. As clay soils can adsorb much more phosphate than light soils, they generally have less readily available to plants than sands or loams which usually have quite high levels. A measure of soil available phosphorus (Index) reflects the reserves of sparingly soluble phosphates in the soil and their capacity to replenish soil solution with water soluble phosphate. Phosphorus may also be released into the soil solution from the mineralisation of organic matter, the amount released being dependent upon:

- soil organic matter content
- soil conditions—temperature, pH, moisture status
- microbial activity, which depends on soil conditions.

4.3.2.2 *Loss of phosphorus*

Phosphorus loss from agricultural land can contribute to poor water quality and nutrient enrichment (eutrophication). Symptoms include algal blooms, fish kills, excessive weed growth, poor water clarity and loss of species diversity. Some algal toxins are also hazardous to health. The main source of agricultural phosphorus loss is through soil erosion and surface runoff of soil particles where precipitated, inorganic phosphorus is bound. This phosphorus is largely released and solubilised when soil enters surface waterbodies, particularly running waters. This loss is a major contribution to diffuse water pollution from agriculture. Smaller quantities of soluble phosphorus can be lost through the process of soil drainage when leaching conditions are prevalent on very sandy soils and when soluble P levels are also very high.

Key measures to reduce phosphorus loss include:

- minimising soil erosion from agricultural land
- careful application of fertilisers and organic manures to minimise the risk of direct loss in runoff
- careful nutrient budgeting to avoid unnecessary phosphorus build-up in soils.

4.3.3 *Potassium*

Potassium (K) has a key function in plants in the process of photosynthesis, particularly in regulating the opening of the leaf stomata cells. It also influences the strength of cell walls through cell water content and plants suffering from potassium deficiency are usually associated with weak, limp growth. Potassium is also essential for a number of plant enzyme systems and generally its role is to regulate processes rather than to become fixed in one form. It is often described as the holistic 'quality' nutrient as good cell functioning can help to maintain a plant's ability to have resistance against pest and disease attack and physiological stress associated with water availability and temperature control.

4.3.3.1 Soil potassium

As with other nutrients, crops take up potassium from the soil solution. This is rapidly replenished by the release of potassium held on the cation exchange sites of clay minerals and organic matter (K^+). This is described as exchangeable potassium. K reserves may also be held 'fixed' or trapped between the plate like units within clay minerals. These reserves, through the action of weathering, replenish exchangeable potassium. The largest quantities of soil potassium are contained deep within crystalline, precipitated materials and this insoluble K is classed as inert K. The size and rate of release of the exchangeable potassium fraction, plus the fixed fraction, determines the long-term need for potassium fertilisers. On clay soils potassium fertilisers are very efficiently used.

Potassium will only be lost from the soil in drainage water on light textured sandy soils low in organic matter content and low in cation exchange capacity. There are no key environmental issues associated with the loss of potassium to the wider environment.

4.3.4 Sulphur, magnesium and calcium

Sulphur (S) has some similar functions to nitrogen in plant development and performance, being a constituent of certain amino acids for protein production and having roles in enzyme systems. Plants need similar quantities of sulphur compared to phosphorus. Sulphur fertilisation of crops is now of increasing interest as the majority of UK crops need sulphur supplementation that used to be historically supplied from atmospheric sulphur deposition chiefly via rainfall. Such deposition has reduced greatly due to industrial activities becoming cleaner with reduced S emissions. This has led to increasing risk of sulphur deficiency for a wide range of crops such as oilseed rape, milling wheat and multi-cut silage. Sulphur is mainly contained within soil organic matter and becomes available for plant uptake when microbial mineralisation occurs. It is prone to leaching losses as sulphate in soils which would normally leach nitrogen as nitrate. On mineral soils, S should be applied to all winter oilseed rape crops and to all cereal and grass crops grown in high risk sulphur deficient situations.

Magnesium (Mg) is a key component of plant chlorophyll and therefore plays a vital role in carbohydrate synthesis through photosynthesis. It is also important for protein synthesis and enzyme reactions. Calcium (Ca) is important for plant cell structure and division. Soil solution levels of both, that are available for plant root uptake, are replenished from exchangeable forms retained on cation exchange sites and from weathering of soil minerals. Both calcium and magnesium have important roles in the regulation of soil pH through addition of liming materials. The role of calcium as a nutrient is often overlooked since well-managed soils are usually adequately supplied with Ca through the addition of liming agents.

Table 4.3 summarises the major nutrients associated with crop production.

Table 4.3 Major nutrients - needs and effects.

Plant nutrient	Crops which are most likely to suffer from deficiency	Field conditions where deficiency is likely to occur	Deficiency symptoms	Effects on crop growth	Effects of excess
Nitrogen (N)	All farm crops except legumes; especially important for leafy crops	On all soils except peats, and especially sandy soils and where organic matter is low	Thin, weak, spindly growth; lack of tillers and side shoots; small yellow/pale green older leaves, sometimes bright colours	Speeds up growth of seedlings and roots; hastens leaf growth and maturity	Lodging in cereals; delayed ripening; soft growth susceptible to frost and disease; may lower sugar and starch content
Phosphorus (P)	Root crops, clovers, lucerne and kale	Clay soils; acid soils especially in high rainfall areas; chalk and limestone soils and peats; poor grassland	Similar to nitrogen except that leaves are a dull, bluish-green colour with purple or bronze tints; poorer rooting	Increases leaf size, rate of growth and yield; makes leaves dark green	Might cause crops to ripen too early and so reduce yield if not balanced with N and K fertilisers
Potassium (K)	Potatoes, carrots, beans, barley, clovers, lucerne, sugar beet and fodder beet	Light sandy soils, chalk soils, peat, badly drained soils; grassland repeatedly cut for hay, silage or 'zero' grazing	Growth is squat and growing points 'die-back', e.g., edges and tips of leaves die and appear scorched	Crops are healthy and resist disease and frost better; prolongs growth; improves quality; balances N and P fertilisers	May delay ripening too much; may cause magnesium deficiency in crops and 'grass-staggers' in grazing animals

Continued

Table 4.3 Continued

Plant nutrient	Crops which are most likely to suffer from deficiency	Field conditions where deficiency is likely to occur	Deficiency symptoms	Effects on crop growth	Effects of excess
Magnesium (Mg)	Cereals, potatoes, sugar beet, peas, beans, kale	Light sandy soils, chalk soils; often of a transient period associated with poor soil structure; excessive potash	Interveinal yellowing on leaves (short of chlorophyll)	Associated with chlorophyll, and potassium metabolism	Unlikely, requirements about same as phosphorus and one-tenth that of nitrogen and potash
Sulphur (S)	Most crops but especially brassicas, e.g., oilseed rape, kale; grass cut for conservation; milling wheat	Light sandy soils—especially if low in organic matter and intensively cropped	Yellowing on younger leaves and less vigorous growth	Nitrogen made more efficient	Unlikely to occur

4.4 Micronutrients (trace elements)

These are generally required in small quantities for essential metabolic processes and growth needs in crop and animal systems. Certain elements in this role are essential for animal development and functioning, but may not be similarly required by plant crops e.g., selenium, and growth is therefore not limited if in short supply. Animal deficiencies may result though if nutrient intake is reduced when grazing or fed forage with low nutrient status. In this situation direct nutrient supplementation to the animal is usually the most effective response, though application to grazing pastures may sometimes be used as a suitable alternative.

Crop micronutrient deficiencies are likely to be greatest:

- on very poor soils in terms of depth and structure
- where soil conditions, such as a high pH or a very high organic matter, make them unavailable
- where intensive farming (with high yields and high nutrient offtake) is practised
- where organic products such as farmyard manure and slurry are not used.

Supplying micronutrients to plants is not always easy and care is required to prevent overdosing which may damage or kill the crop. To facilitate their application and crop availability, elements such as copper, iron, manganese and zinc are regularly used in a *chelated* form. These chelates are 'protected' water-soluble complexes of the micronutrient element bound with organic substances such as EDTA (ethylenediaminetetraacetic acid) or EDDHA (ethylenediamine di(o-hydroxyphenylacetic acid)). These organic substances form a protective 'crab's claw' around the element, effectively sequestering them and protecting them temporarily such that availability is not compromised through bonding associated with cationic charges. They can be safely applied as foliar sprays for quick and efficient action, or to the soil, sometimes as a supplement with a compound fertiliser, for root uptake without wasteful soil 'fixation' because they do not ionise. Foliar micronutrients are often crop-applied when tank-mixed with many spray chemicals, such as fungicides. Compatibility is generally good but product labels should be checked to ensure successful mixing. Micronutrients may also be applied as *frits* in bespoke, often horticultural, situations. These are produced by fusing the elements with silica to form glass which is then broken into small particles for distribution on the soil or growing medium, or as salts in solution.

One important scenario for micronutrients in plant nutrition is when plants are grown in culture solutions circulated around their root systems (*hydroponics*). This is a form of horticultural crop production in fresh produce and salad cropping (often in such developments as urban farming and vertical farming) especially where the culture solutions can be replenished automatically with nutrients (see Chapter 17).

It is important that whenever micronutrients are applied that there is a confirmed current, or likely future, need. This is because crops that are not growing satisfactorily may be suffering for reasons other than a micronutrient deficiency, e.g., acidic or alkaline soil, deficiency of a major nutrient, poor drainage, drought, frost, mechanical damage, viral or fungal diseases. Micronutrient deficiencies can be diagnosed by either nutrient analysis of leaf or other tissue, harvestable product or soil depending on the nutrient under investigation. Visual symptoms on crop tissue can also be an indication of specific nutrient disorders in certain situations. Care must be taken though in using this system as visual symptoms are sometimes transient and not necessarily bespoke and specific to a single nutrient in a deficiency or excess situation e.g., interveinal yellowing of crop tissue linked to manganese deficiency can also be a symptom of magnesium or sulphur deficiency in a number of cropping scenarios. Visual symptoms in crops may also only occur once nutrient levels are very deficient. Once symptoms have developed, crop growth and development are compromised and it can often be too late to correct the deficiency in that growing season.

Manganese (Mn) is the micronutrient element most frequently found to be deficient in field crops often associated with high pH levels and high soil organic matter, both of which reduce its availability for plant uptake. Mn deficiency can also be an issue on loose, unconsolidated soils. Crop symptoms can include pale, limp, new leaves with developing necrotic streaks or lesions. Deficiency can be remedied using foliar sprays of manganese sulphate or chelated manganese, usually applied just before periods of rapid crop growth and repeated throughout the season in severe situations. Seed treatments are also used for manganese deficiency.

Boron (B) deficiency is most often seen in root vegetables and sugar beet where it causes a problem of growing-point necrosis and developing crown rot. Soils can be tested for boron levels and remedial action should be taken before crops are grown. Boron sprays or boronated fertilisers can be used where necessary.

Copper (Cu) deficiency is a problem of peaty, very light sandy and thin organic chalky soils. Symptoms are most often seen in wheat and barley where it produces plants with pale, twisted or curled leaves and shrivelled leaf and ear tips. Chelated copper can be foliar applied directly to crops or copper treatments such as copper sulphate can be applied to soils.

Other micronutrient issues in field cropping can be linked to zinc (Zn) deficiency in cereal and forage situations and molybdenum (Mo) with certain field vegetables especially cauliflower.

4.5 Assessment of soil nutrient content

Plant available phosphorus, potassium and magnesium levels are typically measured via laboratory analysis from sampled soil and are quoted as a concentration and as an equivalent Index or status level. Indices indicate the relative amounts of nutrients in the soil that are available for plant uptake and range from deficient to very large. Values also reflect the past history of fertilisation and soil type. The most commonly used Index system places results in scoring bands ranging from 0 to 9. The lowest deficiency level is indicated by Index 0 and an excessively high level by Index 9 (never reached under field conditions). The Index 2 band for potassium is split into 2− and 2+. For arable, grassland and forage crops, the target soil Index is 2 for phosphorus and 2- for potassium, whilst for vegetable cropping the target soil Index is 3 for phosphorus and 2+ for potassium. Target soil indices represent soil nutrient levels that match optimal plant nutrition and any nutrient inputs aim to maintain these levels when working alongside nutrient removal in crop offtakes. As Indices drop below 2 more nutrient application is recommended to 'build up' the Index and soil nutrient level over a period of time back towards optimum, and as they get above 2 or 3 the recommendations decrease until, at Index 4 and over for all crops, apart from certain field vegetables, no applications are required and a period of 'running down' the Index is being instigated. Raising and lowering of Indices in this way can take many years to achieve, often 10 or more per Index level.

Together with pH, the quantities of these plant available nutrients are quoted as part of a routine soil sample nutritional test. These results are used in the decision-making processes leading to nutrient applications in the form of fertilisers and organic manures if needed to raise soil nutrient levels necessary to achieve optimum crop production; likewise, if liming materials are necessary to raise soil pH from acidic conditions to an optimum level for particular crops being grown in the rotation. Such pH optimums, that are partly related to nutrient availability, have been detailed in Chapter 3.

The field sampling for P, K, Mg and pH analysis must be done in a methodical manner if a reasonably accurate result is to be obtained. Generally, sampling should be carried out every 3–5 years, although pH may need to be checked more frequently, especially on light soils where calcium is easily leached. At least 25 soil cores, 15 cm deep (7.5 cm on long-term grassland), should be taken, bulked and sub-sampled to make up a representative amount weighing at least 0.5 kg. Unless the area is very uniform, samples should be taken for every four hectares. A 'W' pattern transect should be walked across the area with cores being taken frequently and regularly along this line. Field headlands, gateways and minor areas of obvious soil differences should be avoided. The analysis is carried out on the fine earth part of the soil, and this must be borne in mind when interpreting the results for very stony soils.

Options are available for more detailed assessment of soil nutrient variability within fields or cropping areas. The use of soil conductivity scanning and passive, gamma-ray detection technology allow zones and mapping layers to be established where detailed soil sampling can be targeted. GPS-based precision tools can also be used in order to establish sampling grids which allow more detailed assessment of nutrient variability. More details of precision sampling and nutrient assessment are provided in Chapter 9.

As has already been indicated nitrogen behaviour within soil follows a rapid flux with pathways of conversion and loss being of key importance. A description of the Soil Nitrogen Supply (SNS) - amount of soil N available for crop uptake from establishment to end of growing season, taking account of N losses -has already been discussed and its formulation is based on either:

- a *measurement method* involving assessment of soil mineral nitrogen (SMN) through laboratory analysis of sampled soil-combined with assessment of N in crop tissue at the time of soil sampling (by laboratory analysis or through a number of prediction systems) and assessment of additional N released by mineralisation (usually by estimation related to soil type)
- or a *field assessment method* using previous cropping, previous fertiliser and manure use, excess winter (or annual) rainfall and soil type. With this method there is no requirement for any sampling or analysis. The categories used for rainfall and soil type are given in Table 4.4

Table 4.4 Variables used to determine Soil Nitrogen Supply by Field Assessment Method.

Rainfall category	Soil category
Low rainfall areas: 500–600 mm annual rainfall, or up to 150 mm excess winter rainfall	Light sand soils
Moderate rainfall areas: 600–700 mm annual rainfall, or 150–250 mm excess winter rainfall	Shallow soils
High rainfall areas: over 700 mm annual rainfall, or over 250 mm excess winter rainfall	Medium soils
	Deep clayey soils
	Deep silty soils
	Organic soils
	Peat soils

The field assessment method for calculating SNS is the most popular (largely due to time and money), but the measurement method is particularly advocated when the SNS is likely to be large or uncertain such as where large amounts of organic manures have been applied. It is not recommended though for peaty soils or when grass leys have been recently ploughed out, as the mineralisable N can be very variable in these cases. Practically, the SNS of a field following an arable, forage or vegetable crop will depend on what soil type and in what rainfall area the crop was grown, hence the use of the field assessment method. For example, cereals can leave a field anywhere between SNS Index 0 and 6 depending on where they were grown. A winter wheat crop grown on a light sand soil in a low rainfall area will mean an index 0 field; the same crop grown on a deep silty soil in a moderate rainfall area will mean an index 1 and grown on a peat soil in all rainfall areas will leave the field at between index 4 and 6. The nitrogen requirements for the next crop would then be linked to the determined SNS value.

SNS Indices for sequences of crops following grassland are more complicated and take additional factors into account that are related to how the grassland has been managed. For long-term grassland there is a different system with three SNS categories—low, moderate and high—and three Grass Growth Classes—poor, average and good. The SNS status is based on previous nitrogen fertiliser use, organic manure use and grass management history in the last 2–3 years.

In addition to the major nutrients routinely assessed in soil for nutrient management planning purposes and fertiliser and manure use, soil assessment can also be used for micronutrient analysis and therefore more detailed nutrient interpretation. This is more likely to be considered in high value cropping situations or when potential nutrient deficiencies may be apparent or need to be investigated. As there are more soil extractant reagents needed for the bank of micronutrients this can increase the costs of soil analysis considerably.

4.6 Assessment of crop nutrient content

Plant tissue analysis of leaf or shoot material can be used for nutrient assessment and to assist with achieving optimum crop yield and quality. Suspected nutrient deficiencies based on plant visual symptoms can be confirmed by laboratory analysis of tissue material. The method can also be useful for crop nutrient deficiencies ahead of visible symptoms or where crops can show no obvious deficiency symptoms, yet the nutrient content is not sufficient to give optimum yield ("hidden hunger").

Crop material analysis can be used for most of the nutrients of interest, whether bespoke or broad spectrum. Just as with soil sampling, it is important to collect tissue material that accurately represents the crop under investigation. This could be by using diagnostic sampling in certain areas of a field or by taking samples from representative transects. If it is the latter, then a 'W' shaped pattern with at least 25 sample points as with soil sampling is recommended. At each point samples from 2 to 3 plants should be taken. Samples should be protected from damage and sent for analysis without a lengthy delay to avoid deterioration occurring.

It is important to note that whilst the nutrient analysis obtained will reflect the plant's nutrient status at the time of sampling this could be influenced by the stage of maturity of the crop, or preceding soil or weather conditions. Samples taken at a later stage of crop growth and that show potential nutrient deficiencies may mean that it is then too late to effectively replenish deficits for that particular crop, but could be useful for rotational considerations. Analysis of crop harvestable material is also a potential consideration for determining whether there has been effective nutrient uptake and whether corrective changes need to be made for the future. There is increasing interest in using cereal grain analysis for nutrient budgeting work.

4.7 Terminology and formulation of fertiliser products

Straight fertilisers supply only one of the major plant nutrients in their formulation. *Compound* fertilisers typically supply two or more of the three primary plant nutrients, nitrogen, phosphorus and potassium. Other major nutrients or micronutrients, as well as pesticides, can also be added, although this is not commonly done now. The exception would be sulphur which is increasingly being offered as part of compound fertilisers.

Approximately 75% of all fertilisers now used are *complex* compounds or *blended* compounds. They can exist as either solid or fluid materials. Complex compound fertilisers are normally made by drying a wet slurry, containing the appropriate raw materials, on a fluid bed system to produce granules, each containing the declared nutrients in the correct ratio, and size being 2–5 mm diameter. Blending of the straight fertilisers to make a compound mixture is becoming more popular again with a number of UK companies offering bespoke blends tailored for individual farms. In modern blends it is important that the individual single ingredients are, as far as possible, matched in physical characteristics (granular size and density). This is to avoid segregation out (separation) of the ingredients and uneven spreading patterns. They are dry mixes.

Table 4.5 Example compound fertilisers and nutrient supply to crops.

Compound ($N:P_2O_5: K_2O$)	Crop	N (kg/ha)	P_2O_5 (kg/ha)	K_2O (kg/ha)	Applied rate of product (kg/ha)
12:12:18	Potatoes	150	150	225	1250
20:10:10	Spring cereals	100	50	50	500
0:24: 24	Autumn cereals	0	60	60	250

High quality blended fertilisers can be made to specific plant nutrient ratios. They are cheaper than the complex fertilisers, but generally the handling and, most importantly, spreading qualities are not as good. However, new blending technology, involving blending towers and computer control systems, means that high quality blends are now being produced without the need for fillers and with accurate matching of ingredients.

Some examples of compound fertilisers and possible spreading uses are shown in Table 4.5.

4.8 Fertiliser products and nutrient supply

4.8.1 Nitrogen fertilisers

The nitrogen in many straight and compound fertilisers is in the form of ammonium (NH_4^+ cation) but, depending on soil conditions including aeration, temperature and drainage, it is quickly changed by bacteria in the soil to nitrate (NO_3^- anion) through nitrification. Many crop plants, e.g., cereals, take up and respond to the NO_3^- anions quicker than the NH_4^+ cations, but other crops, e.g., grass and potatoes, are equally responsive to NH_4^+ and NO_3^- ions. The ammonium cation is held in the soil complex at the expense of calcium and other loosely-held bases which are lost in the drainage water. Hydrogen ions are also produced as ammonium is nitrified. Both of these actions can have an acidifying effect on the soil.

Ammonium nitrate and urea fertilisers are produced by spraying a solution of the fertiliser from a vibrating shower head into a 'prilling tower'. As the droplets fall down the tower against a stream of cold air they become round and solid, producing prills 1—3 mm in diameter.

Nitrogen fertilisers in common use on-farm are:

Ammonium nitrate (AN) (33.5%—34.5% N). This is a very widely used fertiliser for applying on to the growing crop (top-dressing). Half the nitrogen (as nitrate) is very readily available. It is marketed in a special prilled or granular form to resist moisture absorption. Ammonium nitrate fertiliser is a hazard due to its detonation potential and ability to act as an oxidising agent fuelling a fire. Risk is managed through safe storage and handling and being kept well away from combustible organic matter or incompatible materials. Because of the ammonium present, it has an acidifying effect.

Calcium ammonium nitrate (26%—28% N). This granular fertiliser is a mixture of ammonium nitrate and calcium nitrate. It has a lower acidifying potential when added to the soil.
Urea (46% N). This is the most concentrated solid nitrogen fertiliser and it is marketed in the prilled form. It is sometimes used for aerial top-dressing. In the soil, urea changes to ammonium carbonate which may temporarily cause a harmful local high pH. Nitrogen, as ammonia, may be lost from the surface of chalk or limestone soils, or light sandy soils, when urea is applied as a top-dressing during a period of warm, often windy, weather. When it is washed or worked into the soil, it is as effective as any other nitrogen fertiliser and is most efficiently utilised on soils with adequate moisture content, so that the gaseous ammonia can go quickly into solution. In dry conditions in the height of summer it is probably better to use ammonium nitrate. Chemical and bacterial action changes it to the ammonium and nitrate forms. If applied close to seeds, urea may reduce germination. Future use of urea fertilisers in England has been subject to a recent government consultation. The context was a need to achieve a 16% reduction in ammonia emission by 2030 in comparison to the 2005 level. Certain future restrictions on use and timings of solid and liquid forms, with and without urease inhibitors, are expected to be implemented from 2023 onwards.
Ammonium sulphate (21% N). At one time, as a fertiliser, this was the main source of nitrogen. However, sulphate of ammonia is seldom used now. It consists of whitish, needle-like crystals and has a greater acidifying action on the soil than other nitrogen fertilisers. Some nitrogen may be lost as ammonia when it is top-dressed on chalk soils. It also contains 60% SO_3.
Sodium nitrate (16% N). This fertiliser is obtained from natural deposits in Chile and is usually marketed as moisture-resistant granules. The nitrogen is readily available and the sodium is of value to some fresh produce crops. It is expensive and is not widely used. It also contains 26% Na.
Calcium nitrate (15.5% N). This is a double salt of calcium nitrate and ammonium nitrate in prilled form. It is mainly used in continental Europe.
Aqueous nitrogen solutions (18%—37% N). These are usually solutions of mixtures of urea and ammonium nitrate, and are commonly used on farm crops (liquid fertilisers), often with other nutrients in formulated compounds. Foliar applications of liquids can often be targeted in later crop growth stages.

Various attempts have been made to produce slow-acting nitrogen fertilisers. Such products have included resin- or polymer-coated granules of ammonium nitrate (26% N), sulphur-coated urea prills (36% N), urea condensates and urea formaldehydes (30%—40% N). These types of fertiliser are used more in amenity and production horticulture than farm agricultural crops, but may have more of a future as a way of improving nitrogen use efficiency through reduced nitrogen losses and emissions.

It should also be noted that a significant proportion of nitrogen supplied to farm crops comes from compound fertilisers in which it is usually present mainly as mono-ammonium phosphate (MAP) or diammonium phosphate (DAP), as described in the next section on phosphorus fertilisers.

4.8.2 Phosphorus fertilisers

These can be classed as:

- those containing water-soluble phosphorus
- those with no water-soluble phosphorus. The insoluble phosphorus is variably soluble in the weak soil acids.

By custom and by law, the quality, grade and phosphorus content of phosphorus fertilisers are declared and, in the UK, expressed as a percentage of phosphorus pentoxide (P_2O_5) equivalent. The rest of Europe uses a declaration based on elemental phosphorus. The main fertilisers used in agriculture that supply phosphorus (being termed phosphate fertilisers) are:

Single superphosphate. This contains 18%–21% water-soluble P_2O_5 being produced by treating ground rock phosphate with sulphuric acid. It also contains a small amount of unchanged rock phosphate in addition to gypsum ($CaSO_4$), giving a sulphur content of 27%–30% SO_3. It is suitable for all crops and all soil conditions, but is not widely used now.
Triple superphosphate (TSP). This contains approximately 46% water-soluble P_2O_5 in a granulated form. It is produced by treating rock phosphate with sulphuric acid, followed by phosphoric acid. It also contains a small amount of elemental sulphur. TSP is the most common form of farm phosphate fertiliser.
Ammonium phosphates (monoammonium phosphate (MAP) and diammonium phosphate (DAP)). These are produced by treating rock phosphate with phosphoric acid and ammonia. They range from 46% to 55% P_2O_5 and also contain useful amounts of nitrogen.
Nitrophosphates. These contain 20% P_2O_5 and the same amount of N. They are made by treating rock phosphate with nitric acid.
Ground mineral phosphate (ground rock phosphate). This contains 25%–40% insoluble P_2O_5. It is the natural sources of rock that are ground to a fine powder. Its practical farm use is really only on acid soils in high rainfall areas and then preferably for grassland. A softer rock phosphate obtained from North Africa is available as 'Gafsa'. It can be ground to a very high degree of fineness. This means that it will dissolve more quickly in the soil, although it should still only be used under the same conditions and for the same crops as ordinary ground mineral phosphate. To be classified as a soft rock phosphate under EU Regulations, at least 55% of its total P_2O_5 must dissolve in a 2% solution of formic acid. Ground mineral phosphate fertilisers are now granulated.
Basic slags. These contain 5%–22% insoluble P_2O_5. They are by-products from the manufacture of steel. However, because of improved methods of steel manufacture, and the shrinkage of the industry itself, very little of this valuable fertiliser is now available to farmers in the UK. Small quantities of low-grade slag are sometimes imported.

There has been recent speculation about 'peak' fertiliser use and in particular whether rock phosphorus supplies are running out. Estimates of our remaining world phosphorus reserves vary between 100 and over 500 years. However, reserves are not the same as resources, and the estimated resource levels of phosphorus would allow us to keep using them for thousands of years, albeit at a higher cost. There are also advances in well-managed and controlled recycling of phosphate, in particular from sewage.

4.8.3 Potassium fertilisers

These are also known as potash fertilisers, which is short for potassium ash. The quality, grade and potassium content of these fertilisers are declared and, in the UK, expressed as a percentage of potassium oxide (K_2O) equivalent. The main potassium fertilisers used in agriculture are:

Muriate of potash (MOP) (potassium chloride). As sold, it usually contains 60% K_2O. It is the most common source of potash for farm use and is also the main potash ingredient for compound fertilisers containing potassium. As a straight fertiliser it is normally granulated, but some is marketed in a powdered form. KCl is found in large quantities all over the world and is mined from rock deposits left by dried-up oceans. It is nearly always found in conjunction with salt (NaCl) and the two are separated and purified by a flotation process.

Sulphate of potash (potassium sulphate). This is made from the muriate (chloride) form of potash and so is more expensive than MOP. It contains 48%–50% K_2O and, ideally, is targeted for quality production of high value crops such as potatoes, tomatoes and other fresh produce crops. However, the cost limits its use. It also contains 45% SO_3.

Polyhalite. This is a naturally occurring evaporite mineral (sulphate of potassium, calcium and magnesium) found deeper under the potash seam in North Yorkshire and is mined and purified to produce a commercial granular or powder fertiliser product that is 14% K_2O, 48%SO_3, 17%CaO and 6%MgO.

Kainit, sylvinite and potash salts. These are usually a mixture of potassium and sodium salts and, depending on the source, magnesium salts. They contain 12%–30% K_2O and 8%–20% Na. They are most valuable for sugar beet and similar crops, for which the sodium is a beneficial plant nutrient, partly replacing some of the potassium requirements.

4.8.4 Sulphur fertilisers

Ammonium sulphate (60% SO_3) and *Kieserite* (52%–55% SO_3) provide soluble sulphur in immediately available forms. They have quick effects after a top-dressing application. Kieserite can be useful for peas and beans where N is not applied.

Gypsum (typically 40% SO_3). This supplies sulphur as calcium sulphate. It is reasonably soluble once in soil and is an effective source.

Polyhalite (Polysulphate) (48% SO_3). This naturally occurring mineral supplies potassium, calcium and magnesium in addition to sulphur.

Epsom salts (33% SO_3) supply sulphur as magnesium sulphate with a 16% MgO content.

Elemental sulphur forms (typically 200%–225% SO_3 (80%–90% S)) are slower to act, but are longer lasting (and less predictable). This is because the sulphur needs to be oxidised to sulphate before it is available for plant uptake. Speed of oxidation is dependent on particle size and only the most finely divided ('micronised') elemental forms will be more rapidly acting.

4.8.5 Magnesium fertilisers

Kieserite (25% MgO). This fertiliser is quick acting and is particularly useful on severely magnesium-deficient soils where a magnesium-responsive crop such as sugar beet is to be grown. It also contains 52%–55% SO_3.

Calcined magnesite (typically 80% MgO). This is the most concentrated magnesium fertiliser, but it is only slowly available in the soil.

Epsom salts (16% MgO). This is a soluble form of magnesium sulphate and is used as a foliar spray where deficiency symptoms may have appeared on a high-value crop. It also contains 33% SO_3.

4.9 Fertilisers and nutrient supply

4.9.1 Plant nutrient requirements

Nutrient recommendations are typically quoted in terms of kilogrammes per hectare (kg/ha). For example, a recommendation for a spring barley crop may be to apply 110 kg N/ha, 45 kg P_2O_5/ha and 95 kg K_2O/ha in a particular field situation. To convert this into quantity of fertiliser product(s) per hectare, it is necessary to use the percentage analysis figures for the materials in question. This is clearly stated through the product nutrient content declaration, e.g.,. a 20:10:10 compound fertiliser would mean that the material contains 20% N, 10% P_2O_5 and 10% K_2O, always given in that order. Therefore, as an example, 100 kg of 20:10:10 fertiliser would supply 20 kg N, 10 kg P_2O_5 and 10 kg K_2O. In some situations, a compound fertiliser supplying N, P_2O_5 and K_2O may also have other nutrients in its formulation, such as sulphur declared as SO_3 e.g., 8:24:24 + 5%SO_3

Dependent upon certain factors including soil type, soil nutrient content, crop use or market, nutrient applications to crops will have important differing timings to ensure most effective uptake and utilisation of the nutrients. This will mean that important decisions need to be made on the choice and combination of products, possibly also including farm or non-farm produced organic materials. This may mean that nutrient recommendations are not matched exactly and a compromise has to be accepted.

In the case of liquid fertilisers, the amount applied will depend on the type of declaration of the manufacturer. Some declare their fertilisers on a weight per volume basis and some on a weight per weight basis. The weight per volume declaration is relatively easy to work out, i.e., so many kilogrammes of plant nutrient per 100 litres of water. A weight per weight declaration means that the user must know the specific gravity (SG), of the product, e.g.,

100 L of water weigh 100 kg,
but a 28% N solution has a specific gravity of 1.29 kg/L,
therefore 100 L of 28% fertiliser weighs 129 kg.
28% fertiliser means 28 kg in 100 kg product (weight per weight),
therefore, there are 28 kg of N in 100/1.29 = 77.5 L of product,
or, put another way, the weight of N in 100 L is 36 kg.

4.9.2 Nutrient cost comparison

The cost of plant nutrients per kilogramme can be calculated from the cost of fertilisers which contain only one plant nutrient such as nitrogen in ammonium nitrate, phosphate in triple superphosphate, and potash in muriate of potash.

Example:

If 1 tonne of ammonium nitrate (34.5% N) costs £680 and contains 345 kg N, then

1 kg of N costs 68 000p/345 = 197 pence

If 1 tonne of triple superphosphate (46%) costs £520 and contains 460 kg P_2O_5, then

1 kg of P_2O_5 costs 52 000p/460 = 113 pence

If 1 tonne of muriate of potash (60%) costs £515 and contains 600 kg K_2O, then

1 kg of K_2O costs 51 500p/600 = 86 pence

1 tonne of a 9:24:24 compound fertiliser would contain:

N	P_2O_5	K_2O
90 kg	240 kg	240 kg

The value of this in comparison, based on the costs of a kilogramme of straight nitrogen, phosphate and potash, is therefore:

N	90 × 197p = £177.30
P	240 × 113p = £271.20
K	240 × 86p = £206.40
Total	£654.90

Normally, the well-mixed granulated compounds cost more than the equivalent in straights. Bulk blends of straights are usually cheaper. Fertiliser prices have varied considerably in recent years; actual costs at any time should be substituted in the calculations shown.

It is also possible to compare the values of straight fertilisers of different composition on the basis of cost per kilogramme of plant food, e.g., the cost of a kilogramme of nitrogen in ammonium nitrate (34.5% N) is 197 pence from the calculation above. This could be compared with the cost of a kilogramme of nitrogen in urea (46% N) e.g.,

460 kg N/tonne at £670 per tonne = 67 000p/460 = 146 pence

Therefore, apart from lower handling costs because of the higher concentration of N, the urea is the lower-priced fertiliser although, compared with ammonium nitrate, there may be certain limitations with use.

4.10 Fluid (liquid) fertilisers

Across the UK as a whole the fluid fertiliser share of the market is around 10%. However, in the typical arable areas where the system has generally been concentrated, liquid nitrogen has a share in the range of 10%–15%, and fluid compounds account for between 5% and 10% of the total tonnage of compound fertiliser used. Fluid fertilisers are generally separated into solutions and suspensions.

4.10.1 Solutions

These fertilisers are non-pressurised, fully dissolved, solutions of the same raw materials that are used for solid fertilisers. They should be distinguished from pressurised solutions, such as aqueous ammonia and anhydrous ammonia. Fluid compounds are based on ammonium polyphosphate or ammonium phosphate, urea and potassium chloride, whilst ammonium nitrate and urea are the main constituents for liquid straight nitrogen fertilisers.

The concentration of liquid nitrogen is, for all practical purposes, the same as solid nitrogen fertiliser as long as both urea and ammonium nitrate are used together in the same solution. Fluid phosphorus and potassium compounds are only about two-thirds the concentration of solids. However, fluids have bulk densities in the range 1.2−1.3 kg/L compared with 0.9−1.0 kg/L for solid fertilisers (i.e., 100 L = approximately 125 kg). The higher bulk density, plus quicker handling and application of fluids, will compensate for the lower concentration. It is an obvious consideration when fertiliser work rates are assessed on the weight of nutrients applied in a given time and not the weight of the product. However, the cost of storage may be higher and special care must be taken during handling, storage and application to avoid spillage and environmental pollution.

Fluid solutions are stored in steel tanks (up to 60 t capacity) on the farm. A cheaper form of storage—glass-reinforced plastic tanks—is available as an alternative to mild steel. Fluid fertiliser storage tanks should be as far away as possible from any surface water, drains or ditches. Ideally, secondary containment (bunding) should be provided and the Environment Agency (or equivalent in the devolved administrations) should be consulted on the suitability of fluid fertiliser storage sites.

There is a range of equipment available for applying fluid fertiliser. The broadcaster types, which range from a 600 L tractor-mounted to 4000 L capacity self-propelled applicators, have spray booms 12−40 m wide. Although the broadcaster is specially designed for fertiliser, by changing the jets it can be used for other agricultural chemicals, and so the fluid system fits in well with cereal tramlining. Equipment is also available for the placement of liquid fertiliser for sugar beet, potato and brassica crops, and a combine drill attachment can be used for sowing the cereal and applying the fluid together. For top-dressing, and to minimise scorch, stream jets can easily be fitted which produce larger droplets for a better foliar run-off. Alternatively, dribble bars, which trickle the fertiliser on to the soil, can be used.

Volume of product to apply:

> e.g., if 100 kg N/ha is recommended, liquid nitrogen, N 37 kg per 100 L, is used, then $100/37 \times 100$ L/ha = 270 L/ha.

Units of sale:

> Whilst solid fertilisers are applied by weight (kg/ha) and sold by weight as £ per 1000 kg, fluid fertilisers are applied by volume (L/ha) and sold by volume as £ per 1000 L (1 m^3).

4.10.2 Suspensions

These are fluids containing both mixed solids and dissolved nutrients, and they can be produced with a concentration almost up to that of a granular solid, thus meaning suspensions typically have higher nutrient concentrations than solutions. Suspensions are achieved either by crushing the solid raw material and adding up to 3% of a semi-solid attapulgite clay to it to help minimise the settling of the soluble salts, or by reacting phosphoric acid and ammonia followed by the addition of potassium chloride and extra nitrogen. For long-term storage, crystallisation may be a problem and occasional agitation is necessary to prevent the crystals from growing whilst in store. Special applicators are also needed and this is why suspensions are mostly applied by a contractor. Suspensions use is developing in North America (where, in fact, fluids are more widely used) and to some extent on the European Continent.

4.10.3 'Distressed' fertiliser

This is fertiliser material that has been damaged, usually in transit from overseas. It is bought by distributors and sold at a heavy discount to farmers who normally make it into a liquid fertiliser, simply by dissolving it in water.

There is no essential difference in plant growth following the application of fertiliser in solution, suspended or solid form. The popularity of fluid fertilisers when used is associated with a belief that liquids are easier to handle on farm and can be spread more accurately. Also, that nutrients can be delivered more quickly and accurately to the plants needing them and that urea in fluid form is not as volatile as the solid product, which is valuable when dealing with drying conditions.

4.11 Application of fertilisers

Accurate and uniform fertiliser application is essential to maximise the effective utilisation of nutrients by crops and avoid nutrient loss to the wider environment that could cause contamination and pollution of water and air. For solid fertilisers, the main distribution methods used are *broadcasting* where materials are added to the soil surface before or immediately after planting, *top-dressing* on to a growing crop or *placement* within the soil very close or near to the seed during sowing or crop planting. For fluid fertilisers, materials are either *sprayed* on to the soil surface in solution or suspension form, *top-dressed* as a *foliar* application on to a growing crop as solutions (but not as suspensions due to crop scorch risk) or *placement* in the soil.

Fertiliser distribution is mainly associated with the following application systems: *Broadcast distributors* using various mechanisms such as:

(a) *Pneumatic types*. The principle involved is that the fertiliser is metred into an airflow which conveys it through flexible pipes to individual outlets placed over deflector plates to distribute the fertiliser evenly. Hopper capacity is from 1 to 6 t with a spreading width of 12–36 m (selected to match tramline widths). Application rates vary from 5 kg/ha (for broadcasting seeds) to 2500 kg/ha.

(b) *Spinning disc types.* These consist of a hopper placed above a single or twin rotating disc, each with one or more deflector vanes. The fertiliser is fed on to the disc from the hopper. Wide areas can be covered, but the accuracy of distribution can often vary somewhat. This system of application is used for most fertiliser spreading.

(c) *Oscillating spout type.* A large nozzle situated at the back of the hopper moves rapidly from side to side throwing the fertiliser to the required width. This type accounts for about 20% of fertiliser application in the UK.

The spread patterns of all broadcast fertiliser spreaders should be checked regularly. This can be done by driving the spreader through a line of collection trays and then measuring the amount of material delivered to each one (*'tray-testing'*). A number of commercial companies now offer this service to farmers. The quality and uniformity of the fertiliser material also has an effect on spread pattern. For nitrogen fertilisers, a standardised Spread Pattern (SP) rating was adopted measuring the 'throw and flow' characteristics of these materials. Ratings have been shown on a scale of 1–5 with SP5 being the best.

Inaccurate, uneven spread patterns of fertilisers reduce crop economic margins, crop quality, harvesting efficiencies and is often manifested by crop striping in the field. Better targeted application is achieved through the use of headland or border systems on the spreader reducing fertiliser contamination of non-cropped areas and the use of variable rate application providing limits or cut-offs in order to match soil and crop need. These systems can be controlled electronically or manually from the tractor cab.

Combine drills. Fertiliser and seed (e.g., cereals) from separate hoppers are fed down the same, or an adjoining, spout. A star-wheel feed mechanism is normally used for the fertiliser and this usually produces a 'dollop' effect along the rows. In soils low in phosphorus, this method of placement of the fertiliser is much more efficient than broadcasting and can require less fertiliser to be used per hectare. It is known as combine drilling and is sometimes referred to as 'contact placement'. Because of possible scorch, combine drilling should only be used for cereal crops.

Placement drills. These machines can place the fertiliser in bands 5–7 cm to the side and 3–5 cm below the row of seeds. It is more efficient than broadcasting for crops such as peas and sugar beet. Other types of placement drills attached to the planter are used for applying fertilisers to the potato crop and some brassicas.

Aircraft. Used for aerial broadcasting, but rarely used except in inaccessible hill land and forests. If used it is usually undertaken by helicopter.

Liquid spray boom applicators (non-pressurised). These are used with stream jets or dribble bars for broadcasting or placing fluid solutions or suspensions.

Liquid applicators (pressurised). Materials are injected under pressure into the soil, e.g., anhydrous and aqueous ammonia.

4.12 Organic materials

There are a diverse range of organic materials that are applied to agricultural land and which can provide valuable sources of plant nutrients and organic matter for crop nutrition and soil improvement. These can be classified in a number of ways such as farm and non-farm produced and can include animal manures, human sewage sludge (marketed as biosolids), composts, anaerobic digestates and waste-derived materials from the food, drink, industrial and textile industries. In the description given here there is a distinction drawn between farm animal (livestock) manures and other organic materials mainly because significantly higher quantities of livestock manures are produced and applied to land than other organic materials even when they are considered together as a separate category to inorganic manufactured fertiliser products.

4.12.1 Effective nutrient utilisation

The recycling of organic materials to land should be undertaken in such a way that benefits achieved outweigh the risks involved and that the latter are managed effectively. Organic materials can pose a significant risk to the wider environment if not used well. Applications should be made in a targeted way such that materials are effectively and evenly distributed for soil and crop use. Solid materials are applied using rear and side discharge spreaders and liquid manures or slurries using tanker or umbilical systems ideally with precision capabilities. These can include low level spread through '*dribble bars/hoses*' and '*trailing shoe*' adaptors or the use of shallow or deeper injectors. Applying materials in this way will improve nutrient use efficiency in the soil and crop system and help to reduce odours and gaseous losses of nitrogen. However, tankers and spreaders when loaded can place pressure on the soil and so due care and timeliness are needed to avoid soil compaction and damage. Suitability of cropping, and growth stages, is also of practical relevance such that most applications are geared to grassland or as part of cultivation practices ahead of crop establishment. Although it is not so usual, slurry can be used on cereals. Band spreaders have been developed which make it possible to spread liquid materials across the whole width of tramlines.

Benefits of organic materials recycled to land will be mainly associated with nutrient additions and soil conditioning value such as organic matter increase, dependent on the material being spread. It is highly desirable that these nutrient additions are effectively accounted for as fully as possible. This will need to consider a number of factors including:

Total and readily available nutrient (RAN) content of the material - often large proportions of the nutrient content of organic materials are not in chemical forms available for plant uptake at the time of application, these being organically bound. Soil microbial mineralisation can convert these forms to readily available forms over a period of months to years depending on soil and weather conditions. It is important that readily available nutrients equate to being crop available nutrients as much as possible such that crop uptake is maximised and losses to the environment minimised. Applications can often be limited by total nutrient content e.g., total nitrogen as a field limit in Nitrate Vulnerable Zone (NVZ) legislation.

Application timing - this is best achieved when growing crops can most effectively use the nutrients applied, most likely being spring and early summer. This will need to be managed with effective (and often expensive) manure and slurry storage when materials are not applied to land due to lack of crop need at that time often matched with inappropriate ground conditions for application (e.g., frozen or saturated soil) and/or legislative restrictions.

Application technique - to target efficient and effective spreading value as indicated above.
Soil type - this will affect acceptability of materials in terms of weight and volume, suitability of incorporation and nutrient retention. Lighter, sandier soils could be prone to losses via leaching whilst heavier, clayey soils could suffer runoff issues of liquid materials.

Detailed consideration of all these factors in ensuring effective nutrient utilisation from the addition of organic amendments in soil-crop systems will be closely reliant upon determining the total and readily available nutrient content of the materials being used. This is most accurately gathered through laboratory analysis of dry matter, total nitrogen, phosphorus, potassium, sulphur, magnesium and readily available nitrogen (ammonium-N and, in some materials, nitrate-N and uric acid-N) in representative samples. If sampling on-farm stores of livestock manures it is important to consider factors that can affect nutrient distribution and make-up such as animal bedding or litter material, settling of solids or surface crusting. Rapid on-farm testing kits (e.g., Quantofix, Agros) are available for assessing ammonium-N content of liquid slurries and digestates. Slurry hydrometers can be used for quantifying dry matter contents which can then be related to nutrient levels through correlation calculations. Nutrient contents of typical organic materials can also be obtained from published materials such as *AHDB's Nutrient Management Guide.*

4.12.2 Livestock manures

4.12.2.1 Farmyard manure (FYM)

This consists of dung and urine, and the litter used for stock bedding. It is generally derived from cattle and pig production in the main, but with sheep, horse, goat and duck manure also of importance with those livestock situations. Typically, it will take on a solid, stackable appearance due to the dung and bedding litter content. It is not a standardised product, and its value depends on a number of factors including:

Stock type and age. If animals are fed strictly according to maintenance and production requirements, the quality of dung produced by various classes of stock within each animal category will be similar. But in practice there are differences e.g., it is generally found that dung produced by cows and young stock is poorer in nutrient content than that produced by fattening stock as there is better utilisation of the nitrogen and phosphate from their food intake.
Livestock diet and feeding system. The more proteins and minerals in the diet, the richer will be the dung. Wet or dry feed consistency will alter the dung dry matter and nutrient output.
Quantity of bedding. The less straw used, the more concentrated will be the manure and the more rapidly it will break down to a 'short' friable condition. Straw is the most likely bedding used, although sawdust, wood shavings or other materials may be possibilities. About 1.5 tonnes of straw per animal are needed in a covered yard for 6 months, and 2–3 tonnes in a semi-covered or open yard.
Manner of storage. There can be considerable losses from FYM because of poor storage, for example through rain water contamination. Dung from cowsheds should, if possible, be put into a heap which is protected from rainfall to prevent the washing out and dilution of a large percentage of the plant nutrients which it contains. Farmyard manure made in yards should preferably remain there until it is spread on the land, and then, to prevent further loss, it is advisable to incorporate it as soon as possible.

Farmyard manure is important chiefly because of the valuable physical effects on the soil from the organic matter it contains. It improves soil friability, structural stability and water-holding capacity. It is also a valuable source of nutrients, particularly available potassium in relation to available nitrogen and phosphorus. Most of the nitrogen content will be in crop unavailable form as organic-N so will be reliant on soil mineralisation to make it available over time in the future. An average dressing of 25 t/ha of cattle FYM will provide about 15 kg N, 50 kg P_2O_5 and 200 kg K_2O to the next crop after application. Other nutrition elements are also supplied in smaller amounts. Care should be taken to ensure that a high potash supply does not increase risk of stock hypomagnesaemia (grass staggers) through high potassium levels being antagonistic to magnesium and therefore resulting in reduced herbage Mg levels. All the plant food in FYM is less readily available than that in chemical fertilisers.

4.12.2.2 Poultry manure

Poultry manure derives from a range of poultry systems including laying hens, broiler chickens and turkeys. Examples of different poultry manures include:

- Fresh poultry manure
- Broiler manure
- Deep litter manure
- Dried manure.

Table 4.6 shows a typical available nutrient value of two of these poultry materials.

Fresh poultry manure from battery colony cages or from slatted or wire floors is free from litter. It is semi-solid, but rather sticky and, to reduce any public nuisance from smell, it should be spread as soon as possible after removal from the building and ideally soil incorporated. To reduce nitrogen losses, the preferred application time is in spring and early summer. Application rates are lower than those for farmyard manure due to a higher nutrient concentration.

Broiler litter manure is the bird droppings mixed with litter. This produces a bulky manure, relatively dry and friable and easily handled. It should be spread like farmyard manure.

4.12.2.3 Slurry and liquid manure

Management and housing of cattle and pigs on farms have increasingly led to increased volumes of liquid animal slurry being produced as opposed to solid farmyard

Table 4.6 Comparison of poultry manures.

Crop available nutrients (year of application) (kg/tonne)				
	Dry matter (%)	**N[a]**	**P_2O_5**	**K_2O**
Layer manure	35	9.5	8.4	8.6
Broiler litter	60	12	15	16.2

[a]Assumes spring application and soil incorporation within 24 h of spreading.

manure from strawed yards with high litter content. Slurry is more associated with increased use of cattle cubicle housing with reduced bedding, self-feed silage clamps and slatted pig housing. Larger, and more intensive livestock units with mechanised waste management systems often find it easier to deal with liquids rather than solids. Slurry is classed as excreta produced by livestock (other than poultry) while in a yard or building, (including any bedding, rainwater and washings mixed with it), that has a consistency that allows it to be pumped or discharged by gravity.

Slurry can be separated into solid and more dilute liquid fractions for further management through mechanical (e.g., screening or centrifugal action) or non-mechanical means (e.g., weeping-wall sleepers or stanchions). The solid part (12%–30% dry matter (DM)), can be handled like FYM, and the liquid portion is easier to pump greater distances, is much less likely to cause tainting of pastures and will infiltrate soils more easily after spreading. The effect of slurry management on total nutrient content of cattle slurry is shown in Table 4.7.

By their very nature farm cattle and pig slurries will not be adding large quantities of organic matter when applied to soil. Benefits are not generally associated with improving soil physical characteristics such as structural stability or organic matter rise, but more linked to chemical or biological properties such as raising nutrient indices and biological activity. Cattle slurry can be a good source of nitrogen and potassium for crop use with generally over double the content in comparison with phosphorus. Care must be taken to ensure that the most effective nutrient utilisation as possible is achieved to avoid these nutrients moving to the wider environment, together with other practical concerns such as hypomagnesaemia (grass staggers) as previously mentioned with FYM.

Slurries are classed as high readily available-N manures where the proportion of the total N content that is in an available form is greater than 30%. This can mean more legislative restrictions on application timings (e.g., closed periods in NVZs) when compared with low RAN materials such as FYM and compost. Other high RAN manures with such restrictions include poultry manures and digestates.

The composition of pig slurry will partly depend on the animal feeding system, i.e., dry meal feed has a 10% DM; liquid feed has a 6%–10% DM and whey 1%–2% DM. A comparison of pig slurry analysis is shown in Table 4.8. Supplementation to the diet can mean that pig slurry can contain large amounts of copper and zinc. These elements will build up slowly in the soil, but eventually there could be crop toxicity problems. It should be noted that sheep are particularly susceptible to copper poisoning.

Table 4.7 Typical total nutrient content of cattle (dairy) slurry (kg/m^3).

Unseparated slurry (6% dry matter)			**Mechanically separated slurry (4% dry matter)**			**Weeping-wall separated (3% dry matter)**		
N	**P_2O_5**	**K_2O**	**N**	**P_2O_5**	**K_2O**	**N**	**P_2O_5**	**K_2O**
2.6	1.2	2.5	3.0	1.2	2.8	2.0	0.5	2.3

Source: AHDB (2021) Nutrient Management Guide (RB209) Section 2.

Table 4.8 Comparison of pig slurry.

	Dry matter (%)	Total N (kg/m^3)	Available nutrients (year of application) (kg/m^3)	
			P_2O_5	K_2O
Pig slurry	2	3.0	0.4	1.6
	4	3.6	0.8	2.0
	6	4.4	1.1	2.3

Note: The availability of nitrogen will depend on the timing and method of application, the winter rainfall and the soil type.
Source: AHDB (2021) Nutrient Management Guide (RB209) Section 2.

4.12.3 Other organic materials

4.12.3.1 Sewage sludge (biosolids)

Treated residual sewage material is available from water companies and local councils for land amendment. Benefits include soil conditioning through structural improvement and increasing soil organic matter together with application of valuable nutrients for soil and crop use. The recycling of sewage sludge to agricultural land is seen as the Best Practicable Environmental Option (BPEO) for its recovery. These materials are typically marketed as *biosolids* and have undergone treatment processes to create products for agricultural use. A Code of Practice and Sludge Regulations support and control applications and, together with a Biosolids Assurance Scheme, provide food chain and customer reassurance that sewage sludges can be sustainably and effectively used in agriculture. A Safe Sludge Matrix and a Biosolids Nutrient Management Matrix have introduced tighter controls to those set out with the Code of Practice and are particularly concerned with creating barriers for potential pathogen transfer and soil nutrient rise.

Untreated sewage sludge is now not allowed to be spread on agricultural land for food or energy crops. Necessary standard conventional treatment is associated with anaerobic digestion. Enhanced treatment is used for certain sludges where higher pre-pasteurisation temperatures are reached. This will ensure almost entire pathogen kill (99.9999%) and is a necessity with applications on raw-eaten crops such as fresh produce salads, vegetables and fruit. In addition to this requirement enhanced treatment applications need to observe a 10-month harvest interval for produce eaten raw. Sewage sludge is typically spread as dewatered solids (*cake*) or as liquor. Further treatment processes for certain materials include lime treatment, thermal drying and pelleting to aid use.

Nutrient value of biosolids is mainly linked to their phosphorus and nitrogen contents. There can be useful quantities of sulphur and magnesium supplied in addition, but only very limited potassium inputs. Many biosolid applications can apply more phosphate than is removed in the following crop's harvest and as such soil phosphate levels can rise above target Index level. Regular soil testing for available phosphorus is therefore needed and care used in selecting appropriate fields for application.

Biosolids can contain a range of potentially toxic elements (PTEs), particularly metals such as copper, zinc, lead or nickel, though concentrations are now significantly lower than historically in the past. Critical toxic levels for both crops and grazing stock need to be observed and as such maximum permitted concentrations of soil PTEs are controlled through sewage sludge legislation and testing of both sludge and receiving soil.

4.12.3.2 Digestate

Digestate is a physical output of the anaerobic digestion (AD) industry together with biogas used for energy value. Feedstock materials for digestion typically include farm livestock manures and crop material or other non-farm organic materials such as food waste. These are biologically decomposed in a controlled process in the absence of oxygen. Digestate is typically produced 'whole' in a consistency similar to slurry, but can then be separated into digestate fibre and liquor.

Digestate nutrient value is linked to initial feedstock with food-based digestates typically having higher nutrient concentrations than crop and animal manure-based materials. Digestate can supply significant quantities of crop available nitrogen when applied in whole or liquor form as in these materials most of the total nitrogen content is in readily available forms such as ammonium N, more so than other organic materials including slurry. Separated fibre has much lower percentage of total nitrogen being in readily available forms, but a higher phosphate content than whole and liquor digestate.

4.12.3.3 Compost

Compost is produced by controlled biological degradation of organic feedstock in the presence of oxygen. It is applied to land as a soil conditioner and as a source of soil and crop nutrients. As with digestate, the nutrient content of compost materials will vary depending on the feedstock, the process of production and certain treatment considerations. It can supply a mixed range of both major and micronutrients, but usually contains low amounts of crop available nitrogen as most of the total nitrogen content is as organic N. Repeated use of compost can increase soil organic matter content which can aid soil trafficability as well as improve workability and water holding capacity.

Typical compost feedstocks are green wastes from municipal collections with or without organic food waste. If the latter is used in compost production then sealed ('in-vessel') composting systems are necessary as opposed to open windrow operations. In some situations, animal manures can be a suitable feedstock such that they are stacked and controlled aerobic composting is encouraged. Manure volume is reduced through the composting and there is a suppression of pathogens through high temperature control.

Typical nutrient analysis of sewage sludge, digestate and compost is shown in Table 4.9.

Table 4.9 Total nutrient and dry matter content of other organic materials.

		Total nutrient content (kg/t or kg/m^3)		
	Dry matter (%)	**N**	**P_2O_5**	**K_2O**
Digested sewage sludge cake	25	11	11	0.6
Thermally dried sewage sludge	95	40	55	2.0
Lime stabilised sewage sludge	25	8.5	7	0.8
Whole food-based digestate	4.1	4.8	1.1	2.4
Food-based digestate separated liquor	3.8	4.5	1	2.8
Food-based digestate separated fibre	27	8.9	10.2	3
Green waste compost	60	7.5	3.0	6.8
Green/food waste compost	60	11	4.9	8.0

Source: AHDB (2021) Nutrient Management Guide (RB209) Section 2.

4.12.3.4 Waste-derived materials

There are a host of materials classed as industrial wastes which can be recycled to agricultural land for benefit through soil conditioning or nutrient supply. Application of these wastes are undertaken under an environmental permit, waste licence or an exemption where a proven benefit must be demonstrated and appropriate risk assessments used. If the permitting or licencing regime is used then these wastes are *deployed* to land in controlled situations. Composts and digestates are applied in similar ways if classed as waste materials. However, PAS100 certified compost and PAS110 certified digestate (biofertiliser) (where PAS is the British Standards Institution's Publicly Available Specification) do not normally need an environmental permit or waste exemption to be in place for land spreading. This is also the case for purely farm produced feedstocks in digestate and compost production.

Examples of industrial waste materials include paper crumble, gypsum, poultry litter ash, waste wool and textile cotton (shoddy), abattoir and lairage wastes, spent mushroom compost and food and drink processing wastes. These have very different nutrient analyses and variable agricultural benefits are claimed.

4.12.3.5 Cereal straw

Cereal straw is lignified material and can be a useful source of soil organic matter for maintaining or improving soil fertility. It is particularly beneficial for light sandy and silty soils in which organic matter breaks down rapidly.

The average yield of straw (4 t/ha) supplies about 15 kg N, 5 kg P_2O_5 and 35 kg K_2O, and so the soil should benefit when the straw is ploughed in or otherwise worked into the soil. However, with the sudden influx of straw as a carbon source the soil bacterial population will start to multiply as the straw is decomposed. In so doing, much of the available nitrogen in the soil gets used up for the protein needs of the bacteria. The carbon:nitrogen ratio increases to 50 or 60:1 from the normal 10:1. The protein is released as the straw decomposes and eventually the ratio returns to 10:1. In the meantime, unless extra nitrogen is added (20–25 kg/ha), the following crop can suffer from shortage of nitrogen. This extra nitrogen is only necessary under naturally low nitrogen conditions.

Incorporation of straw is easier to achieve if it is chopped and spread from the combine. Over the long-term regular incorporation of straw will stimulate soil biological activity such that nitrogen is made more available through the nitrogen cycle.

4.12.3.6 Seaweed

Seaweed is sometimes used instead of farmyard manure for crops such as early potatoes in coastal areas, e.g., Ayrshire, Cornwall and the Channel Islands. However, it is expensive to handle which is why it is not extensively used. It can be a useful source of many nutrients: 10 tonnes contain about 50 kg N, 10 kg P_2O_5 and 140 kg K_2O; this would also contain about 150 kg salt. The organic matter in seaweed breaks down rapidly because it is mainly cellulose. It should therefore be collected, spread and worked in the soil without a long delay. In this way the loss of potassium, particularly, will be reduced.

4.13 Fertilisers, organic materials and the environment

When applying nutrient materials, it is economically and environmentally important to use the optimum amounts and not just those required for a maximum yield. Optimum quantities should be considered with both farm economics and with the farm and wider environment in mind. Accuracy of application, timing and handling are vital in order to achieve crop nutrient use efficiency and in conjunction with good agricultural practices. Nutrients lost from soil-crop systems can move to the connected water and air environments causing nutrient increases above critical loads and potential contamination, pollution and damage.

Applications of materials should be considered in such a way to reduce nutrient leaching, runoff or gaseous loss, to avoid eutrophication of both water and land and reduce harm to habitats and species. Issues are associated with nitrogen and phosphorus enrichment of soil and water, potentially toxic elements (PTEs) that can be toxic to plants and animals when present above a critical load in soil (and which have historically been raised through applications of industrial wastes and sewage effluents) and organic material loading to water which can cause oxygen depletion through its microbial degradation.

There is appropriate guidance to support better practice in nutrient management including Codes of Good Agricultural Practice for the Protection of Water, Soil and Air and Codes of Good Agricultural Practice for Reducing Ammonia Emissions. Compliance with appropriate legislation and regulations help to protect soil, water and air quality when linked to nutrient management and the use of fertilisers and organic materials. Further detailed information of examples is provided in Chapter 8. It is important to note that there are different interpretations across the member countries of the United Kingdom. In more recent years the manufacture of synthetic fertilisers has also come into focus in connection with greenhouse gases. This is particularly the case with the high energy requirements necessary for nitrogen manufacture. New technologies associated with *abated* fertiliser are helping to reduce nitrous oxide emissions in particular.

4.14 Key points

- Good nutrient management is an essential aspect of effective crop production that aims to achieve optimum crop yield and quality in respect of both economic and environmental considerations.
- Crop requirement is mostly supplied from soil pools of soluble, readily available nutrients that can be assessed using appropriate tools at recommended times. Assessment may involve soil sampling with laboratory analysis and/or field assessment. A range of factors will affect these nutrient pools including soil physical, chemical and biological factors, nutrient additions through the use of fertiliser products, organic materials or crop residues and atmospheric additions.
- Plant nutrient analysis, such as in leaf tissue and harvestable product, can provide useful, additional information for fine-tuning nutrient management planning and effective crop production.
- Organic materials, whether produced on- or off-farm can supply valuable quantities of plant nutrients for crop production, together with a number of soil conditioning benefits. Soil improvements can include organic matter inputs, and improvements to soil structure and water holding capacity.
- Manures, waste products and other organic materials can vary considerably in physical characteristics and nutrient contents. Care needs to be taken in assessing nutrient contributions and the proportion of those nutrients that are readily available for crop uptake.
- Nutrient products and materials, both manufactured fertilisers and organic materials, are valuable plant food sources, but without appropriate management and good practice can also be lost to the wider environment and contribute to diffuse and point source water and air pollution.
- Using an integrated, broad-ranging approach to nutrient use in crop production will ensure effective nutrient use efficiency and reduced environmental concerns of losses contributing to poor water and air quality. Nutrient management needs to consider inputs/outputs from cropping cycles; application quantities, forms and timing; effective and targeted spreading and nutrient interactions.

Sources of further information and advice

AHDB, 2022. Nutrient Management Guide (RB209) Section 1: Principles of Nutrient Management and Fertiliser Use; Section 2: Organic Materials; Section 3: Grass and Forage Crops; Section 4: Arable Crops; Section 5: Potatoes; Section 6: Vegetables and Bulbs; Section 7: Fruit, Vines and Hops. Agriculture and Horticulture Development Board.

Bennett, W.F. (Ed.), 1993. Nutrient Deficiencies and Toxicities in Crop Plants. APS Press, Minnesota.
Context, 2003. The Crop Nutrition Directory. Context Publications.
Defra, 2009. Protecting Our Water, Soil and Air. The Stationery Office, Norwich.
Tried & Tested, 2017. Think Manures: A Guide to Manure Management. Tried & Tested.

Websites

www.agindustries.org.uk/sectors/fertiliser.html.
www.ahdb.org.uk/nutrient-management-guide-rb209.
https://assuredbiosolids.co.uk/.
https://bfsfertiliserservices.uk/media/content/files/nutrient-deficiency-guide/guide.pdf.
www.biofertiliser.org.uk.
www.gov.uk/government/collections/fertiliser-usage.
www.gov.uk/government/collections/nitrate-vulnerable-zones.
www.nutrientmanagement.org.
www.nutrientmanagement.org/tried-and-tested/3-think-manures-2017/.
www.pda.org.uk.
www.teagasc.ie/crops/soil–soil-fertility/nutrient-deficiencies/.
https://wrap.org.uk/resources/report/digestate-and-compost-agriculture-dc-agri-project-reports.
www.yara.co.uk/crop-nutrition/deficiencies/.

Weeds of farm crops

5

5.1 The impact of weeds

To suggest that weeds are plants that are growing where they are not wanted is an over-simplification. If a non-crop plant is not causing significant problems in a farm crop, and will not cause future problems in that crop, it should not be considered to be a weed. Removal of such plants will cost more, in terms of time and money, than the plants will cost to the crop in terms of yield loss. Also, non-crop plants contribute significantly to agrobiodiversity and should only be removed where crop production systems are at risk. A better definition of a weed, for the purpose of this chapter, is **'a plant that is adapted to an arable or grassland habitat that causes significant problems in the use of those habitats'**. Weeds are not only native plant species. Even crop plants, when present in other cropping systems, can be serious weeds in other crops, e.g., volunteer potatoes or weed beet. Where thresholds are exceeded, weeds can significantly affect crop yields by competing for light, space, nutrients, and water. The competitiveness of weeds varies both between species and at different times of the year. Some crops are also more affected by weed competition than others. Potatoes produce a dense crop canopy which can be very effective at competing with weeds compared with sugar beet which is precision drilled at a low population, and has many weeks when there is little crop cover. As well as affecting crop yield, weeds also affect crops in other ways:

- They can spoil the quality of a crop and so lower its value, e.g., wild oats in cereal seed crops, black nightshade berries in vining peas.
- They can act as host plants for various pests and diseases of crop plants, e.g., couch grass is a host for take-all and eyespot of cereals.
- The presence of weeds can require the implementation of extra cultivations, which in addition to being time and energy consuming, can result in increased erosion and water loss.
- Weeds such as black bindweed, cleavers, couch and thistles can affect combine harvesting. All the green material other than grain going through the harvester reduces the work rate. These weeds can also cause a crop to lodge, which again can affect the speed and efficiency of harvesting.
- Weed contamination at harvest can increase the cost of drying and cleaning of grain crops.
- Thistles, buttercups, docks, ragwort and bracken in grassland can reduce the grazing area and feeding value of pastures. Some grassland weeds may taint milk when eaten by cows, e.g., buttercups and wild onion.
- Weeds such as ragwort, horsetails, nightshade, foxgloves and hemlock are poisonous and if eaten by stock are likely to cause poor growth, health problems, or even death. Fortunately,

Lockhart and Wiseman's Crop Husbandry Including Grassland. https://doi.org/10.1016/B978-0-323-85702-4.00010-8

most stock animals normally do not eat poisonous weeds in the field. However, it should be noted that they cannot avoid these species where they are present in silage or hay.

Conversely, non-crop plants play a number of important positive roles on farms. These include:

- Weeds can be very important for farmland biodiversity (agrobiodiversity); they provide a source of pollen and nectar for bees and butterflies or seed for many bird species. They can also provide a favourable microclimate for crop development, can improve soil biology and structure, can help to reduce erosion by providing ground cover, and alleviate monocultural aspects of cropping systems.

When planning a weed management strategy it is important to correctly identify weeds, understand their life cycle, assess how competitive they are in the current crop, determine if incidences of herbicide resistance are present, as well as assessing the importance for biodiversity and other positive roles that the plants may play. All this information should be considered in order to make a balanced and informed decision on whether management is needed and what management should be carried out.

Some of the most important weed species such as cleavers, common chickweed, wild oats and black-grass are annuals. Only a small number of biennials are important weeds such as spear thistle and ragwort and they are more of a problem in permanent grassland. There are a number of important perennial weeds such as couch grass and field bindweed, although other perennial species such as dock, thistles, and nettles, are again more common in grassland. Seed production is not necessarily the main method of propagation in perennial species, as they may utilise underground stems (rhizomes), above ground stems (stolons) or roots to spread asexually.

5.1.1 Success of plants as weeds

There are a number of reasons why some plant species have been more successful as weeds than other plants. Plants that are successful as weeds are likely to possess at least some of the following characteristics:

- Seed can survive in the soil for a long time and still germinate
- Seed will germinate in a variety of soil types
- After germination, the plant will quickly grow and produce seed
- The plant will produce very high seed numbers and can do this in a variety of environmental conditions
- Seeds will possess methods of both short and long distance dispersal
- For perennial species, the ability to regrow from fragments (of root or rhizome for instance)

5.1.1.1 Seed production

Some weed species are able to produce thousands of seeds per plant. Examples of prolific seed producers include corn poppies and mayweed species. The seed reservoir (soil weed seed bank) in some soils can be as high as 100,000/m^2 in arable soils. In grassland soils this number can be significantly lower (below 20,000/m^2). Not all the seed produced in a year will germinate the next year. Many species have some form of seed dormancy mechanism that has to be broken before they will germinate. Once dormancy has been broken environmental conditions must also be correct for germination; this accounts for some of the variation in weed populations between years. Losses of seed and seed viability are taking place all the time. Depth of burial in the soil, number and type of cultivations, soil type, and weed species affect the rate of decline. The seed viability of some species such as corn poppy decline very slowly compared with the more rapid decline of some grasses such as barren brome and black-grass.

5.1.1.2 Seed spread

As well as being able to multiply in a single field, weeds can spread from field to field, and from farm to farm. Before the implementation of some of the current very strict certified seed standards one important method of spread of some weeds such as wild oats, corn cockle and cornflower was in crop seed. These weeds set their seed at the same time as the cereal crop was harvested. The weed seed was then resown with the cereal seed. Commercial seed is now virtually weed free but weed seed can still be spread around fields and between farms by farm machinery or in straw, and also in the sowing of farm-saved seed. Other methods of spread include wind, water and birds, depending on the characteristics of the seed. Some seeds, such as those of docks, can remain viable in slurry for many weeks and so can be spread around the different fields where the slurry is applied.

5.1.1.3 Time of germination

Only a few weed species such as annual meadow-grass, common chickweed and mayweeds can germinate throughout the year and so can be a problem in a wide range of crops. Most other species have specific germination periods (Table 5.1); some are mainly autumn germinating such as black-grass whereas others such as black-bindweed, redshank, knotgrass and fat hen germinate in the spring.

Continuous cropping of the same crop, sown at the same time each year, will tend to encourage those weeds that germinate at the same time. Examples include autumn-germinating bromes and black-grass in winter cereals, and late spring/early summer germinating black nightshade in forage maize.

5.1.1.4 Same family as crop

Some important weed species are in the same family as the crop plant, e.g., charlock in oilseed rape or fat hen in sugar beet. Weed management using herbicides is then often difficult as there is not enough selectivity in the herbicide activity to kill the weed without injuring the crop.

5.1.1.5 Weed competitiveness

Weed species differ in how they compete with a crop. Some weeds such as speedwell compete early in the growing season and then die back before harvest. Others, such as

Table 5.1 Germination times for some common weed species.

	Calendar month											
Weed species and main period of germination	**Jan**	**Feb**	**Mar**	**Apr**	**May**	**Jun**	**Jul**	**Aug**	**Sep**	**Oct**	**Nov**	**Dec**
Early spring		*	***	***	**	*	*	*				
• Knotgrass												
Spring			*	***	***	**	*					
• Black bindweed												
• Orache												
Early summer				*	***	***	*	*				
• Black nightshade												
Mainly spring		*	***	***	**			*	*	*		
• Wild oat (spring)												
Mainly autumn	*	*	*	**	**			*	***	***	**	*
• Black-grass												
• Cleavers												
Autumn								*	***	***	*	
• Barren brome												
All year	*	*	**	**	**	**	**	**	**	**	*	*
• Annual meadow-grass												
• Common chickweed												
• Mayweed												

Key: * small amount of germination possible; ** some germination; *** main germination period.

Table 5.2 Weed populations (plants/m^2) resulting in a 5% yield loss in winter cereals.

Weed species	Number of plants/m^2 resulting in 5% yield loss
Very Competitive	
Cleavers	1.7
Wild Oats	5.0
Barren Brome	5.0
Moderately Competitive	
Italian ryegrass	8.3
Mayweeds	12.5
Black-grass	12.5
Corn poppy	16.7
Common Chickweed	25.0
Least Competitive	
Annual Meadow-grass	50
Common field speedwell	62.5
Ivy-leaved speedwell	62.5
Groundsel	83.3
Field pansy	250

cleavers, compete later and carry on growing until harvest. Cleavers is a very vigorous plant and much more competitive than the speedwells. Examples of weed competitiveness in a cereal crop are given in Table 5.2.

5.1.2 Assessing weed problems in the field

- Monitoring of the weeds present in the field in recent years can be a very good guide of likely weed problems, so accurate record-keeping is vital. Cropping, time of drilling, weather, soil texture, structure and condition will all affect the weed species and populations that emerge each year.
- Assessing the numbers of weeds present, e.g., per m^2 or per ha, and correct identification. Some plants are very important for wildlife and at low populations can have a negligible effect on crop yields, so where possible threshold data should be used to make management decisions.
- Weed populations tend to be very patchy over the field. It is very time-consuming to map weeds manually. Automated systems for weed mapping have been developed, leading to Site Specific Weed Management and semi-autonomous spray systems. One option for some of the major weeds such as wild oats (because weeds spread by only a few metres per year) is to assess the weeds in early summer before harvest and to produce a map. This map can then be used for patch spraying in future years.
- Effectiveness of control needs to be assessed each year especially with the increase in cases of weed populations that are resistant to herbicides. Again, accurate record-keeping is vital here.

The effects of weeds on yield will depend on the type and number of weeds, how competitive they are as well as on the density and vigour of the crop. Other effects, which must be considered, include ease of management in the different crops being grown in the rotation.

5.1.3 Weeds and biodiversity

Weeds vary in their importance for in-field biodiversity; some species are very valuable for invertebrates and others for seed eating birds. Weeds that are considered valuable for wildlife include annual meadow-grass, charlock, common chickweed, and the spring germinating fat hen, knotgrass, redshank and black-bindweed. Many of the environmental stewardship options include management options where these species may be encouraged (see Chapter 8).

5.2 Weed types and identification

For optimum management of weeds, both culturally and chemically, it is very important to be able to identify the weeds correctly, often at an early growth stage. There are two main groups of weeds, the monocotyledons or grass weeds and the dicotyledons or broad-leaved weeds.

5.2.1 Grass weeds

Grasses in the vegetative stage can be difficult to identify but it is at this stage that management is most effective. There are several features which can be studied in order to identify grasses correctly. Usually each grass has one or two distinct characteristics that aid identification (Table 5.3):

- *Stem/root characteristics.* Annuals spread by seed, whereas perennials can spread by seed and by pieces of vegetative material such as above ground creeping stems *(stolons)* or below ground creeping stems *(rhizomes)* (Fig. 5.1). Couch grass has rhizomes whereas rough meadow-grass has stolons.
- *Leaves.* Some grasses have folded leaves in the leaf sheath though most are rolled (Fig. 5.2); this characteristic is used as the starting point in many grass identification keys; perennial ryegrass and meadow-grasses are examples of grasses with leaves folded in the stem. A third kind of leaf arrangement, bristle-like, is found in sheep's and red fescue.
- *Auricles* (Fig. 5.3) are leaf extensions where the leaf sheath meets the stem. Auricles are only present in a few grass species so a very important structure to look for when identifying grasses.
- *Ligules* (Fig. 5.4), the membrane at the back of a leaf, are usually only assessed when identifying grasses in the same family such as the meadow-grasses or the brome grasses. Rough meadow-grass has a large ligule and smooth meadow-grass has a short ligule. A hand lens is useful in examining these, and other, features.
- Presence or absence of *leaf/stem hairs.* A few grasses have very hairy leaves, such as barren brome and Yorkshire fog. Wild oats only have hairs on the leaf margin. Most other grasses are hairless.

Table 5.3 Guide to the recognition of some common arable grass weeds by their vegetative characters (for cultivated grasses also see Chapter 19).

1. **Shoots flattened**	
A. Leaves with boat-shaped tips and 'tramlines' on upper surfaces	
(i) Young leaves with non-glossy under-surface; non-stoloniferous; non-rhizomatous; ligule tall and rounded.	Annual meadow-grass
(ii) Leaf sheath rough; stoloniferous; ligule pointed; heading mid-May onwards.	Rough meadow-grass
(iii) Leaf sheath smooth; rhizomatous; ligule short and non-pointed; heading May—early June	Smooth meadow-grass
Shoots flattened	
B. Leaves without boat-shaped tips and 'tramlines' on upper surfaces	
(i) Leaf and leaf sheath strongly keeled; each tiller appears flattened; heading June to September	Cock's-foot
(ii) Lower surface of leaf glossy; base of leaf sheath often red; small auricles sometimes present	Perennial ryegrass
2. **Shoots cylindrical**	
A. Leaves and leaf sheaths hairy	
(i) Auricles and rhizomes present; heading June	Couch grass
(ii) Auricles and rhizomes absent	
Basal leaf sheath with pink/red colouration restricted to veins; dense short hairs; heading May	Yorkshire fog Barren brome
Basal leaf sheaths often red coloured; leaves hairy; jagged ligule; heading from mid-May	
(iii) Auricles absent, bulbous base present; slightly hairy leaf blades; tufted habit	Onion couch
(iv) Auricles absent; no stem-based colouration; hairs present on leaf margin; long blunt ligule	Wild oats
B. Leaves, leaf sheaths and stems non-hairy	
(i) Prostrate habit; stoloniferous; internodes with strong colour; ligule long; heading July	Creeping bent
(ii) Upright habit; rhizomatous; ligule long and blunt; heading June	Black bent
(iii) Upper surface of leaves with distinctly marked veins; no rhizomes or stolons; leaf sheaths purplish; ligule blunt; heading from May	Black-grass
(iv) Annual; ligule long and oblong; mainly found in the east and south east of England	Loose silky bent
(v) Similar to black-grass; leaf sheath base pinkish; red sap oozes out of stem base when squeezed	Awned canary grass
(vi) Lower surface of leaf glossy; base of leaf sheath often red; small auricles sometimes present	Italian ryegrass
(vii) Broad, rough leaf blades; base of leaf sheath slightly swollen; ligule long, blunt and white	Timothy

Figure 5.1 Stolons and Rhizomes.

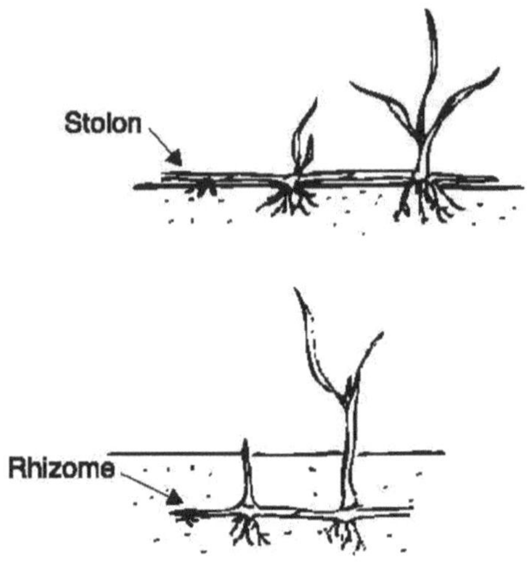

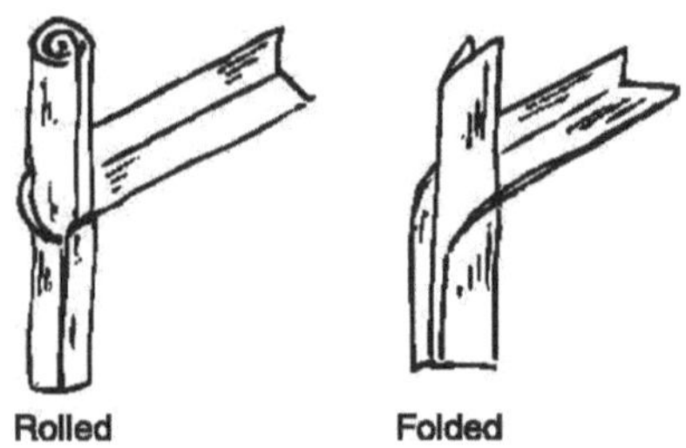

Figure 5.2 Leaves rolled and folded in stem.

Figure 5.3 Auricles.

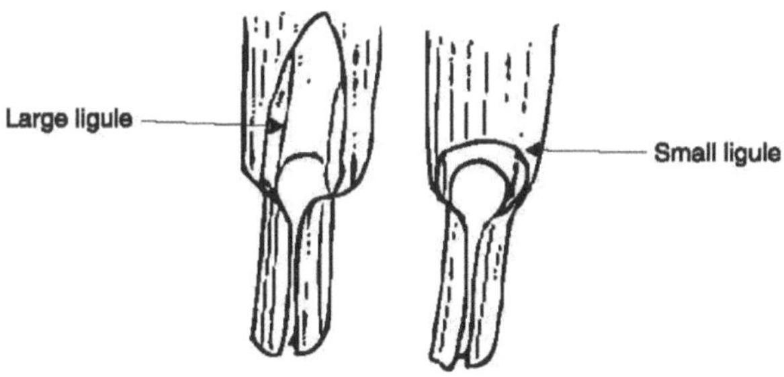

Figure 5.4 Ligules.

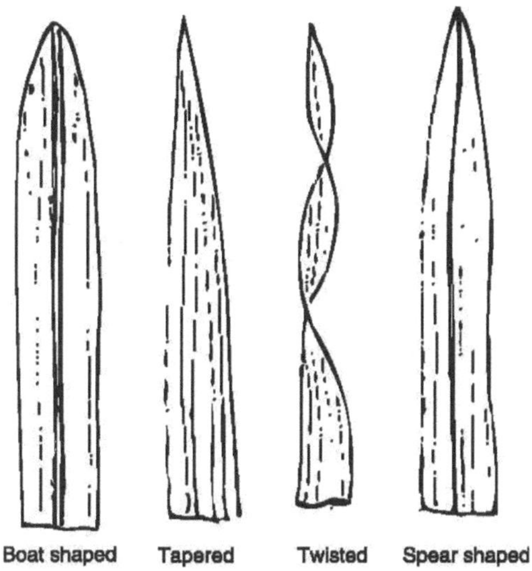

Figure 5.5 Leaf shape.

- *Leaf shape* (Fig. 5.5). Leaf shape and size can be very different, from short with distinct boat-shaped leaf ends of the meadow-grasses, to long and twisted in the brome grasses. Rye-grasses have spear shaped leaves and the bent grasses taper to a sharp point.
- *Stem base colouration.* A few important grasses have coloured stem bases/base of leaf sheath; perennial and Italian ryegrass are deep red, black-grass has a purplish blotch and awned canary grass is pink. Yorkshire fog has pink stripes on the leaf sheath. Timothy has a swollen stem base, and onion couch has bulbils present.
- *Ear characteristics* are distinctive for most of the grasses as detailed below. Normally management is carried out before ear production. Ear number is important when assessing effectiveness of management that year; weed mapping for patch spraying and checking for development of resistant plants.

The following are some important characteristics of the major grass weeds.

Couch grass (Fig. 5.6a and b) spreads rapidly by rhizomes, especially on light soils. Black bent also spreads by rhizomes, but the leaf characteristics are very different. Couch grass leaves are usually slightly hairy, have auricles unlike the bent grasses, and the ear is a spike not an open panicle, as in the bent grasses. Cultivations such as ploughing and rotovating can cause an increase in populations of couch grass due to cutting up of the rhizomes. Each small bit of rhizome that includes a node is

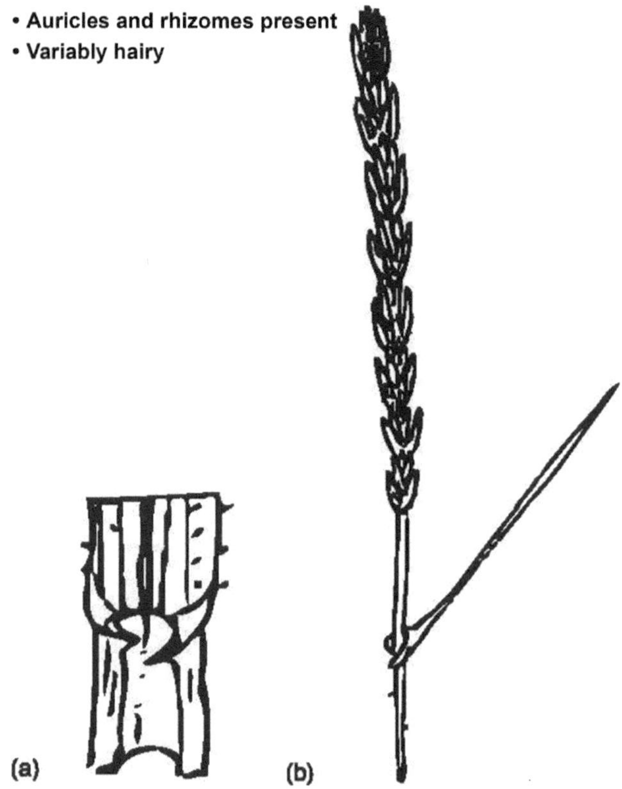

Figure 5.6 Ligule (A) and ear (B) of couch grass.

able to produce a new plant. If there is an opportunity for a fallow period then rhizomes can be dragged to the soil surface with harrows and left to dry out, or be killed by frosts over winter. Stubble treatment or pre-harvest use of a glyphosate-containing product has been very successfully used in many crops so that the weed is not now so important. If there are several clones of couch in a field then viable seed may be produced, although this is not the main method of propagation for this weed. Many of the black-grass herbicides will control couch grass seedlings.

Onion couch (false or tall oat-grass) (Fig. 5.7a and b) can be a more serious problem to deal with than couch grass, though occurring less frequently. It is mainly found in cereal-growing areas on the thin alkaline soils. It has swollen bulbous nodes at the stem base, each of which can produce a new plant. Tall or false oat-grass is the same species as onion couch; it is commonly found in hedgerows but does not have bulbils and is not often a weed in field crops. The ear of onion couch looks like a small upright wild oat, although the spikelets are much smaller. Onion couch can produce much viable seed. Continuous winter cereals and the use of minimum cultivations have encouraged this weed as there are few effective herbicides. Onion couch senesces too early for the pre-harvest glyphosate technique to be effective in some crops, e.g., wheat. Re-growth in stubble can be slow, so management in the stubble can also be poor.

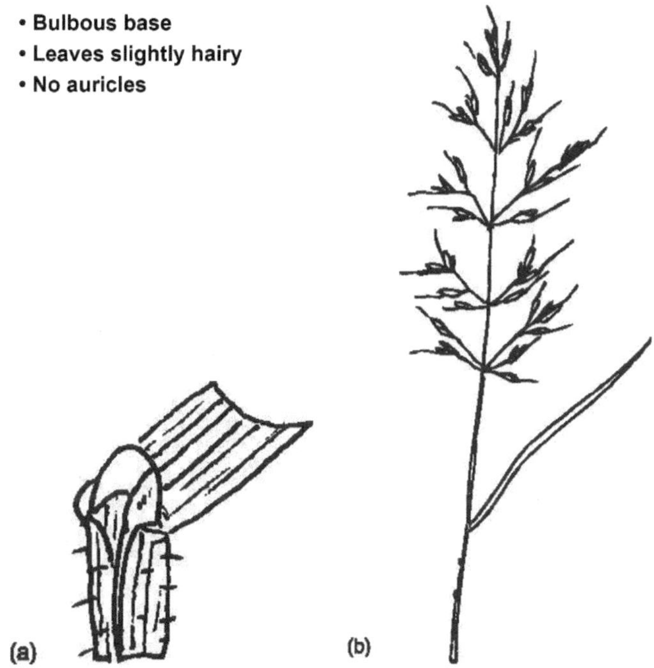

Figure 5.7 Ligule (A) and ear (B) of onion couch.

Creeping bent (Fig. 5.8a and b) is a common grass especially in grassland. It is a perennial that is spread by seed and surface rooting stolons which can be controlled by good ploughing or spraying when it is green in the stubble. Black bent is very similar but has rhizomes like couch grass and management is the same. The bent grasses have characteristic tapered leaves with no leaf hairs or auricles and an open panicle. Sometimes stems are red coloured.

Meadow-grasses (Fig. 5.9a and b) are some of the most commonly occurring grass weeds and can cause problems when present in large numbers. Rough meadow-grass is much more competitive than annual meadow-grass. It is a common weed on mixed arable and grass farms. All meadow-grasses have their leaves folded in the stem and lack hairs or auricles, and the leaf tip is boat-shaped. Rough meadow-grass possesses stolons, whereas smooth meadow-grass has rhizomes. Both are perennial species. Annual meadow-grass has neither stolons nor rhizomes. All meadow-grasses have open panicles. As the seed is small it germinates in the top 5 cm of soil, so ploughing can help reduce populations.

Wild oats (Fig. 5.10a and b) are found on a large percentage of cereal-growing farms and is one of the most important weeds economically due to the high competitive ability of this species. There are two main types of wild oat in the UK. The spring wild oat is the most common, germinating mainly in the spring with a small amount of seed germinating in the autumn. The winter wild oat is found mainly in southern and eastern areas; it mostly germinates in the autumn. These annual weeds look just like

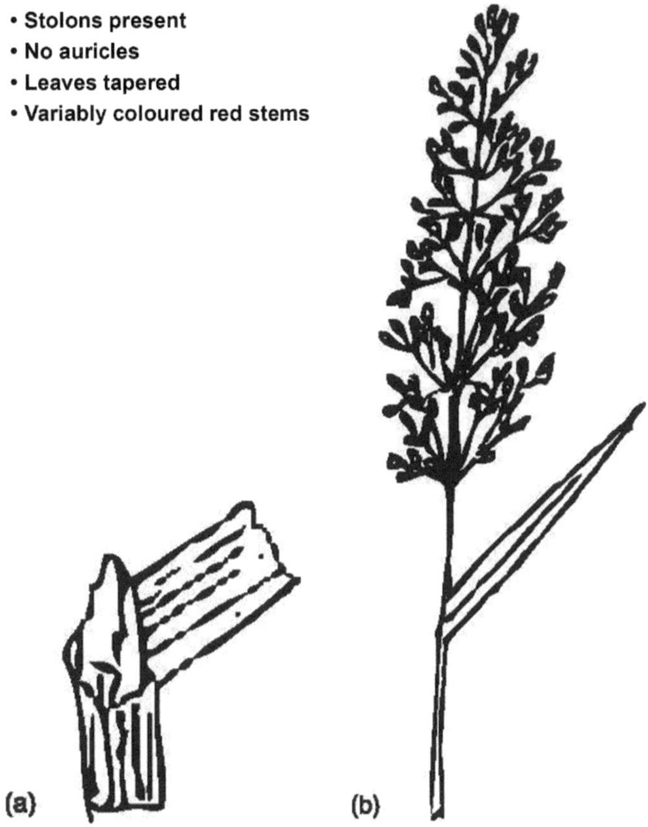

Figure 5.8 Ligule (A) and ear (B) of creeping bent.

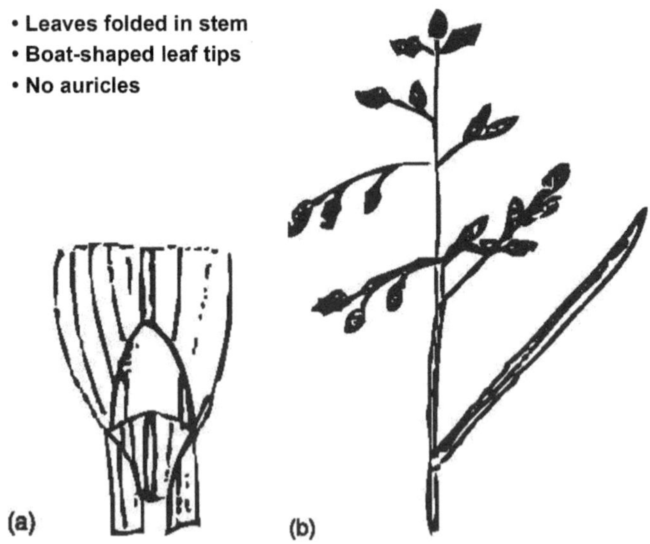

Figure 5.9 Ligule (A) and ear (B) of annual meadow-grass.

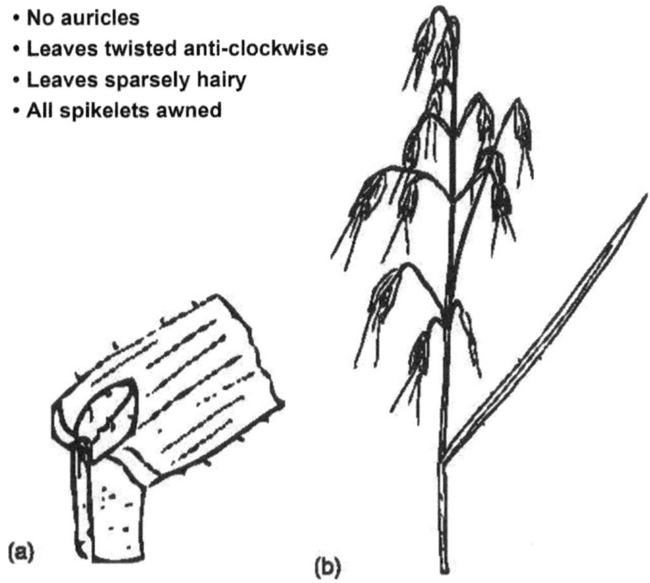

Figure 5.10 Ligule (A) and ear (B) of wild oats.

tame oats in the vegetative state with no auricles but they do have a few hairs along the side of the leaf blade. The leaves often have a characteristic anti-clockwise twist. The spikelets on the open panicle have long, kinked awns. Once established in a field, no matter what management measures are taken, wild oats are likely to persist for a very long time because of dormant seed in the soil. This dormancy problem is made worse by stubble cultivations after harvest which buries the seed. Most of the shed seeds, if left on the surface, are predated by birds and mammals. Deeply buried seed, which may have fallen down cracks on clay soils, for example, can remain viable for over 20 years and germinates from up to 25 cm depth of soil. The seed can arrive in a field in many ways, e.g., in the crop seed, dropped by birds, in farmyard manure made with infested straw, and from the combine. The first wild oat to appear in a field should be removed by roguing. Hand roguing can be justified when numbers are fewer than 500 plants/ha. Later, if numbers are allowed to increase, it may be necessary to use herbicides. It is difficult to decide when this can be justified. It can be based on yield response or likely grain contamination or to prevent a build-up of numbers in the future. Left uncontrolled, wild oats can increase threefold each year. There are now many populations of wild oats that have been found to be resistant to commonly applied herbicides. Cereal yield reductions are very high from wild oats as they are one of the most competitive grass weeds. Premiums can be lost for seed, malting and bread-making if there is contamination with wild oats.

Black-grass (Fig. 5.11a and b) economically is the most important annual grass weed on many cereal farms (although 12.5 plants/m^2 are required to reduce yield by 5% populations in excess of 500 plants/m^2 are often encountered) in the UK. This species used to be associated with heavy, wet soils where winter cereals were grown, but

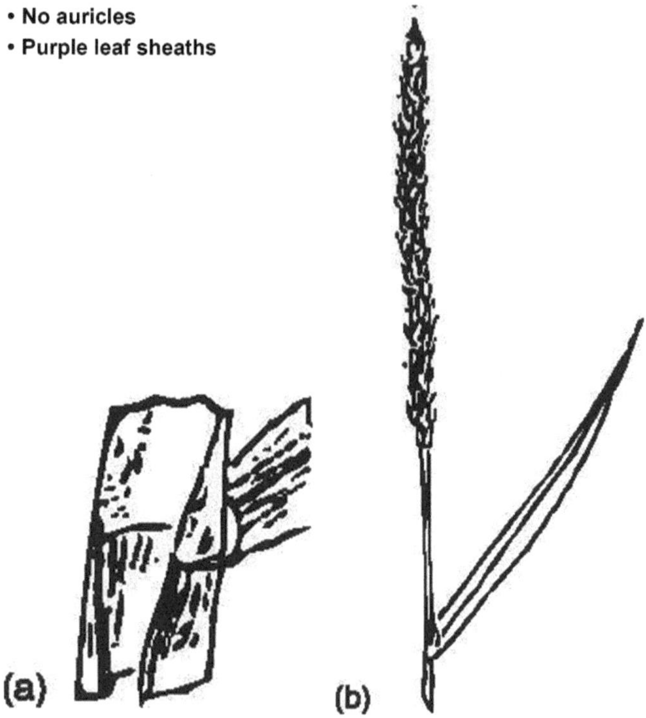

Figure 5.11 Ligule (A) and ear (B) of black-grass.

is now widespread on most types of soil where autumn-sown cereals predominate. As a seedling black-grass is not very distinct, with no auricles or hairs on the leaves. The stem base is often partially coloured purple. The ear is a long narrow spike with awns and is coloured purple/black, hence the name black-grass. Black-grass produces a very great number of viable seeds (typically, individual plants can produce 2–20 ears with between 10 and 300 seeds/ear). Most seed germinates in the first 3 years but some can remain dormant for up to 7 years. The seeds germinate mainly in early autumn, but in heavily infested fields spring germination can also be important. If uncontrolled, populations can increase significantly over a single year, so a high percentage control must be achieved to contain the problem; this usually means that cultural practices are necessary to supplement herbicides. Rotational ploughing, to bury the seed deeply can be a useful management method. Black-grass only germinates in the top 5 cm of soil. An alternative method that is now being used more widely is to undertake zero tillage. This method leaves seeds from previous years buried, and herbicides are used to control any seedlings that appear. Over a few years the number of viable black-grass seeds in a field should be effectively reduced using this method of management. Spreading the seeds to clean fields should be avoided by using weed-free crop seed and by thoroughly cleaning the combine. Vigorous crop competition is very important. Delaying drilling or even including a fallow can be very important in helping reduce weed populations. One of the most effective cultural management methods is to carry out spring cropping, as 80% of black-grass germinates in the autumn. A number of herbicides are available and the cost can be justified in cereals where economic thresholds are met (Table 5.2). Herbicide resistance is now considered to be endemic in black-grass populations in the UK, so programmes involving herbicide tank mixtures and sequences should always be used.

Barren or sterile brome (Fig. 5.12a and b) is a serious problem on many cereal farms. It spread from hedgerows and uncropped areas of fields, and is encouraged by early sowing of winter cereals, established using minimal cultivations. The seeds have little dormancy and germinate very readily in the autumn. There are five different bromes that can cause problems in field crops; barren brome and great brome are in the *Anisantha* family. This group has large open drooping panicles with long awns. Great brome has awns that are twice the size of those found in sterile brome. Rye brome, meadow brome and soft brome are in the *Bromus* family; they have more upright heads with broader spikelets. It is very difficult to distinguish between the different bromes at the seedling stage, but this can be done by looking at the nature of the hairs on the leaf and also the size and shape of the ligule. All the bromes have hairy leaves with no auricles. Ploughing can be very effective in managing the *Anisantha* bromes if the seed is buried more than 13 cm. Stubble cultivation, in a damp autumn, can also aid control of the *Anisantha* bromes, but not the *Bromus* bromes, whose seed should be left on the soil surface for at least a month prior to any cultivations. Field margins need to be managed to ensure that viable brome seeds are not produced. It is much better to have grass strips at field edges rather than a bare areas which can quickly be colonized by brome. There are only a limited number of herbicides for use in cereals and control can be variable, although a number of herbicides used in other crops are effective.

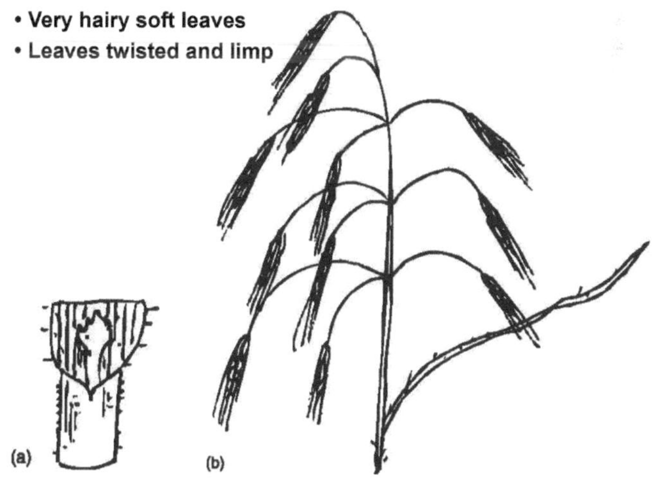

Figure 5.12 Ligule (A) and ear (B) of barren brome.

Italian ryegrass (See Chapter 19) can be a problematic weed in winter cropping situation, both because of seed carryover from grass leys and also from native populations of the grass. It is fast growing, and can form large populations that will result in significant yield loss in affected crops. Its ability to produce many tillers, and the presence of herbicide resistant populations in the UK, both contribute to its competitive ability and its increased presence in cereal crops. Management of this species should include stale seedbeds, although these might be less successful for this species than for some other grasses due to the long period over which Italian ryegrass seed can germinate. Ploughing seed to a depth from which they will not germinate and emerge also offers management option where seed are already present in the system. Greater range of crops will offer a greater range of herbicide options. If unmanaged, populations can rise quickly due to the high numbers of seed this species produces. Management of rotational grassland within rotations to reduce Italian ryegrass seed production, through timely grazing and cutting, can also help to reduce the amount of seed being returned to the soil.

5.2.1.1 Other grass weeds

Awned canary grass is a relatively new problem in the UK although it is more common in other parts of the world. It is not widespread but it can cause serious yield losses and harvesting problems where it does occur. The heads resemble large timothy heads (See Chapter 19). At the seedling stage the plant looks very like black-grass. The seedling has a pink stem base and when broken a red sap is often released. The seed has a limited amount of dormancy. Deep ploughing and spring cropping can help reduce weed incidence. Care must be taken with herbicide choice as only a few of the commonly used grass weed herbicides control awned canary grass.

Loose silky bent is a local problem on sandy or light loam soils in eastern and southern England and many parts of Europe. As with many other annual grass weeds, deep ploughing and spring cropping can be very good management methods to aid control.

Rat's tail fescue is a competitive and invasive annual grass which has been found in greater numbers in the UK in recent years. Once populations have become established they can have significant effects on both cereal yields and on quality of forage. This species is more often a problem is no-till winter cereals, and rotational ploughing can help to reduce populations. Spring cropping or delayed drilling of winter cereals, along with stale seedbeds, may also offer some management too.

Occasionally in maize crops some of the important tropical grass weeds such as common millet and green bristle grass have emerged but rarely at populations that would affect yields. Unusual grass species are also introduced with game bird feed. Climate change, with associated milder winters, may increase the survival of such species though, so it is important that farmers and agronomists remain vigilant.

5.2.2 Broad-leaved weeds

There are a wide range of broad-leaved weeds that can be important in field crops in the UK; within annual crops most of these are annuals although there are a few perennial broad-leaved weeds that can cause problems, and these may occur more often in minimum tillage and direct drill situations. Perennial weeds are more of a problem in grassland and in other perennial crops. How important economically a broad-leaved weed species is depends on both the weed species and the crop being grown. Cleavers are as competitive in cereals as some of the grass weeds. Numbers of species found tends to depend on the rotation; rotations including spring crops usually have the widest range, but may have lower numbers of some of the most problematic species. Some broad-leaved weed species favour different soil types; redshank is often associated with the heavier wetter soils whereas the mayweeds are often associated with the lighter soils. Corn spurrey is found on the light to medium textured soils, especially those that are naturally acidic. Populations of broad-leaved weeds are often higher where soils are ploughed rather than where only shallow cultivations are used.

Management, both chemical and mechanical, of annual broad-leaved weeds is usually most effective at the seedling stage. In order that the correct product is chosen it is important to identify the weed correctly. Many broad-leaved weeds have very characteristic cotyledons (seed leaves) and true leaves which aid identification. Size, shape, colour, if stalked, presence of hairs or prominent leaf wax all help in identification. Fig. 5.13a–d shows the characteristics of some common broad-leaved weeds.

Cleavers (Fig. 5.13a) is the most competitive weed in winter cereals and winter oilseed rape, even more competitive than wild oats. It can be confused at the seedling stage with ivy-leaved speedwell (Fig. 5.13a). Note that cleavers has a notched cotyledon. Cleavers germinates at fairly low temperatures mainly in the autumn and winter, with some in the spring.

Charlock (Fig. 5.13a) is a brassica showing the typical large kidney-shaped cotyledons. It is the cotyledon shape and roughness and presence or absence of hairs on the true leaves that identifies the different brassica species. Charlock used to be a major weed problem in cereals but with the introduction of more effective herbicides it is now less important. Management in brassica crops, including oilseed rape, can still prove difficult though. From an agrobiodiversity point-of-view it is an important plant for a number of different invertebrate species.

Speedwells (Fig. 5.13b) are also commonly found in arable crops. Ivy-leaved speedwell is mainly autumn germinating whilst common field-speedwell germinates all the year round. Common field speedwell has spade-shaped cotyledons whereas ivy-leaved speedwell has large oval cotyledons ending in a knob. It is important to be able to distinguish between these weeds, as their different germination periods will dictate whether changes in rotations will help in their management. Speedwells are low growing and are not very competitive (Table 5.2).

Fat hen (Fig. 5.13b) is in the same family, Amaranthaceae, as sugar beet and so can sometimes be difficult to control in beet crops. It germinates in the spring and is often associated with the application of manure. It has a very distinct waxy/mealy appearance on the cotyledons and first true leaves. The stalk and underside of the cotyledons are red. It is an important plant for invertebrates and seed eating birds.

Polygonums (Fig. 5.13b), including knotgrass, redshank, pale persicaria, and black bindweed, are a very important group of spring germinating weeds. Of these, knotgrass

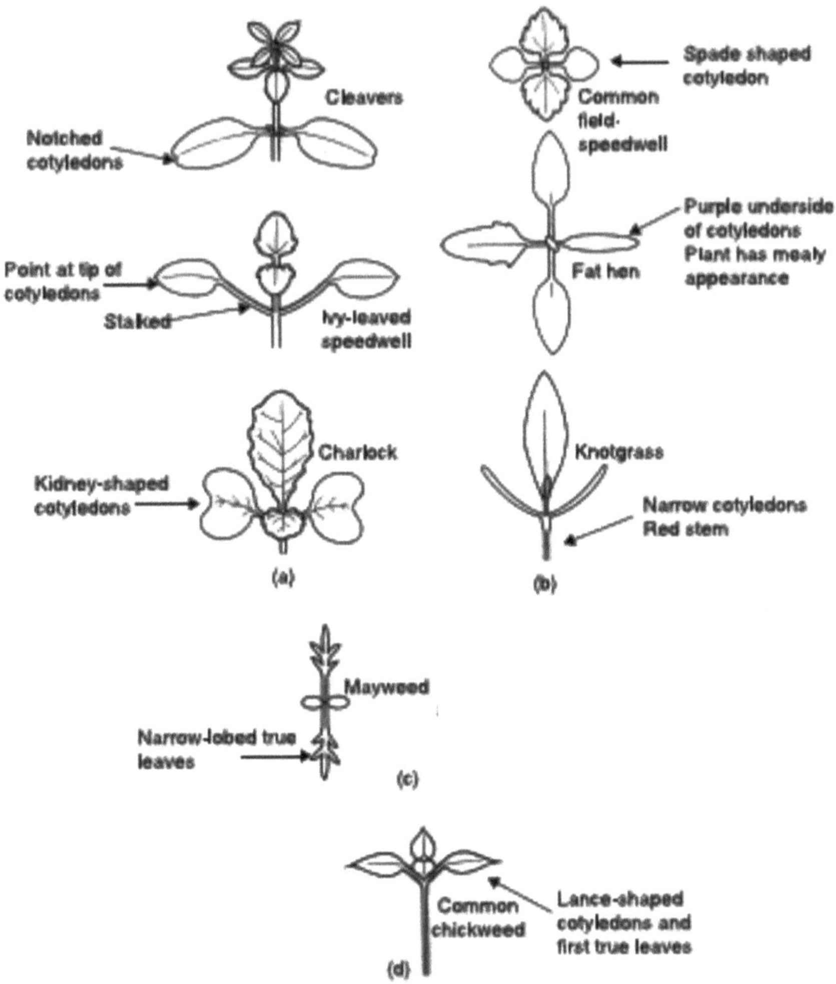

Figure 5.13 Identification of common broad-leaved weeds: (A) large cotyledons; (B) medium sized cotyledons; (C) small cotyledons; (D) similar-shaped cotyledons as true leaves.

germinates earliest in the year; it has narrow cotyledons arranged in the shape of a 'V'. All the polygonums seedlings have a red stem. They are important for in-field biodiversity.

Mayweeds (Fig. 5.13c) germinate in most months, although spring is the peak period. Unlike many broad-leaved weeds, mayweeds remain green all season and so compete with the crop for a long time. They are prolific seed-producers and the seed can remain dormant for many years. Mayweeds have very small cotyledons and very distinctive shaped first true leaves. They are also important for invertebrates. Populations of mayweeds possessing herbicide resistance against ALS-inhibiting herbicides are now present in the UK and poor control should be investigated.

Field pansy only cause significant problems in cereals at populations above 250 plants/m^2, so their management is rarely required. The seedlings have distinct small, and hairless, first true leaves.

Common chickweed (Fig. 5.13d) is a very common broad-leaved weed found in virtually all arable fields in the UK. Chickweed favours fertile soils, germinating all the year round and it readily overwinters. This weed is very successful, partly because it can produce more than one generation of plants a year and thus a lot of seed. At the seedling stage it is a bright green colour and has true leaves with the same shape as the cotyledons. As with mayweeds, populations of chickweed possessing herbicide resistance to ALS-inhibiting herbicides are now present in the UK.

Common poppy is another common weed found in cereal and oilseed rape crops. Seed of this species can survive in soil for up to 100 years, far longer than other arable weed species. This means that ploughing gives only very limited, short-term control. As with mayweeds and chickweed, populations of common poppy possessing herbicide resistance to ALS-inhibiting herbicides are now present in the UK.

Red and henbit dead-nettles are fairly uncompetitive species that will colonise bare patches on fertile, sandy loam soils. They are easily recognizable at cotyledon stage by the notch in each cotyledon where the stalk is joined.

5.3 Management of weeds: general

Although the management of weeds with herbicides is standard practice on many farms it is worth remembering that other weed management methods also play an important part. This is especially important with the increase in herbicide resistant weed populations, loss of herbicides from the market, and the limited introduction of new products. An integrated weed management (IWM) approach should be employed, incorporating where appropriate manual, mechanical, cultural, biological and chemical methods. IWM is the incorporation of all of these methods in a mutually supportive manner, and can have both environmental and economic benefits. Prevention of weed problems should underpin any IWM programme. If a problem is prevented then management will not be required. Non-chemical methods of weed management are the only ones available to organic farmers, and the methods used in these systems are very important in informing the design of IWM programmes.

5.3.1 Methods of weed management

- Crop hygiene. It is particularly important to ensure that only clean seed is sown when using home-saved seed. Hand roguing can be a very effective method of preventing weeds becoming a problem in the first place. This can be especially useful for managing wild oats. Avoidance of machinery contamination can also help to prevent weed increase and spread both within and between farms.
- Cultivations. Ploughing can be very effective at containing or reducing some annual grass weed problems such as black-grass and barren brome; ploughing can bury freshly shed seed to 15/20 cm and will normally bring to the surface fewer viable weed seeds. Most seeds

only germinate from the top 5 cm of soil. Black-grass populations can be reduced by approximately 70% after ploughing, although this figure does vary greatly between field trials. With heavy clay soils, where minimum cultivations are used, it is recommended to rotationally plough once every 3 or 4 years to help reduce the grass weed burden. Weeds such as wild oats that have large seeds that can germinate from greater depths are unaffected by ploughing. In the case of wild oats it can be beneficial to leave the seeds on the surface in an overwinter stubble. They will then be predated by birds and mammals.

In some years, if there is enough soil moisture, stubble cultivations and the 'stale' seedbed technique can be a useful aid to weed management. The soil is cultivated ready to sow the next crop; weeds are allowed to grow and are then killed before planting the crop. Inter-row cultivations are still used in some crops, particularly if there is a difficult weed to control, such as weed beet in sugar beet, or where there are limited herbicides approved for use, as in field vegetables. There is increasing interest in the use of in-crop weeders as a method of reducing herbicide inputs in such crops.

- Cutting, e.g., bracken, rushes, ragwort and creeping thistles. This weakens the plants, causes depletion in the carbohydrate reserves in roots and rhizomes, and prevents seeding. This often needs to be repeated three or four times a year to be successful.
- Drainage. This is a very important method of controlling those weeds which thrive in waterlogged soils. Lowering the water table by good drainage will help to manage weeds such as rushes, sedges and creeping buttercup.
- Rotations. Growing grass leys and various arable crops that are planted at different times of the year usually leads to a different weed flora. There is also an opportunity to use different types or groups of herbicides. This method is useful if there are difficult weeds such as barren brome, volunteer potatoes, herbicide resistant black-grass and weed beet to control.
- Maintenance of good fertility. Arable crops and good grass crops require a high level of fertility, i.e., the soil must be adequately supplied with nitrogen, phosphates, potash, lime and organic matter. Under these conditions crops can compete strongly with many weeds.
- Crop seed rates. High seed rates and good crop establishment all help to reduce the impact of weeds.
- Time of drilling. Delaying drilling cereals from September to October can significantly reduce the number of black-grass and brome seedlings that emerge. Later drilling will also allow more stubble treatments to be undertaken.
- Competitive crop/variety. Some crops, such as winter oilseed rape and potatoes, are very competitive and can help smother any weeds present. There can also be a difference in competitiveness of different crop varieties due mainly to differences in crop height and leaf angle.
- Burning. The use of propane burners can be an effective method of weed control; it is mainly used in organic systems. When farmers were able to burn crop stubbles this was very effective at reducing the weed seed burden, but is non-selective and will also kill non-target organisms.
- Mulches. There are a number of mulching materials that can be used to prevent weed growth in some small-scale plantings, e.g., amenity plantings, fruit and vegetable crops. Black polythene is the most commonly used material. As well as preventing weed growth it also helps to conserve soil moisture.

- Cover crops. The use of cover crops, along with the use of catch crops and intercropping, is continuing to increase in popularity. Such crops have multiple benefits to cropping systems. Weed suppression, either by competition or by allelopathic means, is one of these benefits.
- Chemical control. There are many different herbicides and products on the market, although a lot have now been withdrawn recently due to EU and UK legislation. New herbicides cost a lot to develop and to get approval for use, so currently few new products are coming on to the market. The following is a summary of modes of action of some of the currently approved herbicides.

5.3.2 Herbicides—modes of action

Most of the herbicides used have a *selective* effect, i.e., they are substances which stunt or kill weeds and have little or no harmful effects on the crop in which the weeds are growing. A severe check of weed growth is usually sufficient to prevent seeding and to allow the crop to grow away strongly. There are a limited number of *non-selective* or *total* herbicides such as glyphosate. They kill or check all vegetation and are usually used in non-cropped areas, before the crop emerges, or as crop desiccants pre-harvest.

Herbicides are often sold under a wide range of proprietary names which can be very confusing, but the common name of the active ingredient (ai) must always be stated on the container. In the text of this book the common name of the active ingredient is used when referring to herbicides.

The chemicals now commonly used as herbicides can be classified according to their chemical group and biochemical mechanism of action. Most simply they can be classified by their basic mode of action as follows:

- *Contact-acting herbicides.* These will kill most plant tissue by a contact action with little or no movement through the plant; shoots of perennials may be killed but re-growth from the underground parts usually occurs. Effective control relies on good coverage of green plant material by the herbicide. Some examples of contact-acting herbicides are carfentrazone-ethyl and phenmedipham.
- *Soil-acting or residual herbicides.* These chemicals act through the roots or other underground parts of the plant after being applied to the soil surface. Some examples are propyzamide, pendimethalin, tri-allate, flufenacet and prosulfocarb. These herbicides vary in their persistency in the soil from weeks to months, and can be affected by soil organic matter levels. Some residual herbicides also have some contact activity, such as metribuzin.
- *Translocated herbicides* are those which can move through the plant before acting on one or more of the growth processes. Some of these herbicides can be very effective at controlling perennial weeds, e.g., the control of couch by glyphosate. Good spray cover is not so important with this type of herbicide, but it is vital that they are used at the correct plant growth stage to ensure maximum translocation to the roots and rhizomes.

Herbicides work by affecting one or several different biochemical pathways or physiological processes within the plant. The speed of action reflects the mechanism of action. Some of the earliest developed herbicides are similar to substances (hormones) produced naturally by plants. Susceptible plants initially show distorted growth and then will take a few days to die back. They are mainly used for controlling broad-leaved weeds in cereals and grassland. Examples include 2,4-D and MCPA.

A large number of herbicides affect photosynthesis. Symptoms are usually seen as a rapid yellowing or *chlorosis* of the leaves followed by leaf death or *necrosis*. Other herbicides limit or stop cell division and elongation of the growing point. These chemicals can be very slow acting as seen with propyzamide (a residual amide) in oilseed rape.

Several of the newer chemicals, such as the sulfonylureas, are very specific and only affect the production or synthesis of single compounds such as specific amino acids. Again, these herbicides tend to be slower acting than those that affect photosynthesis. Examples include amidosulfuron, thifensulfuron-methyl and rimsulfuron. Further details on herbicide modes of action are given on the Herbicide Resistance Action Committee website. Here, different modes of action are given different numbers, to try and help growers to select good mixtures and sequences to reduce the development of herbicide resistance.

5.3.3 Herbicide choice

It is vital to check the manufacturer's recommendations before buying and applying pesticides and to follow the Code of Practice for Using Plant Protection Products. Only products approved by the Chemicals Regulation Directorate (CRD) can be sold or applied in the UK. It takes many years for new chemicals to be approved as the CRD has to have adequate data to ensure the products are safe, do what they are supposed to do and have limited environmental impact. The European directive (91/414/EEC) was introduced to harmonise all the different pesticide regulations within the EU. Following withdrawal from the EU, the UK is developing its own legislation based upon this directive. The aim of the legislation is to further reduce risk and hazard to both human health and the environment. This legislation has led to the loss of a number of commonly used pesticides in the UK.

Herbicide choice will be affected by a number of factors including:

- *Weeds present and their growth stage.* The efficacy of herbicides is often affected by the growth stage of the weed. Control of annual broad-leaved weeds is usually most effective when the weeds are small. Control of annual grass weeds is usually more effective before the 2-leaf stage. Herbicide recommended rates are often lower when the weeds are small. Descriptions of weed and crop growth stages are included in the UK Pesticide Guide and are useful when considering herbicide label recommendations.
- *Crop/variety and crop growth stage.* Some herbicides are only recommended for use on a limited range of crops. Other herbicides are only recommended for use on some varieties of crops as there can be problems with crop damage, i.e., metribuzin in potatoes.

  The crop growth stage is important partly because of crop damage. The hormone herbicides such as mecoprop-P are not recommended after the early stem extension stage of cereals. Crop growth stage is also important with the contact-acting herbicides as they usually work best when the crop and weed are growing well. Note, however, that if the crop is too advanced, and is shading the weeds, then it is very difficult to get good spray coverage on to the weeds. It should also be noted that if a crop is under environmental stress (temperature stress, drought stress, nutrient stress) then herbicide damage to that crop may be more likely to occur.

- *Soil type and condition.* In order for the residual chemicals to work effectively the soil tilth should not be too cloddy. Activity is affected by the amount of soil moisture present as this will affect movement of the chemical to the germinating weeds. Persistency of the chemical in the soil will be affected by the rate of chemical applied, and speed of breakdown. The main method of breakdown is by microbial activity, which is affected by soil moisture and temperature. The rate of breakdown varies considerably between fields and is difficult to predict.

  The rate of application of some residual chemicals is affected by the soil texture. Some are not recommended on sandy or stony soils as there can be too much leaching down to where the crop is growing, which can lead to crop damage.

  Herbicide choice is restricted on soils with a high organic matter or those that have a high adsorption coefficient. Residual activity is reduced on soils with a high organic matter (usually greater than 10%). Trash on the soil surface can also have this effect; the chemical becomes attached to the charged sites on the organic matter *(adsorbed)* and is then unavailable for weed uptake. The activity of residual herbicides can also be affected by soil pH.
- *Weather conditions.* Weather conditions, including rainfall and temperature, can affect activity and/or efficacy of a treatment, as well as crop damage. Control of many weeds using foliar applied chemicals is most effective when the weather conditions are optimal for weed growth. Some herbicides are rain-fast fairly soon after application whereas glyphosate requires at least 6 hours dry weather after application to give the best results.
- *Tank mix compatibility.* Some chemicals only have limited compatibility with other pesticides; this can be due to a problem with the formulations or with the activity of the chemical being affected. To get optimum weed control in some crops mixtures of chemicals are required or sometimes the use of sequences/stacking (products applied in close succession) is more effective.
- *Cost.* Pesticides (including herbicides, fungicides and insecticides) are expensive when they are first marketed, partly to cover the very high research and development costs. Once they are off patent costs usually fall. Cost can significantly affect product choice and rate used, especially if crop prices are low.
- *Following crops.* There are a few restrictions on following crops and intervals between applying certain herbicides and following crops, especially with residual chemicals; always check the label. Some chemicals also have recommendations to plough before planting the next crop.
- *Water buffer zone requirements.* Many pesticides have a buffer zone requirement when spraying next to water courses. There are not as many herbicides as other pesticides that are affected in this way, but it can influence choice of chemicals. If required, a Local Environmental Risk Assessment for Pesticides (LERAP) should be carried out; this takes into account width of water course, rate of chemical used and type of nozzles on the sprayer. Undertaking a LERAP can reduce the buffer zone requirement.
- *Resistance.* Herbicide resistance is one of the key problems faced by arable farmers both in the UK and worldwide. This can be particularly problematic on cereal farms, where herbicide resistance in certain grass weed species is now very common. Herbicide resistance is the inherited ability of a weed to survive rates of chemicals that normally control the weed. Black-grass, wild oats and ryegrass are grass weeds that have developed a problem of resistance to some of the most commonly applied grass weed herbicides. Once herbicide resistance has been diagnosed then chemical choice will be severely affected, and it is vital that Integrated Weed Management methods are employed to tackle these weed populations.

5.4 Herbicide resistance

Cases of herbicide resistance have developed more slowly than insecticide and fungicide resistance mainly because most weeds only have one generation each year, unlike diseases and some pests like aphids. Cases of resistance started being found in arable weeds in the UK in the early 1980s. This was in black-grass populations on farms where populations were very high, continuous winter cereals were grown (often established using minimum cultivations) and the same herbicide was often applied both in the autumn and spring every year. Currently in the UK, herbicide resistance in black-grass populations is described as endemic, meaning growers should assume it is present on their farms. Resistant Italian ryegrass is also fairly widespread, wild oat resistance is less common and there are a growing number of confirmed cases of broad-leaved weed resistance (common poppy, mayweed, chickweed). Poor results after applying herbicide may not necessarily be due to herbicide resistance. Seed samples can be collected and tested if resistance is suspected. Other indicators for resistance are good control of other susceptible species and healthy plants next to dead plants of the same species.

5.4.1 Mechanisms of herbicide resistance

There are two main mechanisms of herbicide resistance

1. *Enhanced metabolism resistance (EMR).* This is the most common mechanism and develops slowly over many years. Herbicides affected by enhanced metabolism are broken down (detoxified) more rapidly in resistant than susceptible plants. Each time the same herbicide is sprayed more of the resistant plants will remain so that gradually the weed population will be dominated by the more resistant plants.
2. *Target site resistance (TSR).* Most herbicides work by targeting a single active site in a plant, often inhibiting a biochemical pathway. In TSR this active site is slightly altered, and the herbicide no longer 'fits' the active site. Because of this the biochemical pathway is not inhibited and the weed survives. In the UK, TSR is reported against the grass weed herbicides (the 'fop', 'dim' and 'den' graminicides) and also the ALS-Inhibiting herbicides (including the sulfonylureas). This type of resistance develops very quickly.

5.4.2 Management of herbicide resistant weeds

Once resistance is confirmed it will always be present on the farm. Recently there has been a reduction in the number of different herbicides available and few new herbicides are coming on the market, so management of resistant weed populations is becoming more difficult.

- Non-chemical management methods must be used in combination with herbicides. Rotational ploughing, delayed drilling and spring cropping can help reduce grass weed populations.
- Use of pre-emergence treatments reduces risk of developing resistance compared with many of the post-emergence products applied in cereals.

- Use mixture or sequences of herbicides with different modes of action. Do not rely on the same chemical or chemical group.
- Growing crops other than cereals will give an opportunity to apply totally different and effective chemicals.
- Continue to monitor effectiveness of management measures to ensure resistance problems do not increase.

5.5 Spraying with herbicides: precautions

This is a skilled operation and should only be carried out by trained operators. A recognised certificate of competence is required by all who store, sell, recommend or apply pesticides. It is vital that users are familiar with the HSE Code of Practice for Using Plant Protection Products. Method of application and timing can significantly affect herbicide activity. The following precautions should be taken when spraying:

1. Make a careful survey of the field to determine the weeds to be controlled. Choose the most suitable and safest chemical and the best time for spraying. The risks associated with the use of any herbicides should be assessed as defined in the HSE regulations (COSHH—*Control of Substances Hazardous to Health*). Appropriate measures should be taken to control any risk.
2. Wear the correct personal protective equipment (PPE). Read carefully the instructions issued for that product when handling the concentrate and when spraying.
3. Check carefully the amount of chemical to be applied and the volume of water to be used. Droplet size is very important when using post-emergence herbicides, whereas with pre-emergence herbicides medium size droplets are just as effective as small droplets. When weeds are small, droplet size needs to be small to ensure the highest deposition on the plants. Use of angled nozzles can increase deposition on small and vertically growing weeds and is also preferred for good coverage of seedbeds with pre-emergence residual herbicide applications.
4. Only use approved tank mixes. Follow the manufacturer's recommendations for mixing.
5. Control spray drift. Wind speed, boom height, speed of application, atmospheric conditions, and water volume and spray quality can all affect the amount of spray drift. Fine sprays are most affected by drift. The optimum wind speed is when there is a light breeze and tree leaves are just rustling. Often, air-induction sprayers produce smaller droplets than conventional sprayers, but these sprayers can give as good results as the medium droplets produced by conventional sprayers and produce less spray drift.
6. Make sure that the boom is level. The lower the boom the less the drift.
7. Keep pesticides out of water courses and other environmentally sensitive areas. Due to UK water quality legislation this is a major issue particularly with some herbicides such as MCPA and propyzamide. Pesticides can get into a water course from spray drift, spillages of the concentrated chemical, sprayer washings or from surface run-off and, importantly, through drains. Establishing grass buffer strips of at least 6 m next to water courses can significantly reduce the risk. Not spraying when field drains are running or when the soil is very dry and cracked can also help stop herbicides entering water courses.

8. Some estimates suggest that in excess of 40% of surface water pollution by pesticides results from spillages in pesticide handling areas. Use of biobeds or bunds can help in reducing pollution from these sites.
9. Check buffer zone requirements for all herbicides, not just the sensitive ones, when spraying next to water courses. (If necessary undertake a Local Environmental Risk Assessments for Pesticides (LERAP) assessment.) A LERAP assessment is not required if there is a 5 m no-spray zone or buffer strip next to the watercourse.
10. Wash out the sprayer and pesticide containers thoroughly after use. Follow the guidelines for safe disposal.
11. Record keeping. By law, records must be kept of all sprays applied to crops grown for human or animal feed. Details of operator, field, date of spraying, chemical used, and rate of application and water volumes plus weather conditions when sprayed must be kept. It is also good practice to note effectiveness of the treatment, especially with the increases in pesticide resistance.

5.6 Key points

- A plant should only be considered to be a weed if it has been determined that it will cause significant problems to a cropping system.
- The positive roles that non-crop plants play in agroecosystems should also be considered when deciding if management is required.
- Accurate identification of plants is a key component of all weed management strategies.
- A detailed knowledge of the biology and ecology of a plant plays an important role in designing integrated weed management strategies.
- Herbicides are a very useful tool in weed management but should be used with care and should not be used as the only method of management due to the risk of herbicide resistance developing.
- Integrated Weed Management should assess physical/manual, mechanical, cultural and biological methods of management alongside the use of herbicides.
- Management strategies should also focus on prevention of future weed problems developing.

Sources of further information and advice

AHDB, 2014. Identification and Control of Brome Grasses IS31.
AHDB, 2021. Managing Weeds in Arable Rotations—A Guide.
AHDB, 2018. The Encyclopaedia of Arable Weeds. AHDB.
AHDB, 2016. Distribution and Management of Herbicide Resistance in Broad-Leaved Weeds IS54.
Anderson, W.P., 2007. Weed Science, Principles and Applications. Waveland Press.
British Crop Protection Council, 2016. Field Scale Spraying.
Cobb, A.H., Reade, J.P.H., 2010. Herbicides and Plant Physiology. Wiley-Blackwell.
Defra, 2011. Identification of Injurious Weeds.

Garthwaite, D., Ridley, L., Mace, A., Parrish, G., Barker, I., Rainford, J., MacArthur, R., 2019. Pesticide Usage Survey Report 284, Arable Crops in the UK 2018.

Hanf, M., 1983. The Arable Weeds of Europe. BASF.

Hester, R.E., Harrison, R.M., 2016. Agricultural Chemicals and the Environment: Issues and Potential Solutions: Edition 2. RSC.

HSE, 2006. Pesticides—Code of Practice for Using Plant Protection Products. HSE/DEFRA.

Hubbard, C.E., 1992. Grasses. Penguin Books Ltd.

Lainsbury, M.A., (published annually) The UK Pesticide Guide. BCPC/www.cabi.org.

Lutman, P.J.W., Moss, S.R., Cook, S., Welham, S.J., 2013. A review of the effects of crop agronomy on the management of Alopecurus myosuroides. Weed Research 53 (5), 299–313.

Marshall, E.J.P., Brown, V.K., Boatman, N.D., Lutman, P.J.W., Squire, G.R., Ward, L.K., 2003. The role of weeds in supporting biological diversity within crop fields. Weed Research 43 (2), 77–89.

Mathews, G., 2016. Pesticides, Health, Safety and the Environment, second ed. Wiley-Blackwell.

Naylor, R.E.L., 2008. Weed Management Handbook, ninth ed. Wiley.

Wallace, H., 2021. Grasses: a guide to Identification using vegetative characteristics. FSC.

Wilson, M.F., 2003. Optimising Pesticide Use. John Wiley and Sons Ltd.

Websites

www.agricentre.basf.co.uk.

www.cropscience.bayer.co.uk/.

www.croplife.uk/.

www.ewrs.org.

www.fera.co.uk/.

www.ahdb.org.uk/.

www.hracglobal.com.

www.hse.gov.uk/pesticides/.

www.voluntaryinitiative.org.uk/.

The Weed Resistance Action Group (www.ahdb.org.uk/knowledge-library/the-weed-resistance-action-group-wrag).

www.weedscience.org/Home.aspx.

Diseases of farm crops

6

6.1 Introduction to plant diseases and disorders

The main living organisms causing diseases of crop plants are fungi, oomycetes, bacteria and viruses (biotic diseases). Mineral deficiencies, physiological disorders and symptoms associated with environmental conditions are often classified as diseases (abiotic diseases or disorders) and are included in this chapter. Diseases, like pests, annually cause millions of pounds worth of damage and loss to the agricultural industry. Diseases can significantly reduce crop yields and quality, as well as increasing losses of produce in store.

6.1.1 Key terminology

Any organism that causes plant disease is known as a **pathogen**. Pathogens are sometimes referred to as **parasites** as they obtain nourishment from the infected crop plant. Plant pathogens fall into a number of categories

(a) *Biotrophs* (or obligate parasites): These organisms can only survive and reproduce in the cells of living hosts including the crop plant, its volunteers and some wild plant species. Biotrophic fungi use special organs known as haustoria to extract nutrients and do not use enzymes during invasion. When biotrophs move on to volunteers and/or wild plants, this is referred to as 'the green bridge'. Examples of biotrophs include the rust and powdery mildew fungi, and viruses.

(b) *Necrotrophs* (or facultative pathogens): These are pathogens that survive on dead plant tissue. They use an array of cell wall degrading enzymes and toxins to kill the cells of their host and can survive on dead or decaying plant matter in the absence of their plant hosts. Unlike biotrophs, some necrotrophs can have wide host ranges. Examples of necrotrophs include the fungi that cause diseases Sclerotinia stem rot (oilseed rape, peas and beans, carrots and others) and take all in wheat and barley.

(c) *Hemibiotrophs*: These pathogens have biotrophic and necrotrophic phases to their infection cycle. An example of a hemibiotroph is the pathogen *Phytophthora infestans*, which causes the disease late blight in potatoes and tomatoes.

(d) *Saprophytes*: These organisms live on decaying or dead organic matter; either plants that have reached maturity and died or the residues of these plants in the soil. All necrotrophs have a saprophytic stage. Saprophytes play an important part in helping to break down plant remains into organic matter.

Lockhart and Wiseman's Crop Husbandry Including Grassland. https://doi.org/10.1016/B978-0-323-85702-4.00017-0

Figure 6.1 'The disease tetrahedron' summarises the main factors involved in plant disease epidemics.

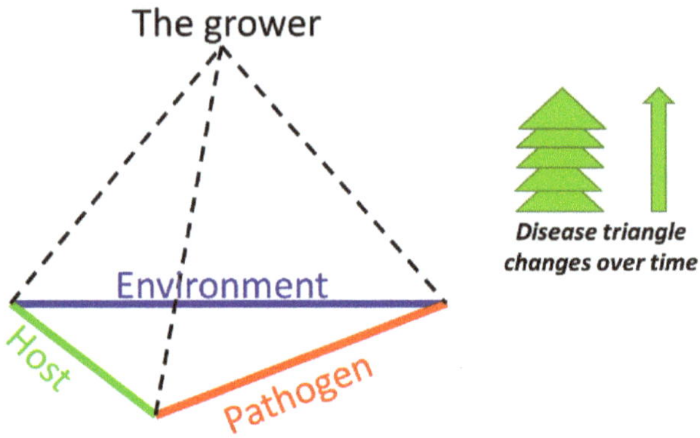

Most pathogens only attack one crop species or plant family (i.e., biotrophs), but a limited number have a wider host range (i.e., necrotrophs). For a pathogen to be successful, it must be able to invade and colonise a plant, grow and reproduce, be able to spread effectively and have some method of survival. This is called the disease cycle (lifecycle) and it is important to appreciate that pathogens have distinct cycles. A thorough understanding of disease cycles, allows effective methods of control to be selected and applied at an optimum time.

Knowledge of a particular disease cycle can also help when predicting likelihood of a disease epidemic. Conditions have to be right for diseases to develop, so not every year/season has the same disease importance. Disease development is affected by many factors and they are all interlinked (see Fig. 6.1):

- *The host plant.* The vigour and growth stage of the crop can affect the amount of infection that is likely to occur. Crop variety resistance is also important.
- *The pathogen.* Amount and type of pathogen infection material, e.g., bacterial cells, fungal spores, and ability to colonise the host plant.
- *The environment.* Each pathogen has specific environmental requirements (temperature, moisture/rainfall and humidity) for infection, colonisation, reproduction and spread. Because of the environmental differences across the country, disease incidence is often different. For example, there is a much higher risk of Septoria tritici (Septoria Leaf Blotch) in winter wheat grown in the west of the UK, whereas yellow rust is a greater risk in the east of England and Wales.
- *Farming activities.* Rotation, variety choice, time of sowing, seed rate, fertiliser use, irrigation and cultivations can influence the incidence and severity of disease.

6.2 Disease symptoms

Symptoms of many diseases are characteristic for that disease and crop. Often the disease can be identified without having to send them on to a plant health clinic for diagnosis using microscopy, culturing or even molecular diagnostic tests. Some diseases can be identified visibly, i.e., by observing the fungal growth or fruiting bodies, as with powdery mildew and the rusts. For other diseases it is the type of lesion that helps identification, as follows:

- *Yellowing or chlorosis.* Many of the viruses, such as beet virus yellows, cause foliar chlorosis (loss of chlorophyll) as do some mineral deficiencies. Where the yellowing is on the leaves, the shape of the infected area may help with identification.
- *Death of tissues/necrosis.* Many diseases can cause the death of the plant tissue (necrosis), including leaves and stems. Again, size and shape of the lesions are important. Leaf spots are well-defined lesions and sometimes the dead tissue falls away to leave a shot-hole effect e.g., advanced stages of Phoma leaf spot in oilseed rape. Leaf blotches tend to be of variable size. Blights can cause total death of leaves and stems and finally to all above-ground parts of the plant.
- *Abnormal growth.* Some diseases cause the infected tissue to enlarge either by an increase in cell numbers (hyperplasia) or size (hypertrophy) e.g., brassicas with club root or sugar beet with Rhizomania.
- *Cankers.* Typically associated with stems. Cankers are diseased collars (girdles) that may be raised, sunken, swollen or cracked (e.g., stem canker in oilseed rape).
- *Stunted growth.* Stunted growth is a common symptom associated with plant viruses e.g., barley yellow dwarf virus. However, root diseases, such as take-all in cereals can also result in stunted growth.
- *Wilting.* There are a number of bacterial and fungal diseases that cause wilting, normally late on in the development of the disease. Wilting is caused due to blockages in the vascular system of the plant leading to reduced uptake of water. An example is Verticillium wilt of potatoes.
- *Tissue disintegration.* This type of damage is associated with many of the root or foot rots and storage rots. The affected cells are broken down and release liquid (wet rot) or become dry and brittle (dry rot).

6.3 Some important types of plant pathogens

6.3.1 Fungi

Around 70% of all plant diseases are caused by fungi. Over 100,000 different species of fungi have been recorded, with most being invisible to the naked eye and approximately 10% being important as plant pathogens. Fungal diseases are given a common name plus an internationally agreed scientific name for the causal pathogen. As an understanding of fungal pathogens and their life cycles has developed, some names have been changed, see Appendix 3. Naming can get confusing as sometimes the original scientific name has been incorporated into the common name, then the scientific name changes as has happened with Septoria tritici in wheat. A typical fungal disease cycle is provided in Fig. 6.2.

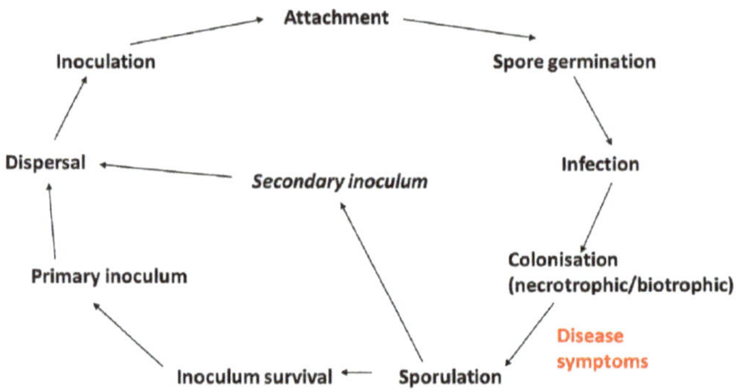

Figure 6.2 Outline of a disease cycle for fungal plant pathogens.

6.3.1.1 *Fungal infection*

A typical fungus is comprised of long thin filaments (made up of single cells) termed *hyphae*. Collectively, these are known as the *mycelium*. Fungi enter the host plant surface either by direct penetration of surface cells (main form of entry) or through wounds or natural openings. Not all fungi that enter plants produce disease symptoms. Sometimes the fungus is killed within the host plant, leaving a visible necrotic spotting or hypersensitive reaction. This reaction is seen in plants that possess resistance genes to the pathogen or genotype (strain) of the pathogen.

Once within a susceptible plant the fungus can grow and spread between cells. As previously discussed, biotrophic fungi produce extensions to the hyphae which enter the host cells and act as nutrient-absorbing organs (*haustoria*). In the case of most parasitic fungi, the mycelium is enclosed within the host (only the reproductive parts protruding), although some fungi are only attached to the surface of the host, e.g., powdery mildew. Sometimes, even after a fungus has entered the host, plant symptoms may not be visible at first. This lag time is called the *latent period* and once the symptoms have appeared it is too late to treat with a fungicide.

6.3.1.2 *Reproduction*

Fungi can reproduce simply through fragments of hyphae, but reproduction typically occurs through production of spores. Spores can be compared to the seeds in ordinary plants, but they are microscopic and occur in vast numbers. Asexual spores are usually important for the spread and increase of a population or epidemic. Sexual spores are often important for fungal survival and can be specialised for airborne dispersal. Some fungi produce tight masses of mycelium, known as *sclerotia*, as the host begins to senesce. Sclerotia allow fungi to persist in the soil until a suitable host becomes available.

6.3.1.3 Fungal survival and dispersal

Not all fungi have the same mechanisms for spread and survival; this knowledge will help in deciding on disease prevention and control methods.

- *The seed.* Some fungi are carried from one generation to the next by surviving on the seed coat or inside the seed itself, e.g., loose smut diseases of cereals.
- *The soil.* The spores or even resting bodies such as *sclerotia* drop off the host plant and remain in the soil until another susceptible host crop is grown in the field. A suitable rotation will go a long way to check these diseases such as those caused by *Sclerotinia* spp.
- *The air.* Spores of many foliar diseases are dispersed on air currents. Spores of some of the cereal rusts have been known to travel hundreds of kilometres in air currents.
- *Water.* Rain droplets are important for the dispersal of some fungal spores over short distances. Additionally, motile spores use moisture films on plants and in soil to reach their host e.g., the zoospores of *Phytophthora infestans* (late blight in potatoes).
- *Alternate hosts.* Some diseases have the ability to overwinter on alternative hosts as seen with some of the rust fungi.
- *Infected plant material.* Many fungal pathogens survive from one season to another either on plant debris (necrotrophs) or volunteer crops (biotrophs).

6.3.2 Oomycetes (water moulds)

Oomycetes also known as water moulds are frequently and incorrectly categorised as fungi. Agriculturally important examples include *Phytophthora infestans* that causes late blight in potatoes and *Peronospora pisi*, the causal pathogen of downy mildew in peas. Oomycetes are similar to fungi in that they produce hyphae and different types of spores (sexual and asexual). The spores of oomycetes can be motile e.g. zoospores and require moisture films (free water) for movement. Because they are related to organisms such as brown algae, they have cellulose in their cell walls. Specialised fungicides have been developed to target the synthesis of this molecule and are used to combat late blight.

6.3.3 Plasmodiophoromycetes (slime moulds)

The slime moulds are a distinct group of protozoan plant pathogens with the most well-known being *Plasmodiophora brassicae* (club root) which infects brassica crops such as oilseed rape. As well as producing motile zoospores, these pathogens use feeding structures, known as plasmodia, to extract nutrients from within plant cells. A number of slime moulds are vectors for important soil-borne viruses such as Rhizomania (Beet Necrotic Yellow Vein Virus) and Barley Yellow Mosaic Virus. Slime moulds frequently produce thick walled resting spores that enables long term survival of 10 or more years.

6.3.4 Viruses

The virus was discovered in the nineteenth century. It is a very small sub-cellular organism that cannot be seen using a normal light microscope; only by using an electron microscope can it be seen and identified. Serological and molecular techniques are now commonly used to aid identification. Viruses exist as particles consisting of nucleic acids (DNA or RNA), usually surrounded by a coat of protein. Some viruses can cause the host cell to produce various crystalline structures which can also be used to aid identification, as can shape of the outer virus coat.

Once a virus is in a host plant, it causes the host cell to produce more nucleic acids and proteins from which more virus particles are formed, which can then move around the plant in the phloem. All viruses are obligate parasites. They are not known to exist as saprophytes. The virus is generally present in every part of the infected plant except the seed, pollen and often the apical meristem. Therefore, if part of that plant, other than the seed, is propagated, the new plant is itself infected, e.g., the potato. The tuber is attached to the stem of the infected plant, and infection is carried forward when the tuber is planted as 'seed'.

In most plant virus diseases, the infection is transmitted from a diseased to a healthy plant by vectors; insects are the main vectors although nematodes and mites are known to transmit some virus diseases. As discussed previously, slime moulds can also be vectors, e.g., the mosaic viruses in cereals are transmitted by the soil-borne slime mould *Polymyxa graminis*; rhizomania in sugar beet is transmitted by *Polymyxa betae*. Viruses can overwinter on perennial infected crops that might not show any visible symptoms and on volunteer crops. One of the main methods of survival is in alternative host plants such as weeds. Viruses tend to have a fairly wide host range. The host species must also be suitable for the vector. The only way viruses can survive in the soil is in a living organism such as a nematode or fungus.

Viruses cause a variety of symptoms from plant yellowing to distortion of growth e.g., stunting. The name of a virus usually describes the symptoms and a main host plant such as barley yellow dwarf virus. Virus names are abbreviated using the first letter of each name, e.g., barley yellow dwarf virus is BYDV.

6.3.5 Bacteria

Bacteria are microscopic single cell organisms, only visible under a light microscope. They are of a variety of shapes, but those that cause plant diseases are all rod-shaped. About 12% of the identified bacteria are plant pathogens (c. 100 species). Like fungi, bacteria feed on both live and dead material. Although they are responsible for many diseases of humans and livestock, in arable crops in the UK they are usually of minor importance compared with fungi and viruses.

Bacteria overwinter in a similar way to fungi except that no resting spores or bodies are produced. They can survive in infected plants, seeds and tubers, in plant debris and, just a few, in the soil. They mainly enter plants through wounds or occasionally through openings like the stomata. Once inside the plant host, bacteria use pectinolytic enzymes to breakdown pectin, which is a major component of the cell walls. Bacteria, causing wet rots, often gaining entry into their host root crops after mechanical damage during harvest. Once in the host plant bacteria can reproduce themselves simply by the process of splitting into two (binary fission). Under favourable conditions this division can take place about every 30 min. Bacterial pathogens can be spread via soil movement, machinery, water and through infected seed.

6.4 Abiotic disorders

6.4.1 Lack of essential plant foods (mineral/nutrient deficiency)

When essential plant foods become unavailable to particular crops, deficiency symptoms will appear. These symptoms, such as chlorosis and necrosis, are characteristic for that nutrient and crop but can often look like a foliar disease (see Chapter 4). Distribution of the symptoms in the crop can be useful in deciding if it is likely to be a biotic disease causing the problem or a nutrient deficiency. Fungal diseases often start with a primary focus (diseased patch) and then spread out, whereas nutrient deficiency symptoms are more widespread or confined to certain distinct soil types or farm operations. Soil and or leaf tissue analysis can be used for diagnosing a nutrient deficiency. If a crop is suffering from a nutrient deficiency, it can make it more susceptible to attack by some pests and pathogens.

6.4.2 Physiological diseases (stress)

Physiological diseases are often triggered by adverse environmental conditions which can upset the normal physiological processes of the plant. Usually this is only temporary, but there may be occasions when the effect is more permanent.

- Temporary conditions, e.g., a high-water table in the early spring. This will cause yellowing of the cereal plant as its root activity is restricted, considerably reducing its oxygen and plant food intake. When the water table falls, the plant is able to grow normally once more, assuming a healthy green colour.
- Permanent conditions, e.g., where the soil has become compacted the root activity of the plant can be restricted. This will result in poor stunted growth with the plant far more vulnerable to pest and disease attack. The yields will be reduced. Periods of very dry conditions followed by wet weather can cause root cracking and splitting in crops such as carrots and potatoes; it can also cause secondary tuber production in potatoes. Hail damage can often be confused with a disease. It can cause leaf and flower spotting and even damage the flower parts. In cereals, hail and heavy rain at flowering have been known to increase the number of blind grain sites in the ear.

6.4.3 Herbicide damage

Herbicide drift or application of a spray contaminated with a herbicide not recommended for that crop can occasionally cause crop damage. Symptoms vary depending on the chemical involved. Usually it is very obvious that the problem is not caused by a disease because of distribution in the field.

6.5 The control of plant diseases

Before deciding on control measures it is important to know what is causing the disease. Once the pathogen is identified, then knowledge of its life cycle and factors affecting the outbreak and spread (*epidemiology*) will help decide on the most appropriate method of control. Depending on the disease (see Table 6.2 at the end of this chapter) there are many non-chemical methods of control that will help reduce the risk before any agrochemicals are required. With the increase in number of cases of fungicide resistance, it is more important than ever that alternative control methods are included in the programme (Integrated Disease Management or IDM) in order to maintain the effectiveness of currently available chemicals.

6.5.1 Non-chemical methods of control

- *Crop rotation*

  A good crop rotation, avoiding growing too many years of the same crop or related species, can help to avoid an accumulation of disease inoculum, particularly those that are soil borne. In many cases the organism cannot exist except when living on the host, so if the host plant is not present in the field then the pathogen will decline. Rotation is most effective with pathogens that are weakly saprophytic. For example, the organism that causes take-all (*Gaeumannomyces graminis* var. *tritici*) of wheat and barley only persists for a year without these cereal hosts. Some pathogens take years to decline as they have resting spores in the soil waiting for the next susceptible crop to be planted again, e.g., club root that affects plants in the brassica family.
- *Soil fertility*

  A crop under stress from low nutrient status can be more prone to disease attack e.g., winter wheat with take-all. Alternatively, overly lush growth, associated with high nitrogen applications, can also encourage diseases such as powdery mildew and rusts. Early application of nitrogen in winter cereals can cause increased infection from foliar diseases. The acidity or alkalinity of the soil can affect both the availability of nutrients as well as some diseases. Maintaining a high soil pH of seven or above is very effective at reducing incidence of club root in brassicas. In potatoes, common scab is usually worse if the pH is high so liming a soil should always be done after growing this crop.
- *Crop hygiene*

  Diseases should be discouraged by avoiding sources of infection on the farm, e.g., late blight from old potato clamps, and virus yellows where sugar beet was stored prior to being sent to the factory. Good stubble cleaning will minimise certain cereal diseases being carried over from one cereal crop to the next year's cereal crop. Some biotrophs use weeds as alternative hosts. By controlling the weeds, the disease can be reduced, e.g., brassica weeds such as charlock are hosts to the organism responsible for club root.

- *Seedbed conditions and cultivations*

  Puffy seedbeds can increase the risk of take-all in cereals, whereas over-consolidation or compaction can lead to poor growth and other diseases. Wheat grown in land cultivated with minimal tillage has been shown to have lower incidences of eyespot than in ploughed land. On the other hand, *Fusarium graminearum,* causal pathogen of fusarium ear blight and major producer of mycotoxins, can be a greater risk when the stubbles of wheat or maize are not buried through inversion tillage. This is because *Fusarium* spp. are excellent saprophytes - colonisers of crop debris.
- *Clean seed*

  Seed Certification Schemes ensure that seed lots meet strict standards for presence of seed-borne diseases. These schemes have been very successful at reducing the incidence of some diseases such as bunt or stinking smut in cereals. Home-saved seed should be tested to avoid poor crop establishment and growth due to seed-borne disease.

  With potatoes it is essential to obtain clean 'seed', free from virus. In some districts where aphids are very prevalent, potato seed may have to be bought each year. Under the Seed Certification Scheme, it is mandatory to have the field inspected and the seed virus tested.
- *Resistant varieties*

  While the breeding of resistant varieties has advanced significantly, it is not by any means simple, and in most cases it takes at least 10 years to develop a new variety. For some years, plant breeders concentrated on what is called single or major gene resistance. However, with few exceptions, this resistance is overcome by the development of new races/genotypes of fungi to which the gene is not resistant e.g., yellow rust of wheat.

  Some breeding programmes are now concentrating on partial (polygenes) or 'field resistance', which means that a variety has the characteristics to limit infection from a wide range of races with little lowering of yield. Varietal resistance is the main method of control for slime mould-transmitted viruses such as barley yellow mosaic virus and Rhizomania.
- *Variety diversification*

  The risk in any year of a serious disease infection can be reduced on farms growing cereals using a range of varieties that have different disease resistance ratings and resistance to different races of the disease. NIAB monitors races of yellow rust, brown rust and powdery mildew in an ongoing survey known as the UK Cereal Pathogen Virulence Survey. The aim of the survey is to make growers and the cereals industry aware of current and emerging races of pathogens, which might threaten the currently available resistant varieties. Previously, the scheme produced tables showing variety diversification groups for yellow rust. Unfortunately, these are not currently available due to the complexity of the current yellow rust races.
- *Variety mixtures (blends)*

  This is a natural extension of diversification in that varieties from different diversification groups are grown together in the same field. In this way a disease carrying spore from a susceptible variety, but within a blend of varieties (two or three varieties) making up the crop, has less chance of successfully infecting a neighbouring plant than in a pure crop. Yields of blended crops can be more reliable than pure crops, and this may be achieved with the use of fewer fungicides. The problem comes when selling a mixed crop (different end market specifications), and in practice usually only grass seed is sold as mixtures. However, recent interest in regenerative agriculture has resulted in some seed merchants offering blends of wheat in their product ranges.
- *Time of sowing*

  Time of drilling can affect disease progression in many crops, partly due to the amount of inoculum present and environmental conditions that favour the disease. Early-drilled winter cereals are more likely to be infected by eyespot, take-all and some foliar diseases, while early-drilled spring barley is more susceptible to *Rhynchosporium*, but late drilled barley is more susceptible to powdery mildew. The pathogen *Rhizoctonia solani*, causing stem and stolon canker in potatoes, causes more damage in young immature plants. Consequently, a later planting date, when soil conditions are warmer and more conducive to faster emergence results in milder disease symptoms.

- *Crop density*

  Many rain-splash diseases, and those that require a high humidity, are more common where there is a high plant population compared with a more open crop with a low population. For instance, lush and thick crops of winter wheat are far more susceptible to powdery mildew.

- *Control of insects/virus vectors*

  Some insects are important vectors of viruses and need to be managed. For example, the peach potato aphid (*Myzus persicae*) that transmits virus yellows in sugar beet (BYV and BMYV) can be reduced by using tolerant varieties, cleaning up old sugar beet clamps and using a limited selection of insecticide sprays if the aphid threshold is met (one aphid per four plants up to the sixteen-leaf stage). Other virus vectors such as slime moulds cannot be targeted by chemical controls, and use of resistant varieties is the main option for arable farms.

- *Legislation*

  A number of countries have introduced legislation to ensure that some diseases do not spread and become a serious threat. In the UK potato wart disease, brown rot and ring rot in potatoes are three examples of notifiable diseases which are subject to quarantine measures. Potato wart disease has been controlled by these statutory measures and some contaminated fields where potatoes were prohibited from being grown are now being found to be free from the disease and are being de-scheduled.

- *Irrigation*

  Irrigation can often encourage disease development by altering the crop environment so that it is more conducive for disease spread and survival; one exception is in potatoes where irrigation applied as the tubers are initiated can be very effective at controlling common scab.

- *Biological control*

  Biological control is the use of other organisms to control the disease organisms. Very few treatments are currently available commercially as often in the open field results can be very variable. One successful treatment is the fungus *Coniothyrium minitans* for control of the soil-borne sclerotia of *Sclerotinia sclerotiorum*.

- *Biostimulants and biofumigation*

  Biostimulants encompass a wide variety of microbial (e.g., plant growth promoting bacteria or mycorrhiza) and non-microbial (seaweed extracts or chitin/chitosan) products that are used to stimulate plant growth by increasing nutrient uptake or increasing resilience to abiotic stresses. Some of these products may also have potential to reduce infections caused by pathogens by stimulating plant defence systems or through other means. For instance, applications of chitin or chitosan can elevate soilborne microorganisms which break down chitin via secretion of chitinase. Since chitin is found in the wall of fungi and pests such as insects and nematodes, chitin/chitosan applications can lead to reductions in the populations of these organisms.

  Biofumigation involves the use of brassica cover crops such as Indian mustard to suppress soil-borne pests and pathogens. These crops are grown during the summer months and are mulched in during the early autumn to release volatile, biocidal compounds.

6.5.2 Chemical control—fungicides

Over the last 50 years there has been a large increase in both the number of fungicides available and their subsequent use on crops. They are used in the following ways:

- The dressing of seed with a fungicide; this is carried out to prevent certain soil-borne and seed-borne diseases. Various fungicides can be used, depending upon the disease to be controlled and the crop.
- Foliar application to the plant. Different treatment programmes involving the use of fungicides are now considered as an essential part of many crop production programmes.

Fungicides are typically used when a specific disease has developed to a point (the economic threshold) which will cause a loss of yield that will justify the cost of treatment and application. There are now many established thresholds for application of fungicides. Often the rate of the fungicide applied will be adjusted depending on the disease risk and varietal resistance. This is called an appropriate dose rate. Some computerised models have been or are being developed to aid decision making.

With potato late blight, treatment needs to be applied before symptoms are seen and is based on late blight warnings known as The Hutton Criteria. These warnings rely on defined periods of temperature (>10°C) and humidity (>90%) recorded over a 2-day period.

6.5.2.1 Protectant fungicides

The first chemicals developed for control of fungal diseases were inorganic compounds such as those based on sulphur and copper. These chemicals do not move in the crop plant (non-systemic), they simply protect the crop plant from pathogen infection by providing a protective layer. They are appropriately referred to as protectant fungicides or protectants. Good crop coverage is essential for protectants to be effective, hence why they are often used earlier in the season when crops are less developed. Protectants may be multisite inhibitors meaning that they bind to a number of biochemical sites within a fungus. Some of the first fungicides produced after inorganic compounds, such as the dithiocarbamates, have very similar characteristics, e.g., they are protectant and multi-site fungicides. Although these chemicals are not as effective as some of the newer systemic compounds they still have some uses today, particularly in programmes where there is a high risk of disease resistance, e.g., control of potato blight.

6.5.2.2 Systemic fungicides

Systemic fungicides have been developed since the 1960s and are now the most commonly used fungicides. On entry into the crop plant, they can move to a certain extent within the crop, usually in the xylem vessels, to the site of infection. These fungicides tend to affect a single biochemical pathway within the pathogen and are called single site inhibitors.

Table 6.1 Summary of some commonly used fungicide groups.

Family group (based on mode of action (MOA))	Abbreviated family group name	Chemical family	Example active ingredient (ai)
Demethylation inhibitors	DMI	Triazole	Prothioconazole
Quinone outside inhibitors	QoI	Strobilurin (strob)	Azoxystrobin
Succinate-dehydrogenase inhibitors	SDHI	Carboxamide	Fluxapyroxad
Methyl benzimidazole carbamates	MBC	Benzimidazole	Thiabendazole

Many systemic fungicides can be applied after the initial infection period, before symptoms appear (within the latent period); these treatments are called curative or eradicant. Eradicant fungicides have the ability to prevent further colonisation and development by fungal pathogens and then protect the plant for a certain time after application. Persistency and eradicative activity vary between chemicals.

Fungicides are grouped together according to their mode of action and chemical structure. Some frequently used fungicides are included in Table 6.1, which also includes commonly used abbreviations. It is important to know the family group of the fungicide when considering fungicide programmes in order to reduce risk of fungicide resistance (see the individual crop chapters for further details on disease control programmes). In recent years, widely used fungicides such as chlorothalonil, pencycuron, mancozeb, epoxiconazole and thiram have been banned from use in the UK. Concerns over environment, pollinators and human health mean that further fungicide bans or restrictions are highly likely.

6.6 Fungicide resistance

Disease resistance to fungicides is now widespread. It is a problem with the systemic products that act on one site only of the fungus. When a fungicide controls a fungal disease effectively, the fungus is 'sensitive' to the chemical. However, other strains of the fungus can and do occur over a period of time, and some of these may be resistant ('insensitive' or 'tolerant') to the fungicide which means that the disease is then not controlled adequately. In some cases, once there is resistance it is total and the fungicide is not effective. An example of this type of resistance (single step) is seen in the control of Septoria leaf blotch with strobilurin fungicides. In other cases, there is a shift in the sensitivity of the fungus population to the fungicide but there is still some control. This has been found with control of mildew in cereals using many of the triazole fungicides, it is called multi-step resistance.

Resistance builds up through the survival and spread of the resistant strains and it is speeded up by repeated application of the same fungicide treatment or fungicides with the same mode of action. There is an increased risk of this happening with fungicides which are site-specific in the fungus compared with multi-site fungicides. Likelihood of resistance developing is also affected by the biology of the target disease.

There are several ways of avoiding a build-up of resistance by a fungus or reducing the risk. It is very important that farmers follow these guidelines in order that currently available chemicals remain effective as few new products are coming on the market.

- Combine application of fungicides with non-chemical methods to reduce disease risk.
- Avoid growing large areas of very susceptible varieties in areas where disease incidence is usually high.
- Where possible, use fungicides with different modes of action (i.e., from different family groups) especially when more than one fungicide application is to be used on the same crop.
- Use approved tank mixtures of fungicides with different modes of action, rather than always relying on single fungicides.
- Apply fungicides only when necessary; use disease forecasts and thresholds to avoid unnecessary treatment. Walk and monitor crops regularly.

Unsatisfactory disease control following the use of fungicides is not always due to fungicide resistance; there are several other reasons, the main ones being wrong timing, wrong chemical, use of too low a dose rate and poor application.

6.7 Management of specific diseases of field crops

Please refer to Table 6.2 for a guide to some of the main diseases affecting field crops in the UK and options for their management. Note that the table is intended to provide an overview and further reading may be required (see suggestions for further reading at the end of the chapter).

Table 6.2 Major plant diseases affecting UK field crops and their management options.

Crop attacked	Disease	Causal agent	Symptoms of attack	Life-cycle	Methods of control
Cereals	(1) **Bunt, covered** or **stinking smut** of wheat (2) **Covered smut** of barley. **Leaf spot** of oats (3) **Covered** and **loose smut** of oats	Fungus	Brown or black spore bodies with distinct smell (decaying fish) replace the grain contents.	Infected grain is planted; seed and fungus germinate together and thus young shoots become infected. The spores are released when the skin of the grain breaks, and so combining contaminates healthy grain. With use of seed dressings and certified seed, diseases are now rarely seen.	(1) Seed dressing, virtually all the currently available fungicide seed dressings control these diseases (2) Seed tests (treat if more than one spore per seed)
	(1) **Leaf stripe** of barley (2) **Leaf stripe** of oats	Fungus	The first leaves have narrow yellow or brown streaks. Subsequently, brown spots appear on the upper leaves.	Infected grain is planted; seed and fungi germinate together and thus young shoots are infected. From the secondary infection, spores are carried to developing grain.	(1) Seed dressing, virtually all the currently available fungicide seed dressings control these diseases (2) Seed tests (treat if more than 2% of seed is infected—leaf strip of barley)
	(1) **Loose smut** of wheat (2) **Loose smut** of barley	Fungus	Infected ears become a mass of black spores. They do not remain enclosed within the grain as with the covered smuts.	Similar to the covered smuts, but the fungus develops within the grain. The spores are dispersed by the wind to affect healthy ears.	(1) Sow certified seed (less than 0.5% or less than 0.2% at a higher voluntary standard) (2) The seed can be dressed with a broad-spectrum fungicide

	Yellow rust	Fungus	Yellow-coloured pustules in parallel lines (stripes) on the leaves, spreading in some cases to the stems and ears. In a severe attack, the foliage withers and shrivelled grain results.	The fungus mainly attacks wheat though there are distinct species that attack barley and oats. Initial primary foci may be found in early spring. From the pustules, spores are carried by the wind to infect healthy plants. During winter, spores are normally dormant on autumn-sown crops. Cool (c. 15°C), humid conditions favour disease. Temperatures above 25°C inhibit growth.	**(1)** Resistant varieties, although new races appear against which these varieties can soon have reduced resistance **(2)** Fungicides, e.g., a triazole, strobilurin or SDHI **(3)** Control volunteer plants between harvest and emergence of next crop to reduce carry-over of inoculum
	Powdery mildew	Fungus	On winter cereals, grey-white or light brown fluffy pustules on lower leaves in February. Infection spreads to other leaves and plants. Disease common between May and August. Early infected leaves go yellow. Toward the end of season black spore cases (Cleistothecia)form among the pustules.	From self-sown cereals in stubble, winter and spring cereals can be infected. Specific species attack wheat, barley, oats and rye. Warm, humid (but not wet) weather favours disease. Dense crops and those where high rates of N have been used tend to encourage disease development.	**(1)** Clean-up old stubble **(2)** Resistant varieties **(3)** Fungicides available, although field resistance to some fungicides is now common. Better control with some of available fungicides when applied early in disease development **(4)** Optimum N and seed rates

Continued

Table 6.2 Major plant diseases affecting UK field crops and their management options.—cont'd

Crop attacked	Disease	Causal agent	Symptoms of attack	Life-cycle	Methods of control
	Rhynchosporium—leaf blotch of barley	Fungus	Initially symptoms look like a water soaked area on the leaf. When fully formed diamond shaped blotches (light blue/ green/grey with dark brown margins) are seen on the leaves. As disease progresses, blotches coalesce.	A disease of barley, rye and triticale. Fungus over-winters on self-sown barley plants and on crop debris. From here spores are carried to planted barley crops. Disease is spread by rain splash. Early sown crops most at risk.	**(1)** Clean stubble of all self-sown barley plants **(2)** Many fungicides are available including some triazoles, strobilurins and SDHIs **(3)** Select resistant varieties in high risk areas
	Net blotch of barley	Fungus	Short brown necrotic stripes on older leaves; on younger leaves in addition to the striping, irregular-shaped dark brown blotches 'spots' or 'netting' symptoms. Can spread to ears.	Can be seed-borne but mainly spread from previously infected crop or volunteers. Disease encouraged by cool, wet weather.	**(1)** Seed dressing **(2)** Crop hygiene to clear stubble of previously infected crop **(3)** Fungicides, e.g., SDHI mixtures **(4)** Resistant varieties
	Halo spot of barley	Fungus	Small pale oval lesions with a dark brown border, scattered across the leaf. Black fruiting bodies (pycnidia) found in lesions.	Fungus survives in seed, plant debris and on volunteer cereals. Cool moist conditions favour the disease only found in south west England.	**(1)** Many of currently available fungicide seed treatments and foliar sprays can help suppress the disease

	Brown rust of barley and wheat	Fungus	Numerous, very small individual orange-brown pustules on the leaves. Normally does not develop significantly until the summer. A severe attack causes shrivelled grain.	Different species in wheat and barley. The resting spores over-winter on volunteers. From here the pustules are airborne to infect healthy plants. Encouraged by warm (15–20°C), humid weather; spores can be spread by wind.	**(1)** Crop hygiene to clear stubble of volunteer plants **(2)** Resistant varieties **(3)** Many systemic fungicides, e.g., a triazole, strobilurin or SDHI mixture
	Crown rust of oats	Fungus	Orange-coloured pustules spread mainly on leaf blade. Later in season black pustules are produced. Severe attack prior to, and including, milk-ripe stage, causes shrivelled grain.	Only affects oats, a distinct species affects other grasses. Spores are air borne from over-wintering volunteer plants, and winter oats, to the spring crop. Disease encouraged by warm (20–25°C), humid conditions.	**(1)** Crop hygiene to clear stubble of volunteer plants **(2)** Keep winter and spring oat crops as far apart as possible **(3)** Some varieties show reasonable resistance **(4)** Some of the triazoles, stro-bilurins and SDHI mixtures give good control
	Ramularia leaf spot of barley	Fungus	Brown rectangular spotting on upper side of leaf that can be confused with net blotch. Symptoms also seen on awns and stems.	A seed borne disease. Symptoms often appearing when crop stressed. Spores spread following wet weather. A disease mainly found in the north of the UK.	**(1)** Try to ensure that the crop is not stressed **(2)** In high risk areas, grow more resistant varieties **(3)** Apply a fungicide, e.g., triazole/SDHI mixture at booting stage

Continued

Table 6.2 Major plant diseases affecting UK field crops and their management options.—cont'd

Crop attacked	Disease	Causal agent	Symptoms of attack	Life-cycle	Methods of control
	Septoria tritici—leaf blotch of wheat	Fungus	Mainly a foliar disease. Oval, bleached/grey discoloured blotches of varying sizes and shapes (on which appear minute black dots or fruiting bodies (pycnidia)) seen from late autumn to early summer. Infection of the ear is rare.	Leaf spores are liberated in wet weather, and they over-winter on volunteer crops and then transfer to winter cereals. Symptoms can develop 3–4 weeks after the infection period. Optimum temperatures for the disease to develop are between 15 and 20°C. Early drilled crops most at risk.	**(1)** Crop hygiene to clear stubble of volunteer plants **(2)** Clean seed **(3)** Partially resistant varieties **(4)** Rotation and time of drilling (early drilled crops more at risk) **(5)** Many systemic fungicides, e.g., triazole and SDHI mixtures typically applied at GS30-31, 32 and 39
	Septoria nodorum—leaf and **glume blotch** of wheat	Fungus	Disease (other than seedling blight) does not develop until midsummer. Leaf lesions start as oval yellow blotches. Fruiting bodies are difficult to see without magnification. Glume blotch becomes prominent in July and August. Irregular, chocolate spots or blotches on glumes, beginning at the tips; later ears become blackened with secondary infection.	Glume blotch has a similar life cycle as Septoria tritici but the fungus can also be carried on the seed. The disease develops rapidly in warm (20–27°C) and humid weather. Symptoms develop 10–14 days after an infection period.	Control as for septoria tritici except fungicides applied later at the emergence of the ear (ear wash)

	Tan spot of wheat	Fungus	Causes brown lesions very similar to Septoria nodorum. Spots tend to be oval with a yellow halo and a dark centre.	Tan spot is a trash borne disease. High temperatures (20–28°C) and wet weather encourage development and spread. Mainly a disease of wheat though can affect barley and rye.	**(1)** Plough stubble **(2)** Control volunteers **(3)** Fungicide treatment, some triazoles are effective
	Barley yellow dwarf virus (BYDV)	Virus	Stunted plants in patches or scattered as single plants. Poor root development. Affected leaves in oats are red, yellow in barley and bronze in wheat. Late heading and reduced yield. Yield losses greatest from autumn infection.	Cereal aphid vectors carry the virus, e.g., bird cherry-oat aphid and grain aphid. Volunteers and grasses act as source of infection. Early drilled cereals after grass are most susceptible. Grain aphids less affected by frosts than bird cherry-oat aphids.	**(1)** Spray off previous crop/green stubble and leave at least 5 weeks between primary cultivations and sowing this reduces aphid carry-over **(2)** Apply an insecticide end October/early November. High–risk areas may require earlier treatment in warm autumns. Use of T-sum (day degrees) calculation to determine spray timing **(3)** Do not drill too early in high risk areas **(4)** Oats most susceptible, wheat the least **(5)** Resistant variety available for wheat

Continued

Table 6.2 Major plant diseases affecting UK field crops and their management options.—cont'd

Crop attacked	Disease	Causal agent	Symptoms of attack	Life-cycle	Methods of control
	Soil-borne mosaic viruses (barley yellow mosaic virus (BaYMV), barley mild mosaic virus (BaMMV), oat mosaic virus (OMV), oat golden stripe virus (OGSV) and **soil-borne wheat mosaic virus (SBWMV))**	Virus	Appears in patches in field. Pale green streaks later turning brown, particularly at leaf tip; leaves tend to roll inwards, remain erect to give plant a spiky appearance. Symptoms can disappear during warmer weather but the plants are still stunted and late to mature.	Soil-borne slime mould. *Polymyxa graminis* is the vector. Disease encouraged by hard winter. Symptoms appear from January. Found only in autumn sown cereals. May survive in soil for many years (up to 25).	**(1)** No chemical control **(2)** On affected fields grow resistant varieties **(3)** Reduce the spread of contaminated soil around the farm **(4)** Widen the interval between barley crops to reduce the risk of BaYMV and BaMMV
	Eyespot (wheat, barley, oats, triticale and rye)	Fungus	Often starts as a brown lesion at base of stem later developing into eye shaped lesions on stem about 75 mm above ground. Edge of the lesion is diffuse (not well defined). Infected stems have grey mycelium inside stem; can cause crop to lodge. White-heads present at harvest in severe cases. Two strains of eyespot present in the UK.	The fungus can remain in the soil, on old stubble and some species of grasses for several years. It usually attacks susceptible crops at the earlier stages (late tillering/early stem extension). Early drilled crops and those sown using minimal tillage are most susceptible. Cool moist conditions favour this disease. Disease spread by rain splash.	**(1)** Use risk assessment key based on region, soil type, previous crop, cultivations and sowing date **(2)** Avoid growing continuous cereals/have a 2-year break from cereal crops **(3)** Sow varieties with good resistance in high risk fields **(4)** Monitor crop for eyespot in spring **(5)** Systemic fungicides, such as triazole/SDHI and triazole/strobilurin mixtures are very effective when applied at GS32

	Sharp eyespot (all cereals)	Fungus	Lesions are irregular and more sharply defined than eyespot. Brown/purple border followed by cream-coloured area and brownish centre. Lesions more numerous and occur further up stem than eyespot. Little penetration of the stem, although the first layer may be shredded. Can cause lodging, white-heads and shrivelled grain but not generally considered to be a threat to yield.	The fungus is soil-borne but is also found on plant debris. Tends to be more severe on light soils, with a lower pH and on early drilled winter crops. Mainly a problem in winter wheat and barley; though oats and rye can be affected.	**(1)** Dispose of crop residues **(2)** Delay drilling winter cereals **(3)** Some of the eyespot fungicides will give some control
	Fusarium—brown foot rot and **ear blight** (all cereals)	Fungus	**Brown foot rot**: Emerging seedlings can be killed. Undefined brown discolouration at base of tillers and lower leaf sheaths, dark brown or blackened nodes. Interior of stem and or grain shows pink fungal growth with some species. Premature ripening and 'whiteheads' or 'blind' ears. **Fusarium ear blight:** Water soaked lesions and partial bleaching of the spikelets. Ears with pink and shrivelled grain. Some species can cause the production of mycotoxins in the grain.	There are a number of *Fusarium* spp. but infection is either seed-borne or from previously infected stubble and crop remains. Wet weather during flowering increases risk of Fusarium development. Growing maize in the rotation also increases the risk.	**(1)** Seed treatment using e.g., SDHI or triazole **(2)** Foliar fungicides for Fusarium ear blight e.g., triazoles **(3)** Plough maize and winter wheat stubbles **(4)** Timely harvest (late harvested crops at greater risk from mycotoxin contamination) **(5)** Varietal resistance (Fusarium ear blight)

Continued

Table 6.2 Major plant diseases affecting UK field crops and their management options.—cont'd

Crop attacked	Disease	Causal agent	Symptoms of attack	Life-cycle	Methods of control
	Take-all	Fungus	Black discolouration at base of stem and roots. Infected plants ripen prematurely, in severe cases the ears are bleached and contain little or no grain (whiteheads).	Wheat is most affected though barley, rye and triticale can be infected. Oats are immune. The fungus survives in the soil in root and stubble residues and on rhizomatous grasses. Third wheat crops in a rotation are usually most affected but on light, acid soils in wet areas the second wheat crop can also be seriously infected After 4 years of continuous wheat growing, infection appears to lessen due to take-all decline.	**(1)** Rotation (1-year break from susceptible cereals) **(2)** Extra nitrogen in February helps the growth of new roots. Ensure adequate potash **(3)** Ensure good drainage **(4)** Direct drilling appears to lessen the intensity of the disease as does a consolidated seedbed **(5)** Drill crops at risk in October rather than September **(6)** Limited fungicide seed dressings can help reduce take-all **(7)** Control grass weeds such as barren brome

	Pink snow mould (mainly of wheat and rye)	Fungus	Patchy crop in autumn. Stunted seedlings occasionally with white/pink mycelium at base. After snow has thawed withered plants in patches temporarily covered by white-pink mould. Thereafter, infected plants stunted, weak root system and shrivelled grain.	A seed-borne fungus, but contamination of crop is also possible from the old stubble and other plant debris. Fungus is particularly favoured by low temperatures. Mainly a disease of winter barley.	**(1)** Good stubble hygiene **(2)** Clean seed **(3)** Seed treatment
	Snow rot (Typhula rot) mainly of winter barley	Fungus	Thin, poorly tillered crop; plants yellowing and withered in patches. Old leaves often covered by white and pink mycelium and brown resting spores. Young leaves standing erect but eventually yellowed. Weak root system.	Soil-borne fungus which can remain dormant as sclerotia for years; when active, infects emerging cereal plants, developing rapidly in dark and humid conditions, i.e., under snow. Continuous winter barley increases the risk of this disease.	**(1)** Sow early in the autumn **(2)** Some varietal resistance **(3)** Seed treatment **(4)** Foliar spray with azole fungicide in the autumn
	Ergot	Fungus	Hard black curved bodies up to 20 mm long replacing the grain, and protruding from the affected spikelet. Although disease has little effect on yield, ergot is poisonous to mammals (but it does contain medicinal properties).	Ergots fall to ground and remain until next summer when they germinate and produce short stems with globular heads containing the spores which are then air-borne to the cereal flowers, and certain grasses, depending on the species. Open-flowering cereals most affected, e.g., rye. Infection in oats is rare.	**(1)** Crop rotation (non-cereal crops) **(2)** Control of grass weeds (especially black-grass) **(3)** Sow certified seed (less than 3 pieces of ergot per 500g grain or 1 piece per kilogramme for a higher voluntary standard)

Continued

Table 6.2 Major plant diseases affecting UK field crops and their management options.—cont'd

Crop attacked	Disease	Causal agent	Symptoms of attack	Life-cycle	Methods of control
	Manganese deficiency of cereals	Manganese deficiency	Yellowing on leaf veins followed by development of brown lesions. With oats the spots enlarge and can extend across leaf. Thus the leaf can bend right over in the middle. Older leaves wither and die. Can lead to shrivelled grains.	Manganese deficiency associated with high pH soils.	**(1)** Apply manganese sulphate, treatment may need repeating **(2)** Analyse plant material to confirm deficiency
Oilseed rape	**Phoma—stem canker** and **leaf spot**	Fungus	Beige coloured circular spots with distinct brown margin (0.5–1.0 cm diam.) on leaves and clear black fruiting bodies (pycnidia) in the lesions; fungus migrates through leaf petiole to the stem base to produce a brownish/black canker. Stem splits and rots causing lodging; rapid stem elongation and premature ripening.	Airborne spores produced from the stubble carry infection to young crops in the vicinity. Fungus spreads from the leaves to the stem where cankers develop. The fungus can affect all parts of the plant including the seed. Leaf symptoms seen from October to April.	**(1)** Destroy/bury stubble debris soon after harvest **(2)** Rotation—avoid close cropping with oilseed rape (2-year break needed) **(3)** Grow resistant varieties **(4)** Treat with a fungicide when 10%–20% of plants have leaf spot in the autumn and or spring
	Alternaria—dark leaf and **pod spot**	Fungus	Circular small brown to black leaf spots with zonal pattern (target spots) sometimes coalescing on leaves. Small, black, penetrating spots on pods. Leaf spots cause premature ripening and loss of seed.	Seed-borne disease although spores can be carried through the air from other infected brassica crops. Favoured by warm, humid conditions.	**(1)** No resistant varieties **(2)** Stubble hygiene **(3)** Foliar treatment for other diseases will aid control of Alternaria—otherwise treat mid to late flowering if necessary

	Light leaf spot	Fungus	From November onwards, light green/bleached lesions surrounded by small white spore clusters. Lesions become crusty causing the leaves to curl. Infections occur on the stem but are not penetrating. Infections of flowers and pods can lead to abortion. Plant population and seed yield can be severely affected.	A trash-borne fungus. Disease is spread from plant to plant by rain splash during wet weather. Flower buds can become infected and killed at early stem extension stage.	**(1)** Dispose of infected crop residues **(2)** Spray with a fungicide at first sign of the disease in the autumn and in the spring if 15% of plants are affected at stem extension **(3)** Use resistant varieties
	Stem rot—sclerotinia	Fungus	Bleached stem lesions which contain black resting bodies (sclerotia).	Sclerotia left in soil after harvest can survive for 7–10 years. They germinate in spring and produce spores which infect susceptible crops during flowering. Encouraged by warm (c. 20°C), wet weather.	**(1)** Wide rotation of susceptible crops (E.g. peas/beans, carrots, potatoes, lettuce) **(2)** Fungicides such as some triazoles and strobilurins—at early flowering
	Club root	Slime mould	See club root in brassicas below.		
	Verticillium wilt	Fungus	Initially yellow/brown stripes on stem then causes pods to turn pale green and to ripen prematurely. Mainly found in eastern and southern counties of England.	Soil-borne sclerotia germinate and fungus invades susceptible host roots and spreads to xylem vessels. Microsclerotia are then produced on outside of vascular system.	**(1)** Widen rotation (at least 4 years between susceptible crops) **(2)** Plant clean seed **(3)** Sow resistant varieties
	Turnip yellows virus TuYV	Virus	Leaf margins turn red. Reduces crop growth and subsequently reductions in pod numbers and oil content. Wide range of plant species affected.	Virus spread by peach potato aphid. Mild winters favour survival of aphid and disease spread. TuYV is a persistent virus.	**(1)** Foliar insecticide treatment if large number of aphids found in the autumn **(2)** Resistant varieties

Continued

Table 6.2 Major plant diseases affecting UK field crops and their management options.—cont'd

Crop attacked	Disease	Causal agent	Symptoms of attack	Life-cycle	Methods of control
Linseed	**Alternaria**	Fungus	Damage affects seedlings as they emerge. Brick red lesions are found on stems and roots. Can also affect mature plant.	A seed-borne disease. Can spread up mature plant during periods of wet weather.	**(1)** Some evidence of varietal resistance **(2)** Foliar application of a number of fungicides have been effective against late infection
	Botrytis—grey mould	Fungus	Attacks leaves, stems and seed capsules. Reddish browning on stem base. Plants then become covered in grey mould.	A seed-borne disease encouraged by warm, moist conditions.	**(1)** Sow certified seed **(2)** Fungicide spray for spring linseed **(3)** Low nitrogen rates and plant populations tend to reduce problem
	Pasmo	Fungus	Grey/black spots with black pycnidia found on leaves and stems.	A disease of winter linseed. Can be seed borne or spread from plant debris.	**(1)** Sow clean seed **(2)** Some triazole fungicides are effective when applied at mid-flowering
Peas	**Downy mildew**	Oomycete (water mould)	Infection on young seedlings can cause plant death. A secondary infection shows as isolated yellow/green spots on the upper surfaces of the leaves with greyish-white mycelium on underside. Can considerably reduce yield with fewer or no seeds per pod.	The disease is soil and seed-borne from where spores infect the growing point of seedlings. A secondary infection can follow, with spores spreading by air currents and rain splash to the developing foliage of other plants. Cool moist conditions favour this disease.	**(1)** Widen rotation **(2)** Seed treatment if growing susceptible varieties **(3)** Some varieties are more tolerant **(4)** Foliar fungicides give limited control

	Ascochyta—leaf and **pod spot**	Fungus	Brown sunken spotting on stems, leaves and pods. The lesion is usually surrounded by a dark brown margin. Can cause seedling loss. Peas can become stained.	Disease is mainly seed-borne. Infected seed germinates and lesions develop on first leaves from where spores spread to rest of foliage, including the pods, and to other plants by air currents and rain splash.	**(1)** Sow healthy seed **(2)** Use a fungicide seed dressing
	Botrytis—grey mould	Fungus	Grey mould develops where petals have fallen on to pods and leaves.	Fungus grows on many host plants and crop debris. Disease spread by damp and humid conditions during flowering.	**(1)** Treat with a suitable fungicide at early flowering **(2)** Avoid over fertilising the crop
	Mycosphaerella—foot rot and leaf spot	Fungus	Small brown spots (tea leaf like lesions) on leaves and pods. Infections can girdle stems and cause whole plants to die.	Spread on seed and from sclerotia in the soil or on crop residues. Wet weather favours disease.	**(1)** Widen rotation **(2)** Use healthy seed **(3)** Apply a fungicide at mid flowering
	Bacterial blight	Bacteria	Dark brown lesions on leaves, stems and pods.	Spread on seed. Wet windy weather favours disease.	Only sow healthy seed
	Pea wilt—Fusarium wilt	Fungus	In late May/June, lower leaves tend to turn grey before rolling downwards. All leaves eventually affected. Death of plant either before podding or before pods have swollen. White mycelium appears on stems after death.	The pathogen is soil-borne. It invades the plant and only returns to the soil from the infected dead plant material.	**(1)** Rotation **(2)** Use resistant varieties *Note*: There are other Fusarium species which cause other diseases in peas notably root rot
	Pre-emergence damping--off	Fungus—complex of different species	Poor germination and seedling establishment. The seed rots, or if not the stem is soft and dark brown in colour.	Soil-borne pathogens invade seeds and/or plant stems at or just below soil level. Worst when peas sown in cold, wet soil.	Avoid planting into cold wet seed beds

Continued

Table 6.2 Major plant diseases affecting UK field crops and their management options.—cont'd

Crop attacked	Disease	Causal agent	Symptoms of attack	Life-cycle	Methods of control
	Marsh spot	Manganese deficiency	Yellowing of leaves between veins which remain green; with severe deficiency growth is restricted and yield reduced. Brown discolouration (marsh spot) in centre of pea.		Apply manganese sulphate as soon as in full flower. May need repeating 7 days later
Field beans	**Chocolate spot**	Fungus	Starts with small circular chocolate coloured discolourations on leaves and stems; with an aggressive attack symptoms move to flowers and pods. Spots can coalesce and cause total plant death.	Fungus carried over from previous year on debris of old bean haulm and on volunteer plants. Disease favoured by warm, humid and wet weather. Autumn-sown beans, especially early sown, are more liable to attack than those that are spring sown.	**(1)** Clean up stubbles containing remains of old crop of beans **(2)** Use optimum seed rates **(3)** Apply foliar fungicide. First treatment in winter beans at early flowering stage if disease present on lower leaves. Treatment should be repeated 3 weeks later
	Ascochyta—leaf spot	Fungus	Leaves affected by regular brown to black spots, some up to 2 cm in diameter; spots have slightly sunken grey centres (in which can be seen small black spots with brown margins). Pods and seed also affected, the latter covered with sunken brownish-black lesions.	Infected seeds when sown may produce seedlings with characteristic disease symptoms on stem at soil level or on lowest leaves. In cold moist conditions the disease will move up the plant and on to other bean plants.	**(1)** Healthy seed—seed can be tested **(2)** Hygiene—destroy any volunteer beans in other fields **(3)** Grow resistant varieties

	Sclerotinia—stem rot	Fungus	Rotting of shoots and roots. Can cause death in severe attacks on winter beans.	Soil-borne. The strain that attacks spring beans also attacks peas, red clover and oilseed rape. The sclerotia or resting bodies can persist in the soil for 7–10 years.	**(1)** Limited fungicide recommendations **(2)** Use wide rotations
	Bean rust	Fungus	Red/brown rust pustules surrounded by a yellow halo appear on leaves.	Can be spread on seed, trash or volunteers. More of a problem on spring beans and on potassium deficient soils. Disease encouraged by warm temperatures and high humidity.	**(1)** Hygiene—kill any volunteer beans **(2)** Apply a foliar fungicide—e.g., some triazoles and strobilurins can be very effective
	Downy mildew	Oomycete	Pale green water soaked lesions. Grey fungal growth on underside of leaf.	Seed- and soil-borne. Spring beans most affected. Disease encouraged by cool, wet weather.	**(1)** Rotation **(2)** Hygiene—destroy debris **(3)** Foliar fungicide (metalxyl-M) - treat if risk high
Sugar beet and fodder beet	**Virus yellows (beet mild yellowing virus—BMYV, beet yellows virus—BYV and beet chlorosis virus—BChV)**	Virus	First seen in June/early July on single plants scattered throughout the crop—yellowing at the tips of the leaves and between the leaf veins. This gradually spreads over all but the youngest leaves. Infected leaves thicken and become brittle. The yield is seriously reduced by an early attack.	The crop is infected by aphids which have over-wintered mainly in clamps. Several green aphids carry the virus, particularly peach potato aphid. After a mild winter and warm spring aphid migration is early and the chances of an epidemic are increased. BMYV and BYV have a wide host range including some common weeds.	**(1)** All beet clamps should be cleared by the end of March **(2)** Development of tolerant (partially resistant) sugar beet varieties **(3)** Aphid warnings issued by British Sugar to aid timing of aphicide **(4)** Foliar aphicides applied if threshold of one aphid per four plants exceeded. Care must be taken with insecticide choice as there is resistance to some of the foliar products

Continued

Table 6.2 Major plant diseases affecting UK field crops and their management options.—cont'd

Crop attacked	Disease	Causal agent	Symptoms of attack	Life-cycle	Methods of control
	Powdery mildew	Fungus	Powdery greyish-white mycelium on foliage seen in dry weather in late summer, early autumn.	The spores are air-borne and move from diseased plants found at loading sites and from roots left in the field and also from weed beet to infect the new crop. Disease favoured by dry, warm (25°C) weather. The number of frosts recorded in the winter affects the incidence of powdery mildew.	**(1)** Spray with a fungicide. Several triazoles/strobilurins mixtures recommended. Spray as soon as disease is seen but before mid-September **(2)** Some varieties are partially resistant **(3)** Use warnings given by Broom's Barn
	Rhizomania—root madness (beet necrotic yellow vein virus—BNYVV)	Virus	Wilted plants (sometimes in patches in field) showing pale yellowing of veins; development of elongated, strap-like leaves often protruding above surrounding plants. Infected roots smaller than healthy roots, constricted below soil level, usually fanging with a proliferation of small lateral roots (bearding). Inside of root shows brown-streaked tissue from tip of tap root upwards.	The virus is transmitted by the soil-borne slime mould, *Polymyxa betae,* which is present in UK soils. Widespread in Europe and now in the UK. *Polymyxa* can survive in the soil for 25 years. Germination of resting spores occurs in moist and warm conditions (>12°C) and latent periods are shortest at high temperatures (25°C).	**(1)** Partially resistant varieties are the main method for control against the common pathotypes (A and B). A minority of varieties are available for the less common AYPR strain **(2)** Extend the rotation to reduce the risk **(3)** Reduce movement of infected soil. Give special attention to sugar beet harvesters

	Ramularia—leaf spot	Fungus	Pale brown circular spots with a pale appearance in centre of lesion. Can eventually cause death of all leaves.	Mainly soil-borne disease. Wet weather encourages infection and spread.	**(1)** Bury infected crop residues, manage volunteers and extend rotation **(2)** Apply an approved fungicide, e.g., a triazole/strobilurin mixture
	Cercospora—leaf spot	Fungus	Similar spots as those caused by ramularia but smaller and darker.	A seed and soil borne disease. Initial infection can come from seed, alternate hosts and infected crop residues. High temperatures (23–25°C) and humidity favour disease development	**(1)** Bury any infected crop residues and manage volunteers **(2)** Some triazole/strobilurin mixtures are effective
	Stemphylium	Fungus	Irregular yellow spots between July and August. Older spots are irregular and brownish (necrotic) and up to 1–3 cm.	The pathogen survives in soil and is dispersed during wet weather from soil and from sporulating lesions. There are a range of *Stemphylium* species; some infect plants such as garlics, onions and leeks, while others are found on spinach, fat hen, mustard and asparagus.	**(1)** Bury any infected crop residues **(2)** Limited evidence that triazole/strobilurin mixtures are effective
	Downy mildew	Oomycete	Beige to purple, clumpy mould found on the underside of the central young (heart) leaves. Rolling/distortion of the heart leaves.	The pathogen survives in the soil as resting spores (oospores). The disease develops on young plants (seedlings) under cool humid conditions.	**(1)** Bury any infected crop residues **(2)** Extended rotations

Continued

Table 6.2 Major plant diseases affecting UK field crops and their management options.—cont'd

Crop attacked	Disease	Causal agent	Symptoms of attack	Life-cycle	Methods of control
	Violet root rot	Fungus	Patches of wilting plants seen in late summer/early autumn. Roots have a superficial rot that is violet/black in colour. Dark mycelium can be seen within the rot. Infected beet develop secondary rots in the clamp.	The pathogen is soil-borne, surviving saprophytically and as sclerotia. Survival time is estimated at 4 years. Disease mainly seen from September—November and is active when the soil temperature is above 13°C.	**(1)** Avoid susceptible crops in the sugar beet rotation (e.g., carrots, parsnips and potatoes). Use wider rotations **(2)** Avoid feeding infected beet to cattle **(3)** Some broadleaf weeds can act as alternative hosts e.g., creeping thistle, nettles and shepherd's purse
	Speckled yellows	Manganese deficiency	Small yellowish areas between leaf veins, later turning to buff-coloured angular, sunken spots (speckled yellows) which eventually coalesce.	Symptoms often disappear as root system develops. But disease can be a problem on near-alkaline soils or those soils with high organic matter content.	Foliar spray with manganese sulphate in May and June which may have to be repeated.
	Heart rot	Boron deficiency	In young plants the youngest leaves turn a black/brown colour and die off. A dry rot develops in the root and spreads from the crown downwards. The growing point is killed, being replaced by a mass of small deformed leaves.	This deficiency is more apparent on dry and light soils and can be made worse by heavy liming.	**(1)** Deficiency diagnosed by soil analysis **(2)** Apply boron if required

Potatoes	**Late blight**	Oomycete	Irregular dark brown (necrotic) areas on leaves. Whitish mould occurs around the margin of the necrosis on the underside of leaves ('the white halo'). Leaves and stems become brown with lesions and die off prematurely. Affected tubers have a superficial rot (granular and reddish brown in colour) and may break down in-store due to secondary infection with soft rotting bacteria.	Infected tubers (either planted, ground keepers, or from dumps) are the main source of late blight. Asexual spores (sporangia) are carried by the wind or rain to infect the haulms. Recently sexual resting spores (oospores) have been found, although their importance is not fully understood. From the infected haulms, the spores are washed into the soil to infect the tubers. Infection can also take place at harvest. The oomycete cannot survive on dead haulm. Risk of late blight encouraged by certain weather conditions—Hutton Criteria—e.g., two consecutive days with min. temp above 10°C and relative humidity at 90% for at least 6 h/day. Disease more likely to develop once crop meets between rows. New more aggressive strains (genotypes) have been recorded in recent surveys. Some genotypes have reduced sensitivity to fungicides.	**(1)** Ensure that potato dumps do not sprout. Spray off or cover with plastic sheeting and control volunteers **(2)** Late blight spreads rapidly—warnings of blight periods are given by various organisations. Early preventative spraying, followed by repeated sprays every 7–10 days, is advisable **(3)** A wide range of fungicides are available, both protectants and systemics. Due to resistance problems, care should be taken with chemical choice and mixtures **(4)** The haulms should be desiccated before harvest to prevent the tubers becoming infected whilst being lifted **(5)** There is some difference in susceptibility to foliage and tuber blight between varieties

Continued

Table 6.2 Major plant diseases affecting UK field crops and their management options.—cont'd

Crop attacked	Disease	Causal agent	Symptoms of attack	Life-cycle	Methods of control
	Early blight	Fungus	Discrete necrotic lesions occur between the leaf veins of the lower/mature compound leaves, typically during the onset of flowering. Lesions have a characteristic zonal pattern to them—'target spot'.	The fungus survives saprophytically in the soil. Crops are infected by spores that are released from the soil. Alternating wet and dry periods (overnight dew) favours sporulation from the lower leaves. The pathogen prefers warm conditions (20°C).	**(1)** Some indeterminate varieties are more susceptible to the disease e.g., those used for processing **(2)** Bury any infected crop residues **(3)** Avoid crop stress (drought, PCN damage)
	Potato leaf roll virus—PLRV	Virus	Lower leaves are rolled upwards and inwards; they feel brittle and crackle when handled. The other leaves are lighter green and more erect than normal. Yield is lowered.	The virus is transmitted by aphids from plant to plant. Peach potato aphid is the main vector. Infected tubers (which show no signs of the disease) are planted and thus the disease is carried forward from year to year.	**(1)** Resistant varieties **(2)** Use high grade 'seed' **(3)** Ensure no aphids in the chitting house **(4)** Apply a suitable systemic insecticide to reduce the spread of the virus **(5)** Reduce sources of infection—dumps etc
	Mosaics—potato virus Y (PVY)	Virus	Symptoms range from a faint yellow mottling on leaves to a severe distortion of the leaves and distinct yellow mottling. Yield can be seriously reduced by the severe forms.	Peach potato aphid is the main vector of PVY. The virus is non-persistent within the aphid. Symptoms may not appear for 4 weeks after infection.	**(1)** As for PLRV **(2)** Some varieties resistant **(3)** Using aphicides can give variable results

	Common scab	Filamentous bacteria (several species)	Skin-deep irregular-shaped scabs on tuber: these can occur singly or in masses. With a severe attack cracking (star like) and pitting takes place with secondary infection by insect larva and millipedes.	The soil-borne organism attempts to invade the growing tuber (the lenticels) which responds by development of corky tissue to restrict the parasite to the surface layers. Organism re-enters soil when infected seed is planted. Disease is particularly prevalent on light sandy, alkaline soils, in dry conditions.	**(1)** Avoid liming just prior to planting potatoes **(2)** Irrigation is main method of control. Irrigate when soil moisture deficit reaches 12 mm during the 6 weeks from the start of tuber initiation **(3)** Some varieties are more resistant than others
	Powdery scab	Slime mould	Appearance can be similar to common scab, but the spots are rounder and formed as raised pimples under the skin which burst; sometimes cankers and tumours develop at the rose end of the tubers.	Spore balls can remain in the soil for many years (7+) or may be planted on infected tubers and the zoospores attack the new tubers by way of the lenticels, eyes or wounds, resulting in scab development. Usually more troublesome in wet seasons.	**(1)** Avoid using infected seed or contaminated FYM **(2)** Some varieties are very susceptible **(3)** Check zinc levels. Soils with a high level of soil zinc may have a lower disease risk
	Dry rot	Fungus (several species)	Infected tubers are usually first seen in January and February. The tuber shrinks and the skin wrinkles in concentric circles. Blue/pink or white pustules appear on the surface.	The soil-borne fungus enters the tuber from adhering soil. Infection can only enter through wounds and bruises caused by rough handling at harvest. Favoured by high storage temperatures.	**(1)** If the potatoes are handled carefully, infection is considerably reduced **(2)** Dry cure tubers before moving into store **(3)** Use of fungicides at lifting or first grading (seed potatoes) **(4)** Varieties vary in their susceptibility

Continued

Table 6.2 Major plant diseases affecting UK field crops and their management options.—cont'd

Crop attacked	Disease	Causal agent	Symptoms of attack	Life-cycle	Methods of control
	Spraing—tobacco rattle (TRV) or **potato mop-top virus (PMTV)**	Virus	Foliage—very variable; TRV stem mottling; PMTV yellow blotches and bunching of leaves on short stems—like a mop. Tubers primary (after soil infection): wavy or arc-like brown, corky streaks in flesh of cut tuber. Secondary—from infected tubers: PMTV—badly formed and cracked tubers. TRV—brown spots in flesh.	TRV spread by stubby root (Trichodorid) nematodes in soil, especially in light sandy soils. PMTV spread by powdery scab pathogen and can remain in the soil for years as resting bodies.	**(1)** Plant only resistant varieties on infected soils (TRV) **(2)** Control stubby root nematodes with a nematicide (TRV) **(3)** Avoid planting seed with powdery scab (PMTV)
	Blackleg and **tuber soft rots**	Bacteria	Plants stunted and pale green or yellow foliage; easily pulled out of the ground and stem base is black and rotted. Infected and neighbouring tubers develop a wet rot in the field or in store, especially in damp and badly ventilated (warm) conditions.	The bacteria move to tubers via the stolons and in wet soil to healthy tubers and enter via lenticels or damaged areas. Carried on seed tubers.	**(1)** Rogue or reject seed crops where the disease shows on foliage **(2)** Do not plant infected tubers **(3)** No resistant varieties, although some are less susceptible **(4)** Monitor store regularly. Ensure good store ventilation

	Gangrene	Fungus	A serious tuber rot which develops in storage, usually late; it shows as grey 'thumb-mark' depressions on the tubers and the flesh beneath rots also, pin-head black spore cases.	The fungus remains alive in the soil and on trash, and can infect tubers in the soil and from tuber to tuber when handling. Cold and wet growing conditions favour this disease. More commonly found in cool stores (seed potatoes) due to slow wound healing.	**(1)** Do not plant diseased seed **(2)** Reduce mechanical damage **(3)** Dry cure tubers before moving into store **(4)** Tuber treatment with a fungicide reduces incidence in store (seed potatoes) **(5)** Varieties vary in their susceptibility
	Silver scurf	Fungus	A superficial disease. Causes well defined silvery patches with a sooty centre. Can affect quality for pre-pack market.	Mainly spread from air-borne spores during storage. Symptoms develop in store especially in warm conditions and when condensation is not managed.	**(1)** Plant disease free seed **(2)** Apply a fungicide (treatment for seed tubers) **(3)** Fungicides available for soil treatment **(3)** Cool, dry storage will reduce risk **(4)** Some varieties are very susceptible
	Skin spot	Fungus	Tuber symptoms develop during late storage and appear as pimple-like, dark brown, shrunken spots with raised centres. The worst damage is the destruction of the buds in the eyes of seed tubers.	Mainly spread by infected tubers. Tuber infection occurs at lifting and is worse in cold, wet seasons.	**(1)** Apply a fungicide (treatment for seed tubers) **(2)** Dry storage conditions **(3)** Dry cure tubers before moving into store **(4)** Varietal resistance
	Black dot	Fungus	Similar symptoms on tubers as silver scurf but the lesions are less well defined. Small black sclerotia present on the lesions. A problem in potatoes for pre-packing.	Soil- and seed-borne disease. Warm and moist conditions in store increase the amount of infection.	**(1)** Some varietal differences in susceptibility **(2)** Plant clean seed **(3)** Apply a recommended fungicide at planting **(4)** Store potatoes in cool dry conditions

Continued

Table 6.2 Major plant diseases affecting UK field crops and their management options.—cont'd

Crop attacked	Disease	Causal agent	Symptoms of attack	Life-cycle	Methods of control
	Stem canker and black scurf	Fungus	Black sclerotia (black scurf) present on seed tubers can lead to poor emergence due to the development of stem canker. On young plants, the fungus may cause sunken brown stem lesions below ground. Stem infection can lead to yellowing and rolling of leaves. In severe cases shoots may die due to pruning damage. Pruning damage to stolons (stolon canker) results in a reduction of the marketable fraction. Infection of developing tubers can cause an unsightly netted scab sometimes referred to as 'giraffes' neck' or elephants' hide' White sporulation may be found at base of stems.	Seed and soil-borne disease. Low soil temperatures encourage disease development especially on light soils. Pathogen survives in soil as sclerotia (limited survival), as a saprophyte and on wild plant species e.g., couch grass.	**(1)** Widen rotation **(2)** Avoid planting into cold and dry seedbeds **(3)** Plant clean seed (certified seed or use seed tests) **(4)** Plant chitted seed to encourage rapid emergence **(5)** Limit time between burning off and harvesting to reduce black scurf developing on progeny tubers **(6)** Fungicide tuber treatments available **(7)** Fungicide soil treatment
	Brown rot	Bacteria	Symptoms only seen in warm conditions. Initially leaves wilt leading eventually to plant death. Brown streaking on the stem. Tubers have brown vascular discolouration and ooze watery droplets when cut.	Outbreaks linked to contaminated irrigation water. Woody nightshade is secondary host. Potentially a very serious disease.	A notifiable disease in the EU and UK. Legislative control measures

	Ring rot	Bacteria	Similar symptoms to brown rot in the tuber, except watery droplets do not appear from freshly cut tubers. Instead, creamy white discharge is released from the vascular tissue when pressure is applied (cheese like rot). Cracking symptoms are seen externally on tubers.	Pathogen spread mainly through infected seed. Can survive in soil for a limited period of time.	A notifiable disease in the EU and UK. Legislative control measures
Brassicae (Brussels sprouts, cabbage, kale, swedes and turnips)	**Club root** or **finger and toe**	Slime mould	Swelling and distortion of the roots. Stunted growth. Leaves pale green in colour.	A soil-borne slime mould. The pathogen grows in the plant roots and causes the typical swellings. Resting spores can pass into the soil, especially if diseased roots are not removed. They can remain in the soil for up to 15 years, releasing motile spores when the host crop is again grown in the field. The spores are more active in acid and wet conditions.	**(1)** Rotation. With a serious attack advisable not to grow the crop for at least 5 years in the same field **(2)** Liming (try to achieve a soil pH of 7.2 or above) and drainage **(3)** Manage brassica weeds such as runch, charlock and shepherd's purse **(3)** Resistant crops. Kale is more resistant than swedes or turnips **(4)** Some varieties of swedes and turnips are more resistant than others. Resistant varieties available for oilseed rape. **(5)** Test soil if a possible risk

Continued

Table 6.2 Major plant diseases affecting UK field crops and their management options.—cont'd

Crop attacked	Disease	Causal agent	Symptoms of attack	Life-cycle	Methods of control
	Black rot	Bacteria	V shaped lesions at edge of leaves. Lesions start yellow then eventually become necrotic. Can lead to leaf loss. Symptoms vary depending on brassica species.	A seed-borne disease. Spreads during propagation and from infected weeds and crop residues. Warm humid conditions favour the disease.	(1) Plant healthy seed (2) Rotate crops with non-host crops
	Powdery mildew	Fungus	Upper surface of leaves show blue/black discolouration; sprouts turn black.	Spores over-winter on infected plants and are carried by air currents to infect the following year's crop. Particularly susceptible crops are Brussel sprouts, swedes and turnips.	(1) Varieties differ in their susceptibility to this disease (2) Apply an approved fungicide at first signs of disease (3) Widen rotation (4) Remove infected volunteers
	Downy mildew (e.g., cauliflower)	Oomycete	Pale green spots on leaves. Fungal growth/spores can be found on underside of leaves. Badly infected plants may die.	A soil-borne problem very important at the seedling stage especially when producing plants in modules.	(1) Ensure good ventilation of the transplants (2) Apply an approved fungicide at the first sign of disease (3) Variety resistance
	Ring spot (sprouts, cabbages and cauliflower)	Fungus	Small spots on leaves initially with a well-defined edge. As the disease develops the lesions enlarge to produce several concentric brown rings.	Infected debris is the main source of infection. Encouraged by cool wet weather.	(1) Rotation (2) Destroy crop debris (3) Apply a recommended fungicide (4) Grow resistant varieties

	Brown heart of swedes and turnips (Raan)	Boron deficiency	No external symptoms, but when the root is cut open a browning or mottling of the flesh is seen. Affected roots are unpalatable.		See heart rot of sugar beet
	Stem rot in kale	Boron deficiency	Brown rot in the stem pith followed by stem collapse.		See heart rot of sugar beet
Grass	**Barley yellow dwarf virus (BYDV)**	Virus	Leaves turn yellow, red or brown, discolouration starting at tips and going down leaves. Plants generally stunted, but can produce more tillers. Disease more conspicuous in single plants than in whole sward.	Spread by several species of aphids. Ryegrass and the fescues are the most susceptible.	Some ryegrass varieties more resistant than others
	Ryegrass mosaic (RMV)	Virus	Yellowish/green mottling or streaking of leaves. Severe infection can show more general browning of leaf.	Number of different strains of virus. Spread by wind-borne mite vectors which are favoured by hot, dry weather. Spring sown Italian ryegrass is most susceptible.	**(1)** Some ryegrass varieties are more resistant than others **(2)** Autumn sowing **(3)** Early grazing can reduce mite population
	Crown and **brown rust** of perennial ryegrass	Fungus	Crown rust usually seen in late summer. Pale yellow leaf flecking, followed by bright orange/yellow oval pustules. Brown rust is similar in appearance but is found in spring/early summer.	Spores are air borne and can quickly infect a clean sward especially in warm, dry weather.	**(1)** Frequent grazing and cutting to remove diseased tissue **(2)** Some grass varieties are more resistant than others **(3)** Limited off-label fungicides available for seed crops **(4)** Ensure nitrogen inputs are optimal

Continued

Table 6.2 Major plant diseases affecting UK field crops and their management options.—cont'd

Crop attacked	Disease	Causal agent	Symptoms of attack	Life-cycle	Methods of control
	Rhynchosporium—leaf spot	Fungus	Irregular scald-like blotches on leaves and is most apparent in spring and early summer.	Spores are air borne and move from a diseased to a clean sward in the spring. Favoured by cool, wet weather.	(1) Choose resistant varieties in susceptible areas (2) Early cutting
	Drechslera—leaf spot	Fungus	Causes necrotic spotting or streaking of leaves which looks similar to net blotch in cereals. Can also cause yellowing of leaves around necrotic area. Grass yield and quality can be severely affected.	Can be seed-borne. Encouraged by wet weather and lush crops. Occurs throughout the year though more commonly seen in the spring and autumn. There are several different species of *Drechslera*.	(1) Ensure adequate levels of potassium in soil. High nitrogen inputs can increase disease (2) Limited off-label fungicides available for seed crops (3) Grow resistant varieties (4) Frequent grazing and cutting to remove diseased tissue
	Powdery mildew	Fungus	Disease found throughout the country especially in Scotland. Present in early spring onwards, particularly in dense swards of short duration ryegrass. Greyish-white mycelium on leaves.	Spores are air borne, and move from an infected to a clean crop. There are many strains of this disease.	(1) Resistant varieties (2) Limited off-label fungicides available for seed crops

Red and white clover, lucerne, sainfoin	**Clover rot (sclerotinia)**	Fungus	In autumn necrotic spots found on leaves then spreads to stems. Foliage turns olive-green and then in early spring turns black and eventually dies. The root can also die.	Soil-borne disease. Resting bodies of fungus produced on affected plants in winter and spring. They are small (size of clover seed), white at first and then turning black. Bodies remain dormant in summer but in autumn produce spores which affect other plants.	**(1)** Clean seed **(2)** Use resistant varieties where possible **(3)** Rotation at least an interval of 4–5 years between susceptible crops
Lucerne	**Verticillium wilt**	Fungus	Usually seen in fairly isolated patches in first harvest year; in next 2 years spreads to many parts of the field. Normally after first cut, lower leaves turn pale-yellow colour and then white, and eventually shrivel from base upwards. Whole plant finally dies.	Disease can be introduced by contaminated seed; spores can also be transported by air, as well as being spread by contaminated fragments of the crop moving from plant to plant and then from field to field by machine.	**(1)** Where suspected use tolerant/resistant varieties **(2)** Use clean seed **(3)** Harvest healthy crops first before moving onto older infected crops **(4)** Extend interval between lucerne crops
Maize	***Fusarium* spp.—stalk rot**	Fungus	Base of plants attacked in August/September; foliage grey/green colour and wilting. Pith brown/pink at base of stem; leads to premature senescence. Some *Fusarium* spp. produce important mycotoxins.	Soil- and seed-borne disease. Can build up in the soil. Grain maize more likely to be affected because disease develops most rapidly in mature crops.	**(1)** Differences in varietal susceptibility **(2)** Fungicide treatment

Continued

Table 6.2 Major plant diseases affecting UK field crops and their management options.—cont'd

Crop attacked	Disease	Causal agent	Symptoms of attack	Life-cycle	Methods of control
	Kabatiella—eyespot	Fungus	Round cream spots on leaves surrounded by a dark brown ring and a yellow 'halo'. Causes crop to die prematurely.	Overwinters on crop debris. Airborne spores can be important for spread between crops. Disease encouraged by cool wet weather and growing successive crops.	**(1)** Widen crop rotation **(2)** Plough previous crop residues in the autumn **(3)** A limited number of fungicides are available **(4)** Some variety differences in susceptibility
	Smut	Fungus	Large black galls on any of the above ground parts of the plant, including the cob.	A seed-and soil-borne disease. Spores from galls can re-infect other maize plants, or they can remain dormant in the soil, surviving for several years.	**(1)** Crop rotation—more serious when maize is cropped frequently **(2)** Resistant varieties

The scientific (Latin) nomenclature for the pathogens associated with the above diseases can be found in Appendix 3.

6.8 Key points

- Plant diseases are important because they affect yield, produce quality and crop storage.
- The vast majority of diseases in the UK are caused by fungi, oomycetes, slime moulds, bacteria and viruses. Organisms causing disease are referred to as plant pathogens.
- Diseases can be biotic (caused by living organisms and therefore infectious) or abiotic (caused by environmental extremes, nutrient availability and physiological disorders).
- Plant pathogens have unique and individual lifecycles and respond to different environmental conditions. It is useful to refer to the disease triangle or tetrahedron when investigating disease.
- Knowledge of pathogen lifecycles and epidemiology (factors affecting disease) enables the development of suitable crop protection plans.
- Disease management can be achieved through cultural management (rotations, seed rates, planting/harvest dates, irrigation, nutrient inputs), varietal resistance, disease forecasts, legislation, biological control and appropriate use of fungicides.
- Many fungicides are susceptible to the development of resistance by plant pathogens. To avoid resistance, it is important to use fungicides mixtures with different modes of action and to rotate products within a fungicide spray programme.

Sources of further information and advice

Agrios, G.N., 2005. Plant Pathology, fifth ed. Elsevier Academic Press, London.

Alford, D.V. (Ed.), 2000. Pest and Disease Management Handbook. Blackwell Science, Oxford.

Biddle, A.J., Cattlin, N., 2007. Pests, Diseases and Disorders of Peas and Beans. Academic Press, London.

Carlile, W.R., Coules, A., 2012. Control of Crop Diseases. Cambridge University Press, Cambridge.

FRAG, U.K., 2021. Fungicide Resistance Management in Cereals 2020-21. Available at: https://ahdb.org.uk/frag-cereals. (Accessed 10 September 2021).

Kavanagh, K., 2005. Fungi, Biology and Applications. John Wiley & Sons Ltd, Oxford.

Koike, S.T., Gladders, P., Paulus, A.O., 2007. Vegetable Diseases. Manson Publishing, London.

Lainsbury, M.A., 2021. The UK Pesticide Guide. BCPC*, Alton.

Lucas, J.A., 2020. Plant Pathology and Plant Pathogens, fourth ed. Wiley-Blackwell, Oxford.

Mathews, G., 2006. Pesticides, Health, Safety and the Environment. Blackwell Publishing, Oxford.

Mulder, A., Turkensteen, L.J., 2005. Potato Diseases. NIVAP, AC Den Haag.

Murray, T.D., Parry, D.W., Cattlin, N.D., 2008. Diseases of Small Grain Cereal Crops. Manson Publishing, London.

Scott, P., Strange, R.N., Korsten, L., Gullino, M.L., 2021. Plant Diseases and Food Security in the 21st Century. Springer International Publishing, New York.

Tronsmo, A.-M., Collinge, D.B., Djurle, A., Munk, L., Yuen, J., Tronsmo, A. (Eds.), 2020. Plant Pathology and Plant Diseases. CABI, Wallingford.

Thind, T.S., 2012. Fungicide Resistance in Crop Protection, Risk and Management. CABI, Wallingford.

Wale, S., Platt, H.W., Cattlin, N., 2008. Diseases, Pests and Disorders of Potatoes. Manson Publishing, London.
Walters, D. (Ed.), 2009. Disease Control in Crops—Biological and Environmentally Friendly Approaches. Wiley-Blackwell, Oxford.
*UK Pesticide Guide published annually.

Websites

Levy bodies provide useful information on diseases (identification and crop protection).
AHDB Cereals and Oilseeds: https://ahdb.org.uk/cereals-oilseeds.
British Beet Research Organisation: https://bbro.co.uk/.
Processors and Growers Research Organisation: https://www.pgro.org/.

Agrochemical and breeding companies often have useful factsheets on disease and their fungicide/variety products

BASF: https://www.agricentre.basf.co.uk/en/.
Bayer Crop Science: https://cropscience.bayer.co.uk/.
Corteva: https://www.corteva.com/.
KWS: https://www.kws.com/gb/en/products/.
LG Seeds: https://www.lgseeds.co.uk/expertise/?crop=maize.
Syngenta: https://www.syngenta.co.uk.

Disease forecasts and surveys

Blight watch (Hutton Criteria forecasts for late blight): https://blightwatch.co.uk/.
Crop Monitor: https://secure.fera.defra.gov.uk/cropmonitor/.
CropProtect (useful information on key UK diseases): https://croprotect.com/.
Phoma forecast (AHDB): https://ahdb.org.uk/phoma-leaf-spot-forecast.
Sclerotinia forecast (AHDB): https://ahdb.org.uk/sclerotinia-infection-risk-alerts-for-oilseed-rape.
UK Pathogen Virulence Survey (yellow rust, brown rust, powdery mildew): https://www.niab.com/research/agricultural-crop-research/research-projects/uk-cereal-pathogen-virulence-survey.

Fungicides and fungicide resistance

Chemicals Regulation Division: https://www.hse.gov.uk/crd/ AND https://www.hse.gov.uk/pesticides/index.htm.
FRAC (international committee): https://www.frac.info/.
FRAG UK (national committee): https://ahdb.org.uk/frag.
Voluntary initiative: https://voluntaryinitiative.org.uk/.

Plant Health Clinics

Fera: https://www.fera.co.uk/crop-health/plant-clinic.
NIAB: https://www.niab.com/.

Pests

7

7.1 Introduction

Animal pests are vertebrate or invertebrate species that have a negative impact on humans, which can range from being a domestic nuisance to transmitting diseases such as malaria. Within agricultural crop production systems, animal pests are responsible for economic losses totalling millions of pounds per year. Animal pests that directly impact agricultural productivity are referred to as crop pests hereafter. Despite overwhelming attention being given to crop pests, most animals are either benign or beneficial to crop production. Species considered benign have no discernible impact on crop productivity while beneficial species provide key ecosystem services, such as pollination or pest control, that enhance crop productivity.

Insects are the most important group of crop pests, followed by nematodes, molluscs, birds, mammals and mites. Some crop pests are a reoccurring problem, though not always at population levels that cause economic damage. Other crop pests are sporadic and only occur when environmental or crop conditions are optimal. Several crop pests are specific to individual crop species (i.e., specialists) whereas others damage many different crops (i.e., generalists). Many crop pests are successful because they have many generations per year, can effectively disperse to new areas and are adapted to changing environmental conditions. This Chapter provides a basic introduction to the biology and identification of key crop pests before discussing their control methods.

7.2 Invertebrate pests

7.2.1 Insects

Insects are the most successful group of organisms on earth, found in almost every conceivable environment. There are approximately one million described insect species at present, however it is estimated that this number is more likely to be between 6 and 10 million. Given this huge diversity it is unsurprising that there are many different insect groups considered crop pests. Common insect groups that contain species considered to be crop pests include true bugs (Hemiptera), beetles (Coleoptera), moths/butterflies (Lepidoptera), thrips (Thysanoptera), sawflies (Hymenoptera) and true flies (Diptera).

Insects have several defining physical characteristics: three pairs of jointed legs, three body sections, a pair of antennae and compound eyes. As insects are classed

Lockhart and Wiseman's Crop Husbandry Including Grassland. https://doi.org/10.1016/B978-0-323-85702-4.00001-7

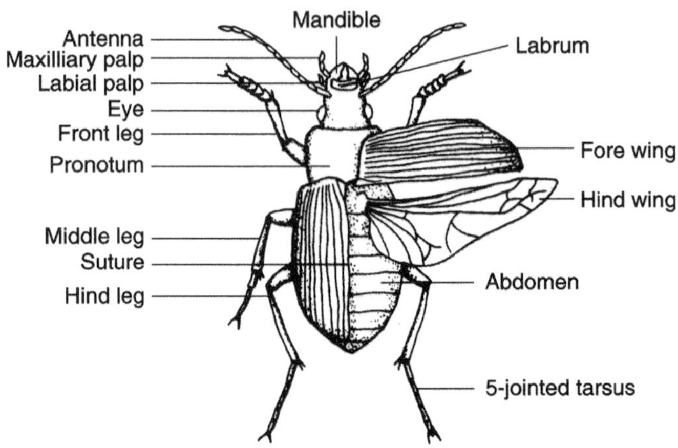

Figure 7.1 Basic insect structure.

as invertebrates, they do not have an internal skeleton, instead relying on a hard external covering called the exoskeleton to provide support for their bodies as well as protection from the environment. This exoskeleton is predominantly formed from chitin and is segmented to allow insects to move freely. These segments are grouped into three main parts (Fig. 7.1):

1. Head – this segment contains important sensory organs that allow an insect to perceive and interact with their environment:
   a. Antennae (sensory organ) for perceiving chemical stimuli (equivalent to smell in humans) within the environment;
   b. Eyes (sensory organ) for perceiving visual stimuli. Most insects have compound eyes as well as simpler eyes known as ocelli;
   c. Mouthparts for perceiving chemical stimuli (equivalent to taste in humans) and feeding. There are two main mouthpart types found in insect crop pests (Fig. 7.2):
      i. Mandibulate (biting and chewing) for grazing on foliage, stem and root tissues;

Figure 7.2 Insect mouthparts.

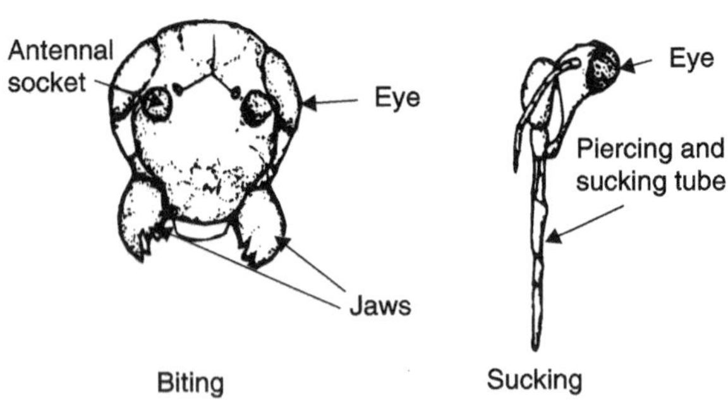

   ii. Haustellate (piercing) for accessing plant vascular tissue to feed on plant phloem sap. Insects with these mouthparts do not feed on foliage, stem or root tissues directly.
2. Thorax – this segment gives rise to structures required for movement:
   a. Legs for walking. All adults have three pairs of jointed legs;
   b. Wings for flying. Most, but not all, insect species have two pairs of wings. True flies (Diptera) only have a single pair of wings as their hind wings are modified into club-shaped structures (halteres) used for balance during flight. Aphids may also be wingless.
3. Abdomen – this segment generally has no structures of note except for in some female adults where an egg-laying organ (ovipositor) may protrude from the end and in aphids where the abdomen gives rise to two distinctive structures, the cauda and siphunculi, which can aid identification.

Understanding insect life-cycles is important when identifying a pest and deciding on the most appropriate control method. Most insects begin life as an egg laid by a female (oviposition), however some insects (e.g., aphids) can give birth to live offspring (larviposition) in certain circumstances. There are two main types of insect life-cycle:

1. Complete metamorphosis – each different life-cycle stage is morphologically distinct from one another (i.e., differ in appearance). This type of life-cycle is found across a range of different insect groups: true flies, beetles, moths/butterflies, thrips and sawflies. There are four stages associated with complete metamorphosis (Fig. 7.3):
   a. Egg;
   b. Larva (singular) or larvae (plural) are morphologically different from adults of the same species. This is the main growing stage of the insect life-cycle, where individuals shed their exoskeleton, in a process called moulting, to replace it with a larger one. The stage between each moult is called an instar. Most larvae moult at least three or 4 times (go through three or four instars). Larvae usually possess mandibulate mouthparts and it is at this stage, in many insects, that they are most destructive to crops;
   c. Pupa (singular) or pupae (plural) is the resting stage of the insect life-cycle. Once larvae are fully grown, they pupate and undergo complete metamorphosis;
   d. Adult.
2. Incomplete metamorphosis – each life-cycle stage is morphologically similar to one another (i.e., similar in appearance), except the egg stage, and lacks a pupal stage. This type of life-cycle is found in true bugs and grasshoppers. There are three stages associated with incomplete metamorphosis (Fig. 7.4):
   a. Egg;
   b. Nymphs are morphologically similar to adults, though are typically smaller and lack wings. This is the main growing stage of the insect life-cycle and there are several instars before reaching maturity. Nymphs usually possess haustellate mouthparts and are active feeders that can cause significant crop damage;
   c. Adult.

Insects vary in the length of time it takes to complete their life-cycle. For example, click beetles (larvae – wireworm) may take 5 years, whereas bean seed fly can complete up to six generations in a single year. The time of year when larvae are found also varies between species. There are distinct differences in larval

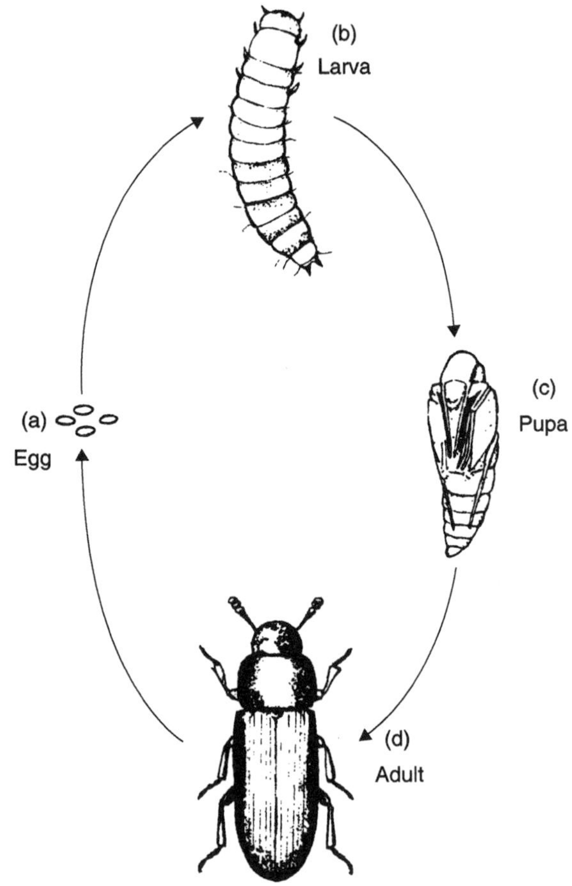

Figure 7.3 Complete metamorphosis. There are four stages associated with this type of life-cycle: (a) egg (b) larva or the immature growth stage, (c) pupa and (d) adult.

appearance between different insect groups (e.g., beetles and butterflies) as well as within insect groups. Identification characteristics can include size, colour, presence or absence of legs on the thorax and presence or absence of prolegs (false legs) on the abdomen (Fig. 7.5).

Not all insects are harmful to crops. Some species are beneficial as they predate on or parasitise crop pests. Some of these, especially at the larval stage, can look like some crop pests so it is important to be able to identify accurately. Ladybirds, hoverflies, ground beetles and lacewings are key predatory insects.

7.2.2 Nematodes (eelworms)

Nematodes are microscopic, non-segmented, worm-shaped invertebrates enclosed within a tough cuticle. They vary from 0.1 to 2.0 mm in length with an average size of 1.0 mm. Most nematode species are free-living and beneficial, but there are several important species that are either parasitic within crop plants (endoparasitic) or feed on

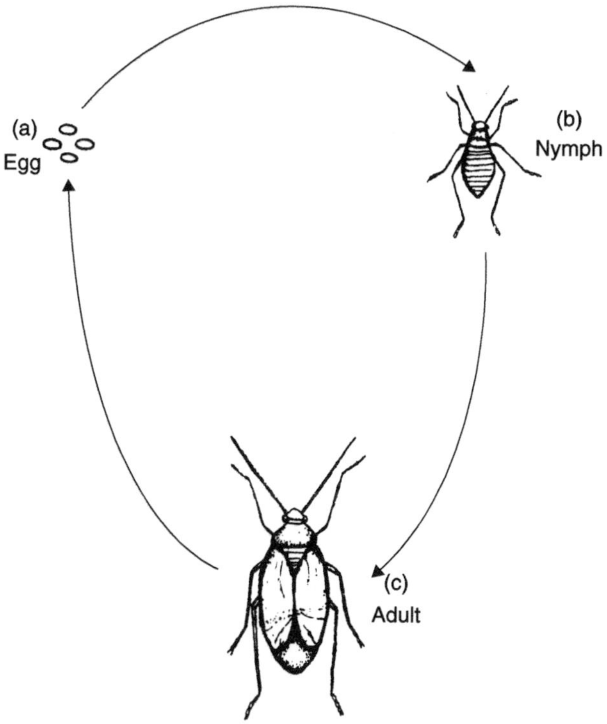

Figure 7.4 Incomplete metamorphosis. There are three stages associated with this type of life-cycle: (a) egg (b) nymph and (c) adult.

the surface tissues of crop roots (ectoparasitic). Most nematode pests only attack or cause damage to specific host species. Nematodes themselves have a limited ability to move far, the most important means of spread is by movement in water or soil. Farm machinery and contaminated planting material can be especially important in translocating nematode pests between fields or farms.

The nematode life-cycle is relatively simple, consisting of an egg hatching into a larval form. There are several juvenile stages (shedding a cuticle each time) before reaching the sexually mature male or female stage. In most cases, part of the life-cycle takes place in the soil. With many pest species there is a life-cycle stage (e.g., a cyst containing eggs) that may remain dormant, sometimes for several years, to only become active again when environmental conditions are suitable. Some species, however, have more than one generation per year and are only inactive during winter months. Both juvenile and adult life-cycle stages cause host plant damage. Plant-feeding nematodes have simple mouthparts that consist of a cavity (stoma) that contains a needle-like stylet (Fig. 7.6). It is this which pierces the cellular tissue for sucking out the cell contents of the parasitised plant.

Identifying nematodes can be challenging due to limited morphological variation between species. Due to these difficulties, identification typically relies on using plant symptoms with plant or soil analysis. Advances in molecular barcoding have enabled laboratory tools to be developed to help differentiate between closely related nematode

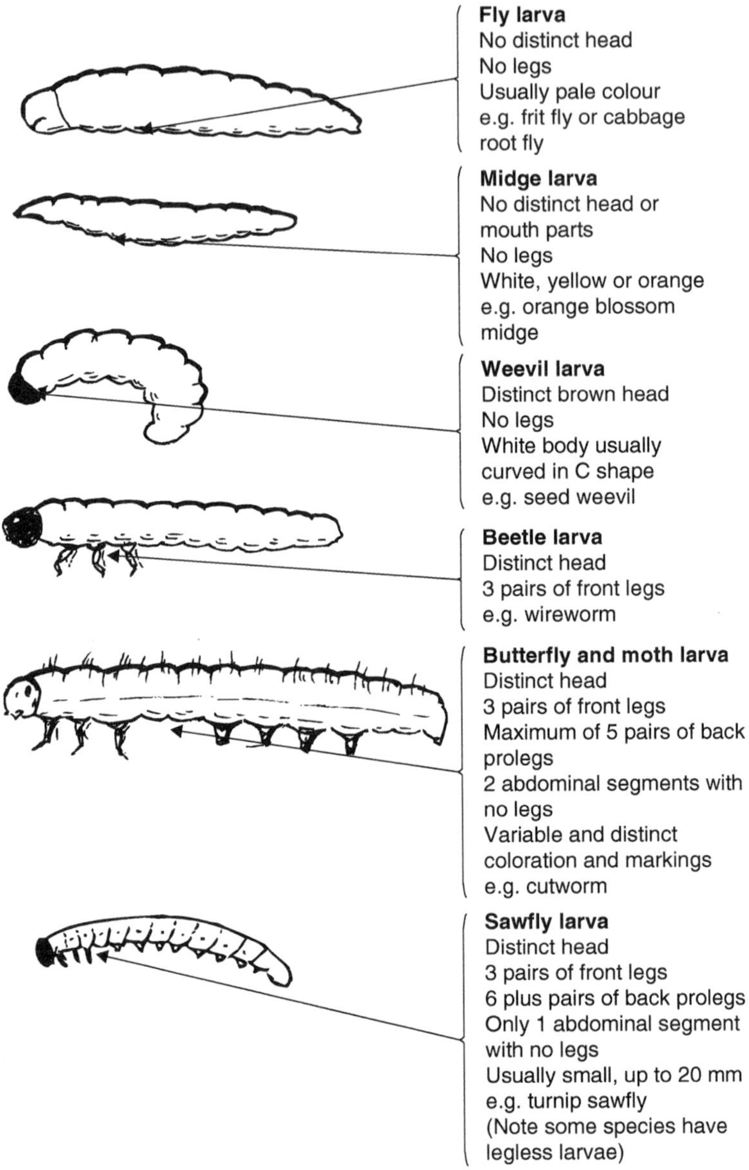

Figure 7.5 Characteristics of insect larvae.

species. Identification is important as it allows preventative control strategies to be implemented. The major nematode pests are the root-attacking cyst nematodes and stem nematodes. These can have a wide host range but on each host plant they cause varying forms of necrosis and deformity, which negatively impacts plant health and yield.

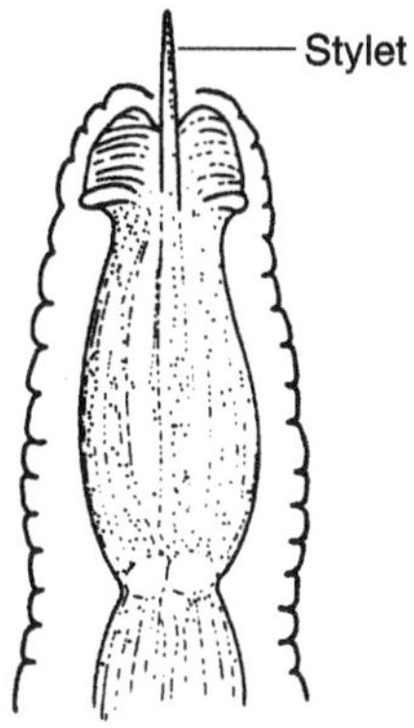

Figure 7.6 Plant-feeding nematode mouthparts.

7.2.3 Molluscs

Slugs (shell-less gastropods) are the most important mollusc pests of field crops and can negatively impact most crops if present in significant numbers. There are several defining features for slugs – they are legless, unsegmented and boneless. Unlike insects, slugs do not have an exoskeleton and are prone to desiccation without protection from their mucus. To further minimise the risk of desiccation, slugs are mostly active on humid nights when the soil is moist. There are three slug groups of concern in agriculture: grey field slugs, garden slugs and keeled slugs. Feeding activity can occur throughout the year, though slugs are particularly damaging during autumn. Periods of heavy rain, frost or drought often reduce slug damage as they become inactive and stop feeding. Slugs impact a wide range of crops, including cereals, potatoes, field brassicas (oilseeds and vegetables) and carrots. Crops grown on heavy clay soils are most at risk; previous crop and soil cultivations and soil consolidation can affect activity. Each slug can lay up to 500 eggs over several weeks, which usually hatch either in early autumn or late spring depending on weather conditions. On hatching, the young slugs resemble the adults, except in size. Initially the young feed on organic matter in the soil before moving to above and belowground plant material.

7.2.4 Mites

Mites are very small invertebrates typically less than 0.1 mm in size. Many mites have an important role as decomposers by breaking down organic matter in soil, but some species are major crop pests. Mites morphologically differ from insects as they are wingless, have two body sections and four pairs of legs (adults only) (Fig. 7.7). Crop pests have mouth parts (chelicera) adapted for piercing plant tissues to feed on their cell contents. Most mites have an incomplete metamorphosis life-cycle, with a 6-legged nymph emerging from an egg that develops into an 8-legged nymph before maturing into an adult. There is no pupal stage in the mite life-cycle and both nymph stages are morphologically similar to adults. Mites can reproduce very quickly (approximately 7 days from egg to adult under optimum conditions), so large populations can establish before feeding damage is recognised.

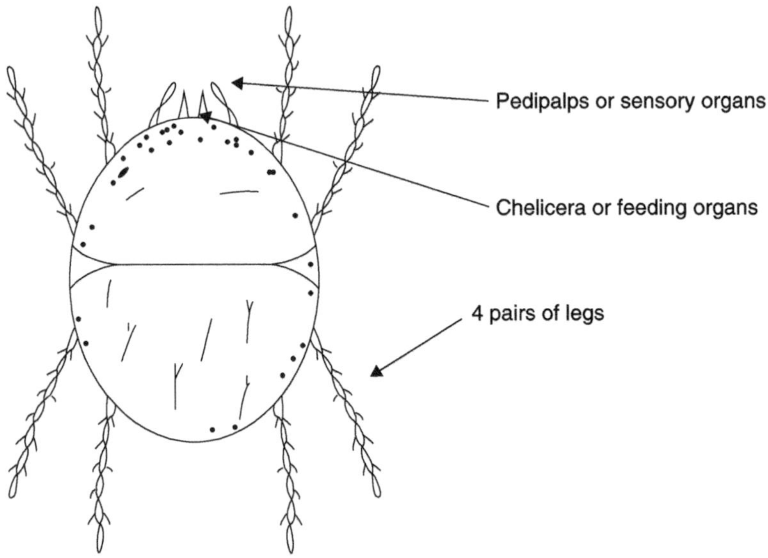

Figure 7.7 Basic mite structure.

7.3 Vertebrate crop pests

7.3.1 Birds

Birds are widely considered to be more helpful than harmful as they consume a range of pestiferous animal species (insects, rodents, etc.) and plant seeds (including those of weeds). Even beneficial birds can cause some issues, often inflicting accidental damage to emerging cereal crops while searching for insects (e.g., leatherjackets). There are exceptions to the beneficial status of birds. Wood pigeons cause significant crop damage by feeding on cereal seed and grain of lodged crops as well as on young pea and brassica crops. Corvids may also feed on the seed of a range of different crop plants. The only effective methods of limiting bird damage to crops are through organised shoots or use of bird scarers.

7.3.2 Mammals

There are several mammals that can cause crop damage. Rabbits and deer can be very serious pests, consuming many growing crops, particularly young cereals. Mice and rats can eat and damage both growing and stored crops. Some mammals, however, are beneficial to farmers as they will consume other pest organisms. Good examples of these beneficial mammals include hedgehogs that eat insects and slugs and foxes that kill rabbits and rodents.

7.4 Damage types

Crop damage is more often noticed in the field before a pest is found and identified. The type of damage, distribution in the field, time of year and the crop are important aids when identifying a pest and deciding on whether control measures are required. There are several key damage types to be aware of:

- *Total plant loss.* Some pests eat out the seed before it even germinates. For example, slugs can hollow out the grain of cereals, or mice can eat all the seeds in a row of peas. At other times the seed germinates but then subsequently dead plants are found eaten at ground level or just below the soil surface. Slugs as well as leatherjackets, cutworm, swift moth larvae, vine weevil larvae and the cockchafer grubs can cause this type of damage.
- *Loss of the central growing shoot.* A few pests feed on the central growing shoot causing it to die, producing dead hearts. Frit fly in grass and cereals and wheat bulb fly produce these symptoms. The dead stem can be split open to find the larva.
- *Holes and/or leaf damage.* Leaf damage can severely reduce crop vigour and even result in total crop loss during periods of slow growth. Adult weevils cause characteristic notching around the outside of crop leaves whereas slugs cause leaf shredding. Flea beetle adults cause shot holing in the leaves of brassicas. This type of damage may also be superficial and not worth controlling if the crop is growing faster than it is being eaten.
- *Distorted leaves.* Aphids feeding on crop leaves can cause the crop leaves to become distorted.
- *Loss of flowers and/or buds.* A few pests enter flowers and/or flower buds and their feeding damage can cause the buds to abort. Pollen beetle causes this type of damage especially in spring oilseed rape.
- *Stunted growth.* Nematodes such as potato cyst nematode and stem nematodes can cause severe crop stunting.
- *Virus transmission.* Some animals act as vectors for some important virus diseases. Examples include free-living nematodes transmitting tobacco rattle virus that causes spraing in potatoes, aphids transmitting barley yellow dwarf virus (BYDV) in cereals and mites that transmit ryegrass mosaic virus.
- *Damage just before or after harvest.* Several pests can damage the crop just before harvest or even after harvest. Grain aphids can affect wheat grain filling so that the harvested grain is shrivelled. Slugs and wireworm can eat into potatoes so that the quality is severely affected, and the crop becomes unsaleable. Grain weevils and mites are pests that can cause significant damage to grain in store. See Table 7.1 at end of chapter for further details on crop pests, their life-cycles, typical damage and control methods.

7.5 Integrated pest management (IPM)

Conventional agricultural crop production systems have depended on plant protection products, more commonly referred to as pesticides, to minimise crop losses to pests and secure profitable harvests. It has, however, become increasingly clear that pesticides can negatively impact both human and environmental health. Beyond these impacts there an increasing number of cases of resistance in populations of the target organism to key active ingredients. Greater awareness of these issues has led to widespread acknowledgement that crop protection needs to be more sustainable. Integrated pest management (IPM) is an environmentally and economically sustainable crop protection approach that aims to reduce pesticide use. This is achieved by implementing all available pest control measures within a framework that emphasises prevention, monitoring and thresholds. If pest populations grow above economic injury thresholds, then control measures, including pesticides, are used to reduce populations below this threshold. An economic injury threshold is the population of a pest that, if controlled, will give a yield return that will pay for the cost of the pesticide and application. Economic injury thresholds have been calculated for many pests and are an important IPM tool.

Determining if economic injury thresholds have been exceeded is important when deciding whether or not to implement controls to reduce pest populations. Another important tool to help with decision making for application of a pesticide is the use of forecasts of pest populations. Several organisations (e.g., Agriculture and Horticulture Development Board (AHDB), Processors and Growers Research Organisation (PGRO), Rothamsted Research) run forecasting services for some important crop pests such as cutworm, pea moth, carrot fly and several aphid species. These forecasts are often combined with crop monitoring for pests such as pollen beetle in oilseed rape and use of in-field traps. Examples of in-field traps include use of chicken layers mash covered with a plastic saucer or similar for slugs, sticky yellow traps for aphids and whitefly, yellow water traps for flea beetles and pheromone traps for pests such as orange wheat blossom midge. Use of pest forecasts, crop monitoring and in-field traps all help to inform whether pesticide applications are required or not. When pesticides must be applied the most selective should be chosen and applied at the correct time and dose rate. It is important to vary the mode of action of pesticides used in order to slow down the development of resistance.

7.5.1 Cultural control

A cornerstone of integrated pest management is to prevent crop pest populations from establishing and growing above economic injury thresholds by using cultural control methods (i.e., non-pesticide control methods). There are many cultural control methods available to farmers:

- *Rotations*. Changing the sequence of crops grown in a particular field to prevent crop pest populations from building up. Successfully implementing rotations requires knowledge of crop pest biology to minimise unintended consequences. For example, introducing grass into an arable rotation can increase the risk of frit fly and leatherjackets, whereas continuous cereals encourage wheat blossom midges. Adding unrelated crops (e.g., field beans, brassicas, potatoes) into this rotation can minimise pest population build up.
- *Sowing time*. Altering sowing time can have a significant impact on pest populations. If a crop is sown early enough it may be able to develop sufficiently to withstand pest attack. It is also possible to delay sowing to avoid peak pest attack (e.g., drilling winter cereals late to avoid aphid-transmitted BYDV).

- *Cultivations.* Ploughing and other cultivations can expose pests such as wireworms, leatherjackets and caterpillars, which are then eaten by birds. Well-prepared seedbeds encourage rapid germination and growth which will often enable a crop to grow away from the pest attack. However, ploughing can have a negative effect on beneficial organisms. A fine consolidated seedbed can help reduce the impact of some pests. Rolling is a useful control method for slugs and leatherjackets in crops growing in cloddy seedbeds.
- *Hygiene.* It is important to practice good hygiene in all aspects of crop production. Control volunteer crop plants and those weeds that are alternative hosts to pests in other crops so that the pest life-cycle can be broken. To reduce the spread of pests such as cyst nematodes, contaminated soil should be returned to the same field after grading and cleaning crops such as potatoes or sugar beet. Care should be taken when moving machinery from heavily infested land to clean land to try and reduce movement of contaminated soil. Grain stores should be thoroughly cleaned before harvest to reduce the incidence of grain pests.
- *Resistant and tolerant varieties.* Some crop varieties show resistance to certain pests. Such resistance often manifests by slowing pest growth and development, therefore limiting populations from growing to economically important levels. Perhaps the best-known example of a resistant variety being successfully implemented on-farm can be found in wheat, where resistance has been bred for orange wheat blossom midge. Tolerant varieties can produce good yields even when they become infested with a crop pest. Resistant varieties have a negative impact on pest biology while tolerant varieties have no such impact.
- *Barriers.* Fleece or net crop covers prevents some pests, such as carrot flies, from laying their eggs on to a crop. On a field scale it is expensive although a very effective option. In swedes this is the only method available to growers to control cabbage root fly as there are currently no suitable recommended chemicals.
- *Trap cropping.* Growing a susceptible plant species before the main crop or growing it in strips around a crop protects the main crop. This sacrifice plant/crop is used to attract mobile predators before they attack the crop; these pests can then be killed so that the pest life-cycle is not completed. An example of a trap crop is sticky nightshade (*Solanum sisymbriifolium*), which is grown to reduce populations of the potato cyst nematode.

7.5.2 Biological control

Many insects are predators (e.g., ladybirds) or parasitoids (a type of parasite that eventually kills its host e.g., parasitoid wasps) of crop pests. For example, a hoverfly larva can eat 400 aphids in its lifetime while an aphid parasitoid may parasitise up to 100 aphids. Farmers can exploit these beneficial insects to naturally control pest populations within their crops, which is widely known as biological control. This approach to pest control was predominantly developed for use in protected cropping systems (e.g., glasshouses, polytunnels) and relies on releasing large numbers of mass-reared beneficial insects into the crop (augmentation biological control). Although this approach to biological control has been successfully used in protected cropping systems for over 60 years it hasn't been widely adopted in field crops. There are many reasons for this but it is predominantly a result of financial pressures. Typically, each hectare of a field crop has a lower value than the same area of a protected crop while mass-reared beneficial insects are more expensive than their synthetic pesticide alternatives. However, beneficial insects can be encouraged into field crops by ensuring that there are suitable diverse habitats such as flower rich margins around field perimeters. These habitats provide alternative food sources (e.g., floral nectar) and shelter. Increased numbers of predators and parasitoids of crop pests helps to reduce the requirement for chemical controls by preventing pests from reaching economic injury thresholds.

7.5.3 Biopesticides

Biopesticides are plant protection products derived from natural sources. These products are increasingly being used as a component of IPM programmes. Reasons for the increased use of biopesticides include improved operator safety, greater specificity, reduced persistence in the environment and improved compatibility with biological controls than many of their synthetic counterparts. There are two main biopesticide groups in use within UK agriculture:

- *Microbial.* Fungi, bacteria, protozoa and viruses are natural enemies of invertebrate organisms that can be used to suppress invertebrate pest populations. These microbial organisms are widespread within nature and do not negatively impact non-target or vertebrate organisms, so are considered environmentally benign. There are many examples of microbial biopesticides but two widely used species are the bacteria, *Bacillus thuringiensis* (often simply referred to as *Bt*) and the fungus, *Beauveria bassiana. Spinosad* is another example of a biopesticide derived from a microbial source (bacterial) as a by-product of sugarcane fermentation and is predominantly used for Lepidopteran pest control in the UK. Although entomopathogenic (i.e., insect-killing) nematodes are not strictly microbes as they are multi-cellular, they are often grouped within microbial biopesticides due their microscopic size.
- *Naturally occurring chemicals.* These chemicals may originate from a variety of sources, but most biopesticides based on naturally occurring chemicals originate from plant material. Such chemicals are normally produced by plants as anti-herbivore defences that can be extracted, formulated and applied to crops to reduce pest populations. Biopesticide products based on neem oil are commercially available for use in the UK and these act as anti-feedants to reduce pest feeding on crops. Besides plant-derived biopesticides, insect-derived semiochemicals (e.g., pheromones) may be used to facilitate specificity and improve pest monitoring.

Although biopesticides account for 20% of all plant protection products registered for professional use in the UK, these are currently almost exclusively used on high value horticultural crops. There are several reasons for their restricted use. Biopesticides are relatively expensive compared to conventional synthetic pesticides, they often need to be applied repeatedly to be effective due to poor residual effects (i.e., they are only effective for a limited period of time before requiring further applications) and they are typically applied in large volumes of water. Environmental conditions during application also determine product efficacy, limiting their use to periods of dry weather or during the evening when humidity is higher and UV radiation is lower.

7.5.4 Chemical control

If prevention using cultural control methods is inadequate, there are a wide range of pesticides available for pest control. Many of them achieve control by acting on target organism's nervous system. Currently the Insecticide Resistance Action Committee (IRAC) recognises 29 different chemical modes of action groups, although in the UK the number is smaller. Pesticides are applied in several different ways and formulations. Most are applied as sprays or as seed coatings, though a small number are applied in granular form or as baits. Fumigants are used in grain stores for the control of beetles and weevils. There are two main ways that pesticides kill their target organisms:

1. Contact – the pest is killed when it comes into contact, such as:
   a. Directly hit by a spray during application;
   b. Walking over treated areas;
   c. Absorbing vapour;
   d. Passing through treated soil.
2. Ingestion – the pest is killed when it consumes treated material. Stomach poisons need the pest to feed on pesticide treated materials (plant or bait) to be effective. Many pesticides that kill by ingestion are therefore systemic. A systemic pesticide is one that is applied directly to crops via the foliage or seed and is translocated within the plant through their vascular structures (xylem and phloem). Systemic pesticides may also be applied to the soil around the base of plants.

Most pesticides kill by more than one method, which makes them very effective. But many active ingredients within pesticides are extremely toxic to animals and humans and, by law, certain precautions must be observed by the persons using them, both when handling and applying. It is important to remember that a harvest interval must be observed between the last application of the pesticide and the harvesting of edible crops, as well as a re-entry period for the access of humans and animals to treated areas.

7.6 Pesticide classification

Pesticide is a broad term reserved for chemicals used to control any pest organism (insects, nematodes, molluscs, mites, rodents, weeds, diseases, etc). Strictly speaking we should refer to specific groupings rather than this broad term. For example, insecticides are used for insect control, nematicides are used for nematode control, molluscicides are used for slug control and acaricides are used for mite control. For up-to-date information on pesticides authorised for use in the UK, as well as regulations and advice on the use of these chemicals, reference should be made to the *UK Pesticide Guide*, published annually by CABI/BCPC or through the Health and Safety Executive (HSE) Website. Pesticides are typically grouped according to their chemical properties:

- *Fatty acids or soaps.* These compounds are used in horticulture against soft bodied insects by applying directly to foliage. Pests must be sprayed directly to achieve any control.

- *Organophosphates.* As a group, the organophosphates are dangerous to use. This group of insecticides mainly affects the pest's nervous system. They are generally non-persistent and do not accumulate in the environment. They should be handled strictly in accordance with the manufacturer's instructions. There are few approved products still being used but one example is *pirimiphos-methyl*, which is used as a fumigant for stored grain pest control.
- *Carbamates.* These compounds also affect the pest's nervous system. They work quickly and have a reasonable rate of breakdown. *Pirimicarb* is used for aphid control in pea and bean crops. This insecticide is quite specific and has less of an effect on beneficial insects.
- *Synthetic pyrethroids.* These chemicals have a particularly high insecticidal activity while generally being safe to humans and farm animals. They are efficient contact insecticides with a rapid knockdown action and some are thought to act as antifeedants. These properties are seen as being valuable in the chemical control of plant viruses by controlling the insect vectors (e.g., *deltamethrin* and *lambda-cyhalothrin*).
- *Neonicotinoids.* The newest group of insecticides developed from nicotine. Mammalian toxicity is lower than for the organophosphates. This group of insecticides are systemic and very effective against sap feeding insects. These insecticides work by affecting the pest's nervous system. In the UK most treatments were applied as seed dressings until a total ban on outdoor use in 2018 due to concern regarding their impact on pollinators and other beneficial invertebrates.

7.6.1 Insecticides

A wide range of insecticides have been developed for controlling insect pests within crops since the end of the Second World War. Most of the early synthetic insecticides (e.g., DDT; *dichlorodiphenyltrichloroethane*) are no longer permitted for use in UK cropping systems due to their negative impacts on environmental and human health. Other insecticides developed more recently have similarly been withdrawn, such as several neonicotinoids, for the same reasons. Consequently, there is a diminishing portfolio of active ingredients approved for use in the UK. The most widely used class of synthetic insecticide in the UK are pyrethroids. This group of broad-spectrum synthetic insecticides are based on natural chemicals called pyrethrins that are extracted from chrysanthemum flowers. Pyrethroids, such as *lambda-cyhalothrin* and *deltamethrin*, are contact-acting pesticides used to control a range of insect crop pests. In recent years, however there has been an increasing number of cases of resistance to these insecticides (e.g., cabbage stem flea beetle, peach-potato aphid, pollen beetle) meaning that in these cases pyrethroid insecticides are no longer effective.

7.6.2 Nematicides

Nematicides have been developed for controlling nematode pests within crops. Like insecticides, many nematicides have been withdrawn from the market due to their negative impacts on human and/or environmental health. *Fosthiazate* is one of the few nematode control options remaining and is a soil-applied organophosphate, contact-acting nematicide used to control potato cyst nematodes.

7.6.3 Molluscicides

A number of molluscicides have been developed for controlling slugs and snails within crops. This group of pesticides tend to be deployed as pellets around the plant base. The molluscicide *metaldehyde* has recently been withdrawn from the market due to concerns around the impact of this chemical on human health, particularly as it has been found in catchments where water was abstracted for human consumption. *Ferric phosphate* is now the only molluscicide used in UK agriculture and works by interfering with stomach calcium metabolism and causing molluscs to stop feeding, which then typically die below ground.

7.6.4 Acaricides

Acaricides have been developed for the control of mite pests within crops. Some insecticides may also be used against mite pests as both insects and mites are arthropods and so have many similarities. Some products are, however, used only as acaricides, such as *tebufenpyrad*. This acaricide is effective against a wide range of different mite pests.

7.6.5 Pesticide resistance

Pest populations resistant to insecticides have developed due to overuse of products with the same mode of action, especially in species that have many generations a year and have high multiplication rates. After introduction of new active ingredients, it usually takes from 2 to 20 years for cases of resistance to appear. Resistant populations develop as susceptible insects are killed leaving those that are resistant or less sensitive to increase by passing on their resistance to their offspring. There are several mechanisms of resistance:

- Metabolic resistance – this is the most common mechanism. Resistant pests can detoxify or destroy the pesticide faster than susceptible insects. Pyrethroid resistance in UK pollen beetle populations has been identified as being facilitated through enhanced metabolism by P450 monooxygenases;
- Target-site resistance – this is the second most common mechanism. Resistant pests have a modified activity site so that the pesticide does not work. Pyrethroid resistance in UK cabbage stem flea beetle populations has been identified as being facilitated through a mutation in the target-site of this pesticide;
- Penetration resistance – resistant pests have developed a cuticle that restricts pesticide absorption into their bodies. This mechanism offers limited protection and is typically observed with one of the other resistance mechanisms. Pyrethroid resistance in UK diamondback moth populations has been partly attributed to reduced penetration efficacy;
- Behavioural resistance – this is the least common mechanism. Resistant pests can detect the pesticide and leave the treated area to avoid coming into contact with it.

When pests exhibit resistance to pesticides it is often due to more than one of these mechanisms. Peach potato aphid is an excellent example of this as resistance in UK populations is facilitated through multiple mechanisms. Organophosphate and carbamate resistance is facilitated through enhanced enzyme expression that degraded the pesticides while pyrethroid resistance is facilitated through mutations in the pesticide target-site.

Table 7.1 Economically important crop pests, their damage symptoms and control methods.

Crop attacked	Pest	Description	Life-cycle	Symptoms of attack	Control	Notes
Cereals	**Slugs** (mainly grey field slug, *Deroceras reticulatum*)	*Grey field slug:* grey or fawn in colour and up to 50 mm in length.	Active throughout the year when conditions are warm and damp. Breeding occurs mainly in the spring and autumn. Eggs are laid just below the soil surface and slugs tend not to move far from these sites. Slugs often have a patchy distribution within fields and are typically more damaging if the seedbed is loose, cloddy and trashy. Populations and damage are highest on heavy textured soils.	Slug feeding causes seed hollowing in cereal crops where the embryo is destroyed. In addition, young cereal plants can be completely 'grazed off' if slug pressure is high.	Establishing a fine consolidated seedbed can help to reduce the problem. Use a molluscicide such as ferric phosphate if there is a risk of damage, indicated by use of refuge traps baited with chicken layers mash. Rolling and early nitrogen applications can help a damaged crop to recover in the spring.	Winter wheat chiefly attacked. Often worse after oilseed rape or other broad-leaved crops. Use 9 slug bait traps per field, however in fields greater than 20 ha increase trap number to 13. Leave overnight and count slugs following morning. There is a risk of damage to winter cereals if at least 4 slugs are caught/trap.

	Wireworm (adults known as click beetles and larvae known as wireworms) - key species include *Agriotes sputator, A. obscurus* and *A. lineatus*	*Larva:* Initially white but turn golden brown in colour, cylindrical body, three pairs of very small legs, may grow up to 25 mm in length.	Wireworm hatch out during summer from eggs laid just below soil surface, mainly in grassland or around grass weeds. Wireworm actively feed in the spring and again in the autumn and take 4–5 years to complete their development. After pupation in the soil, the adults emerge in late summer and lay their eggs the following spring.	Yellowing of foliage followed by the disappearance of successive plants in a row. This is caused by wireworm moving along the row. Larvae eat into the plants just below soil surface. They are usually found in soil around the base of plants.	Good growing conditions to help the crop grow away from an attack. Insecticide treated seed may be used where high numbers of wireworm present, which is most likely in fields after permanent pasture. Damage is reduced when the soil is well consolidated. Control grass weeds.	Wheat and oats are more susceptible than barley. Seed treatment is recommended if soil counts find more than 750,000 larvae/ha. Recently wireworms have been found in some fields in arable rotations, which may be associated with use of zero or minimum tillage.
	Leatherjackets (adults known as crane flies or daddy long legs and larvae known as leatherjackets) - key species include *Tipula paludosa* and *T. oleracea*	*Larva:* Greyish-black in colour, no legs and a tough 'leathery' skin, may grow up to 40 mm in length.	Eggs laid on grassland or weedy stubble in the autumn from which the larvae soon emerge. Leatherjackets feed overwinter when soil temperature is above 0.5°C and feed more actively in the spring in warmer conditions before pupating in the summer with adults emerging in late summer.	Crops die away in patches, the roots and stems of plants having been eaten just below the soil surface. Leatherjackets found in soil around the base of plants.	Where possible, plough the field in July and early August, covering the old grass sward well can reduce egg laying. Rolling and nitrogen top-dressing may be used to help the crop to recover from leatherjacket damage.	Spring cereals after long-term grassland are most at risk from leatherjacket damage.

Continued

Table 7.1 Economically important crop pests, their damage symptoms and control methods.—cont'd

Crop attacked	Pest	Description	Life-cycle	Symptoms of attack	Control	Notes
	Wheat bulb fly (*Delia coarctata*)	*Larva:* Creamy white in colour, no legs, indistinct head and a blunt tail end to the body. Larvae are found inside tillers and may grow up to 12 mm in length.	Eggs laid on bare soil or on soil between row crops such as potatoes, sugar beet and onions in July and August. Eggs hatch from early January and larvae invade plants. Larvae feed on the crop until May before pupating at the base of plants.	The central shoot of a plant turns yellow and dies in early spring, this is called 'deadheart' symptom.	Late sown wheat after vegetables, potatoes, sugar beet or peas are most at risk. Increase seed rates in susceptible fields and consider use of insecticide seed treatment in high risk situations where there are more than 100 eggs/m^2. Encourage natural enemies, such as ground beetles and rove beetles. Avoid bare ground in July and August where possible and delay drilling until after egg laying.	Wheat bulb fly is a problem mainly of central and eastern counties. Use a seed treatment if late drilled and >2.5 million eggs/ha.

	Frit fly (*Oscinella frit*)	*Larva:* Creamy white in colour, no legs, indistinct head and slightly pointed at head and tail end of body. Larvae may be found inside tillers and can grow up to 3 mm in length.	There are three generations within the year. In the first generation, larvae feed on crops in May and June. The second generation when larvae feed on oat kernels or grasses in July and August. The third generation when larvae feed initially on grasses from October onwards and overwinters on the grass unless it is cultivated for cereals when the larvae will move onto the establishing crop.	In early summer, larvae from the first generation may kill the central shoot of the oat plant, known as 'deadheart' as the outer leaves remain green. Blind spikelets and shrivelled grains are caused by larvae from the second generation. The third generation of larvae cause 'deadheart' in winter cereal crops.	Sow spring oats before mid-April, and try to get these crops past the 4-leaf stage as quickly as possible. Allow a 6-week interval between ploughing grass and sowing winter cereal crops. Delay sowing of winter cereals where possible. In high risk situations use an insecticide seed dressing where permitted. Encourage natural enemies such as ground beetles, rove beetles and predatory flies.	Spring oats and maize are particularly susceptible. Mainly a problem in rotations with grass.

Continued

Table 7.1 Economically important crop pests, their damage symptoms and control methods.—cont'd

Crop attacked	Pest	Description	Life-cycle	Symptoms of attack	Control	Notes
	Yellow cereal fly (*Opomyza florum*)	*Larva:* Yellowish in colour, no legs, indistinct head, slender body with slightly pointed head and tail end. Larvae are found inside tiller and may grow up to 7 mm in length.	Eggs laid near wheat plants in October and November, hatching February and March the following year. Larvae enter the central shoot of the plant which is killed. Larvae pupate inside the shoot or in the soil in May with adults emerging in the summer. Adults migrate to hedgerows before returning to wheat fields in late autumn to lay eggs.	Yellow cereal fly produces a characteristic spiral shaped scar at the base of a tiller as it enters the plant. Larval feeding results in classic 'deadheart' symptoms. Larvae only inhabit one tiller and do not move to other tillers (unlike wheat bulb fly).	Early sown winter wheat in the eastern counties is at greatest risk but because only a single tiller is damaged crop losses are usually low. Establish a good population of at least 200 plants/m^2 and consider use of an insecticide seed treatment in high risk situations.	Winter barley is rarely attacked and spring sown cereals are not considered to be at risk from this pest.
	Gout fly (*Chlorops pumilionis*)	*Larva:* Creamy white in colour, no legs, indistinct head. May grow up to 6 mm in length.	There are two generations a year. In the first generation, eggs are laid in May and larvae complete their development by September. In the second generation, eggs are laid in August and September and larvae overwinter, completing their development in March and April of the	Tillers on which the larvae feed become swollen and stunted. Crops affected usually compensate by producing new tillers. The first generation in late sown spring cereals can cause yield losses of 30%–50%.	Time of drilling is important, September drilled winter crops are most at risk. Encouraging tillering in the spring by application of an early nitrogen treatment may be effective in minimising crop losses.	Barley and wheat are mainly affected in England and Wales. Oats are not attacked. Crops can compensate for up to 25% of plants damage. If 50% of winter cereal plants have eggs present, then consider a foliar insecticide treatment.

			following year for the second generation. Larvae from both generations tunnel into the plant tillers and pupate inside the plant.			
	Cereal aphids (species including English grain aphid (*Sitobion avenae*), bird cherry-oat aphid (*Rhopalo siphum padi*) and rose-grain aphid (*Metopolo phium dirhodum*)	*English grain aphid:* 1.3–3.3 mm in length, green or yellowish to almost black in colour but always with black tubes (siphunculi) at the rear of the insect. *Bird cherry oat aphid:* 1–2 mm in length, olive green in colour with rusty coloured patches at the rear of the insect. *Rose-grain aphid:* 1.6–3 mm in length, pale green in colour with a darker green stripe running down the back of the insect.	*English grain aphid:* Migrates into establishing cereal crops between September and March. Asexual populations build up on cereals and grasses through the spring and summer. This species remains on cereals and grasses throughout the year. *Bird cherry oat aphid:* Migrate to cereal crops and grasses from bird cherry trees in May and June. Asexual populations build up through summer and then from September to November may either return to bird cherry trees to complete a single sexual	Damage is predominantly through transmission of viruses, such as barley yellow dwarf virus (BYDV), however direct feeding may also negatively impact the crop if populations are large enough. The English grain aphid causes empty and/ or small grain; by puncturing the grain at the milk-ripe stage, the grain contents seep out. This also reduces the weight of the grain. Rose grain aphids affect the leaves not the ear	Timing of drilling is important with early sown crops often being at greater risk of viruses such as BYDV. Where appropriate consider BYDV resistant or tolerant varieties of wheat and barley. Encourage natural enemies such as ground beetles, rove beetles, ladybirds, lacewings and parasitoid wasps. Where insecticide applications are necessary use products that are less harmful to aphid natural	Thresholds for direct feeding damage by cereal aphids is 50% of tillers infested before the start of flowering and 66% of tillers infested after the start of flowering and up to 2 weeks before harvest. There are no established thresholds for virus transmission and instead temperature models should be used to determine the need to control and to optimise the timing of control.

Continued

Table 7.1 Economically important crop pests, their damage symptoms and control methods.—cont'd

Crop attacked	Pest	Description	Life-cycle	Symptoms of attack	Control	Notes
			generation and overwinter as eggs or migrate into establishing cereal crops where asexual reproduction continues. *Rose-grain aphid:* Migrates into crops from wild and garden roses in March and April. Asexual populations build up on cereals and grasses through the summer before migrating back to roses to complete a single sexual generation and overwinter as eggs.	and damage is usually less important. Virus transmission is important before the crop reaches stem extension. The main vectors of viruses in cereal crops are the English grain aphid and the bird cherry-oat aphid.	enemies. Many aphid pests are now resistant to key insecticides used to control them, particularly pyrethroids.	
	Orange wheat blossom midge (*Sitodiplosis mosellana*)	*Adult:* Bright orange in colour and up to 3 mm in length. *Larvae:* Bright orange in colour, no legs, indistinct head and may grow up to 2.5 mm in length.	Adults emerge from the soil in June and lay eggs in the florets at dusk. Larvae hatch from the eggs after 4–10 days and then feed on the developing grain for about 2 weeks before	Crops are only susceptible to damage between ear emergence and flowering. When larvae are present during these crop growth stages yield loss occurs due to	Many wheat varieties are now resistant to this pest. Where a resistant variety is not grown a carefully timed foliar insecticide may be considered. Monitoring through	Orange wheat blossom midge are mainly a pest of wheat in the main arable areas of the UK and are more important than yellow wheat blossom midge

			dropping to the soil. Larvae overwinter as a cocoon and can survive in the soil for at least 10 years, only pupating in the spring when soil conditions are warm and moist.	shrivelled grain. Grain quality is also reduced and can be seen in a reduced Hagberg falling number.	the use of crop inspections, pheromone traps or sticky yellow traps are important to correctly time use of insecticides. Encourage natural enemies such as parasitic wasps and dance flies.	(a pest with a similar appearance and life cycle but bright yellow in colour).
	Saddle gall midge (*Haplodiplosis marginata*)	*Adult*: Red in colour and up to 5 mm in length. *Larva* – initially whitish in colour but become almost red in colour as they mature and may reach up to 4 mm in length.	Adults emerge from soil from April to June and lay chains of eggs on cereal and grass leaves. The larvae that emerge from these eggs burrow down to feed underneath the leaf sheath. Once development is complete in July the larvae drop to soil to overwinter, only pupating the following spring if conditions are suitable.	Saddle shaped galls appear on stems where larvae feed. These galls can cause stems to break or lodge and reduces the flow of nutrients to the developing ears.	Crop rotation is very important; including crops other than cereals can be successful at controlling this insect. Avoid late drilling of winter cereals where problems with this pest have previously been noted. Monitor through crop inspections, use of sticky yellow traps or pheromone traps.	Affects wheat, barley and some other grasses. This is an occasional pest most often found in areas of intensive cereal growing on heavy soils in the UK.

Continued

Table 7.1 Economically important crop pests, their damage symptoms and control methods.—cont'd

Crop attacked	Pest	Description	Life-cycle	Symptoms of attack	Control	Notes
	Cereal leaf beetle (*Oulema melanopus*)	*Adult*: 4–5 mm long with a red thorax and blue-green elytra (wing cases). *Larvae*: Dark green/yellow in colour with a black head. Larvae cover themselves in a mixture of mucus/excreta.	Adults invade crops in the spring and feed on leaves. Eggs are laid onto the upper leaves. The larvae feed openly on leaves before pupating inside a cocoon on the leaf surface.	Both adults and larva consume strips from the leaves.	Control is not needed as this species is not considered economically important. Natural enemies keep populations below economic thresholds.	Rarely considered economically important.
Stored grain	**Saw-toothed grain beetle** (*Oryzaephilus surinamensis*)	*Adult:* Dark brown in colour with six 'teeth' along each side of the prothorax and up to 3.5 mm in length *Larva:* White in colour with a broad head and a flattened body. They may be up to 3 mm in length.	Eggs are laid on the stored grain; larvae feed on damaged grains and pupate inside a cell constructed from food particles and other debris.	The grain heats up rapidly when infestations become severe and the stored grain weight will be reduced. Effects on grain quality include tainting, reduced germination and contamination.	Clean stores thoroughly before use and apply an insecticide or use physical pest control using diatomaceous earth. Seal any gaps/cracks in building.	Use traps to identify pest species and numbers present. Check store temperature regularly, maintain grain at < 15% moisture content and below 15°C.

	Grain weevil (*Sitophilus granarius*)	*Adult:* Reddish to dark brown in colour with characteristic elongated snout that is typical of weevils. Typically 2–4 mm in length.	During autumn the weevils bore into the stored grain to lay their eggs. Both the adults and the larvae damage grains.	Damage is seen by the presence of hollowed grains and the rapid heating of the grain within the store. These weevils are typically found 30–60 cm below the surface of stored grain.	As for saw-toothed grain beetle.	As for saw-toothed grain beetle.
	Rust–red grain beetle (*Tribolium castaneum*)	*Adult:* Shiny red in colour with a flattened appearance, characteristically long antennae and up to 2.5 mm in length antennae. *Larva:* White in colour with brown head and up to 4 mm in length.	Eggs are laid on grain. Larvae emerge and burrow into grain. Pupation takes place on grain or other surfaces. Damage is often worse on damaged grain that allows easier access for adults and larvae.	Mainly feed on the grain embryo.	As for saw-toothed grain beetle.	As for saw-toothed grain beetle.
	Mites (species including the grain mite, *Lepidoglyphus destructor*, the flour mites, *Acarus siro*, and *Tyrophagus longior*.	*Adults* and *larval* at 0.5 mm in length are only just visible with the naked eye moving slowing over the surface of infested grains.	Eggs are laid on the stored grain. The larvae hatch from the eggs and then pass through two nymphal stages. The time taken to complete the life-cycle is affected by environmental conditions.	Causes grain to heat up. Can taint flour, produce allergens and spread fungal spores.	As for saw-toothed grain beetle but also consider turning grain to crush mites and physically remove by cleaning.	Check store temperature regularly, maintain grain at < 14% moisture content and 5°C.

Continued

Table 7.1 Economically important crop pests, their damage symptoms and control methods.—cont'd

Crop attacked	Pest	Description	Life-cycle	Symptoms of attack	Control	Notes
Maize	**Wireworm** (as for wireworm in cereals)	As for wireworm on cereals.	As for wireworm on cereals.	Disappearance of successive plants in a row. This is caused by wireworm moving along the row. Larvae eat into the plants just below soil surface. They are usually found in soil around the base of plants.	Good growing conditions to help the crop grow away from an attack. Damage is reduced when the soil is well consolidated. Control grass weeds as these serve as alternative hosts. Implement crop rotations to prevent populations from building up.	Maize crops grown after long-term grass are at greatest risk of damage.
	Frit fly (*Oscinella frit*)	As for frit fly on cereals.	As for frit fly on cereals.	Feeding by larvae may result in twisting of leaves surrounding the growing point. In severe attacks these leaves may be killed, leading to the production of secondary tillers.	Leave at least 6 weeks between ploughing of grassland and the planting of maize to reduce the risk. Encourage natural enemies such as ground beetles, rove beetles and predatory flies.	

Field beans	**Black bean aphid** (*Aphis fabae*)	1.5–3 mm in length, body sooty black in colour, white legs and white antennae.	Black bean aphids migrate from spindle trees to bean crops in May and June. The winged aphids that arrive in the crop give rise to wingless offspring that complete many generations throughout the summer. In the autumn winged migrants are again produced that fly back to spindle trees on which eggs are laid that overwinter.	Colonies of these aphids are seen on the stem leaves, the flowers and developing pods of bean plants. Direct feeding by these aphids results in plants that may be stunted or even killed if the infestation is heavy. Black bean aphid can also transmit viruses that affect the crop.	Application of a foliar insecticide may be required if the action threshold of 2.5%–10% of plants infested (threshold determined by growth stage) is met. Encourage natural enemies such as ground beetles, rove beetles, ladybirds, lacewings and parasitic wasps. Where insecticide applications are necessary use products that are less harmful to aphid natural enemies.	Mainly a pest of spring beans but this species of aphid may also be found attacking sugar beet.
	Bean seed beetle/ Bruchid beetle (*Bruchus rufimanus*)	*Adult:* Mainly black oval body and up to 5 mm in length. *Larva:* Creamy white body, three pairs of legs and a brown head. Up to 6 mm in length.	Adult lays eggs on the surface of the developing pod in May and June. The larvae hatch from these eggs and burrow into the pod and enter developing seed. Larvae pupate inside	Damage to the seeds affects the value of crop for export and for human consumption.	Apply an insecticide if action threshold is reached on susceptible crops where quality an issue. The action threshold for this pest is if there are two consecutive	Careful nozzle selection may be important to ensure good coverage.

Continued

Table 7.1 Economically important crop pests, their damage symptoms and control methods.—cont'd

Crop attacked	Pest	Description	Life-cycle	Symptoms of attack	Control	Notes
			the seed and adults cut a characteristically round exit hole as they leave the seed.		days above 20°C and adult beetles are found then spray and repeat after 7–10 days.	
	Stem nematode (*Ditylenchus dipsaci* and *D. gigas*)	Clear and slender bodies that may be up to 1.2 mm in length.	Young adults can overwinter in the soil before moving in moist soil to invade suitable hosts in late spring. Nematodes then reproduce rapidly within the plant; they can move up the plant and infect the seed.	Stem nematodes cause stunting, malformations such as twisted and thickened stems, staining of the stem and even plant death. Often symptoms only become apparent close to harvest when plants collapse and are difficult to harvest.	No chemical control available, instead ensure that only clean seed is used. Have any home saved seed tested. Avoid growing pulses in close rotation and include non-susceptible crops in rotations such as wheat and barley. Ensure wild oats are controlled as they are also a host for stem nematodes.	Once the soil becomes infested with stem nematodes they may remain present for many years. Wet springs favour stem nematodes.

Peas and beans	**Pea and bean weevil** (*Sitona lineatus*)	*Adult:* Light brown with feint creamy-yellow stripes, up to 6 mm in length. *Larva:* Creamy-white body, brown head but no legs, up to 6 mm in length.	Adults overwinter in leaf litter and become active in early spring when they feed on leaves and lay eggs in the soil close to plants. The larvae hatch from these eggs and complete their development feeding on the roots of plants. Pupation takes place in the soil in mid-summer.	Larvae feed on roots and nitrogen fixing root nodules, which reduces yields and nitrogen content of crops. Adults feed on leaves producing characteristic notches along leaf margins, which again reduces yields.	Foliar applied insecticides may be required when the action threshold of 30 weevils in a pheromone trap on any 1 day is exceeded.	Spring sown crops are more affected than winter sown crops.
	Bean seed flies (*Delia platura* and *D. florilega*)	*Adult:* Greyish-black in colour and up to 5 mm in length. *Larva:* Typical fly larvae, white and up to 8 mm in length. *Pupa:* Reddish-brown in colour and up to 5 mm in length.	Pupae overwinter in the soil and adults emerge between March and September. Eggs are laid into freshly disturbed soil. Larvae feed on buried seed or seedlings and in the absence of fresh plant material on decomposing organic matter. Larvae complete their development in around 14 days but may be found in the soil at any time throughout the year after March. There are three or more generations a year.	Damage is typically seen through the patchy emergence of the crop. Other symptoms include deformed growth of the seedling.	The presence of organic matter is a key factor in stimulating egg laying and so any organic debris should be buried before sowing a new crop. Where possible ensure vigorous early crop growth and where available consider the use of an insecticide seed treatment.	Monitoring traps are available with which to determine pest pressure and the need to apply controls.

Continued

Table 7.1 Economically important crop pests, their damage symptoms and control methods.—cont'd

Crop attacked	Pest	Description	Life-cycle	Symptoms of attack	Control	Notes
Peas	**Pea moth** (*Cydia nigricana*)	*Adult:* Dull brown wings with black and white bars along leading edge of forewings, wingspan of up to 16 mm. *Larva:* Creamy-white to yellowish colour to body, dark head and up to 15 mm in length.	Eggs are laid June to mid-August on the underside of leaves. Larvae hatch from eggs in approximately 1 week and enter pods and feed on peas until fully grown. Larvae leave the pod and make their way to the soil to pupate and overwinter.	Holes in peas caused by larvae reduces crop quality, while the presence of larvae in harvested crops may be a contamination problem.	One or more sprays of a foliar applied insecticide may be required if the action threshold of 10 or more moths in either of two pheromone traps on 2 consecutive occasions (a lower threshold is used for vining peas) is exceeded. Dates for insecticide application can be obtained from PGRO.	Early and very late sown crops suffer less damage. Dry harvested peas for human consumption or seed are most likely to require treating.
	Pea aphid (*Acyrthosiphon pisum*)	A large aphid that is up to 4.5 mm in length, typically green in colour but pink forms also found.	Overwinters as eggs on legumes, such as clover and vetch, and migrates into pea crops in May or June. Several generations are completed throughout the summer and dense colonies may form.	Direct feeding causes distortion of the plant and pods and direct yield loss. This species of aphid can also transmit a number of viruses, such as pea leaf roll virus.	Foliar applied insecticides may be required when the action threshold of 15% plants infested at early flowering of the crop is reached. Encourage natural enemies such as ground beetles, rove beetles, ladybirds, lacewings and parasitic wasps.	

					Where insecticide applications are necessary use products that are less harmful to aphid natural enemies.	
Oilseed rape	**Slugs** (mainly grey field slug, *Deroceras reticulatum*)	As for slugs on cereals.	As for slugs on cereals.	Slug feeding on young oilseed rape plants can reduce growth or kill plants if slug pressure is high.	Establishing a fine consolidated seedbed can help to reduce the problem. Use a molluscicide if there is a risk of damage, indicated by use of refuge traps baited with chicken layers mash.	Monitor in the same way as for cereal crops. There is a risk of damage to oilseed rape if at least 4 slugs are caught/trap in preceding standing cereal crop.
	Cabbage stem flea beetle (*Psylliodes chrysocephala*)	*Adult:* Metallic blue-black in colour, enlarged hind legs and up to 4.5 mm in length. *Larva:* Creamy white body, dark head and tail plate and three pairs of legs. Up to 7 mm in length.	Adults move into oilseed rape crop in September and lays eggs around germinating plants. Larvae invade plants from October to March and migrate back to the soil to pupate.	Adults cause characteristic shot-holing damage to leaves. Larvae burrow into leaf petioles and cause stunting and leaf loss. In severe cases plants may be killed by both adult and larval feeding damage.	Insecticide options are limited due to widespread resistance to pyrethroid insecticides. Early or late drilling may be considered as may the use of cover crops to reduce pest pressure. Badly infested crops may be defoliated to reduce larval numbers.	Monitoring for cabbage stem flea beetle is possible through use of water traps (adults), recording shot-holing damage (adults) and plant dissections (larvae).

Continued

Table 7.1 Economically important crop pests, their damage symptoms and control methods.—cont'd

Crop attacked	Pest	Description	Life-cycle	Symptoms of attack	Control	Notes
	Peach-potato aphid (*Myzus persicae*)	1–2.1 mm in length, typically green in colour but yellow and pink forms may also be found. Tips to the two tubes at the rear of the aphid are black in colour.	Winged adults fly into oilseed rape crops from surrounding areas as crops establish in the autumn or spring. These winged aphids give rise to wingless offspring that may complete several additional generations within the crop.	Most adult aphids migrating into crops carry turnip yellows virus. This is a persistent virus that can reduce yields by up to 26% and produces variable symptoms but may include purple fringing of leaves.	Delayed drilling may be useful in reducing numbers of aphids within crops. There are also now oilseed rape varieties with tolerance to turnip yellow virus. Where insecticide applications are considered necessary it is important to remember that pyrethroid resistance in this pest is very widespread.	There are currently no established action thresholds for this pest.
	Pollen beetle (*Meligethes aeneus* (sometimes referred to as *Brassicogethes aeneus*))	*Adult:* Metallic blue-black in colour, clubbed antennae and up to 2.5 mm in length.	Adults emerge from hibernation during spring to feed on buds and flower parts at temperatures over 15°C. Eggs are laid into damaged buds and larvae hatch to also feed on the buds	Damaged buds wither and die (blind stalks), and the number of pods set is reduced.	Monitoring for pollen beetle required when crop is at the green-yellow bud stage. The action threshold considers both the number of plants per m^2 and	Extreme caution should be taken to ensure that no serious damage occurs to pollinating insects when insecticide applications are made.

			and flowers. When larvae are fully grown they drop to the soil to pupate.		the number of beetles per plant. If more than 70 plants/m^2 then the threshold is 7 beetles/plant compared with 25 beetles/plant with thin crops of less than 30 plants/m^2. Where the threshold is exceeded a foliar applied insecticide may be required. Take account of pyrethroid resistance when selecting the insecticide to use.	
	Cabbage seed weevil (*Ceutorhy nchus assimilis*)	*Adults:* Grey in colour with curved snout and up to 3 mm in length. *Larva:* Creamy-white in colour, brown head but no legs and up to 5 mm in length.	From hibernation sites close to the previous oilseed rape crops, adults lay a single egg in the developing pods. Larvae feed on seeds in developing pods. Once development is complete the larva drops to the soil to pupate.	Seeds destroyed in pods by larva. Damage caused by the adults in laying eggs into pods also allows entry of brassica pod midge.	Using a longer rotation will help reduce the build-up of this pest. Foliar applied insecticides may be required when the action threshold of 0.5 cabbage seed weevil per plant (northern Britain) or 1 cabbage seed weevil per plant (southern Britain) is exceeded.	The headland usually has higher numbers of weevils compared with the rest of the field.

Continued

Table 7.1 Economically important crop pests, their damage symptoms and control methods.—cont'd

Crop attacked	Pest	Description	Life-cycle	Symptoms of attack	Control	Notes
	Brassica pod midge (*Dasineura brassicae*)	*Adult:* Greyish-brown fly that is up to 2 mm in length. *Larva:* Creamy-white larva, without a clearly defined head and no legs that is up to 2 mm in length.	Females can lay up to 60 eggs in a single pod. Eggs are inserted into pods through holes made by cabbage seed weevils, other insects or mechanical damage. Larvae feed on the developing seed and walls of pod; the larvae complete their development before dropping to the soil to pupate.	The feeding larvae cause pods to swell, distort and shatter leading to shedding of seed.	Control of this pest is based on controlling the cabbage seed weevil.	Winter oilseed rape most affected. Most damage is usually found in the crop headland.
	Rape winter stem weevil (*Ceutorhy nchus picitarsis*)	*Adults:* Black in colour with elongated snouts, reddish-brown leg-tips and up to 4 mm in length. *Larva:* Creamy-white in colour, brown head but no legs and up to 5 mm in length.	Adults move into autumn-sown oilseed rape crops in the spring to feed on foliage and lay their eggs in punctures within the plant stalk and crown. Hatching larvae bore into the plant to feed before dropping to the soil to pupate.	Feeding by the larvae may destroy the terminal shoot, either killing the plant or causing the plant to produce secondary shoots which flower and ripen unevenly.	Insecticide applications applied to control cabbage stem flea beetle adults were thought to provide incidental control of this pest.	Risk of infestation is greater when crops are sown near woodland.

	Brassica leaf miner (*Scaptomyza flava*)	*Adults:* Pale brown with faint stripes on the thorax, a large red eye and are 3 mm in length. Wings are approx. 50% longer than head and thorax. *Larvae:* Cylindrical maggot that is light green/yellow in colour and 3 mm in length.	Adults begin to emerge in July but are most abundant during September. Upon emergence and mating, each female can lay over 300 eggs on the lower leaf surface. Hatching larvae feed between the upper and lower leaf surfaces to create mines. When ready to pupate, larvae drop to the ground.	Irregular white/ yellow lines that may merge into large blotches if large pest populations are present.	Controls are rarely employed for this pest in oilseed rape.	Only infests first true leaves, so is not considered economically important.
Vegetable brassicas (cabbage, kale, swedes, turnips)	**Flea beetles** (key species include the cabbage flea beetle, *Phyllotreta cruciferae*.	Similar in appearance to the cabbage stem flea beetle but slightly smaller e.g., adults up to 3 mm in length.	Adults emerge from hibernation during late spring to feed on crops. Eggs are laid around plants but larvae do little damage. Pupation takes place in the soil during the summer.	Very small round holes are eaten in the cotyledons, first true leaves and stems of the plants. In severe cases plants are killed.	Sow the crop either early or late to avoid peak adult activity in April and May. Establish good growing conditions to get the crop quickly past the seed leaf stage. Few effective insecticide options available and there is evidence of pyrethroid resistance in populations of this pest. Crop covers may be considered to provide a physical barrier.	An increasing problem especially in dry springs.

Continued

Table 7.1 Economically important crop pests, their damage symptoms and control methods.—cont'd

Crop attacked	Pest	Description	Life-cycle	Symptoms of attack	Control	Notes
	Cabbage root fly (*Delia radicum*)	*Adult:* Similar in appearance to a house fly, may be up to 7 mm in length. *Larva:* Creamy-white in colour, without a clearly defined head and may grow up to 8 mm in length.	There are usually three generations a year. The first generation of eggs are laid next to brassicas plants during May. Eggs from the second and third generations are laid between July and September but there may be overlap between generations and timings vary depending on region, so no distinct emergence peaks may be observed.	Affected plants are often stunted, may appear bluish in colour and drought stressed. Young plants may be killed.	A limited number of foliar applied insecticides and biopesticides, e.g., spinosad, are approved for use against this pest. Crop covers can be very effective by preventing egg laying around young plants. Control brassica weeds which can act as an alternate host though crop cultivations can be effective.	A potentially serious pest of brassicas. Winter oilseed rape can be attacked by the third generation of flies when drilled early. A national weather-based forecasting system is available.
	Turnip sawfly (*Athalia rosae*)	*Adult*: Characteristic black and orange markings. May be up to 8 mm in length. *Larva*: Greenish-black in colour,	There are 3 generations a year. Adults emerge when warm enough from late spring. Eggs are laid on leaves of host. Larvae hatch and feed voraciously, quickly	Larvae eat leaves and can totally strip plants in a short time.	Crop covers may be effective for preventing eggs being laid onto the crop.	Recently numbers of this pest have increased in the UK. Mainly a problem in turnips, oilseed rape and swedes.

		pale stripe along the side of the body, may grow up to 18 mm length.	skeletonising the plant. Once fully grown the larvae drop to soil to pupate.			
	Mealy cabbage aphid (*Brevicoryne brassicae*)	Aphids are covered in a characteristic greyish-white wax and are up to 2.6 mm in length.	Aphids over winter as adults or as eggs on brassica hosts. Eggs hatch in spring and aphid move to the underside of leaves and shoots. Winged forms migrate from the end of May to other crops.	Damage is caused by direct feeding, which weakens plants. Damage is most serious on young plants which can lead to plant death. Marketability is affected in some vegetable brassicas crops due to the presence of the aphids and wax they produce. Important for transmission of viruses, e.g., cauliflower mosaic virus.	Destroy remains of harvested over wintered brassica crops by beginning of May. Foliar applied insecticides may be required in some cases. Encourage natural enemies such as hoverflies, ladybirds, lacewings and parasitic wasps. Where insecticide applications are necessary use products that are less harmful to aphid natural enemies.	Not all brassicas are affected. Several brassica weeds are hosts to this species of aphid. The action threshold for treating winter oilseed rape is 13% plants infested before petal fall.
	Bean seed flies (*Delia platura* and *D. florilega*)	As for bean seed fly on peas and beans.	As for bean seed fly on peas and beans.	As for bean seed fly on peas and beans.	As for bean seed fly on peas and beans.	As for bean seed fly on peas and beans.
	Brassica leaf miner (*Scaptomyza flava*)	As for brassica leaf miner on oilseed rape.	As for brassica leaf miner on oilseed rape.	As for brassica leaf miner on oilseed rape.	Crop covers may be effective for preventing eggs being laid onto the crop.	Economically important pest of vegetable brassicas. Adult monitoring relies on yellow sticky traps and are often most abundant during September.

Continued

Table 7.1 Economically important crop pests, their damage symptoms and control methods.—cont'd

Crop attacked	Pest	Description	Life-cycle	Symptoms of attack	Control	Notes
	Swede midge (*Contarinia nasturtii*)	*Adults:* Green/yellow/brown fly that is 2 mm in length. *Larvae:* Initially translucent but turn yellow as they develop and are 0.3 mm in length.	Females lay batches of 15–20 eggs on young plant parts. Upon hatching the larvae predominantly feed on the growing point before moving to the ground to form cocoons in the soil to pupate. Larvae pupate in spring and adults emerge from the soil.	Plants typically have scarring on their growing point and leaf petioles as well as flower stalks. The main shoot is often killed, which the plant compensates for by growing more side shoots.	Employing crop rotations are effective at preventing pest populations from establishing. Crop covers may be effective for preventing eggs being laid onto the crop if the mesh size is appropriate.	Pheromone traps can be used to monitor for this pest, however specialist skills and equipment are needed to confirm identification due to their size.
	Diamondback moth (*Plutella xylostella*)	*Adults:* Brownish in colour but often with creamy white markings on the forewings. Up to 6 mm in length. *Larvae:* Light green in colour and up to 16 mm in length.	This species of moth does not overwinter in large numbers and instead adults arrive during the summer from continental Europe. There may be up to three generations each year. Larvae complete their development feeding on a wide range of host plants and pupate inside a silk cocoon on the leaf surface.	Larvae initially feed on the lower leaf surface but leave the upper leaf epidermis intact, creating windows in the leaf. Often however the leaf is completely eaten.	In some crop situations it may be possible to use mesh crop covers in order to prevent egg laying. Where this isn't possible a synthetic insecticide or a biopesticide such as *Bacillus thuringiensis* may be used.	Pheromone traps can be used to monitor for this pest and importantly to allow for a quick response to the presence of this pest within a crop.

Sugar beet and fodder beet	**Beet leaf miner** (*Pegomya hyoscyami*) - adults known as mangold fly	*Adult:* Greyish fly that is up to 6 mm in length. *Larva:* Yellowish-white in colour, without a clearly defined head or any legs. May grow up to 7 mm in length.	White oval-shaped eggs are laid in small batches on the underside of leaves from the end of April. Larvae bore into the leaf tissue and produce mines between the upper and lower leaf surface. Larvae complete development in about 14 days and drop to the soil where they pupate. There are 2 or 3 generations a year but peak egg laying is in May.	Blistering (caused by the leaf mines) of leaf can lead to the leaf becoming withered. Plants may be stunted and in extreme cases killed by this pest.	Good growing conditions to help the crop grow through the vulnerable stages quickly (economically important damage unlikely beyond 8-leaf stage). Foliar applied insecticide may be required if the action threshold is exceeded. The action threshold is where the number of fresh unhatched eggs plus living larvae equals the square of the number of true-leaves (e.g., 16 eggs and larvae on a plant with 4 true leaves).	Control is most likely to be required during May and early June when most eggs are laid.
	Peach-potato aphid (*Myzus persicae*)	As for peach-potato aphid on oilseed rape	Adults of peach potato aphids over winter on a number of host plants. During spring winged aphids migrate to the summer host	The peach-potato aphid does not normally cause direct feeding damage. Instead a small but	Good hygiene is essential in the management of virus yellows. Cleaner-loader clamps, weeds or	Action thresholds have not been used recently but previously when plants have fewer than 12 leaves, one

Continued

Table 7.1 Economically important crop pests, their damage symptoms and control methods.—cont'd

Crop attacked	Pest	Description	Life-cycle	Symptoms of attack	Control	Notes
			crops, including beet crops. Winged aphids give rise to wingless offspring that may complete several generations through the summer.	significant proportion of these aphids transmit virus yellows, a collective term to describe several viruses that cause similar symptoms in beet crops.	sugar beet ground-keepers which aphids feed on before entering the crop are all sources of virus. Seed treatments are no longer available and so foliar applications of insecticides may be considered but it is important to consider widespread pyrethroid resistance.	or more wingless green aphids per four plants justifies a spray. When plants have more than 12 leaves there is no need to spray until there is more than one wingless aphid per plant.
	Beet cyst nematode (*Heterodera schachtii*)	Juvenile nematodes have slender transparent bodies up 0.5 mm in length. Female nematodes swell and form lemon-shaped cysts (0.5 × 1.0 mm) that are initially white but turn brown in colour as they mature.	Cysts remain dormant in soil until a suitable host plant is present. Larvae hatch from eggs held inside the cysts and enter host roots. Once fertilised the females swell to form a cyst on the outside of roots. The cyst may contain up to 600 eggs. There can be two generations a year.	Crop damage is typically seen as areas of stunted plants that wilt in periods of dry weather. Damaged plants are characterised by the presence of many lateral roots (bearding). Yield losses can be as high as 60%.	Wide rotations are effective at reducing the risk as well as having a rotation with few if any brassica crops. Maintain good weed control to avoid presence of potential hosts (e.g., wild mustard, shepherd's purse and purslane).	Most commonly found on organic and lighter soil types. The soil can be tested for the presence of viable eggs and larvae. Other crops such as red beet, fodder beet and oilseed rape can be hosts for this pest.

					Grow tolerant beet varieties. No nematicides are currently approved for use in the UK.	
	Wireworm (adults known as click beetles and larvae known as wireworms) - key species include *Agriotes sputator*, *A. obscurus* and *A. lineatus*	As for wireworm on cereals.	As for wireworm on cereals.	The roots of seedling plants are damaged and the base of the stem may be severed.	Good growing conditions to help the crop grow away from an attack. Damage is reduced when the soil is well consolidated. Control grass weeds.	Wireworm is traditionally a problem in grass rotations. There are an increasing number of cases of wireworm in arable rotations.
	Free-living nematodes (Stubby root nematodes (*Trichodorus* spp. and *Paratrich odorus* spp.) and needle nematodes (*Longidorus* spp.)	Free-living nematodes have slender transparent bodies up 1.0 mm in length and are not visible with the naked eye.	Free-living nematodes move to germinating seedlings and feed on the host plant roots. There can be several generations per year depending on soil temperatures. Nematodes are most active in moist soils.	Causes irregularly stunted plants with multiple taproots known as docking disorder (also known as fanging in carrots). Damage usually in patches or along rows. Most often a problem on sandy soils.	Improve soil structure. Avoid growing other hosts within the rotation e.g., potatoes and carrots.	Use soil sampling services to detect the presence of free-living nematodes.
Carrots	**Carrot fly** (*Psila rosae*)	*Adult:* Shiny black body with a reddish-brown head and yellow legs. Up	Over-winters as larvae or pupae with the first adult generation emerging in spring	Brown and rusty tunnels in tap roots (mines often with larvae protruding)	Use crop rotation to minimise population build-up and avoid growing	Celery, celeriac, parsley and parsnips also attacked by carrot

Continued

Table 7.1 Economically important crop pests, their damage symptoms and control methods.—cont'd

Crop attacked	Pest	Description	Life-cycle	Symptoms of attack	Control	Notes
		to 8 mm in length. *Larvae:* Creamy-white in colour, without a clearly defined head or any legs. May grow up to 10 mm in length.	with most eggs being laid in May. Larvae feed on lateral and tap roots. Larvae pupate in July and a second generation of adults emerges in July, which may give rise to a third generation of adults that emerge in October and November in warmer areas.	becoming progressively worse as season proceeds. In severe cases plants may be killed. Makes the mature carrot unmarketable.	susceptible crops close together. Maintain good crop hygiene around edge of field to reduce available shelter for the adult flies. Some varieties of carrot are partially resistant. Crop covers are an option on small areas. Adjust sowing dates to avoid the first generation or harvesting early to help reduce the damage. Limited number of insecticides available to control this pest.	fly. A weather-based forecasting system is available to optimise timing of in-field monitoring and insecticide applications.

Potatoes	**Potato cyst nematode (PCN)** (*Globodera pallida* (dominant in UK) and *Globodera rostochiensis*)	These nematodes have slender transparent bodies up 1.0 mm in length. Female nematodes swell and form spherical cysts that are 1.0 mm in diameter and initially white (*G. pallida*) or golden (*G. rostochiensis*).	Eggs within the cysts in the soil hatch in April or May when susceptible host plants are grown. The juvenile nematodes invade the roots. New cysts are formed from fertilised females. These cysts can remain viable in the soil for up to 10 years.	The most economically important pest problem in potatoes. Early and severe damage may lead to plant death but surviving plants will become stunted and chlorotic.	An integrated approach needed. Use soil sampling to determine PCN species and numbers. Select tolerant or resistant varieties as appropriate. There are many varieties that are resistant to the golden nematode but only a few varieties that show partial resistance to the white nematode. Use wide rotations to help keep populations low. Plant classified seed. Select available nematicides carefully and consider also use of trap cropping and biofumigation.	Take care to avoid the movement of contaminated soil on machinery.

Continued

Table 7.1 Economically important crop pests, their damage symptoms and control methods.—cont'd

Crop attacked	Pest	Description	Life-cycle	Symptoms of attack	Control	Notes
	Peach-potato aphid (*Myzus persicae*)	As for peach-potato aphid on oilseed rape.	Adults of peach-potato aphids over winter on a number of host plants. During spring winged aphids migrate to the summer host crops, including potato crops. Winged aphids give rise to wingless offspring that may complete several generations through the summer.	Peach-potato aphids are important vectors of potato leaf roll virus (PLRV), potato virus Y (PVY) and potato virus A (PVA) in seed potato crops. In ware potato crops, large populations (5 aphids per compound leaf) may check the growth of the plant.	In seed potato crops a programme of insecticide applications is typically required but this must consider the persistence of the product used, numbers of applications permitted and resistance in peach-potato aphid populations (e.g., to pyrethroid insecticides).	A monitoring programme is available to determine PYV pressure by recording numbers of aphid vectors through the use of yellow water traps.
	Wireworm (adults known as click beetles and larvae known as wireworms) - key species include *Agriotes sputator*, *A. obscurus* and *A. lineatus*	As for wireworm on cereals.	As for wireworm on cereals.	Older larvae tunnel into tubers making tubers unmarketable. Tunnels are characteri stically straight with blackened walls to tunnels.	Sow early maturing varieties. Lift the crop in early September if possible if field at risk.	Wireworm is traditionally a problem in grass rotations. There are an increasing number of cases of wireworm in arable rotations. Pre-planting soil sampling or bait trapping can help identify a potential problem.

	Slugs (mainly round-back, *Arion hortensis* and *A. distinctus* and keeled slugs, *Milax, Tandonia* and *Boettgerilla* spp.)	*Round-back slugs:* Small, often dark upper parts to body with a yellow or orange underside *Keeled slugs:* Larger slugs, grey to dark brown or black in colour with a characteristic ridge running down back.	Active throughout the year when conditions are warm and damp. Breeding occurs mainly in the spring and autumn. Eggs are laid just below the soil surface and slugs tend not to move far from these sites. Slugs often have a patchy distribution within fields and are typically more damaging if the seedbed is loose, cloddy and trashy. Populations and damage are highest on heavy textured soils.	Maincrop potatoes damaged by slugs eating tubers and creating cavities.	Some varieties are more susceptible than others. Use bait traps to assess the need for treatment with a molluscicide. Early lifting will help reduce amount of damage.	Monitor in the same way as for cereal crops but trap over an extended period to develop a better understanding of slug pressure. There is a risk of damage if at least 1 slug is caught/trap.
	Cutworm (mainly *Agrotis segetum*)	*Adult:* brown wings with dark markings. Wingspan up to 40 mm. *Larva:* Up to 40 mm in length; vary in colour from dull greyish brown to green. Curl into a	Eggs are laid on weeds and hatch after approximately 3 weeks. After initially feeding on leaves, larvae drop to the soil after completing the second moult and feed	Plants cut off at the base of the stem, plants wilt and can die. Larvae may also feed on the surface of tubers. Damage worst on light soils in very dry conditions.	Keep land free of weeds to reduce egg laying. Irrigation can be used to knock larvae off plants before the second moult. A number of insecticides are	Also attacks carrots, leeks, turnips, onions, lettuce, parsnips and red beet. A cutworm monitoring programme provides early

Continued

Table 7.1 Economically important crop pests, their damage symptoms and control methods.—cont'd

Crop attacked	Pest	Description	Life-cycle	Symptoms of attack	Control	Notes
		C-shape when disturbed.	(mostly at night) on plant stems above and below ground level as well as on tubers. Most damage is caused in July. Fully grown larvae overwinter and pupate in the spring.		approved for use to control this pest and should be targeted against young larvae while they are still feeding above ground.	warning of this pest.
Grass	**Leatherjackets** (adults known as crane flies or daddy long legs and larvae known as leatherjackets) - key species include *Tipula paludosa* and *T. oleracea*	As for leatherjackets on cereals.	As for leatherjackets on cereals.	Patches of grass dying off where the roots have been eaten away. Larvae found in soil around the base of plants. Secondary damage caused by birds feeding on leatherjackets.	Sowing new seeds before the middle of August and establishing a fine consolidated seed bed reduces the problem.	

	Frit fly (*Oscinella frit*)	As for frit fly on cereals.	The third-generation of larvae can reduce the chances of a successful establishment of some autumn-sown grass seed mixtures.	Larvae bore into stems of establishing grass and cause 'deadheart' symptoms.	When reseeding, allow at least 4 weeks from the destruction of the old sward to the sowing of the new ley.	Seedling Italian ryegrass is more susceptible than perennial ryegrass. Grass following long-term grass creates a higher risk situation. Mid to late August drilling dates tend to be most affected.
	Slugs (mainly grey field slug, *Deroceras reticulatum*)	As for slugs on cereals.	As for slugs on cereals.	Slug feeding destroys seedlings.	Establishing a fine consolidated seedbed can help to reduce the problem. Use a molluscicide if there is a risk of damage, indicated by use of refuge traps baited with chicken layers mash. Rolling can help a damaged crop to recover in the spring.	Direct drilled grass is most at risk.
	Wireworm (adults known as click beetles and larvae known as wireworms) - key species include *Agriotes sputator, A. obscurus* and *A. lineatus*	As for wireworm on cereals.	As for wireworm on cereals.	As for wireworm on cereals.	Good growing conditions to help the crop grow away from an attack. Damage is reduced when the soil is well consolidated. Control grass weeds.	

7.7 Key points

- crops are attacked by a wide range of invertebrate and vertebrate pests.
- Correct identification of a crop pest is essential for effective pest management.
- Adoption of IPM is essential if crops are to be grown in a sustainable way.
- Preventing pest problems should form the basis of any IPM programme.
- Monitoring and forecasting tools should be used to detect pest problems.
- Control of pest problems should employ a wide range of approaches and not be solely reliant on chemical controls.

Sources of further information and advice

Books

Alford, D.V., 2000. Pest and Disease Management Handbook. Blackwell Science, Oxford.

Alford, D.V., 1999. A Textbook of Agricultural Entomology. Blackwell Science, Oxford.

Dent, D., Binks, R., 2020. Insect Pest Management. CABI, Wallingford.

Encyclopaedia of pests and natural enemies in field crops, 2021. Stoneleigh Park, Agriculture and Horticulture Development Board.

Gratwick, M., 1992. Crop Pests in the UK. Chapman and Hall, London.

Gurr, G.M., Wratten, S.D., Snyder, W.E., 2012. Biodiversity and Insect Pests, Key Issues for Sustainable Management. Wiley-Blackwell, Oxford.

Jones, F.G.W., Jones, M.G., 1984. Pests of Field Crops. Edward Arnold, London.

Lainsbury, M.A., 2021. The UK Pesticide Guide. BCPC, Wallingford.

Matthews, G., 2016. Pesticides, Health, Safety and the Environment. Wiley-Blackwell, Chichester.

van Emden, H.F., 2013. Handbook of Agricultural Entomology. Wiley-Blackwell, Chichester.

van Emden, H.F., Service, M.W., 2004. Pest and Vector Control. Cambridge University Press, Cambridge.

Websites

https://www.agricentre.basf.co.uk/en/
https://ahdb.org.uk/
https://cropscience.bayer.co.uk/
https://cropprotection.org.uk/
https://www.fera.co.uk/
https://www.hse.gov.uk/pesticides/
https://www.inrae.fr/en/divisions/spe
https://irac-online.org/
https://www.koppert.com/
https://wiki.pestinfo.org/wiki/Main_Page
https://www.pgro.org/
https://www.rothamsted.ac.uk/
https://www.syngenta.co.uk/
https://voluntaryinitiative.org.uk/

Sustainable crop production

8

8.1 Introduction

Over the last 70 years crop production systems in the UK have intensified to meet the demands of a growing population and remain profitable in a competitive global market. Farmers have had access to an armoury of crop protection and crop nutrition products that have enabled them to perhaps neglect some of the basic principles of 'good agricultural practice' such as maintaining soil health and the use of diverse and resilient rotations. However, there is now a growing realisation and concern amongst policy-makers, consumers and indeed farmers, that some of our current production methods are not sustainable in the longer term, especially in the context of resilience to climate change.

Alongside concerns about the environment, there is also a somewhat conflicting need to provide food, energy and water for a growing global population which is projected to reach almost 10 billion by the year 2050.

These conflicting needs have resulted in a renewed interest in 'sustainable' farming, being led, not just by policy makers and consumers but also by increasing numbers of progressive farmers themselves. Many farmers have recognised that for their businesses to survive in the future a more sustainable model of production and perhaps a return to some of the fundamental biological principles of farming practices will be needed. These principles are not new but the advantage that farmers have today is the precision technologies that are available as described in Chapter 9, alongside a good understanding of biological systems to help them to achieve some of these sustainability aims.

There is no universal definition of sustainable crop production but one which has commonly been accepted by the Department of the Environment, Food and Rural Affairs (DEFRA) is 'ensuring the continuing availability to the consumer of adequate supplies of wholesome, varied and reasonably priced food, produced in accordance with accepted environmental and social standards. The term 'Regenerative Farming' is also worthy of mention here, as one which is receiving significant interest and has been borne out of the need to address the sustainability of our agricultural systems. Definitions of regenerative farming are broad and varied but centre around improving soil health which in turn improves productivity. 'Agroecological' is another increasingly common term to describe farming practices that minimise negative environmental impacts. However, it should also be remembered that sustainable crop production businesses must not only address environmental protection but must also address the other two commonly recognised pillars of sustainability, social equity and economic viability.

Lockhart and Wiseman's Crop Husbandry Including Grassland. https://doi.org/10.1016/B978-0-323-85702-4.00019-4

8.2 Crop production techniques

General techniques used in the preparation of land for cropping, decisions on the sequence of crops to grow, and providing crops with adequate water and protection from weeds, pests and diseases will be described in this section. Production techniques specifically related to individual crop types are addressed in the separate chapters relating to the management of a particular crop.

Some of these techniques have not changed but many are being adapted in response to concerns over the environment as will be highlighted throughout this chapter.

8.2.1 Drainage

Well drained land is the starting point for good soil management and provides the widest window for effective cultivations and good crop establishment. Normally, the soil can only hold some of the rainwater which falls onto it, particularly during the winter months. The remainder either runs off, is evaporated from the surface, is used by the plants if still growing actively, or soaks through the soil to the subsoil. If surplus water is prevented from moving through the soil and subsoil, it soon fills up the pore space; this will kill or stunt the growing crops, and any living organisms in the soil, as air is driven out of the profile.

The water table is the level in the soil or subsoil, below which the pore space is filled with water. This is not easy to see or measure in clay soils but can be seen in open textured soils (Fig. 8.1).

The water table level fluctuates throughout the year and in the UK is usually highest in February and lowest in September. This is because of a higher amount of evaporation, more transpiration from growing crops and usually lower rainfall in the summer.

In chalk and limestone areas, and in most sandy and gravelly soils, water can drain away easily into the porous subsoil. These are free-draining soils. On most other types of farmland some sort of artificial field drainage is necessary to carry away the surplus water and so keep the water table at a reasonable level. For most arable farm crops the water table should be about 0.6 m or more below the surface but for grassland 0.3–0.45 m is sufficient. The withdrawal of grants for drainage work, the greater protection of some wetland environments and the rising costs of specialist contractors have meant a large reduction in the area of land being drained compared with the immediate post-war period and the 1970s and 1980s.

Indications of poor drainage are:

- Machinery is easily 'bogged down' in wet weather.
- Stock grazing pastures in wet weather easily poach the sward.
- Water remains on the surface for many days following heavy rain.

Figure 8.1 Position of water table and the effect on water levels in wells and ponds.

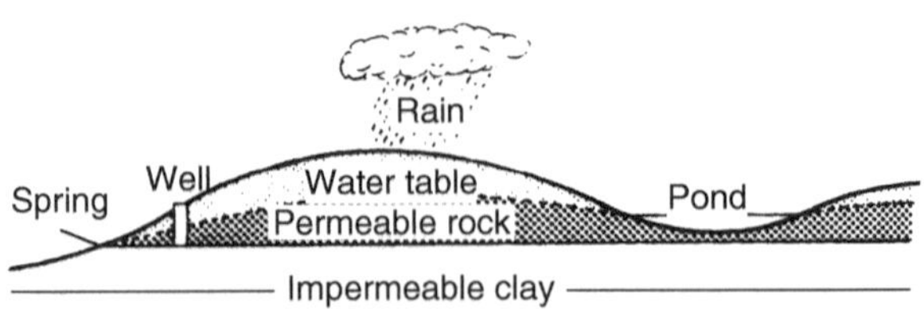

- Weeds such as rushes, sedges, horsetail, tussock grass and meadowsweet become common in grassland. These weeds usually disappear after drainage. Peat forms in places which have been very wet for a long time.
- Young plants are pale green or yellow in colour and have generally poor growth.
- The subsoil is often coloured in various shades of blue or grey (mottled), compared with shades of reddish-brown, yellow and orange in well-drained soil.

The benefits of good drainage are:

- Well-drained land is better aerated and the crops grow better and are less likely to be damaged by root-decaying fungi.
- The soil dries out better in spring and so warms up more quickly, encouraging mineralisation of nitrogen and it can also be worked earlier.
- Plants are encouraged to form a deeper and more extensive root system. In this way they can often obtain more plant food and are better able to survive periods of drought.
- Grassland is firmer, especially after wet periods. Good drainage is essential for high density stocking and where cattle are out-wintered if serious poaching of the pasture is to be avoided.
- Disease risk from parasites is reduced. A good example is the liver fluke. Part of its life cycle is in a water snail found on badly drained land.
- Inter-row cultivations and harvesting of root crops can be carried out more efficiently.
- The soil has more potential to support a wider range of crops in the rotation.
- The soil is able to carry heavier machinery without suffering compaction damage.

The main methods used to remove surplus water and control the water table are:

- Ditches or open channels.
- Underground plastic pipe drains.
- Mole drains.
- Ridge and furrow.

8.2.1.1 *Ditches and open drains*

Ditches may be adequate to drain an area by themselves, but they usually serve as outlets for underground drains. They are capable of dealing with large volumes of water in very wet periods. The size and shape of a ditch varies according to the area it serves and the surrounding soil type (Fig. 8.2). Ditches should be cleaned out to their original

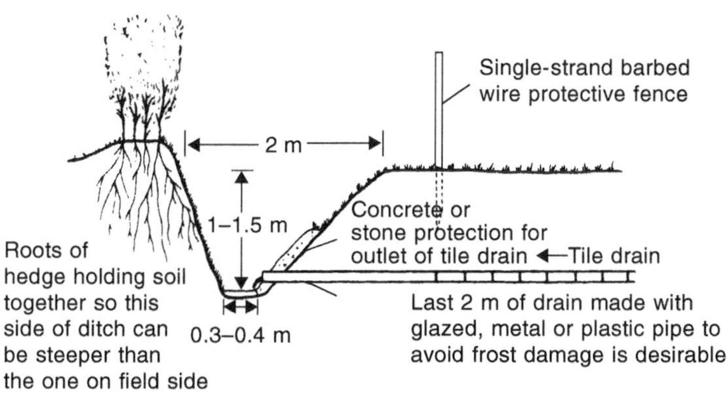

Figure 8.2 Section through a typical field ditch and tile drain.

depth once every 3 or 4 years. The spoil removed should be spread well clear of the edge of the ditch. Many different types of machines are now available for making new ditches and cleaning neglected ones.

Small open channels 10–60 m apart are used for draining hill grazing areas. These are either dug by hand using a spade designed for the purpose, or made with a special type of plough drawn by a crawler tractor. Similar open channels are used on low-lying meadow land where underground drainage is not possible.

8.2.1.2 Underground drains

The distance between drains which is necessary for good drainage depends on the soil texture. In clay soils the small pore spaces restrict the movement of water. Therefore, the drains must be spaced much closer together than on the lighter types of soil where water can flow freely through the large pore spaces (Figs. 8.3 and 8.4).

In the past, many different materials have been used in trenches to provide an underground passage for water (e.g., bushes, straw, turf, stones and flat tiles) before cylindrical clay tiles became popular (concrete pipes have also been used). However, the flexible plastic pipe has replaced the clay tile for all new drainage work. Underdrainage is very expensive and cost must be related to likely benefits.

Figure 8.3 The steepness of a water table (W.T.) varies with different soil types.

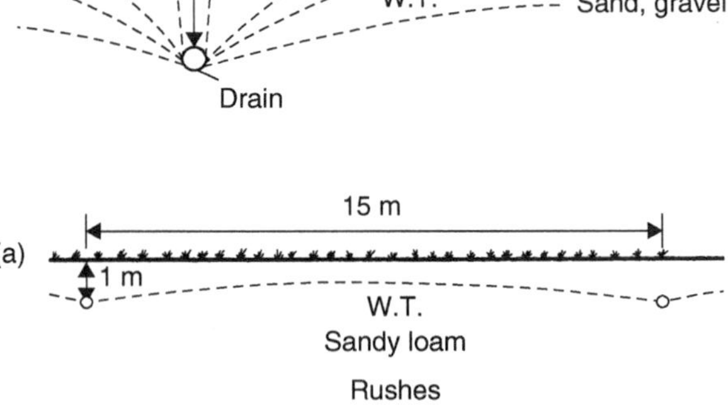

Figure 8.4 Diagram showing the effect of spacing of drains on the water table (W.T.); (a) sandy loam, (b) and (c) silt or clay loam.

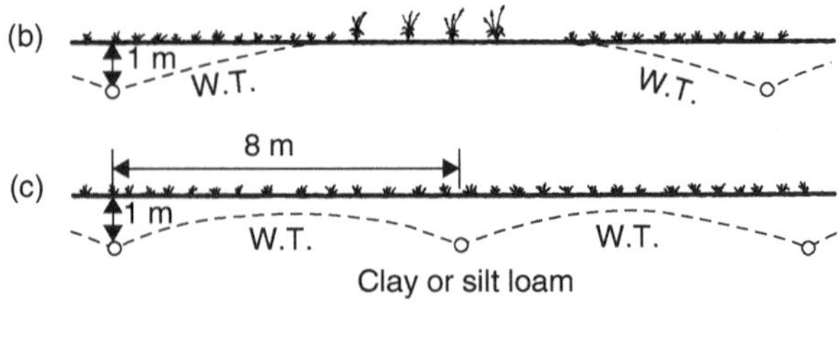

Consequently, when designing a scheme, the spacing of the drains must be carefully considered. Digging some holes and examining the subsoil can be very helpful in deciding how the field should be drained. If cost is a limiting factor, it would, for example, be advisable to lay the drains about 40 m apart in an area where 20 m would be an ideal spacing. This should be quite satisfactory over most of the field and only a few extra drains may be required later in the wetter areas. This should not be difficult if a detailed map of the drainage layout has been made.

The slopes on a field determine the way the drains should run to be efficient in removing water (Fig. 8.5). Water will not run uphill, unless it is pumped, and so the drains must have a gradient. If the slope is more than 2% (1:50) the laterals (side

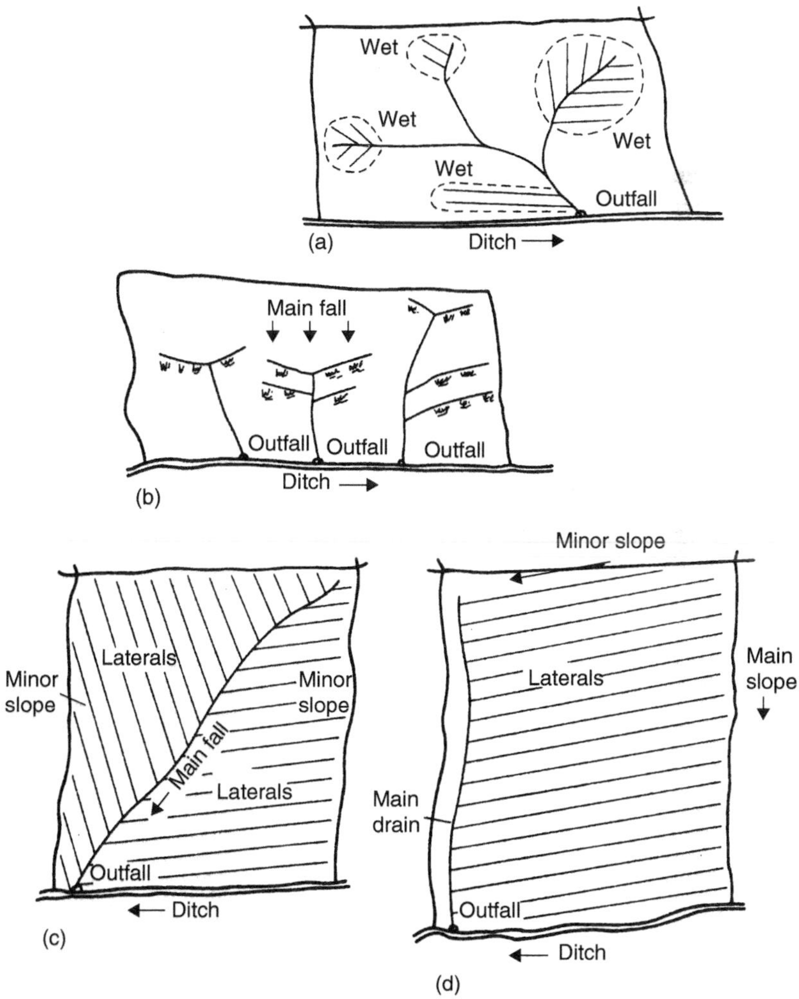

Figure 8.5 Drain layouts: (a) drainage of wet areas; (b) intercepting springs (rushy areas); (c) herringbone system in valley (note separation of junctions); (d) grid system.

drains) should run across the slope. Ideally, the minimum fall on laterals should be 1:250 and, in mains, 1:400. It is usually preferable that the laterals go into a main drain before entering a ditch; this means fewer outfalls to maintain. In flat areas, and where the soil is silty (causing silting up of the drains over a long period), it is preferable to let each lateral run straight into the ditch.

A high-pressure jetting device can be used to clean them periodically. Iron ochre deposits can also be cleared in this way.

Where the land is low lying and there is no natural fall to a river, the area can be drained in the usual way and the outlet water from the drainage system can lead to a ditch from which it is pumped into a river or over a sea wall. Underground drainage is best carried out in reasonably dry weather.

8.2.1.3 Underground piped drains

Plastic pipes come in various configurations. Most of them are corrugated with slits cut at intervals to allow water to enter easily.

The size of pipe required will depend on the rainfall, the area to be drained by each pipe, gradient and soil structure. In the case of main drains, it will depend on the number of laterals entering.

Various types of machine are available for laying plastic pipes, and the porous backfill (e.g., washed gravel, clinker) required on the heavier soils. The depth of porous fill is more important than the width in the trench. However, it is very expensive and may constitute 50% of the total cost of the drainage scheme.

Modern machines can lay pipes very rapidly, especially when a laser beam system is used automatically to adjust the gradient on the drain, and a GPS system allows very accurate measurement of spacing within the field.

8.2.1.4 Underground drains—mole drainage

Mole drainage is a cheap method which can be used in some fields. Although mole drains are sometimes used on peat soils, the system is normally applied to fields which have:

- A clay-rich subsoil with no stones, sand or gravel patches.
- A suitable gradient (a fall of 5–10 cm over 20 m).
- A reasonably level surface.

A mole plough, which has a torpedo or bullet-shaped 'mole' plus an expander attached to a steel coulter or blade, forms a cylindrical channel in the subsoil along which water can travel (Fig. 8.6).

The best conditions for mole draining occur when the subsoil is damp enough to be plastic and forms a good smeared surface inside the mole channel. It should also be sufficiently dry to form cracks above as the mole plough passes (Fig. 8.7). Furthermore, if the surface is dry the tractor hauling the plough can get a better grip. The plough should be drawn slowly (about 3 km/h) otherwise the vacuum created is likely to spoil the mole. Reasonably dry weather after moling will allow the surface of the mole to harden and so it should last longer.

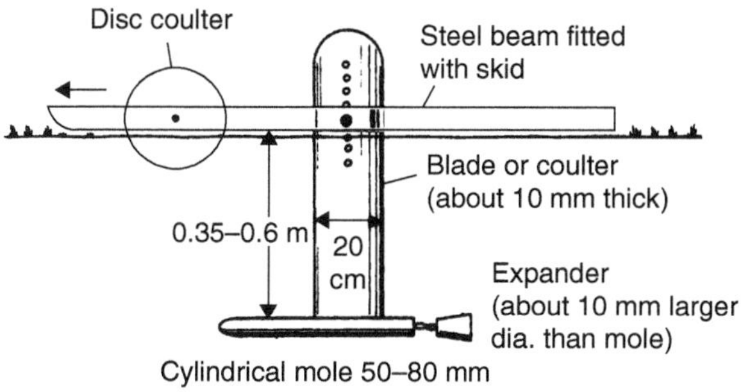

Figure 8.6 Mole plough.

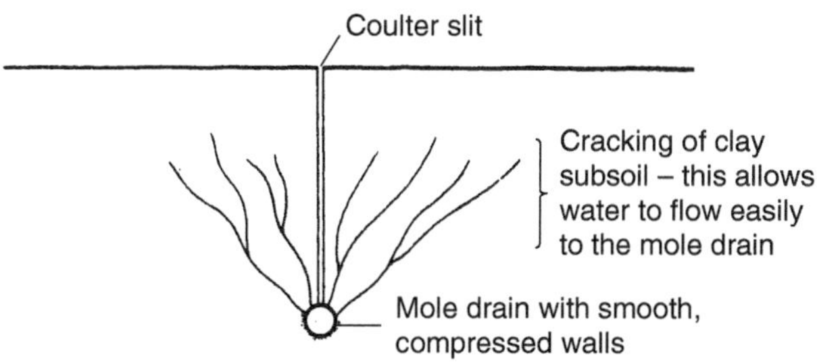

Figure 8.7 Section through mole drain and surrounding soil.

Mole drains are drawn 3–4 m apart usually at right angles to the direction of the plastic drain pipes. For best results the moles should be drawn through the porous backfill of a plastic main or lateral drain (Figs. 8.8 and 8.9). A new set of mole drains can be drawn every 3–10 years as required. They will last much longer in a soil with a high clay content than in one with more sand or silt where the sides of the moles collapse quite quickly.

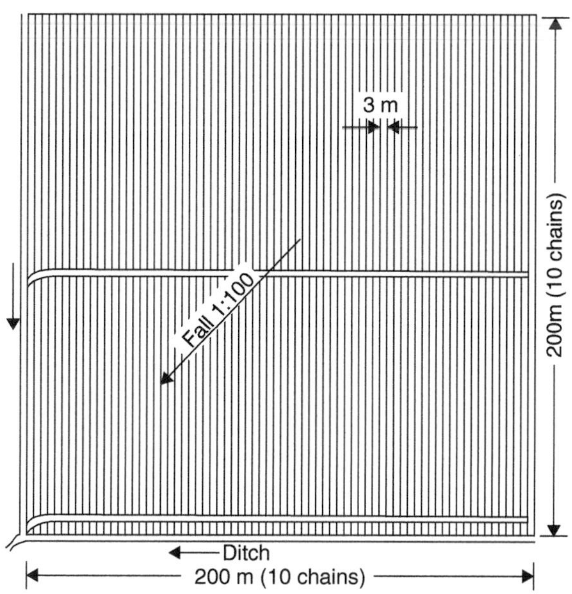

Figure 8.8 Layout of mole drainage (with field main) in a 4 ha area.

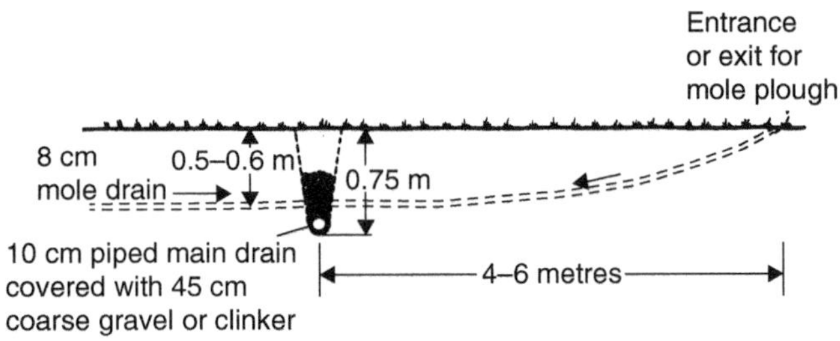

Figure 8.9 Section through the tile drain shown in Fig. 8.8 to show how water from the mole drain can enter the pipe drain through the porous backing.

Installing a new drainage system is the first step in removing surplus water from a field. However, the structure of the soil and subsoil must, if necessary, be improved so that water can move easily to the drains.

8.2.2 Seedbed preparation

Establishing the desired plant populations and conditions that facilitate uptake of nutrients and water is crucial to providing the potential for a high yielding crop. The aim of soil cultivations is to produce a seedbed which is firm enough to provide good soil to seed contact but has spaces between aggregates that allow free movement of water and air with no impedance to root growth. In most cases a good seedbed should be prepared with the minimum amount of working and the least loss of moisture, however, on heavy soil and in a wet season, some moisture loss can be desirable. Overworked soils or inappropriate cultivations for the conditions can result in problems such as compaction, slumping and capping depending on soil texture.

The general principles of seedbed preparation will be described here whilst the seedbed requirements and commonly used sowing methods for individual crop types are described in more detail in their respective chapters.

Seedbed preparation techniques can be categorised into those where no cultivations take place between the previous crop and establishment of the next crop with the seed being direct drilled into the trash left by the previous crop; those where minimal soil disturbance takes place using combinations of discs and tines to a depth of 5–10 cm (non-inversion tillage, minimum tillage) and modified forms of this where minimal soil disturbance takes place only in narrow bands where seed is to be placed (strip tillage); and finally those where soil is fully inverted to a greater depth by the use of a plough (inversion tillage).

If compaction has been identified as a problem then some sub-surface restructuring may also be required (subsoiling) prior to any of the above seedbed preparation techniques.

An individual farmer may well practice more than one method of seedbed preparation on their land depending on variations in soil texture, moisture levels and cropping. Indeed, maintaining some flexibility to respond to different conditions in terms of seedbed preparation is useful but may be limited by the high cost of the machinery.

8.2.2.1 Sub-surface restructuring

A well-structured subsoil is important to allow roots to grow unimpeded and water to drain away freely. To maintain healthy soils that will support high yielding crops it is important to use a spade to regularly examine the subsoil for signs of compaction such as obvious differences in soil moisture or root growth or visibly hardened layers termed 'pans'.

A pan is a hard, cement-like layer in the soil or subsoil which can be caused by ploughing at the same depth every year. Plough pans are partly caused by the base of the plough sliding along the furrow but also occur when the soil is wet, and there is some wheelslip which has a smearing effect on the bottom of the furrow. Plough pans are more likely to form on the heavier types of soil with a high clay content. Pans may also be formed by the deposition of calcium carbonate, silica, iron, manganese or aluminium compounds causing particles to bind together tightly.

Compaction may also occur in situations where shallow cultivations have been used for many years and the soil below has become compacted. This is most likely to occur on headlands or where heavy tractors and cultivation equipment have been used on wet soils in the past.

Subsoiling is the term commonly used to describe the operation to break up pans or compacted layers and loosen the subsoil with minimal disruption of the surface. A subsoiler is a very strong tine which can be drawn through the soil and subsoil at the required depth. Most tines have 'wings' fitted near the base of the tine to produce a lifting and shattering effect. This will only produce satisfactory results when the subsoil is reasonably dry and drainage is good. In most cases a rear packer or roller follows the tines to close and consolidate.

8.2.2.2 Inversion tillage (ploughing)

Traditionally, ploughing has been the primary method of cultivation, indeed, the origins of the mouldboard plough commonly used today began in China over 2000 years ago. The mouldboard plough commonly inverts the soil to a depth of 15–30 cm to bury residues and weed seeds and alleviate surface compaction. However, it is a costly operation, both in terms of the time taken and also the amount of energy required. Today, large tractors are now used to pull eight or nine furrows, so work rate is increased. Most commercially used ploughs are reversible. They have right-hand and left-hand mouldboards and so no openings or finishes have to be made when ploughing which produces a more level seedbed. Concave mouldboards break the furrow slices better and are often used on the lighter soils and slatted mouldboards are recommended for light and fluffy soils such as peat. Deep digger ploughs are used where deep ploughing is required, e.g., for root crops.

The proper use of skim and disc coulters and careful setting of the plough for depth, width and pitch can greatly improve the quality of the ploughing. Presses attached to the plough or as a following operation can also be used to help close the furrow. If possible, it is desirable to vary the depth of ploughing from year to year to avoid the formation of a plough pan.

'Chisel ploughing' is a term used to describe the work done by a heavy duty cultivator with special spring or fixed tines; unlike the ordinary plough, it does not move or invert all of the soil.

Ploughing will usually be followed by a secondary cultivation, in many cases this is carried out by a 'combination' drill where there is a power harrow mounted on the drill.

Rotational ploughing describes a method whereby a heavy burden of grass weed seeds is ploughed under and the plough is not used again for 3–5 years until it is likely that a large proportion of the buried seed is no longer viable.

8.2.2.3 Non-inversion (minimum tillage)

The principal of minimum tillage is to mix the previous crop residues with the soil close to the surface to produce a surface tilth whilst retaining the natural subsurface structure. The depth to which minimum tillage is carried out varies depending on any need to remove compaction and the depth at which it occurs and the amount of residue to be incorporated. Deep non-inversion tillage (20–35 cm) can be used in a broader rotation that may include root crops as an alternative to ploughing but in most combinable crop rotations on well-structured soils, shallow non-inversion tillage (5–10 cm) is used. Minimum tillage equipment usually consists of disc or tine coulters or a combination of both to do the mixing and a packer or roller to level the seedbed. Discs are usually considered more effective in drier conditions whilst tines have a wider soil moisture window. Some management of previous crop residues to ensure their even distribution, such as the use of a straw rake, may be required prior to a pass with minimum tillage equipment.

As well as the obvious economic benefits of moving less soil, reduced forms of tillage used in appropriate conditions can offer benefits to soil health and biodiversity such as an increase in soil organic matter content in the top soil, a reduction in soil erosion, improved drainage and water holding capacity and increased levels of microbial and earthworm activity as well as a reduction in greenhouse gas emissions. Reduced tillage techniques can also improve the timeliness of crop establishment and retain moisture which will aid crop establishment and ultimately improve yields. However, for many of these benefits to be realised in practice, careful management is required and these techniques are often most successful when implemented alongside other measures such as diversifying rotations. Without careful management, compaction, build-up of grass weeds and patchy emergence can reduce yields in reduced tillage systems.

Benefits of reduced or no-till systems in terms of the environment can be complex. For example, if any form of tillage reduces yield then crop production efficiency is reduced and carbon footprint may be affected. Another example is that there is evidence that no-till practices can increase urease levels close to the soil surface and therefore increase the risk of ammonia volatilisation if urea is applied.

8.2.2.4 Strip-tillage

Strip tillage techniques are increasingly common and often viewed as a stepping stone to direct drilling. With strip tillage only the narrow bands into which the crop is sown are cultivated to 5–10 cm depth leaving the strips in between undisturbed. Wider row widths are commonly used for strip tillage so the system also facilitates placement of fertiliser into the bands in which the crop is sown and establishment of companion crops between the bands. There is also the potential to carry out inter-row weeding operations either mechanically or chemically using hooded spray nozzles.

8.2.2.5 Direct drilling

Direct drilling is a relatively low-cost method of crop establishment whereby there is no prior cultivation. A tine or disc opener is used to open a slot into which the seed is placed and some form of closing wheel follows to level the surface. It is easier to adopt in naturally self-structuring soils or those where improvements have already been made over a number of years following minimum or strip tillage. Some cultivator drills allow flexibility by raising the cultivation elements enabling them to be used as a direct drill.

8.2.3 Sowing

Broadcasting seed onto the surface without cultivation can successfully be used for the establishment of small seeded crops such as grass/clover or cover crops. It was also used in the past as a cheap method of establishing oilseed rape, often termed 'autocasting' but it is no longer widely practised for this crop. Establishment can be uneven and seed rates need to be increased to compensate but the advantage is that it maximises the time for a catch or cover crop to establish before the autumn and reduces costs.

In situations where some seedbed preparation has taken place either by inversion or non-inversion techniques then seed of combinable crops is generally sown using a seed drill where seed is discharged in rows in an even trickle and where spacing within the row is not critical e.g., for cereals. Many seed drills of this kind are 'combination' drills where the secondary cultivation in the seedbed preparation process is carried out by a power harrow mounted on the drill. Precision drills are used where accurate seed placing and precise plant populations are required in crops such as sugar beet and maize where even spacing of seeds within the row is also required.

Following drilling, seedbeds will usually benefit from being rolled where conditions allow to consolidate and improve seed to soil contact and optimise the activity of soil applied residual herbicides.

Bed systems are used for root and vegetable crops to eliminate fanging, give quick vigorous establishment, and reduce reliance on good ground conditions for timely machine operations. Crops are grown in strips or beds which the tractor straddles, thus restricting wheelings and compaction to the areas between the beds; all equipment widths must be matched to the bed width (or multiples of the bed width) to use this system effectively. Bed width is usually determined by harvester design.

8.2.4 Crop rotations

The practise of 'rotating crops' or planting different crops sequentially in a field has been an important feature of cropping systems since organised agriculture began. Rotations are well proven to maintain crop yields, by maintaining soil fertility, especially where legumes, which fix atmospheric nitrogen, are included and also by reducing pathogen and pest pressure. Rotations also help to avoid repeated use of certain pesticides where there is a high risk of resistance developing, encourage biodiversity within the farm system by providing different habitats, manage workload, machinery and storage resources and spread financial risk.

First wheats, due to their profitability are generally the cornerstone of an arable rotation and the 'alternative' crops, more commonly known as break crops fall into two categories: (a) combinable break crops such as oilseed rape, linseed, peas and beans; and (b) non-combinable break crops such as sugar beet, potatoes, field scale vegetables and grass. Choice of crops grown will depend on the constraints imposed by climate and soil type, whether there is a profitable local market, expertise and the availability of equipment and storage facilities. The range of crops grown will also be limited by the fact that small amounts of individual commodities do not put farmers in a strong marketing position. Examples of typical rotations are shown in Table 8.1 but it should also be remembered that many planned rotations are subject to change due to weather conditions and their impact on harvesting and seedbed preparations.

Over recent decades the availability of artificial fertilisers and effective pesticides allowed rotations to become market led 'narrowing' them in some areas to such an extent that they relied on an alternating winter wheat/oilseed rape 'rotation' for a number of years. This cannot be considered a sustainable practice in the longer term and yields will ultimately decline. It also led to repeated use of a limited range of herbicides and contributed to resistance problems e.g., in black-grass. There was some attempt to address this problem by the introduction of the 'three crop rule' under the Basic Payment Scheme in 2015. The three-crop rule specified that farms with more than 10 ha of arable land had to grow at least two crops, whilst at least three crops were required on farms with more than 30 ha and the main crop could not cover more than 75% of the land. This rule has since been abolished as the agricultural transition plan in the UK comes into force.

Table 8.1 Typical rotations.

Year	Combinable crop rotation	Root crop rotation	Mixed farm rotation
1	First wheat	First wheat	Grass (Year 1)
2	[a]Second wheat or winter or spring barley	[a]Second wheat or winter or spring barley	Grass (Year 2)
3	Winter oilseed rape	Potatoes	Grass (Year 3)
4	First wheat	Sugar beet	Forage brassica (sow late summer)
5	[a]Winter or spring barley	Winter wheat	Forage maize
6	Winter or spring beans	[a]Second wheat or winter or spring barley	Wholecrop cereal

[a]These rotations may be shortened by removal of the second cereals.

However, other factors such as widespread pesticide resistance, the loss of key pesticides through legislation, the depletion of organic matter levels in soils and the increasing cost of inorganic fertilisers are now encouraging a return to wider rotations and even the incorporation of grass leys and livestock into previously arable rotations. However, profitability also remains a key factor in break crop choice and despite recent challenges in growing it, due to high demand and therefore relative profitability winter oilseed rape (see Chapter 14) remains the key combinable break crop in many rotations. It is also drilled and harvested at a different time to cereals and therefore sits well alongside them in a rotation. The inclusion of spring crops is increasingly important as a tool in integrated weed management strategies.

8.2.5 The use of cover cropping and companion cropping in rotations

A cover crop is a non-cash crop grown primarily for the purpose of 'protecting or improving' the soil between cash crops which is then incorporated prior to planting the cash crop. There has been renewed interest in cover crops over recent years as a tool to improve soil health and resilience, reduce carbon losses and as part of the solution to pesticide resistance and the loss of key pesticides through legislation.

Depending on the species used potential benefits of cover cropping include:

- nitrogen fixation
- a reduction in run-off and leaching
- the return of organic matter to the soil and associated soil structural and water retention benefits
- habitat value
- agronomic benefits such as the disruption of pest and disease lifecycles and weed suppression as part of an integrated control strategy

Typical cover crops are mustards, radishes and turnips from the Brassica family which are good at establishing quickly and growing vigorously and are useful for suppressing weeds. Vetches and clovers from the Legume family are good at fixing nitrogen and oats, rye and rye-grasses from the Cereal family have similar functional traits to brassicas but also have extensive root systems benefitting soil structure. Other species such as phacelia, buckwheat and chicory may also be included and mixtures of species are frequently used to provide multiple benefits but it must be ensured that they do not out-compete one another.

In some cases, specific cover crop types have very specific benefits in reducing a pest problem such as acting as a trap crop or a biofumigant as part of an integrated pest management strategy for nematodes in root crops.

The duration of the cover crop can vary but it will most often be sown following harvest in late summer and incorporated prior to planting a spring crop. It should be relatively cheap to establish, and should establish and grow quickly thus ensuring rapid ground cover. The choice of cover crop should be considered carefully in relation to the practiced rotation; it is very important that it should not exacerbate potential weed, volunteer, disease or pest problems such as clubroot (*Plasmodiophora brassicae*) where oilseed rape or vegetable brassicas are present in the rotation. Consideration should also be given to the destruction of the cover crop which may be sprayed off with a non-selective herbicide, killed by frost action, grazed or mechanically destroyed.

The use of perennial legume cover crops such as clover or lucerne often termed 'living mulches' is also gaining some interest in reducing fertiliser and pesticide inputs in no-till systems (Fig. 8.10). Potential benefits include nitrogen accumulation, weed suppression, improved soil structure and biodiversity. However, such systems are complex and challenging in that to achieve these benefits a relatively high biomass of living mulch is required which then competes with the 'cash' crop and compromises yield.

Companion planting is another technique which is also gaining some interest in conventional arable systems and involves sowing a different type of plant or secondary crop together with a primary cash crop. In some cases the companion crop is sown separately in rows between the cash crop and in some cases it is broadcast with the cash crop. The aim of companion cropping is to reduce fertiliser and potentially also agrochemical inputs by using nitrogen fixing legumes and species which will suppress weed and pest problems. Typical combinations are oilseed rape and clover and vetches for the purposes of providing nutrition and suppressing weed and pest problems and also maize and ryegrasses for the purposes of reducing the risk of soil erosion.

8.2.6 Integrated pest management

Once a crop has been successfully established the effective control of weeds, pests and diseases is crucial to it fulfilling its yield potential. The term Integrated Pest Management (IPM) was first coined in the 1970's in response to concerns over the social and environmental effects of some pesticides. In practical terms it is a planned approach to the prevention, detection and control of pests, weeds, and diseases aiming to reduce the incidence of pest problems and therefore reduce the reliance on pesticides. Some measures such as the use of rotations are commonly practiced but others have seen slower uptake, primarily due to concerns over the impact of any reduction in pest control on yield and profitability and therefore pesticides have remained the main-stay of pest control in conventional systems. The impact of pesticides is easily measured and

Figure 8.10 A living mulch system with oats as a cash crop.
Source: Organic Research Centre.

cost benefit can therefore easily be demonstrated whereas it is more difficult to measure the impact of some methods of pest prevention and therefore farmers have less confidence in them even though they can potentially be used to reduce input costs.

Over the last 10 years the loss of plant protection products and rising levels of resistance have meant that the pesticide toolbox has shrunk dramatically and further losses are likely which has certainly put greater emphasis on IPM strategies. The use of IPM strategies may help to preserve the limited toolbox of conventional pesticides that are still available. At the same time, many policy drivers emphasise a need for reduced reliance on pesticides. The EU Sustainable Use Directive (Directive 2009/128/EC which was transposed into UK law by the Plant Protection Products (Sustainable Use) Regulations 2012 requires that measures are taken to promote low pesticide input pest management, giving priority to non-chemical methods. The revised national action plan published in 2022 will feature integrated pest management (IPM) at its core and a need for more advice, training and education to increase the uptake of IPM has been identified in a consultation exercise prior to the release. However, there is no doubt that uptake is increasing as farmers themselves are recognising that the 'answer is no longer in the can'.

In practice, the first stage in an IPM strategy is the prevention of pests through measures such as those listed below, further examples are available in Chapters 5, 6, and 7:

- Diverse crop rotations e.g., the inclusion of spring crops for a reduction in those weeds that mainly germinate in the autumn such as black-grass
- Use of appropriate cultivation techniques e.g., stale seedbed technique for grass weeds such as some of the brome species or black-grass or rolling seedbeds to reduce the risk of slug damage
- Use of resistant/tolerant cultivars e.g., orange wheat blossom midge resistance in winter wheat, TuYv resistance in oilseed rape
- Companion cropping e.g., buckwheat in oilseed rape for the reduction of cabbage stem flea beetle
- Trap cropping e.g., sticky nightshade planted prior to potatoes for a reduction in potato cyst nematode
- Variety blends e.g., for the reduction of disease pressure in cereals
- Use of certified seed and planting material e.g., loose smut in barley
- Use of balanced nutrition, liming and irrigation/drainage practices e.g., maintaining high pH for reduction in the risk of clubroot occurring in oilseed rape and vegetable brassicas, irrigating at tuber initiation for the reduction of scab in potatoes
- Preventing the spread of harmful organisms by hygiene measures e.g., by regular cleansing of machinery and equipment
- Protection and enhancement of important beneficial organisms e.g., beetle banks

The second stage in an IPM strategy is the correct detection and monitoring of the pest problem. This can include field observations as well as the use of predictive or forecasting tools. Results of field monitoring may be compared to scientifically derived economic thresholds to decide whether treatment is required.

Before a plant protection product is used any alternative means of control should be evaluated, for example, the use of biological controls or biostimulants or mechanical means such as defoliation of the crop for pest or disease control or the use of mechanical hoes for weeds. In regenerative systems and organic farming (particularly biodynamic farming systems) there is interest in the use of 'compost brews' which are based on feeding the soil biology and therefore improving the resilience of the crop to pest and disease attack reducing the need for pesticide use.

The final stage in an IPM strategy is the application of a crop protection product or pesticide. Once the need for a pesticide has been justified the choice of product, rate and timing are all very important decisions to make by a qualified adviser who holds the statutory BASIS Certificate in Crop Protection. The spray operator is also required to hold a PA2 certificate. These certificates are issued by the National Proficiency Tests Council to show professional competence in calibrating and operating a boom spray applicator.

Accurate records must be kept of any pesticide treatments made to satisfy the requirements of regulations and assurance schemes. It is also good practice to assess the efficacy of each pesticide treatment in the field to help identify any resistance issues that may be developing.

8.2.7 Crop nutrition

The provision of a balanced supply of nutrients at the correct rate and timing to allow a crop to grow to its full potential is very important for economically and environmentally sustainable crop production. Some of the demand is met from mineralisation of soil organic matter, deposition from the atmosphere, weathering of soil minerals and biological nitrogen fixation but the shortfall is provided by nutrients applied to the crop from both organic and inorganic sources (Chapter 4) both of which have important roles to play in sustainability. Organic sources can bring great benefits to soil health by increasing soil organic matter levels and therefore improving other characteristics such as water retention though may not always supply adequate levels of all nutrients at the time a crop needs them. Inorganic materials are an important tool in balancing nutrient levels and providing readily available nutrients but are increasingly costly which is providing incentive to use both inorganic and organic fertilisers more effectively.

Accurate application of both organic and inorganic nutrients is very important for both economic and environmental reasons. The application of excessive rates or applications at times of the year when they are not utilised efficiently by crops, can lead to losses to water and air causing environmental pollution problems. In England around 60% of nitrates and 25% of phosphates in water originate from agricultural land. High levels of these nutrients can cause eutrophication and a reduction in biodiversity and there are also cost implications in their removal from water for human consumption. High losses will also represent an economic loss and therefore affect the profitability of the crop.

The manufacture of synthetic fertilisers, especially nitrogen, is an energy intensive process which is one of the most significant contributors to the greenhouse gas production associated with arable cropping. The application of organic manures and nitrogen fertiliser is also a source of nitrous oxide emissions, a potent greenhouse gas, and their efficient use is important in reducing these emissions. Optimising the use of both organic and inorganic nutrients will be an important factor in agriculture meeting its net zero target.

Accurate nutrient planning should be carried out on a field-by-field basis, tailoring the crop's requirement based on recommendations given in the *Nutrient Management guide* (*RB209*) published and updated by the Agriculture and Horticulture Development Board (AHDB). In the majority of situations estimates of the soil nitrogen supply (SNS) are used to adjust nitrogen recommendations based on rainfall, soil type and previous cropping. However, the soil mineral nitrogen (SMN) content of the soil, which is a measure of the amount of ammonium and nitrate nitrogen available to the crop, can also be tested for on an individual field basis to calculate the need for nitrogen fertiliser (see Chapter 4). Phosphate, potassium and magnesium requirements are determined by regular soil sampling or by calculation of the amount taken off in crops.

Where there is a fixed cropping rotation on a farm, the nutrient management plan can be improved by carrying out nutrient balance calculations. From records of crop yields, the nutrient removal of phosphorus in the form of phosphate and potassium as potash can be calculated from standard values. These are then totalled over the rotation and compared with the recommendations calculated from the Nutrient Management Guide. In many situations there will be imbalances between off-takes and inputs that can be corrected. This should be carried out by farmers or agronomists who hold the Fertiliser Advisers Certification and Training Scheme (FACTS) qualification (FACTS Qualified Advisers - FQAs).

Where organic manures are being used it is very important that their nutrient content is estimated or measured as accurately as possible so that this can be taken into account when calculating inorganic fertiliser requirements. For an individual crop this means analysing the manure to assess the nutrient content or using standard values from the Nutrient Management Guide and subtracting the quantity of available nitrogen, phosphate and potash from the inorganic fertiliser recommendation.

Records should be kept of all operations, which should include the type and amount of fertiliser or organic manure applied and storage details. In Nitrate Vulnerable Zones (NVZs) records of all organic and inorganic nitrogen applications must be kept. All staff carrying out field operations should be fully trained and have access to records including the Nutrient Management Plan and the Codes of Good Agricultural Practice for Air, Soil and Water.

Precision farming tools are increasingly being implemented to provide more targeted nutrient applications based on soil and crop variability. The development of weigh cells on fertiliser spreaders, yield monitors on combines and crop nitrogen sensors has led to the use of variable rate fertiliser application on many farms.

The concept of canopy management for winter wheat and oilseed rape crops where the size of the crop canopy is assessed to determine how much nitrogen is already in the crop is also regularly used to help refine nitrogen applications in the spring.

Under the term of regenerative agriculture there is also a developing interest in using 'Brix' testing to help tailor nutrient applications to crops. A refractometer is used to measure sugar and nutrient levels in a sap sample from the crop and combined with tissue testing this can inform nutrient input decisions.

Under the Governments Clean Air Strategy ammonia emissions must be reduced by 15% by 2030 (compared to 2005 levels). Approximately 88% of ammonia emissions in the UK are from agriculture, and in the arable sector one of the major contributors is from volatilisation associated with the use of urea forms of nitrogen fertiliser. Nitrogen in urea fertiliser is converted into the ammonium form by urease enzymes produced by soil bacteria before it can be taken up by plants. In warm, moist soil conditions this can happen very rapidly before plants can take it up which leads to the loss of ammonia by the process of volatilisation. Some EU countries such as Germany have already banned the use of straight urea fertilisers, only allowing the use of 'inhibited' forms that contain urease inhibitors or nitrification inhibitors that slow down the conversion process reducing the risk of volatilisation and ammonia losses. The use of such inhibitors also means that more nitrogen is made available to the crop improving nitrogen use efficiency. Use of inhibited forms of urea is increasing in the UK but the question of whether the UK will ultimately ban the use of straight urea remains the subject of a DEFRA consultation. There are some concerns around the long-term effects of urease inhibitors on soils and some potential negative consequences of a ban such as increased use of ammonium nitrate forms of fertiliser with greater potential for nitrous oxide emissions.

One further factor to bear in mind in terms of long-term sustainability is the limited supply of the raw materials from which some inorganic nutrients are derived such as rock phosphate. As countries run out of domestic sources it may become a political issue within decades unless other sources can be found and exploited.

8.2.8 Irrigation

Most crops in the UK are produced using natural rainfall and our climate is generally perceived to be 'wet' but rainfall varies significantly across the country, both seasonally and annually and this variability is increasing as some of the impacts of climate change are being seen. In the UK more than 1000 agri-businesses, particularly in the midlands, east, and southeast England need to supplement natural rainfall with irrigation to produce high quality vegetables and fruit that meet the requirements of retailers. Of the total volume of water used for irrigation in the UK, most is used on potatoes (54%) and outdoor field vegetables, such as carrots, onions, parsnips, and salad crops (31%). Some water is used on soft fruit (e.g., strawberries) and orchard fruit (e.g., apples, pears), sugar beet, and very occasionally on cereals and grass.

There is concern over the future availability of water with rainfall uncertainty and increased competition for water resources, focussing farmers on using it as efficiently as possible.

8.2.8.1 Soil water availability

Irrigation is the term used when water is applied to crops which are suffering, or about to suffer, from drought. It is usual to measure irrigation water, like rainfall, in millimetres:

- 1 mm on 1 ha = 10 m^3 = 10 tonnes
- 25 mm on 1 ha = 250 m^3 (a common application at one time)
- 100 mm on 1 ha = 1000 m^3 = 1 dam^3 or 1000 tonnes (about a season's requirement)
- 100 mm on 50 ha = 50 dam^3 (a guide to the amount of water required). [dam^3 = cubic decametre]

Green plants take up water from the soil and transpire it through their leaves (see Section 1.4.1), although a small amount is retained to build up the plant structure. Green crops which cover the ground have a water requirement. This varies from day to day and place to place according to climate and local weather conditions. It is called potential transpiration (Table 8.2).

When the rainfall is well distributed, crops are less likely to suffer than in wet periods followed by long dry periods. During such dry periods the crops have to survive on the available water held in the soil (see Section 3.3.4). The difference between the actual amount of water in the soil and the amount held at field capacity is known as the soil moisture deficit (SMD). It is one of the main factors in determining the need for irrigation.

An approximate soil moisture deficit can, at any time, be calculated from the rainfall and the figures given in Table 8.2. For example, if the soil were at field capacity at the end of April, with a green crop covering the ground, and with no rain in May, then the SMD would be 80 mm at the end of May allowing for losses due to evaporation and transpiration. However, if 50 mm of rain fell during May, then the deficit would drop to 30 mm. The amount of water available to a crop at any one time depends on the capacity of the soil per unit depth, and the depth of soil from which the roots take up water. The figures in Table 8.3 are average rooting depths. They would be greater on deep, well-drained

Table 8.2 Water requirement for green crops, southern UK.

	Per day		Per month	
Period	**mm**	**m^3/ha**	**mm**	**m^3/ha**
April and September	1.5	15	50	500
May and August	2.5	25	80	800
June and July	3.5	35	100	1000

Table 8.3 Rooting depths of common crops.

Crop	**Rooting depth (cm)**
Cereals, grassland	35–100
Potatoes (varietal differences)	30–70
Beans, peas, conserved grass	45–75
Sugar beet	55–100
Lucerne	Over 120

soils and less on shallow, badly drained or panned soils. This can be checked in the field by digging a soil profile. Normally, most of the plant roots are concentrated in the top layers and only a small proportion penetrates deeply into the soil.

More than three-quarters of all the farm soils in the UK have available water capacities between 60 and 100 mm (12%–20% by volume) within the root range of most crops (about 500 mm depth). Sandy, gravelly and shallow stony soils have less than 60 mm (12%), whereas deep silty soils, very fine sands, warpland (formed by letting river water flood agricultural land, leaving behind a layer of fertile silty soil), organic and peaty soils have well over 100 mm (20%) of available water within root range. The available water capacity of the topsoil is often different from that of the subsoil. This should be taken into consideration when calculating the amount present within the root range of a crop.

It is estimated that the roots of a plant cannot take up much more than half of the available water within the depth of its deepest roots. The top half of the rooting zone contains 70% of roots and they do not penetrate all the pore spaces and so, where available, irrigation should be started when about half the available water has been taken up. This is known as the critical deficit.

8.2.8.2 Timing and sources of irrigation

Decisions on when to start irrigating must be influenced by the time it will take to apply 25 mm of water over the area involved, usually 5–10 days. Some companies and organisations provide helpful computerised predictions on when irrigation should be started. These take into consideration all the relevant factors and are tailored for each field and its crop. Farmers can work out their own daily water balance sheet fairly easily, using rainfall records and standard transpiration data to get an SMD figure. The greatest need for irrigation is in the eastern and south-eastern counties where the lower rainfall and higher potential evapo-transpiration means that it would be beneficial about nine years out of ten. The need is much less in the wetter western and northern areas.

Water for irrigation is strictly controlled by abstraction licences issued by the Environment Agency. A licence is required to abstract more than 20 m^3 per day in the UK. The charge for the water varies between areas and it usually costs up to 10 times more for taking water directly from rivers in summer than for taking it to fill reservoirs in winter. Licences are increasingly difficult to obtain and even existing licences are often subject to abstraction restrictions.

Most irrigation water is abstracted from surface water (52%) and ground water (41%) sources with the remainder from public water supply, ponds, and harvested rainwater (7%). Water is generally very scarce in summer in the areas which need it most, but *annual* rainfall is adequate for most irrigation needs provided the winter surplus is stored in farm reservoirs. Above-ground reservoir construction is specialised work and must be designed and supervised by a qualified engineer if they are over 25,000 m^3 in volume and lie above adjacent land. A hole in the ground is simpler and may be partly replenished by groundwater. Lining of a reservoir may be necessary and this could double the cost.

All water used for irrigation should be tested for impurities which may be harmful to crops, particularly salt, heavy metals and bacterial contamination.

8.2.8.3 Application of irrigation

Different textured soils absorb water at different rates (Table 8.4).

When irrigation water or rain falls on a dry soil it saturates the top layers of soil to full field capacity before moving further down. However, if the soil is deeply cracked (e.g., clay) then some of the water will run down the cracks. Very dry soils can absorb water at a faster rate than shown in Table 8.4, but only for a short period.

Irrigation water is applied by:

- *Rotary sprinklers.* Each rotating sprinkler covers an area about 21–36 m in diameter. The sprinklers are usually spaced 10.5 m apart along the supply line which is moved about 18 m for each setting. Special nozzles can be used to apply a fine spray at 2.5 mm per hour for frost protection of fruit crops and potatoes. Icicles form on the plants, but the latent heat of freezing protects the plant tissue from damage.
- *Rain guns.* These are now the most widely used machines. They are used on all types of crops. The droplets are large and the diameter of area covered may be between 60 and 120 m. Some types are used for irrigation of slurry or dirty water.
- *Travelling rain gun with hose reel.* This is similar in many respects to a mobile machine, but the hose reel remains stationary—usually on a headland—while the rain gun and its hose (nylon/PVC) are towed out to the start of the run (up to 500 m), irrigating as it is wound back, using water flow or other mechanical means to turn the hose reel. At the end of the run it switches itself off automatically, and it can then be moved to the next position. About 80% of all irrigation water is now applied using this system. It can be inaccurate and the large droplets can cause soil erosion problems on uncovered ground, but it has relatively low labour requirements. However, pumping costs can be high in comparison due to the higher pressure needed to generate and propel the water jet over considerable distances and subsequently achieve good break-up of the waterjet. Disturbance of the jet and its distribution in windy conditions can result in poor CV% across the travel lane.
- *Boom irrigator with hose reel.* Similar to the travelling rain gun with hose reel but uses a boom often up to 50 m wide mounted on a carriage attached to the hose and moved across the field as the hose is wound back onto the hose reel at the headland. Carriages are often wheeled but smaller units may use skids. The spray nozzles used, operating at lower pressures, vary from rotary or fixed spray nozzles similar to those used on linear and centre pivot irrigators to rotary impact units appropriately spaced on the boom. The boom operates at a height above the crop sufficient for good distribution and overlap of wetted patterns while only experiencing localised disturbance from wind. Small rain guns are sometimes fitted to the ends of the boom to extend its travel lane range. Obstacles in the field such as trees or electric poles present problems for all mobile boom systems and careful planning of travel lanes is needed if used.

Table 8.4 Time of absorption of water by various soils.

Soil type	Time (hr) required to absorb 25 mm of water
Sandy	1–2
Loams	3
Clay	4–5

- *Centre pivot and linear irrigators.* These are large machines which cover up to 160 ha. They are mainly centre pivots which move round a pivot (the water source) in a complete circle or part of a circle. The infiltration rate of the soil often limits the length of centre pivot machines as the application rate at the outer edge of the pivot boom can result in crop damage, runoff and soil erosion. For this reason applications are often limited in depth but more frequent.
- *Spraylines.* These apply the water gently and are used mainly for protected horticultural crops.
- *Surface trickle.* Water is applied through small bore perforated pipes on or near to the soil surface. They give a much more efficient use of water than sprinklers or rain guns and are often used in perennial horticultural crops or vineyards.
- *Sub-surface trickle.* Often used in orchards or vineyards in hot, dry climates like California where water is becoming a scarce resource. The pipes are buried below ground to reduce evaporation waste and make water use even more efficient.
- *Surface channels.* This method requires almost level or contoured land. It is wasteful of land, water and labour, but capital costs are very low.
- *Underground drainage pipes.* On level land, water can be dammed in the ditches and allowed to flow up the drainage pipes into the subsoil and lower soil layers. This is drainage in reverse!

On sloping fields, problems can arise with run-off if water is applied too quickly. It may run down between potato ridges, for example, without really wetting the ridges where the crop roots are growing. There is less run-off if a bed system is used (see 'The bed system' subsection in Section 15.3.6).

8.3 Drivers of change in crop production

8.3.1 UK agricultural policy

As a member of the EU from the period between 1973 and 2020, the UK was bound by the rules of the Common Agricultural Policy (CAP). The basic objectives of the policy were to increase agricultural productivity, ensure a fair standard of living for farmers, stabilise markets and ensure food security and reasonable prices for consumers. Following a review of the Common Agricultural Policy, in 2012 subsidy payments previously targeted solely at production moved to a regional flat rate payment per hectare and became linked to environmental and biodiversity targets for the first time through cross compliance measures. These changes meant that farming systems became fully exposed to global food markets, and as a consequence prices for agricultural products were more volatile. The impact of these changes encouraged farmers to produce food at the lowest unit price, which generally resulted in specialisation of enterprises and an increase in the area farmed to spread overhead costs. The expansion and intensification of agricultural systems across Europe driven by the CAP are increasingly recognised as a contributory cause of soil degradation, loss of biodiversity and climate change.

Having departed the EU in 2020 the UK is no longer bound by the rules of the Common Agricultural Policy, and the nations of the UK are setting their own regimes for agricultural support. Sustainable food production is high on the political agenda and reflected in the UK government's 7 year transition plan for agriculture which started on January 1, 2021. The aim over the next decade is to achieve sustainable and profitable food production without subsidies. Under the Environmental Land Management Scheme (ELMS) in England farmers will be rewarded for sustainable farming practices and direct payments will be reduced. In England, there are three new schemes that will reward environmental land management:

- Sustainable Farming Incentive
- Local Nature Recovery
- Landscape Recovery

The first elements of these schemes will be rolled out from 2022. The Sustainable Farming Incentive will be a universal scheme open to all farmers and will be based on core standards (sets of land management actions) to support sustainable approaches to whole farm management to deliver environmental benefits. At the time of writing, four core standards will be available to all farmers in spring 2022. The two that relate to cropping are the Arable and Horticultural Soils standard and The Improved Grassland Soils Standard. Under each of these standards three levels of payment will be available depending on the measures adopted. Further standards for which tiered payments will be available will be rolled out from 2022 to 2024.

- Hedgerows standard
- Arable and horticulture land standard (different to the soils standard)
- Waterbody buffering standard
- Improved grassland standard (different to the soils standard)
- Low and no input grassland standard
- Farm woodland standard

Other standards still under development:

- Dry stone walls standard
- Heritage standard
- Farmyard infrastructure standard
- Orchards and permanent crops standard
- Peat soils standard
- Agroforestry standard
- Organic production systems

Local Nature Recovery will be a new scheme that replaces existing schemes, focussing on restoring biodiversity to the farmed landscape.

Finally, Landscape Recovery, for which the first round of applications opens in spring 2022, will support more ambitious changes to land use, on a scale from 500 to 5000 ha. Successful applications are likely to be those that restore threatened species or habitats, either by 'rewilding', a recovery process whereby nature is left to itself or by re-instatement, for example, the establishment of new woodlands, reinstating wetlands or restoring the natural course of rivers and re-creating floodplains.

The framework for future funding policies in Wales, Scotland and Northern Ireland is yet to be announced in as much detail but will undoubtedly also be linked to the adoption of more sustainable farming practices as it is in England.

8.3.2 Encouraging biodiversity

As part of its 25 year environmental plan the government has set an ambitious pledge to halt the decline in species abundance by 2030 and with 72% of the UK comprised of farmland, farmers are pivotal to achieving this. Farming has been held responsible for negative impacts on biodiversity through intensification which has resulted in the loss and fragmentation of habitats, such as hedgerows, flower rich meadows, woodland and rough grazing, the drainage of wetlands, and the use of cropping systems dependent on the use of pesticides, fertilisers and increased mechanisation. Substantial reductions in farmland birds are just one of the outcomes often linked with this intensification. Since 1970, there has been a 48% decline in the combined population of 19 species of bird identified by the British Trust for Ornithology as dependent on farmland such as the turtle dove, grey partridge and corn bunting. However, it should also be noted that 6 of the 19 species such as the wood pigeon have actually seen increases. Since the 1930's there has also been a loss of 97% of wildflower meadows, an important habitat for pollinators. In recent decades these and other measurable decreases in biodiversity have focused attention on farming practices in an attempt to find the reasons for and to halt or reverse these declines.

The first formal acknowledgement of this was seen in the development of the Environmentally Sensitive Area (ESA) scheme in 1987 to encourage farmers to protect those areas of the countryside where the landscape, wildlife or historical features were of national importance. There were 22 designated ESAs covering 1.1 million hectares and farmers in these areas voluntarily entered into 10 year agreements with the Ministry for Agriculture, Food and farming (MAFF), under which they received an annual payment for following specific management practices designed to conserve and enhance the landscape. The countryside stewardship scheme was introduced in 1991 for those farmers outside of designated ESA's. Final agreements under both of these original schemes expired in 2014.

At a similar time, LEAF (Linking Environment and Farming) was one of the first organisations to promote environmentally friendly farming and provide practical guidance on Integrated Farm Management (IFM) improving the public perception of modern agriculture. This was implemented through schemes such as LEAF audits, demonstration farms and the LEAF marque engaging a group of farmers with a strong interest in the environment on a voluntary basis.

Following the CAP reform in 2004 some basic measures such as restrictions on the times of years hedges could be cut and no-spray buffer zones next to hedges and watercourses were included in the 'Cross Compliance' measures that a farmer had to meet in order to receive the basic payment. A new agri-environment scheme was also introduced replacing the ESA and Countryside Stewardship schemes. This offered an Entry Level Stewardship scheme (ELS) with flat rate payments over a 5 year period open to all farmers and a discretionary Higher Level Stewardship (HLS) scheme aiming to deliver significant environmental benefits by also providing payment for capital works.

The ELS scheme was based on a wide range of management options to choose from aimed at providing a range of environmental benefits. Examples included the use of fallow plots or uncropped areas for ground nesting birds or overwintered stubbles to provide winter foraging for birds and mammals (e.g., finches and the brown hare), the installation of buffer strips and grass margins to prevent surface water run-off and the planting of margins and field corners with pollen and nectar rich seed mixtures. At the end of 2015 just over five million hectares of farmland were included in the scheme. ELS and HLS agreements closed to new entrants in 2014 and a new Countryside Stewardship scheme with mid-tier and higher-level options was launched in 2015. These schemes have subsequently received some criticism because in many cases farmers have not had to change their farming practices significantly to receive the payments which were received whether benefits were delivered or not and consequently there have been few measurable improvements in biodiversity.

Whilst perhaps not achieving some of the desired aims, these schemes, as well as the understanding and interest that many individual farmers have in protecting the farmed environment have already led to many examples of positive changes in practice to reduce the decline in biodiversity. For example, in the past field margins were often managed by spraying into the hedge bottom with a non-selective herbicide to prevent the spread of pernicious weeds. This destroyed the field margin vegetation and frequently reduced the growth of the hedge and therefore the biodiversity it could support. Under cross-compliance these grass margins were protected and, in many cases, extended to widths varying from 2 to 12 m using options from ELS. In some cases, the biodiversity in these margins has been improved further by the addition of native wild flowers (Fig. 8.11) to encourage beneficial insects, such as bees, hoverflies and lacewings.

The management of hedgerows themselves has also improved with the times of year when hedges can be cut restricted under cross compliance and the ELS options that reduced the frequency with which they could be cut to allow berry development and thus support more birds. Beetle banks have been created in the centre of large arable fields to provide suitable habitat for predatory insects as well as a nesting habitat for partridge or skylark. They are designed to provide a habitat of tussocky grasses (e.g., cocksfoot) where ground beetles can overwinter in relatively dry conditions. The predation of aphids is more effective because the beetles move into the crop from the beetle bank in the centre of the field and the field margin round the outside.

In the future under ELMS in England, the Local Nature Recovery scheme to be rolled out from 2024 will be the successor to the existing Countryside Stewardship scheme, focussing on restoring biodiversity to the farmed landscape. It will pay for a range of options based around the following themes:

- managing, restoring and creating grassland habitats such as species-rich grassland on farms and in the wider countryside
- managing, restoring and creating wetland habitats such as ponds, lakes, reedbeds and fens
- managing, restoring and creating lowland heathland
- managing, restoring and creating coastal habitats such as sand dunes, salt marsh and shingle
- managing and restoring areas of upland and lowland peat and moorland on farms and in the wider countryside
- targeted measures to support the recovery and reintroduction of particular wildlife species, such as creating and managing nesting and feeding habitat, and to tackle non-native invasive species

Figure 8.11 Wild flower rich field margin.
Source: Andy Wilcox, Harper Adams University.

- managing and creating trees and woodlands, including agroforestry, traditional orchards and tree planting on areas of farms
- nature-based solutions for water—such as creating and managing in-field vegetation, buffer strips and swales to reduce and filter runoff and contribute to natural flood management
- restoring rivers, flood plains, streams and riparian habitats

8.3.3 Protection of soils

Good soil management has always been recognised as an essential element of growing high yielding, profitable crops. However, during recent decades of intensification soil health has measurably declined. Intensively farmed soils have lost 40%–60% of their organic carbon affecting their ability to grow high yielding crops and to fulfil the ecosystem services they offer in mitigating against flooding, supporting biodiversity and regulating greenhouse gases. Reversing the decline in soil health by 2030 is one of the aims of the government's 25 year Environment Plan.

Farmers have always taken steps to reduce the physical removal of soil and nutrients from their fields by runoff, erosion and nutrient leaching particularly on the lighter soil types, sloping fields and after particularly high-risk crops such as maize. The decline in organic matter in agricultural soils has also increased the risk of runoff and erosion. Various steps can be taken to reduce the risk e.g., carrying out cultivations across slopes, the use of over winter cover crops so that soil is not left bare, the use of buffer strips, the use of tied ridges in potatoes, intercropping with grass in maize (see Fig. 18.1), and sowing nurse crops such as spring cereals in sugar beet which are selectively killed by herbicides when the crop is established.

The first regulatory tool to improve soil management practices was introduced in 2005 through the completion of a Soil Protection Review, a documented soil management plan required as part of Cross Compliance under the Single Payment Scheme. In the future the plan to reverse the decline in soil organic carbon levels will be implemented through the Sustainable Farming Initiative Soil Standards for Arable and Horticultural soils and Grassland. Under each of these standards three levels of payment will be available depending on the measures adopted. For the arable standard, at the introductory level a soil assessment will need to be completed with a commitment to establish green cover over winter and increase soil organic matter on a small percentage of the farmed area. At intermediate and advanced levels, the percentage of the farmed area on which the same measures must be taken is increased and a commitment to the use of no, low or minimum tillage techniques on a percentage of the farmed area must also be made.

The prevention of runoff and erosion and retention of soil and nutrients on the land is closely linked to the protection of water which will be covered in the next section.

8.3.4 Protection of water

The Water Framework Directive which originated from the EU in 2000 and was transposed into UK law in 2003 has been retained following the UK's exit from Europe. The directive originally required that all inland and coastal waters within defined river basin districts reached at least 'good' status (assessed in relation to biological, chemical and physical parameters) by 2015. In practice this is implemented through river basin management plans based on a six year cycle. However, few water bodies met this original target and the river basin management plans have been extended by two further six year cycles to 2027.

With around 70% of land in the UK being used for agriculture, tackling diffuse pollution from agricultural activities is critical to meeting these targets. In February 2005 the concept of Catchment Sensitive Farming was introduced by a partnership between Defra, Natural England and the Environment Agency. This aimed to deliver practical solutions and targeted support to enable farmers and land managers to take voluntary action to reduce diffuse water pollution from agriculture. Sixty-five priority catchments and nine catchment partnerships were identified, and catchment sensitive farming officers or river basin coordinators were employed to provide advice and guidance to farmers in these areas.

In 2018 a new legislative approach was introduced in the form of 'The Farming Rules for Water'. Broadly the eight rules (listed below) require that soil is kept on the land, that nutrients match crop and soil requirements, and that livestock, fertilisers and manures don't enter watercourses. Some of these rules align with existing NVZ regulation and cross compliance and an advice led approach is being taken before any enforcement action.

1a) Application of organic manures and manufactured fertilisers must not exceed soil or crop need.
1b) Soils should be tested for P, K, Mg and pH at least once every 5 years. Soil Nitrogen Supply (SNS) must also be assessed either using the Field Assessment Method or using Soil Mineral Nitrogen testing.
2) Do not store manures within 10 m of a watercourse or within 50 m of a spring, well or borehole. Additionally, they must not be stored in any location that might lead to a pollution incident (such as slope leading to watercourse/on top of field drains/waterlogged ground etc.).
3) Organic manures or manufactured fertiliser must not be applied if ground is waterlogged, snow covered or has been frozen for more than 12 hours in the previous 24.
4) Organic manures must not be spread within 10 m of a watercourse or within 6 m if using precision application such as dribble bar/trailing shoe/disc injector no deeper than 10 cm.
5) Manufactured fertiliser must not be applied within 2 m of a watercourse.
6) Reasonable precautions must be taken to prevent erosion and run off from organic manures/fertiliser using suitable management and cultivation practices.
7) There should be no livestock poaching within 6 m of a watercourse.
8) Stock feeders should not be placed within 10 m of a watercourse or 50 m of a spring, well or borehole OR anywhere where there is a significant risk of pollution from the feeder (e.g., steep slope over 12 degrees, waterlogged ground etc.).

8.3.5 Climate change and 'net zero'

Climate change is arguably the most urgent issue facing the world and farmers have a crucial role to play in the national effort to achieve 'net zero' by 2050. Indeed, the National Farmers Union (NFU) has set the ambitious goal of reaching net zero greenhouse gas (GHG) emissions across the whole of agriculture in England and Wales by 2040.

Emissions from UK farms currently amount to 45.6 million tonnes of carbon dioxide (CO_2) equivalent a year, about one tenth of UK GHG emissions. Emissions related specifically to crop production arise from:

1. Burning of fossil fuels in machinery during the production of crops;
2. Transport and on-farm processing of crops;
3. Use of fertilisers and pesticides;
4. During the drainage and cultivation of soils.

Expected effects of climate change in the UK are higher temperatures, wetter winters, drier summers and more extreme weather events which are having an impact on crop and variety choice and weed, pest and disease distributions. These impacts on crop production are discussed in more detail in Chapter 2 (Section 2.6).

The Committee on Climate Change (CCC) which acts as an independent advisor on addressing climate change to the government produced its report on reducing emissions in 2018 which was updated in 2020. The report states that the UK's net zero by 2050 target cannot be met without significant changes in land use but there is also scope for more widespread use of low carbon practices in agriculture to reduce emissions.

Producing a carbon footprint using farm specific data where emissions are expressed as carbon dioxide equivalents (CO_2e) on a 'per net unit of food product leaving the farm' basis allows farmers to identify where there is scope to reduce emissions by benchmarking themselves alongside other similar enterprise types.

8.3.5.1 Reducing emissions from arable agriculture

Improved productivity allows farmers to dilute their emissions per tonne of crop produced. Therefore, many of the principles of good crop management to produce healthy crops that make best use of available resources described throughout this book are relevant here but a few specific areas where there is the greatest scope to reduce emissions from crop production are described below.

- The application of organic manures and nitrogen fertiliser is a major source of nitrous oxide emissions, a potent greenhouse gas. Optimising nitrogen use efficiency e.g., by variable rate applications, accurate timing and the use of enhanced efficiency products (urease inhibitors, nitrification inhibitors and slow-release fertilisers described in Section 8.2.7) have the potential to reduce emissions.
- The use of renewable energy sources such as solar PV panels on arable farms also has scope to reduce emissions.
- Precision farming technology, matching tractors and machines to field operations, planning work to minimise journeys, regular machinery maintenance and ensuring the correct tyre pressure is used can improve machinery and equipment efficiency and therefore reduce fuel use and associated emissions. In the future lower carbon farm machinery e.g., electric or hydrogen powered may be more widely used.
- During cultivation, reducing the amount of soil disturbance using min-till or no-till practices decreases fuel consumption and reduces the release of carbon dioxide from the soil which occurs when soil organic matter is oxidised.
- Increased use of over winter cover crops extends the period of active photosynthesis and therefore increases carbon dioxide removal from the atmosphere and organic carbon storage in soils. Mixtures containing legumes can also reduce the need for inorganic nitrogen fertilisers.
- Including more grain legumes (pulses) in arable rotations reduces reliance on inorganic nitrogen fertilisers.

A number of these practices will be incentivised through the recently introduced Sustainable Farming Incentive scheme as described in Section 8.3.1.

As well as taking steps to reduce emissions through their crop management practices, arable farmers have the potential to improve their carbon footprint by being involved in more major land use changes identified by the Climate Change Committee report as necessary to meet the net zero by 2050 target for the UK. These changes could include increasing the area of bioenergy crops grown, planting trees and restoring peatlands. There may be potential for funding for such land use changes through the Landscape Recovery Scheme or through the market which is developing in trading natural capital.

8.3.6 Food safety and security

Food security can be defined as having reliable access to enough affordable and nutritious food. Both domestic and international factors affect food production and prices for consumers in the UK. Around 60% of food consumed in the UK is imported and both international and domestic supply chains are sensitive to economic and environmental factors such as extreme weather events and conflict. This vulnerability was demonstrated recently in the UK when global shortages of gas had knock on effects on food production and the food processing industry leading to some short-term shortages.

The safety of our food is also very important. In the UK, food assurance schemes help to provide consumers and businesses with guarantees that food has been produced to particular standards. These schemes are mainly voluntary arrangements although many food businesses make certification by an assurance scheme a requirement for their suppliers.

The farmer is the first link in the food chain. Determining the way crops are planted, the inputs applied to the growing crop and the method of harvesting and storage. Therefore, the farmer has a significant impact on the safety and quality of a food product, especially where fruit and vegetables reach the consumer with very little further processing. The farm assurance schemes set standards for good agricultural practice, audit growers and ensure traceability in all aspects of crop production, from pesticide use, to effective grain storage and risk assessments, including mycotoxin risk assessments. Assurance standards are at or above the minimum legal requirements to ensure the producers are complying with relevant legal requirements. A further quality control imposed is the maximum residue level (MRL) of a pesticide in food on the supermarket shelf. This is not a safety limit, but indicates if a pesticide has been applied in accordance with the manufacturers' recommendations, and is monitored and tested for by the Food Standards Agencies for each of the UK nations.

The management of food safety in the food chain is being addressed by assurance schemes, with the use of HACCP (Hazard Analysis of Critical Control Points) principles being seen as the preferred approach to management at all stages of food production, including agricultural production. HACCP identifies hazards which can cause the consumer temporary or permanent injury, and assesses risk as the probability of a hazard occurring. The Critical Control Point is the stage in production where the probability of the hazard occurring can be eliminated or reduced to an acceptable level.

The first crop assurance scheme to be introduced in the UK was Assured Produce in 1996. This scheme was for growers supplying vegetable and salad crops and was an initiative developed by the NFU, producers and retailers, putting in place a system which not only recorded farming operations and inputs, but also allowed external scrutiny or verification of the production systems. The voluntary uptake of the scheme aimed to demonstrate to supermarkets that field vegetables, including potatoes, and salad crops were grown to professionally agreed protocols, and provided traceability. This has given rise to several terms, such as from 'Plough to Plate' and from 'Field to Fork' indicating that the whole food supply chain is included. The Assured Combinable Crops Scheme was launched in 1998 to address the same issues for arable crops.

Whilst not placed directly onto supermarket shelves like many vegetables and salad crops, wheat grown for bread making is processed and baked at the miller's prior to being sold in retail outlets. It is therefore just as important for these to be subjected to the same quality assurance standards as other food.

In 2000, the British Farm Standard was formed to bring together the UK crop and livestock assurance schemes using the Red Tractor as its logo. In 2005 this became the Assured Food Standards with the on-pack logo incorporating the Union Jack. In 2020 the scheme moved towards using the term Red Tractor on the logo. The Red Tractor Produce and Red Tractor Combinable Crops and Sugar Beet standards have become the most widely followed assurance standards in the UK with Red Tractor Assurance. Compliance with the standards ensures that good farming practice is followed when growing crops and increasingly are incorporating aspects of environmental stewardship as well as food safety. Assurance schemes have a cost to the farmer, which depends on the area of land being used for crop production. The fee pays for the administration of the scheme and includes a visit by the external auditor.

The Red Tractor scheme only applies in the UK, similar assurance schemes have also been developed in other countries. One of the largest of these is Global G.A.P. which developed out of the EUREP GAP fruit and vegetable protocol that was introduced in 1999 to standardise the requirements of retailers in the EU. This standard was developed further for non-EU countries and the name was changed to Global G.A.P. in 2007 and now covers livestock, aquaculture and crop production systems.

8.4 Whole farm approaches to sustainability

The term 'Integrated Farm Management' was adopted by LEAF in the 1980's as a whole system approach towards more sustainable farming. In reality, the majority of farmers practice elements of an integrated farming approach and only the strongest enthusiasts have joined the LEAF assurance scheme enabling them to use the LEAF marque to demonstrate that their produce had been grown in accordance with LEAF's IFM principles.

More recently interest in this area has grown with the realisation that more sustainable practices are needed to secure the future of many farming businesses, our land scape and our food supply. Opportunities are also developing to benefit financially from such practices e.g., by trading natural capital. One outcome of this is the development and uptake of terms such as 'regenerative farming'.

8.4.1 Regenerative farming

The term 'regenerative farming is often attributed to Gabe Brown, a North Dakota rancher and farmer who after several years of adverse weather and crop failures began to focus on the health of his soils. There is no set definition of regenerative farming but most will follow the principles of moving the soil as little as possible, minimising the use of pesticides and fertilisers (but not eliminating them), diversifying cropping and potentially introducing grazing livestock into arable rotations (see Section 8.4.3) with the aim of improving soil health and resilience through increasing soil organic carbon. To be successful the system should be introduced slowly and perhaps on a designated area of the farm initially so that lessons can be learnt on a local level before broader implementation.

Because of the loose definition of regenerative farming there are no defined standards or certification schemes as there are for organic systems and therefore no premiums available for produce grown in this way. However, farmers following these practices are hoping for longer term benefits in the resilience of their systems to a changing climate and economic factors such as the high price of inorganic fertilisers.

8.4.2 Agroforestry

Agroforestry is a land management approach that combines trees with crop and livestock farming systems. Integrating trees into farming systems can help to provide shelter, protect soils from wind or water erosion, increase biodiversity, improve animal welfare, diversify income streams and capture carbon. The trees themselves may be species that produce a timber, fruit or nut crop thus also providing an income in the longer term.

Silvoarable cropping refers more specifically to a method of growing rows of productive trees among arable crops (sometimes termed alley cropping), for example, apple trees and winter cereals (Fig. 8.12). The rows of trees are most often aligned north to south to reduce the impact of shading on the arable crop and are spaced at an appropriate distance to easily accommodate normal machinery operations in the arable crop. If the objective is to maintain arable cropping for the duration of the tree rotation, the distance between tree rows should be relatively wide to allow for continued profitable arable crop production. Narrower alleys can be used if the ultimate aim is to focus on profit from the tree derived products.

Figure 8.12 Silvoarable cropping being practised in Eastern England.
Source: Stephen Briggs, Abacus Agri.

The trees reduce wind erosion and make use of resources at a different time of year to the cereal crop. The arable crop maintains income during the period of establishment of the trees. Interestingly, the use of arable cropping to improve the cash flow of tree establishment was the basis of a poplar production system developed in Herefordshire and Suffolk by the match manufacturer Bryant and May in the UK in the 1960s and 1970s.

Agroforestry is not currently widely practised in the UK as there is a lack of funding for the capital investment required as well as a need for more advice and guidance and perhaps a perception that yields of arable crops will be reduced and that the trees will interfere with normal farming practices. However, its potential to contribute to achieving net zero has been recognised by the Climate Change Committee. An ELM pilot project is currently underway to inform future incentives and advice for agroforestry which it is hoped will become part of the Sustainable Farming Incentive and Local Nature Recovery schemes.

8.4.3 Integrating livestock into arable rotations

Under the Common Agricultural Policy farming in the UK has become increasingly specialised, with arable enterprises dominating the flatter and drier Eastern regions and livestock farming dominating the wetter, hillier Western regions. However, a number of issues such as declining soil health in arable areas and weed management challenges have renewed interest in incorporating livestock in arable rotations for long term sustainability.

Depending on the objective a range of options can be considered

- Grass and clover leys can be introduced to rotations alongside livestock such as sheep to help manage weed problems e.g., black-grass
- Cover crops in arable rotations can be grazed providing forage, weed management benefits and improving soil health providing site selection is appropriate
- Outdoor pigs can integrate into arable rotations to improve soil health
- Livestock manures can be used to improve soil health

Investing in livestock and the associated infrastructure obviously requires large scale investment and arable farmers do not necessarily have the required expertise with livestock. However, there are opportunities for existing livestock and arable farms to enter mutually beneficial agreements such as muck for straw deals, contract growing of forage crops and grazing or cutting agreements or licences.

8.5 Key points

- Agricultural policies over recent decades have encouraged intensification which has had some negative repercussions on the environment.
- There is an increasing focus on sustainable production techniques at both a policy and on-farm level in order to protect soil and water, mitigate against climate change and ensure future food safety and security and the future profitability of farming businesses.

- Soil drainage, seedbed preparations, rotations, integrated pest management, nutrition and irrigation are the basic building blocks of crop production but much greater consideration is being given to sustainability in the choices that farmers are making.
- In the future available government funding will be used to encourage farm biodiversity, protect soils and water as well as helping to mitigate climate change; this is likely to influence cropping systems and management and cultivation techniques used.
- In some cases, farmers are considering whole farm approaches such as regenerative farming or the introduction of agroforesty or livestock to arable rotations.

Sources of further information and advice

Agriculture and Horticulture Development Board (AHDB), 2021. Livestock and the Arable Rotation. Available at: https://projectblue.blob.core.windows.net/media/Default/Imported%20Publication%20Docs/AHDB%20Cereals%20&%20Oilseeds/General/GS100%20Livestock%20and%20the%20arable%20rotation%20V2.pdf.

Agriculture and Horticulture Development Board (AHDB), 2021. Nutrient Management Guide (RB209). Available at: https://ahdb.org.uk/rb209.

Agriculture and Horticulture Development Board (AHDB), 2020. Arable Soil Management Cultivation and Crop Establishment. Available at: https://ahdb.org.uk/knowledge-library/arable-soil-management-cultivation-and-crop-establishment.

Bailey, R., 1990. Irrigated Crops and Their Management. Farming Press.

Benckiser, G., Schnell, S., 2007. Biodiversity in Agricultural Production Systems. Taylor Francis Group, Boca Raton.

Brown, G., 2018. Dirt to Soil: One Family's Journey into Regenerative Agriculture. Chelsea Green Publishing.

Castle, L., McCunnall, R., Tring, B., 1984. Field Drainage – Principles and Practices. Batsford.

Defra, 2009. Protecting Our Soil, Water and Air. Available at: https://www.gov.uk/government/publications/protecting-our-water-soil-and-air.

Committee on Climate Change, 2020. Land Use: Policies for a Net Zero UK. Available at: https://www.theccc.org.uk/publication/land-use-policies-for-a-net-zero-uk.

Defra, 2021. Environmental Land Management Schemes: Overview. Available at: https://www.gov.uk/government/publications/environmental-land-management-schemes-overview/environmental-land-management-scheme-overview.

Lindenmayer, D., Cunningham, S., Young, A., 2012. Land Use Intensification. Effects on Agriculture, Biodiversity and Ecological Processes. CRC Press.

Raskin, B., Osborne, S., 2019. The Agroforestry Handbook. Soil Association Ltd.

Vandermeer, J.H., 2011. The Ecology of Agroecosystems. Jones and Bartlett.

Wezel, A., 2017. Agroecological Practices for Sustainable Agriculture: Principles, Applications, and Making the Transition. World Scientific.

Useful websites

ABACUS Agriculture https://abacusagri.com.

Agricology https://agricology.co.uk.

British Trust for Ornithology https://www.bto.org.

Defra https://www.gov.uk/government/organisations/department-for-environment-food-rural-affairs.
Farming for a better climate https://farmingforabetterclimate.org.
LEAF https://leaf.eco.
NFU https://nfuonline.com.
Organic Research Centre https://organicresearchcentre.com.
Plantlife https://plantlife.org.uk.
RSPB http://www.rspb.org.uk.
The Woodland Trust https://www.woodlandtrust.org.uk.

Precision crop production

9

9.1 Introduction

Precision farming (PF) is the management of spatial and temporal variation by increased attention to detail. Data collected from the farm can be used to make site and time specific decisions about a crop. The management scale used can be reduced so rather than treating a crop as a 10 ha field, the area can be divided down into smaller areas which may need to be treated differently enabling the farmer to identify yield limiting factors. Precision farming is therefore not a technology, it is a way of thinking.

Automated farming uses machines to apply these management decisions. Work traditionally done by tractors and their operators can now potentially be automated. The Hands Free Hectare and Hands Free Farm at Harper Adams University are demonstrating how automation can be implemented for crop production.

All of this requires georeferenced data, the farmed area becomes a series of coordinates to which data and management decisions can be allocated.

This Chapter will explore the ways that georeferenced data is collected, analysed and used on the farm. The Chapter will also describe the work done at Harper Adams University on the automation of crop production.

9.2 Obtaining georeferenced data

The creation of satellite constellations orbiting the earth means position coordinates can be assigned to any part of the farm. Global navigation satellite systems (GNSS), often referred to as Global Positioning Systems (GPS) which is one of several available systems, are used to obtain this positioning data through the process of trilateration. The resolution or how detailed georeferencing can go is affected by the GPS accuracy which will vary from 0.3 to 0.025 m depending on the system purchased. Location data from mobile phones is also available but will tend to be at a lower resolution of approximately 4.9 m. GPS location data can therefore be obtained by a variety of methods and used to georeference various field information as described below.

9.2.1 Remote sensing

Remote sensing using cameras mounted on satellites, planes or drones can record variable biomass or colour of crops in individual fields (Fig. 9.1). Aerial digital RGB (Red Green Blue) images can be used for basic visual inspection as well as quantitative assessment such as plant population counting. Numerical indices such as Normalised Difference Vegetation Index (NDVI) values calculated from multispectral images

Lockhart and Wiseman's Crop Husbandry Including Grassland. https://doi.org/10.1016/B978-0-323-85702-4.00025-X

Figure 9.1 Agricultural drone.
Credit: Harper Adams University Hands Free Hectare.

(Fig. 9.2) can be used to monitor crop growth and potential yield throughout the season (Fig. 9.3). Satellite or aerial digital images which show soil brightness indicating texture, soil organic matter and soil moisture content are also possible.

Remote sensing images can be interrogated to provide a range of information about the crop such as nutrient status. However, all can be affected by weather conditions, aerial images in particular are susceptible to cloud cover, making the UK fields harder but not impossible to monitor remotely. Satellite imagery of a particular field could be obtained every 2–16 days if necessary depending on the satellite used or drones can be used below cloud level. Indeed, this potential for frequent scans of the field are the most useful aspect of remote sensing for monitoring crop growth and development.

9.2.2 Machine and field based sensors

Machinery regularly travel across arable fields carrying out routine tasks such as cultivation, drilling, applying fertiliser, spraying and harvesting. Each occasion presents an opportunity to gather georeferenced data about the soil, crop growth and yield.

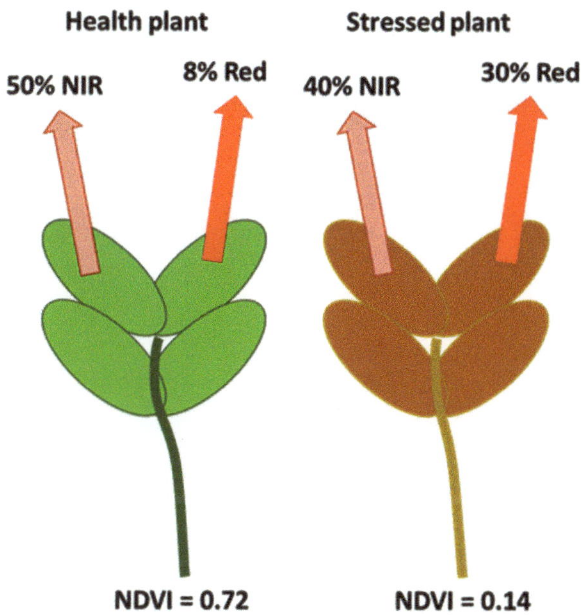

$$NDVI = \frac{NIR - Red}{NIR + Red}$$

Figure 9.2 Normalised difference vegetation index. Credit: Harper Adams University Hands Free Farm.

9.2.2.1 Crop sensing

Sensors can be fitted to the tractor or sprayer which record colour and brightness changes as it passes along the tramlines thereby measuring growth indices such as NDVI which can be used to make decisions on the fertiliser requirement of the crop. Passive sensors pick up natural light and the latest technology allows the adjustment of the sensors to take into account light quality, active sensors have their own light source so can work in low light/night conditions. Developments in technology now can use ultrasound and RADAR to determine crop structure, such as tiller numbers or height of crop.

9.2.2.2 Soil sensors

Soil mapping in its crudest sense has been carried out for many years. pH testing across a field is a form of soil mapping which generates data which allows differential application of lime to correct pH. Taking standard soil samples and testing for P and K levels is another form of soil mapping, although, in the past, samples from across a field were usually combined to give a single field result on which applications were based.

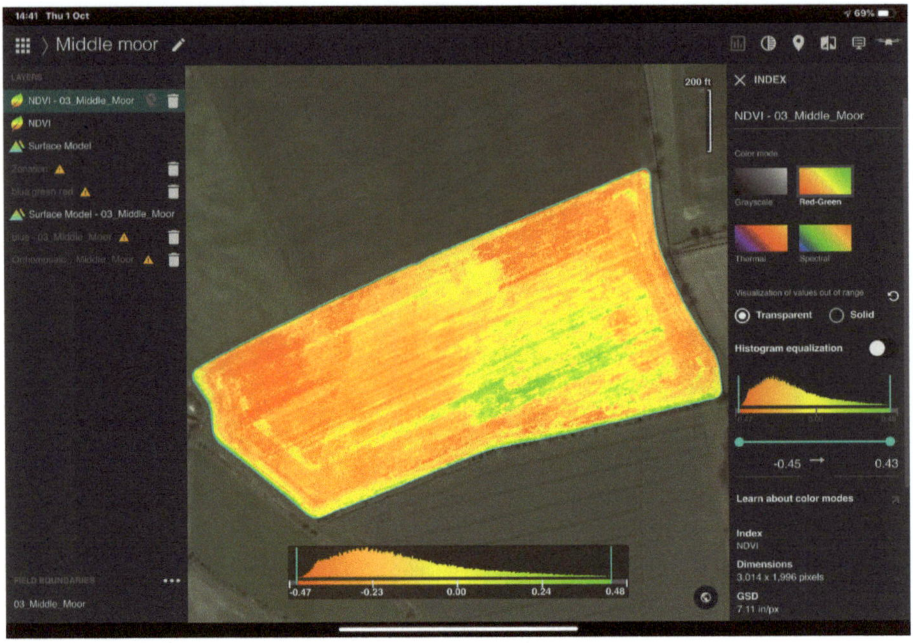

Figure 9.3 Remote sensed normalised difference vegetation index field image.
Credit: Dataset made available by PIX4Dfields.

Now companies offer a mapping service where samples are taken using an All Terrain Vehicle (ATV) fitted with a GPS sensor and a nutrient or texture map can be produced similar to a yield map. Red Green Blue (RGB) images can also be produced which indicate crop health and could point to underlying problems such as soil compaction or salinity.

There are also other measurements that are becoming linked to precision agronomy and soil management. Electrical conductivity metres can be hand held or attached to (ATVs) and the conductivity measurements give an indication of a field's soil texture (sand, silt and clay %) and therefore its ability to hold nutrients and moisture. Hand held compaction metres (penetrometers) are available which measure physical resistance to being pushed through the soil profile and sensors can now be fitted to individual legs of a cultivator to measure soil resistance on the move.

9.2.2.3 Yield monitors

Probably the oldest of the spatial data technology is yield mapping which fuses measurement data collected on the combine harvester regarding grain flow and moisture content along with field position and area covered (header width and combine speed). This technology can be retro-fitted to nearly all types of machine, including sugar beet and potato harvesters. Data can be sent wirelessly in real time direct to computers, or stored remotely for later download and analysis. Yield maps are then produced and used in the development of crop input application plans at a later date. It is important to realise that the more years of data collected, the more useful the maps are. This allows for both spatial and temporal variation within a field and masks the 'odd' years when yields were either very high or very low because of uncontrollable yield limiting factors e.g., the weather.

The accuracy of the maps is only as good as the data generated. The hardware and software are improving with every new generation of machine, but accuracy can still be affected by factors such as sudden speed changes of the harvester, % of the header width used, blockages, temporary loss of GPS signal and then, post-harvest, factors such as data aggregation and simplification, and point 'smoothing' during the generation of the contour yield maps.

9.2.2.4 Field walking

In all cases of georeferenced data collection previously mentioned, it is important to understand that the technology must be used in conjunction with other data, such as field history, weather records and geology maps. Ground truthing is also important—Red Green Blue images may hint at a compaction problem but the farmer will still have to dig a hole to confirm this.

Agronomists can use hand-held GPS units or mobile phone applications to record patches of weeds and incidences of diseases or pests. They can also record areas of poor growth or of damage from larger pests such as rabbits. A specific patch of interest in a field can now quite easily be georeferenced using a pin in Google maps or by using new location applications such as what3words which splits the earths surface into 3 m square blocks and allocates them a unique three word code. Field walking has also been honed with the use of certain drone flight apps enabling the drone operation to cover larger areas faster, capturing images and analysing them for useful information in areas of interest. Skippy scout (TM) can do this and also provide a report allowing the crop walk to focus on areas of interest.

9.2.2.5 Weather

Farmers can buy small automatic solar powered weather stations which can be moved around the farm and automatically record rainfall, soil and air temperature, humidity, sunlight and wind speed and then wirelessly transmit it to a base computer. The roll out of rural 5G and the constant stream of new applications will enable the farmer and agronomist to streamline decision making based on data going from the field to analysis with greater ease such as the use of Hutton Criteria for Blight spraying on an individual field basis. High precision weather services which use mesh technology are also available via subscription providing field level data without the need for on farm hardware.

9.3 Storing and analysing georeferenced data

This part is probably the most difficult to manage when it comes to precision agronomy. Farmers can be overwhelmed by the amount of data being produced, and the numerous interactions that might be taking place in the field. When the precision farming systems were in their infancy, it was often difficult to interpret yield maps, for instance when they were all the farmer had to go on. One common dilemma was: Given an area of low yield in a field, should more fertiliser be used on it to bring the yield up, or accept that it will always be low yielding because of other factors, and therefore to put less on, thus saving money on wasted material? This is often referred to as the Robin Hood versus King John problem, should you steal from the rich to give to poor OR tax the poor to give to the rich. The first strives for uniformity the second aims for peak performance.

The key phrase in the question is 'other factors' and it is these that technology is becoming more sophisticated at measuring and modelling. Companies and agronomists are now able to accurately map differences in soil properties, in incidences of pests, diseases or weeds and have a better understanding of the importance of field history such as old boundaries, manure storage, pipelines, earthworks, spillages and sites of buildings.

Now, the more sophisticated interpretation of data allows what the industry calls 'targeted actions' or 'targeted agronomy'. This targeted response can include variable rate applications of fertilisers or sprays, clearly delineated areas for subsoiling or mole ploughing, alteration of sowing rate or depth, specific rabbit fencing or even requirement for more detailed and careful crop walking (Fig. 9.4).

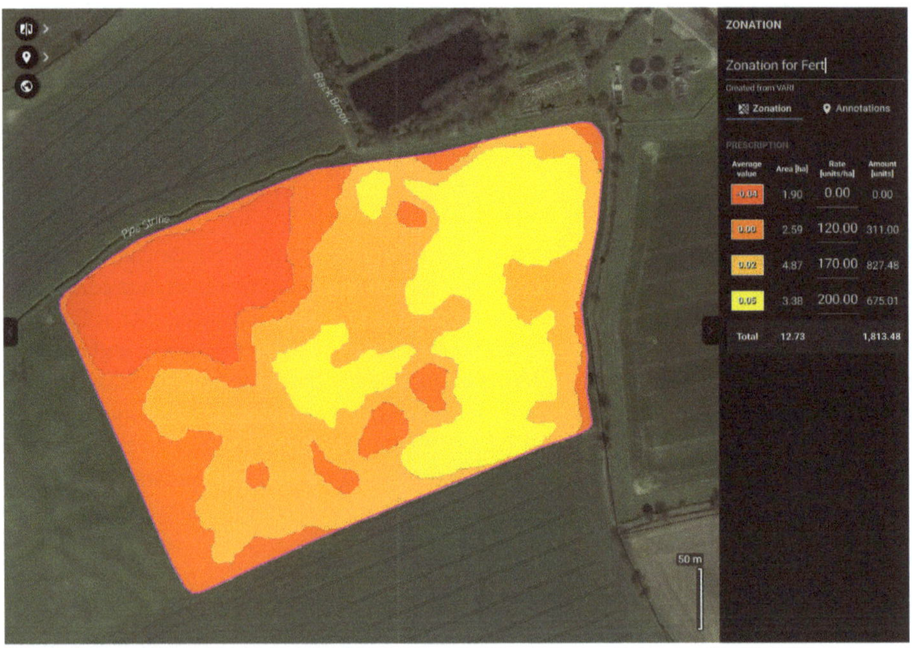

Figure 9.4 Variable application map.
Credit: Harper Adams University Hands Free Farm.

Once data is collected or logged over time it has to be analysed by software, and then used to variably apply inputs or make other management decisions. It is often combined with other data, e.g., weather records, to help the analysis. A mass of individual pieces of data are produced from, say, a combine harvester travelling across a wheat field. Some errors will be amongst this data. To produce a yield map the data must be smoothed or 'kriged' and it is the ability of the programmes used to do this that determines how useful and accurate the map will be. Treatment maps should be assessed by the farmer to ensure any ground truthing issues are also resolved before executing the final map.

Much of the work today is based on modelling or simulation to try and predict what might be the crop response to a range of inputs and other influencing factors such as weed populations, disease pressure, weather factors and soil conditions. These are all being developed to provide an as accurate as possible decision support system for farmers. The emergence of artificial intelligence, fuzzy logic and machine learning in agriculture will almost certainly improve and refine these decision support systems.

9.4 Using georeferenced data

9.4.1 Auto-steering and controlled traffic farming

One of the other main uses for precision farming technology is to make a machine operator's life easier by allowing the machines to steer themselves. There are numerous guidance systems available.

Guidance systems come in many shapes and sizes, as well as complexity and ability. There are two factors involved—positioning of the machine in the field and steering of the machine once in motion. Systems can be used to position and steer fully automatically or can be set up to provide on-board guidance for a tractor driver. Usually a display of some sort will tell the operator if the tractor or harvester is in the right place and heading in the right direction. This can be as simple as a light bar indicator or as sophisticated as a high resolution computer screen with 3D graphics. The simpler systems can be moved from vehicle to vehicle but the more complex ones are inbuilt and stay with the machine.

The operator sets the initial guidance line (sometimes known as the A-B line) and then tells the on-board computer to steer the machinery parallel to it at a predetermined distance away from it. The operator can then concentrate on making adjustments to any implements, watching out for problems or observing the crop if it is an operation during crop growth. Systems are now available that will perform headland turns as well as in-work steering. Reliance on satellites alone can be inaccurate because of atmospheric conditions distorting satellite signals which cause calculated positions to drift, so the most accurate systems use real-time kinematics (RTK) correction signals. These signals are calculated at fixed ground base stations which are continually monitoring the error in GNSS position, the required error correction is then transmitted in real time to the remote GNSS system. This can give accuracy to within ± 2 cm at the remote system. On farm RTK bases are either positioned permanently on a building or for best results on a tripod in the corner of the field transmitting via short range radio. RTK correction subscriptions are also available which are generated locally to the farm and received via cellular communication.

9.4.2 Controlled traffic farming

Guidance systems are at their most useful when there are no tramlines, i.e., before the crop has been drilled and during the drilling or planting process. They can be used in conjunction with the drill to provide very accurate tramlines which can then be used with or without automated steering options. This can also have the effect of more precise and accurate applications and operations as well as making sure that wheelings only occur in a tightly controlled area within the field. This is especially important in crops that are very sensitive to compaction such as sugar beet or carrots. Ensuring that the crop is not grown above a wheeling can increase yield and also improve quality of the final product, e.g., straight or unfanged carrots, which would be accepted for supermarket pre-packs. It also allows subsequent remedial action to correct compaction to be targeted very precisely, thus reducing costs. Controlled Traffic Farming (CTF) takes annual tramlining further by confining all farming operations to permanently fixed traffic lanes, it requires all farm implements to be a standard base span or multiple of this span. An example of this would be 8 m drill and 8 m combine harvester with 24 m crop sprayer. Ideally the track width of the implements would also be matching but this can be challenging due to road transport and differences between tractors and combine harvester axle widths. CTF presents further challenges when considering logistics tasks such as grain handling and bale collection although engineering solutions are emerging to support this. Although CTF could technically be achieved without any digital technology, RTK guidance is necessary for controlled traffic farming as satellite drift is eliminated thereby ensuring tramlines and wheelings stay in the same place from week to week and year to year.

9.4.3 Financial

The most obvious financial benefit is the potential savings in costs of production, although there is an argument to suggest that yields and quality of crops can be improved by more effective use of inputs an example being preventing crop lodging from seed or fertiliser overlap. There can be cost savings on all three major inputs to crops—seeds, fertilisers and agrochemicals. Further, the sustainable farming initiative (part of the environmental land management scheme) requires the use of precision farming techniques for the application of crop inputs to achieve the advanced level standard.

9.4.4 Variable rate sowing

GPS systems are accurate to within a few centimetres so, with the right machinery and set up, seed overlaps can be made a thing of the past. This is especially important when sowing high value crops with expensive seed such as maize and sugar beet. Even with 'cheaper' seed such as cereals or oilseed rape, as long as the individual coulters on the drill can be controlled by the on-board computer, drill overlaps can be substantially reduced, even with the drill travelling at much higher speeds than precision drills.

As well as the physical distribution of the seed, the technology can also allow differential seed rates to be changed on the move to allow for the differences across a field in soil type, condition, nutrition status and weed pressure. This allows optimisation of seed rates to allow for higher yields and less seed wastage.

9.4.5 Variable rate nutrient application

Solid and liquid fertiliser formulations can be adjusted on the move with a spreader or sprayer rate controller via an application map based on previously gathered data such as yield mapping. These variable rate systems are available on newer machines as options but can also be retro-fitted to older machines. When applying nitrogen the technology exists to adjust rates of both liquid and solid N using on the move, real-time on-tractor monitoring of crop colour or near infra-red reflection. These systems aim to improve nitrogen use efficiency reducing over application and waste which has both an economic and environmental benefit. However, application equipment limitations such as the coefficient of variation for fertiliser spreaders will limit the resolution that an application can be made at.

9.4.6 Variable rate pesticide application

Just as with fertiliser, sprays are expensive and costs of purchase, application and disposal are rising all the time. Spray overlaps can be minimised by giving control to the on-board computer to conduct Auto Section Control which automatically turns off overlapping boom sections, rather than relying on the operator's judgement. There may also be savings of labour costs as field operations can be quicker with, for instance, steering guidance. Costly mistakes may also be reduced. Future savings from data collected by agronomists using hand-held GPS systems can be used to target specific problems, with rates and mixtures potentially being changed as the sprayer drives down the tramlines using direct injection sprayer technology, where the sprayer carries clean water and the chemicals are injected into the pipeline as required. Spot herbicide treatment of certain volunteers such as potato volunteers in maize crops using vision systems to distinguish between the volunteer and the crop are now available. However, identifying weeds with similar growth habits and colour to selectively spot spray presents a greater challenge for the technology at present, AI trained 'green on green' vision systems are in infancy but developing quickly to target species such as Docks in grassland. The use of vision systems to guide an automated hoe for weed control in row crops are available too for situations where crop protection via spraying is not the preferred option.

Future developments in the area of precision crop protection might include the use of agricultural drones for pesticide application. These are in use around the world subject to regulatory approval.

9.4.7 Environmental

The new sustainable farming initiative requires Precision Agriculture techniques as one of the actions which will help farmers achieve advanced level standard. A reduced risk of pollution from nitrates, phosphates and agrochemicals is expected if less inputs are used more effectively, but also means savings in the supply chain, from reduced production, fewer road miles for delivery, less packaging production and waste and less pressure on disposal of plastics. It could mean fewer passes through the field if the inputs are used more effectively, saving on fuel and reducing GHG emissions thereby contributing to net zero status.

9.5 Autonomous agriculture

Current farming techniques have characteristics that make precision systems necessary but limit the benefits that can be gained. A machine size such as a 12 m combine header will produce low resolution yield data compared with the individual plants that produced the yield. Application machinery such as 24 m fertiliser spreaders or 24 m sprayers also have the same challenges with respect to resolution. Large machinery, for instance, the average horse power of tractors sold in the UK in 2020 was 171 hp, with a weight around 8 Tonnes can produce soil compaction. Precision techniques using this size and scale of machines may not be able to work on a plant by plant basis.

An autonomous system that will control not only steering and positioning, but also ground speed, engine speed, machine operation such as rates of application without an operator present offers the opportunity to scale down machine size and therefore improve the resolution of the precision farming operations. Groups or 'swarms' of smaller autonomous machines working for longer, such as 24 h shift patterns, could match the work rate of the current large manned machinery.

The Hands Free Hectare at Harper Adams University was the first to have successfully established, monitored, maintained and harvested combinable crops autonomously in 2017 and 2018. Rather than using full sized modern equipment, the Harper approach opted for a smaller 38 hp tractor and associated implements.

The Hands Free Farm is a development on from the hands free hectare. A fully mechanized 35 ha/86 acre driverless, multi crop farming system has been implemented at Harper Adams University Farm. Currently at the end of its second year of production the project continues to use the 38 hp model as the optimum size tractor (Fig. 9.5). Cover crops, wheat and beans have been established, monitored, maintained and harvested autonomously.

Commercial products have been available from disruptive startup companies for a few years and options from the main machinery manufacturers are now becoming available with the latest offerings from John Deere launched in 2021.

The potential to reduce tractor and implement size but increase the number of working hours to 24 h a day or enable an operation to take place in perfect weather conditions is an exciting option for farms to manage the challenges of recruiting staff and reducing soil damage by heavy machinery.

Figure 9.5 Hands Free Farm autonomous spraying.
Credit: Harper Adams Hands Free Farm.

9.6 Barriers to the adoption of precision farming

Precision crop production has some issues to resolve. Communication coverage, system compatibility, GPS accuracy, obsolescence, the right to repair equipment to name but a few. Fully automated, connected systems are open to attack by malicious software. This could have disastrous effects on this industry. Cyber security will need to be at the forefront as we share more of our farming data. Futuristic systems such as Ag 4.0 will need more connectivity and uptake of precision cropping to work.

However, before committing to the investment in the technology it is important to assess the likely benefits to the business by a cost/benefit analysis. AHDB provide a cost: benefit tool which will do this, but, before starting out using such a tool, there are certain factors which should be taken into account.

Yield limiting factors (YLF's) that cannot be controlled in the field will not be affected by precision farming methods. Natural rainfall, sunlight hours, humidity, CO_2 levels and air temperature are variables that cannot be altered. Soil temperature and soil carbon can be affected by husbandry, but only over a long period, sometimes decades.

Soil structure, drainage, acidity, inherent soil fertility, available water content and pests, diseases and weeds are all YLFs that can be managed using precision farming, although most are medium term alterations within the field.

Those YLFs that can be quickly altered are factors such as temporary soil fertility, operator error, timeliness of inputs and the effect of sowing and husbandry on crop architecture.

9.7 Key points

- Technology and agronomy are available to variably apply crop inputs.
- Quick wins to access precision farming is to get machine guidance on the farm tractors.
- Gather field data using a variety of methods.
- The future will certainly have autonomous machines working in UK crops.
- Farms need to train staff in the use of precision technology to get the most benefit from it.

Sources of further information and advice

Anon, 2017. The Precision-Farming Guide for Agriculturists, fourth ed. Deere and Company, USA, ISBN 0-86691-435-8.

Case New Holland, 2017. Autonomous tractor technology shows way forward for farming: enhancing efficiency and working conditions in agriculture. Available at: https://media.cnhindustrial.com/EUROPE/CASE-IH/autonomous-tractor-technology-shows-way-forward-for-farming–enhancing-efficiency-and-working-condit/s/d9d11785-2881-4577-afc2-23e6dadbfc91.

Hands Free Hectare, 2022. Hands free hectare HFH. Available at: https://www.youtube.com/c/HandsFreeHectareHFH/videos.

John Deere, 2022. Future of farming. Available at: https://www.deere.co.uk/en/agriculture/future-of-farming/.

Lowenberg-DeBoer, J., Ericson, B., 2019. Setting the record straight on precision agriculture adoption. Agronomy Journal 111 (4), 1552–1569.

Organic crop husbandry

10

10.1 Introduction

Organic farming is a system that avoids the routine use of readily soluble fertilisers and/or agrochemicals, whether naturally occurring or not. It is a system that aims to use renewable resources where possible and is therefore a good example of a sustainable farming system. Many of the methods used are also being considered by conventional farmers looking for a more sustainable and agroecological approach to their farming system. Most organic farms have a lower energy use than conventional farms, mainly due to no inorganic ammonium nitrate being applied. Organic farming is not a return to pre-1940s farming. It uses current technology and relies heavily on good husbandry. Organic farming aims to maintain soil fertility as well as encouraging biological cycles. Most organic farms include grassland in their rotations which can help maintain soil organic matter and increase carbon sequestration. There is a large reliance on legumes for supplying the crop nitrogen requirements. Weeds, pests and diseases are controlled mainly by non-chemical methods. Animal welfare, pollution and the wider social and environmental issues are also important.

The ideas on organic farming started in the twentieth century. In 1946 a group of like-minded farmers and scientists, including Lady Eve Balfour and Sir Albert Howard, founded the first UK organisation, The Soil Association. It was not until the 1990s that there was a large increase in interest in organic farming both in the United Kingdom and in many parts of Europe and the world. About 16 million ha or 3.3% of land in Europe is now farmed organically compared with 1.5% of global land according to the Research Institute of Organic Agriculture (FiBL). Just over a quarter of the world's organically farmed land is in Europe, with Spain, Italy and Germany being the largest producers. Since 2002 the area of organically farmed land in the UK has slightly increased in many regions (Fig. 10.1). The only decline in area certified was in Scotland where there was a major reduction in area due to withdrawal of certified organic rough grazing and woodland. Some of the other reasons for changes in organic areas were due to changes in support schemes and difficulties in marketing some organic produce partly due to the recession reducing demand for the higher priced organic food. There are now over 485,000 ha of organically managed land in the UK. Currently in the UK there are various stewardship and grant schemes that support organic farming. This will change when the new schemes such as Environmental Land Management Scheme (ELMS) are introduced in 2024 these in turn could affect the area of organically farmed land.

Lockhart and Wiseman's Crop Husbandry Including Grassland. https://doi.org/10.1016/B978-0-323-85702-4.00007-8

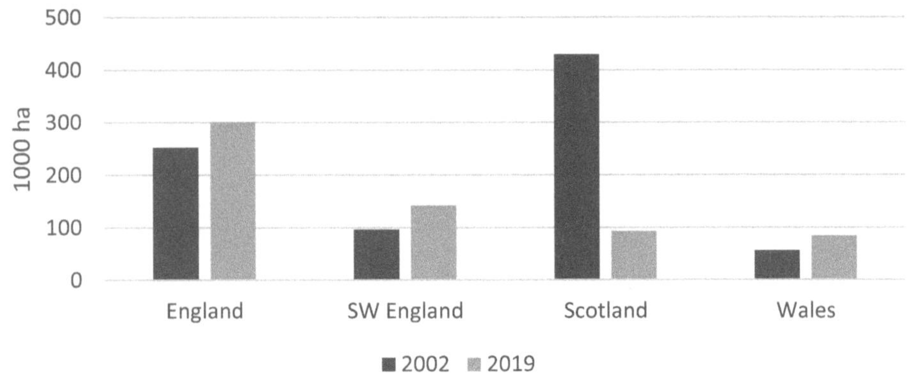

Figure 10.1 Organic land (in-conversion and fully converted) in England, SW England, Scotland and Wales in 2002 and 2019.
From DEFRA statistics.

In other parts of Europe, the picture is different. There has been a much larger shift to organic farming systems for example in Austria, Estonia and Sweden where over 20% of agricultural land is now being farmed organically.

Demand for organic produce has also been much greater in other parts of Europe than in the UK though demand for organic produce in the UK has still been steadily rising since 2011. With the increase in organic livestock enterprises there has been a growing demand for organic feed grain and currently the UK is not self-sufficient and has to import some organic animal feed.

One of the possible reasons affecting interest in organic farming is perceived reduction in farm margins; This has not been the case in some farm management surveys. Organic production is a type of sustainable or agroecological system that is regularly inspected and certified by law. It delivers many known benefits including increased biodiversity and protection of wildlife as well as paying attention to soil quality and pollution control.

Some organic crops such as cereals are cheaper to grow and grain prices are much higher than for conventional crops; but to be viable they cannot be grown too frequently in the rotation. Many field vegetables are much more expensive to grow than conventionally grown crops especially if hand weeding is required, as in carrots. Yields from organic crops can be very variable and are often only 50%–70% of average conventional yields.

For organic farming to work it must be viable financially and the farmer or grower must be committed to the system. There needs to be a change in the approach to the husbandry of growing the crops as agrochemicals and fertilisers cannot be applied if a serious problem arises. The risk of crop loss is much higher than in conventionally grown crops, but the returns can also be greater.

10.2 Achieving organic status

All products sold as organic must be certified. All countries have their own certifying bodies. These regulations are underpinned by the guidelines produced by the International Federation of Organic Agriculture Movements (IFOAM). IFOAM has four major principles to consider.

1. **Health** of soil, plants, animals, people and the planet.
2. **Ecology.** The system should build on and strengthen living ecosystems.
3. **Fairness.** The system must be fair and guarantee equal opportunities.
4. **Care.** The health and well-being of present and future generations need to be considered as well as the environment.

10.2.1 UK organic standards

DEFRA is responsible for overseeing certification of organic produce in the UK. In the UK there are a number of approved certifying bodies including the Soil Association Certification Ltd, Organic Farmers and Growers Ltd, Organic Food Federation, Scottish Organic Producers Association, Quality Welsh Food Certification Ltd and Irish Organic Farmers and Growers. Each farm is inspected annually by its registering body to confirm that the approved guidelines are being followed.

Standards are constantly being amended; it is important that farmers keep good farm records. Current recommendations include: farmyard manure can only come from stock that has not been fed GMOs (genetically modified organisms) and written proof is required; bought-in farmyard manure has to be composted. As organic seed stocks are limited, farmers can currently obtain derogations from their certifying body to plant untreated non-organic seed. In the future the target is for all seed sown to be organic. A limited number of fertilisers and crop protection products are permitted to be applied to organic crops. Some products have restricted use such as meadow salt (a potassium source). Farmers must be able to demonstrate a good reason to apply these restricted products such as a recent soil analysis showing deficiency in potassium if ground potash is to be applied. The permitted and restricted lists are constantly being amended according to UK regulations.

During the annual inspection, input and output records as well as accounts have to be available. Crop records required include: treatments applied for the last 3 years for land in-conversion; crop rotation plans and cropping areas; cropping history including yields; application of farmyard manures and other fertiliser, composting treatments, rates and source; source and types of products used for pest and disease control and source and types of seeds and transplants.

After inspection a compliance form is sent to the farmer with details of the certification decision for each enterprise on the farm, and details of any action to be taken. Each farm must provide up-to-date certification details when selling organic produce.

10.2.2 The conversion period

When a farmer decides to convert to organic production methods a conversion plan is produced in order to get certified by an approved body. Depending on the enterprises on the farm the plan includes details of land areas, dates of conversion, cropping and rotations, stocking rates for livestock, forage management and details of stock housing and produce storage. Conversion can start from the date of application of the last non-permitted product. The ground is not certified as fully organic until 2 years after that date. Some farms convert all of their land at the same time; others find it easier for rotational purposes to convert over a period of a few years. At present, in the UK, a farmer does not have to convert the whole farm, unlike many other European countries.

Support payments for both the 2-year conversion period and subsequent organic maintenance payments are run differently in England, Wales, Scotland and Northern Ireland.

During the first year of conversion, crops can only be sold as conventionally grown non-organic. Second year in-conversion produce can sometimes be sold at a premium which is usually lower than for fully organic products. It is common practice to grow fertility-building crops rather than cash crops during the 2-year conversion period. There is usually a reduction in farm returns during the in-conversion period, hence why organic support grants are particularly important in the first 2 years.

10.3 Rotations

Rotation is the basis of a good organic system. Initially, when changing to this farming method, during the in-conversion period fertility-building crops such as clover are usually grown. A good rotation will aim to balance nutrient requirements, minimise nutrient losses, maintain or even increase soil organic matter, improve soil structure and help eliminate or reduce weed, pest and/or disease problems. Organic rotations use more legumes than conventional cropping; red and white clover, lucerne and even vetches are grown to boost soil nitrogen supply. Usually in a rotation at least 20% of the crops grown will be legumes.

Green manures such as mustard, vetches or rye are commonly grown after a summer-harvested crop and before a spring-planted crop. Green manures reduce the losses of soil mineral nitrogen during the winter; they also provide plant cover so that the soil is less prone or liable to suffer from soil water erosion. When ploughed-in in the spring the green manures release the readily available nitrogenous compounds for the following crop. Having a diverse range of crops means that plants are grown in the rotation with different rooting depths and root biomass which can help to improve soil structure and retrieve nutrients throughout the soil profile.

As well as cereals other crops grown include grain legumes, potatoes and field vegetables. The nitrogen demanding crops such as potatoes or winter wheat usually follow nitrogen fixing crops. There is no one perfect rotation; each farm will have a rotation to suit the enterprises, the soil, and climate of that specific farm. Any potential weed or pest problem may also dictate cropping as well as time of cultivations and planting

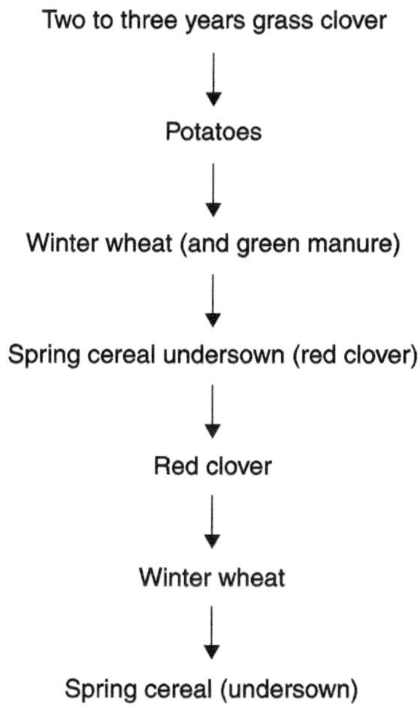

Figure 10.2 An example of a rotation used on a mixed organic unit.

date. Alternating winter and spring crops can help reduce dominance of any specific weed. Continuous cereals are prohibited and there must be at least a 4-year break before cropping again with brassicas or potatoes on the same land. Normally no more than 2 years of cereals are grown in succession and cereals do not make up more than 50% of the crops grown. An example of a rotation used on mixed organic units is given in Fig. 10.2.

Some farms practise a stockless system. In stockless rotation, usually 2 years of cereals will follow a legume. The leguminous crop could either be field beans or even a forage legume for seed. Straw is incorporated and any spring crops grown could follow a green manure. This system has been successfully run for many years on some farms though many have found that livestock needs to be introduced as they are able to utilise the fertility building grass clover leys and can help with nutrient cycling. This may only be feasible on farms where there are adequate livestock facilities or where there are local organic livestock units looking for grass keep.

10.4 Soil and plant nutrition

An important aim in organic farming is to maintain and, if possible, improve the soil structure, soil organic matter, flora and fauna. The ideal is to balance nutrients removed from the farm, in harvested crops or livestock, with inputs from bought-in feed, straw and seed. There is limited data to suggest that organic farming increases soil bacteria and fungal activity and that these are important in mobilising some of the nutrient reserves in the soil.

Soil structure can be affected by crops grown, cultivations and application of organic manures. Some crops which have deep tap roots, such as the vetches and field beans, have been shown to improve soil structure. Shallow cultivations tend to be favoured to try and maintain the level of available nutrients in the topsoil. It is not always possible to cultivate shallowly when ploughing out a ley or after a vegetable crop with large crop residues. When ploughing out grass leys there can be an initial flush of available nitrogen released, and it can be difficult to time cultivations and crop drilling to reduce nitrate leaching. Quite often the use of the stale seedbed technique for weed control and avoidance of some pests and diseases leads to late drilling for some crops, and this means that the soil can be liable to nutrient loss before the crop is able to use the available nutrients.

Grazed leys, application of compost/manure and incorporation of crop remains such as straw and green manure crops help to maintain or even increase the soil organic matter.

10.4.1 Inorganic fertilisers

Nutrient off-takes should be balanced by nutrient inputs. There is only a limited range of fertilisers that are permitted for use on an organically grown crop. Most of these products are not very soluble and tend to become available to the crop over a long period. Production should be planned to minimise the need for bought-in nutrients. Lime, e.g., ground limestone or chalk; phosphorus compounds, e.g., natural rock phosphate; magnesium, e.g., magnesium rock or kieserite as well as some biostimulants are examples of permitted products.

There are limited potassium sources other than farmyard manure (FYM), bought-in straw and animal feeds. Under certain circumstances some restricted products can be applied. Potassium, e.g., meadow salt or Silvinite, is an example of a restricted fertiliser. In the case of Silvinite, the grower has to have available at inspection a recent soil analysis result for the field in question. The soil index must be 0 or 1 for this restricted product to be used. This product is prohibited for use in some other European countries.

Because of the restrictions on fertiliser and manure sources several stockless organic farms have now introduced a livestock system. There are, however, several successful stockless farms, which often have started with high levels of soil potassium or have soils with a reasonable clay content. These soils should have enough potassium mineralised each year for crop production.

10.4.2 Organic manures

As the use of inorganic fertilisers is limited, farmers rely on the application of organic manures including approved sources of amenity waste for both soil improvement and provision of crop nutrients. Some organic manures can be brought onto a holding if they come from an approved livestock system. The use of plant wastes and animal manures from non-organic sources are restricted and must be approved by the certifying body before use. FYM from non-organic sources must be composted for 3 months or stacked for 6 months. As noted earlier, no GM feed should have been fed to the stock producing the manure.

There are several ways of managing compost. The aim is to limit nutrient loss. FYM should be composted to reduce soil-borne diseases and weed seeds. Turning is encouraged to ensure an adequate breakdown process. Turning usually increases aeration and speeds up the microbial activity, but the more the manure is turned the greater the gaseous loss of nitrogenous compounds. The nitrogen in well-composted FYM is usually fairly stable and not prone to soil leaching. During composting the manure will reduce in volume by 50% so reducing spreading costs. There is a lot of debate about whether to cover the manure pile. Covering can help lower the moisture content and leachate losses, but more commonly the heap is outside, uncovered. If so, the aim is to build the heap so that excess water runs off and, if possible, the leachate is collected.

When applying organic manures care must also be taken to ensure that there is no risk of run-off and pollution of water courses. As organic feed can be lower in potassium than in conventional sources, normally organic manure contains less available potassium than standard data would suggest. Use of sewage sludge, effluents and sludge-based composts is prohibited. Manure from some livestock systems such as intensive poultry battery systems is also prohibited.

10.4.3 Nitrogen sources

Most of the nitrogen in well-composted FYM is in an organic form and will be slowly released over the following years. Only a relatively small amount will be available in the first season after application. As FYM can vary considerably in N content (and potassium) between farms due to differences in stock, feed and housing it is worth undertaking an analysis of the manure rather than using estimates from the literature. Timing of application of manure can also affect the amount of soluble nitrogen compounds that are present. Losses are usually lowest from a late winter/early spring application. Application rates per field and overall per farm must follow the strict guidelines laid out by certifying bodies (and by DEFRA in nitrate vulnerable zones—NVZ's) in order to minimise leaching and risk of pollution. For example, for the Soil Association no more than 170 kg/ha of N as FYM can be applied per year.

The major nitrogen source for organic crop production is from nitrogen-fixing legumes. Residues of nitrogen are usually highest after lucerne or red clover. Ploughing-out a legume and planting the following crop can lead to high losses of soluble nitrogen. It is very difficult for an organic farmer to minimise these nitrogen losses as planting date is also governed by weed, pest or disease control requirements.

There are a number of pelleted-approved organic fertilisers. These usually contain a very low percentage of nutrients (N, P and K) and are expensive. Their main use is on specialist horticultural holdings.

10.5 Weed control

In an organic rotation a much broader range of weed species is commonly found than in a more conventional rotation. Some of the most difficult weeds to control in organic crops are perennial weeds such as creeping thistle and docks often associated with mixed farms. The importance of different weeds in organic compared with conventionally grown crops is partly due to the different rotations and the soil nutrient status. Many of the most serious conventional arable weeds, such as cleavers and blackgrass, tend to be very responsive to applied nitrogen and continuous winter cropping and so are less important in an organic system.

Many growers converting to organic production are often seriously worried about potential weed problems. In practice, there can be problems but there are some effective methods of control available.

10.5.1 Rotations and cropping

A mixed rotation, rather than mono-cropping, can have a major effect on weed species. A rotation of winter- and spring-sown crops will tend not to favour any weed species, and the inclusion of good cleaning crops such as potatoes or grass leys will also help to reduce weed populations. Some crops themselves can influence weed emergence and growth (allelopathy). It has been found that different weeds grow in different crops even if planted in the same field at the same time. Oats and red clover are two crops that appear to be particularly good at suppressing weeds. Growing a mixture of crops such as beans and wheat has also been shown to reduce weed biomass compared with the amount found in the individual crops. Weed problems tend to be less in new grass leys if they are established in the spring as an undersown crop rather than after a cereal crop in the autumn.

Green manures can reduce weeds by ensuring good ground cover during the autumn and winter. Ploughed-in green manure crops such as rye and mustard can also help suppress weed (and occasionally crop) germination.

10.5.2 Variety selection

Variety choice should not be underestimated as a method of weed control. Many varieties of the same crop have quite different growth and/or leaf habits. Varieties should be chosen that produce a vigorous leaf canopy. Varieties with horizontal (lax) rather than vertical leaves will be more competitive against weeds. It has been found that the old variety of wheat, *Maris Widgeon* which has a tall straw and lax leaves, can reduce weed biomass by around half that found in commonly grown modern short strawed varieties.

10.5.3 Time of sowing

Delayed sowing is often practised on organic farms usually due to the avoidance of pest and disease problems; it can also aid weed control. Delayed drilling can give time for some weeds to germinate and be killed by cultivations before drilling takes place (the stale seedbed technique). When growing vegetable crops such as leeks and brassicas, transplants are usually used, this gives time to kill weeds before planting the crop. Once the crop is planted the transplants can compete very quickly with many weeds. The combination of transplants, late drilling and in-crop cultivations means that these crops can have a lower number of weeds, so making them excellent cleaning crops.

10.5.4 Seed rates and crop density

High seed rates can help reduce weed competition. In winter cereals for example drilling 500 seeds/m^2 compared with 200 seeds/m^2 can halve the weed biomass. In horticultural crops it is not always possible to plant high populations as this technique can affect crop quality and lead to poor marketable yields.

Having a dense grass clover ley can be effective at reducing weed competition; note though that weeds may be encouraged in grass leys which have been heavily poached or very tightly grazed.

10.5.5 Cultivations

This is one of the major tools that an organic farmer can use. Often on organic farms more cultivations are required than on conventional farms due to the requirements for weed control. It starts as soon as the first cultivations take place. Most organic farms use the plough more than many conventional farmers. When cropping after grass/clover leys the ground is often rotovated before ploughing as this can help kill the grass. Ploughing is a good method for controlling some of the grass weeds and some perennial weeds. The 'stale' seedbed technique is commonly used before sowing the next crop. Once sown or planted there is now quite a range of in-crop weeders that can be highly effective if the timing and weather conditions are right, the most commonly used being the tine or harrow comb weeders (Fig. 10.3). These can be used on a wide range of crops from cereals to many field vegetables such as the *Brassica*. In cereals these harrows can be effective especially if the weeds are very small; results have ranged from 0% to 80% weed control. Mat-forming weeds such as ivy-leaved speedwell and chickweed can be easily pulled up, even late season, compared with weeds with tap roots such as charlock which are not very well controlled. In winter cereals, autumn weeding (when the crop is not competitive) can lead to more weeds emerging. To get the best results with these weeders there will be some crop damage, but usually there is little effect on crop yield. An effective harrowing may cover the crop plants with up to 30% of soil without causing crop damage. Depending on the season, crops may be treated several times with the harrow comb.

As well as whole-field treatments with the harrow comb some crops are also cultivated with an inter-row hoe (Fig. 10.4). Inter-row cultivators are important for weed control in organic field-vegetables and root crops. The use of some of these implements is effective at minimising the amount of hand weeding required. Some of these weeders are self-propelled, some front tractor mounted and others rear mounted with

Figure 10.3 *Harrow comb.*

Figure 10.4 Inter-row cultivator—front mounted on tractor.

an operator on the back (steerage hoe). A wide variety of hoes is used, from goose-foot shares to brushes or even finger tines. Brush hoes (Fig. 10.5) do appear to work better than conventional hoes in wet conditions. The hoes can influence weeds within the crop row by throwing soil into the rows and so smothering the weeds, especially when the weeds are small.

Precision guided weeding systems have been developed which can be much more accurate and work closer to the crop rows than conventional hoes. These weeders use

Figure 10.5 Steerage hoeing using a brush weeder.

video image analysis techniques to distinguish between crop plants and weeds. To work effectively, the crop plants need to be more dominant than the weeds. A precision weeder is being developed for transplanted crops that can weed both inter-row and inter-plants using the same technology.

As in conventional farming it is important to start the weed control programme in time before the weeds start competing. In many crops, such as beans, carrots, swedes and onions, the first in-crop weeding is the important one. It should be timed when the weeds are small, usually only three to 4 weeks after 50% crop emergence. This is when the rows are just visible and the crop plants have their first true leaves. In transplanted crops, the first weeding can take place within a week or two of planting as long as the transplants are not easily uprooted. Weeding with a tine weeder is usually done as soon as the next flush of weeds is at the cotyledon stage. Dry conditions after weeding will ensure better control.

10.5.6 Flame or thermal weeding

This method of weed control is especially important in some horticultural crops. The use of propane gas burners can be effective at controlling a range of weeds. The heat from the burner causes the plant cell walls to burst and the weed normally dries up and dies within 2 to 3 days. These burners do not control all species. Weeds such as chickweed, redshank and cleavers are very susceptible even past the two true leaf stage. Charlock is only susceptible at the cotyledon stage and many grass and perennial broad-leaved species are only checked. In crops such as carrots, onions and parsnips, flame weeding is carried out just before the crop emerges. This gives these small-seeded crops a chance to get established before the weeds start to compete. Flame weeders can also be useful to burn off potatoes before lifting.

There are several tractor-mounted models available. The problem with flame weeding is the high energy consumption and slow work rates, but it can reduce hand-weeding costs by half in some horticultural crops.

10.5.7 Other methods of weed control

Hand weeding, hand hoeing and even digging out can be very important when trying to avoid the development of potential serious weed problems such as docks and even creeping thistle. A tractor mounted dock 'twizzler' has been developed which can be quicker than hand digging and is just as effective. In some horticultural crops such as carrots and onions, even after flame weeding and inter-row cultivations, hand weeding is still needed. At least 200 man hours/ha are commonly required. To make weeding easier some farms use bed weeders (Fig. 10.6). These are tractor drawn or self-propelled frames on which the weeding gang lie so that both their hands are free for pulling weeds; this ensures that the task is not back-breaking and can make the job much quicker and easier than when hand weeding on hands and knees.

Some perennial weeds such as couch, docks and creeping thistle can be difficult to control. One option is to introduce a fallow period and to keep cultivating the ground. If weather conditions are dry, then repeated cultivations can be effective at dragging the weed roots or rhizomes to the soil surface where they will dry out.

A mulch such as black polythene is another excellent method of weed control; the problem is cost. Mulches are only economically viable for use on high value horticultural crops. The film is usually laid using tractor-mounted equipment. Polythene mulches have to be removed after harvest; occasionally they can be reused. Currently several biodegradable mulches are also available.

Other methods of weed control are being investigated including using electrical charges, biological control using fungi and hot water treatments.

Figure 10.6 Bed weeding.

10.6 Disease control

Some diseases common in conventional crops are not so serious in organic crops, this is partly due to a varied rotations and limiting number of times crops are grown in the same field. Also, many organic crops are thinner than the conventionally grown crops due to less N being available which reduces the risk of the spread of rain splash diseases. The thin crops lead to more air movement and less of a microclimate within the crop canopy. Lush crops tend to be those most prone to damage by diseases. Cereal crops are often late drilled to reduce risk of diseases such as take-all, barley yellow dwarf virus (BYDV) and foliar diseases. It is suggested that a balanced soil and crop nutrient status will also help to reduce the incidence of some diseases. Application of manure has been shown to reduce disease possibly due to affecting the soil microflora.

Choice of variety is important. A limited number of organic variety trials have been undertaken in a range of crops; however, recommended lists for conventionally grown crops can still be useful when deciding on the most appropriate variety for organic systems where disease resistance will be of particular interest. Growing mixtures of cereal varieties can help to lower disease incidence and lead to small yield increases. Mixtures are usually only feasible if growing cereals for the feed market, rather than for a specific quality market.

Seed borne diseases can be an important source of disease; organic growers should only sow, or plant, disease-free seed, or transplants. Some vegetable seed is commonly hot water treated which can be effective at reducing some seed borne diseases. Good crop hygiene is also important for organic farmers to help reduce disease incidence. When storing organic crops, the aim should be to only store a disease-free crop.

Even with using all these disease control methods there are still some diseases that can be important in organic crops. One example is potato blight, with only a limited difference in susceptibility between varieties, although there are several eastern European varieties that have good resistance. In the past in the UK growers were able to use a limited quantity of restricted products such as copper oxychloride once blight appears. These copper products are currently not authorised to be used though occasionally there has been emergency authorisation.

10.7 Pest control

There are very few approved products that can be used so organic farmers have to look at husbandry techniques that can help reduce pest pressure. Many organic farmers will tend to accept a higher amount of pest damage than in conventional crops. There are many very good management practices that can help reduce the risk.

Increasing the natural predator population can reduce some pest populations, for example aphids. Organic farmers are encouraged to manage field margins and boundaries sympathetically to encourage predatory insects. On some organic farms higher predatory insect numbers have been found compared with conventional farms where insecticides are frequently used.

Rotations can include a variety of crops and this can help reduce pest numbers.

There has been some success from reduced pest damage using companion planting and mixed cropping.

Time of sowing can help reduce potential damage caused by pests such as carrot fly, frit fly and aphids. Organic cereal growers in high-risk BYDV areas normally drill winter cereals after the middle of October to reduce likelihood of high numbers of virus transmitting aphids. Maincrop carrots can be sown at the end of May or in early June to avoid the first generation of carrot fly. Use of forecasting programmes for emergence of some pests can be important when deciding on planting dates.

There are a few varieties that have good pest resistance such as cereals resistant to orange wheat blossom midge.

On a limited scale, crop covers are used over susceptible crops when there is likely to be damage. These crop covers form effective barriers against pests such as carrot and cabbage root fly and caterpillars. Other than cost the problem with crop covers is that they must be removed before weeding can be carried out.

Pheromone traps are sometimes used as an aid to warn of possible pest problems. There are also a limited number of permitted and restricted treatments approved for use including soap sprays against aphids, the bacteria treatment, *Bacillus thuringiensis* for control of caterpillars such as the cabbage white butterfly and biopesticides for example spinosad against pests such as cabbage root fly. There are several other permitted, naturally occurring, biological control agents mainly for use in glasshouses. Certification bodies have different systems for approving use of restricted products. For example, for the Soil Association farmers and growers need to annually produce a management plan outlying the records that will be kept, crop monitoring and thresholds for treatments as well husbandry precautions taken to reduce potential pest damage.

10.8 Crop management examples

As an organic farmer cannot always control the problems once they appear it is important to get the crop management correct at the beginning of the growing season. The following is an outline of standard husbandry for a range of organic crops.

10.8.1 Winter wheat

Combinable crops such as winter wheat are the most commonly grown crop on organic farms. Organic winter wheat crops usually follow a grass/clover ley or a crop that has been manured, for example potatoes, as wheat is a nitrogen demanding crop. Normally after grass a soil analysis will be undertaken and permitted fertiliser such as lime or major nutrient such as ground rock phosphate applied if required to the soil before drilling the crop. After grass, at least 4 weeks are normally left with bare ground to reduce the risk of frit fly. Ploughing should be as shallow as possible, though adequate grass burial is required.

A disease-resistant variety (or even a blend) should be chosen for the required market. Feed varieties are higher yielding than the quality milling varieties but the end price is lower as in conventional farming. If the farm is growing any conventional cereal crops it is important that a different variety is grown on the organic area. The crop should not be drilled too early, after mid-October is best to reduce the BYDV risk. A fairly high seed rate should be used, 400–500 seeds/m^2; this seed rate will make up for bird and other pest and disease losses as the seed is untreated. Bird problems can be much more serious when using untreated, compared with standard fungicide dressed, seed in conventional systems. In practice, crop establishment can be as low as 40%–50% in organic systems.

Weed control will depend on species present and soil conditions. A stale seed bed pre-drilling can help reduce weed numbers. In the growing crop up to three treatments may be required overall with a harrow comb between November and April. Wild oats and docks will need hand roguing. If the crop is drilled on wide rows, then some weeds can be controlled using an inter-row cultivator.

The crop is harvested at a similar time as conventional crops in August. Speed of combining can be affected if there are high number of tall weeds still present in the crop. If the grain is not going to be stored in a commercial storage facility, then great care needs to be taken to avoid problems with storage pests and diseases. Stored crops need to be stored at 14.5% moisture or less and at a temperature less than 15°C. To reduce insect pest damage stores also need to be thoroughly cleaned and regularly monitored for pests. Only the biophysical product diatomaceous dust is approved for use for treating stores prefilling (See Chapter 13). Yields can vary between 3 and 7 t/ha, depending on the fertility of the site. A major market for the grain has been for human consumption, although currently the feed market is under supplied. Achieving milling quality can be difficult especially obtaining adequate protein levels. Prices can be as much as £100/t more than the conventional crop price.

10.8.2 Potatoes

Potatoes are usually grown after a fertility-building crop such as a grass/clover ley. Following a long-term ley wireworm could be a potential problem. The likely risk can be assessed by soil sampling or bait trapping. Pheromone traps can also be used in the previous grass crop to assess adult populations.

Potatoes are a very responsive crop to composted manure which can be ploughed into the seedbed. Cultivation (and irrigation) systems including destoning are as for conventionally grown crops, the only difference being time of ploughing (See Chapter 15). Autumn ploughing of grass/clover leys can lead to high nitrate losses over winter. In an organic system the aim is to try to limit leaching losses, so cultivations are sometimes delayed.

Varieties should be selected that are suited to the market requirements and with good resistance to foliage and tuber blight. Choose slug-resistant varieties if slugs are a problem. Good quality seed free of disease should be used; organic seed is now readily available. Seed-rate calculations are as for conventional crops, but adjustments should be made for more expensive organic seed and lower yields. To reduce the effect of blight there is an advantage to chitting the seed so that the crop is earlier to mature.

Potatoes are competitive against weeds. Weed control is usually completely mechanised as specialised potato ridge cultivators are available. The other option is to chain harrow the ridges, then re-ridge afterwards. These cultivations are usually done just before crop emergence.

Blight is the biggest problem when growing organic potatoes, especially in high-risk areas. One major source of inoculum is inoculum coming from farm potato dumps and volunteers as well as from using contaminated home saved seed. Good crop hygiene and husbandry is therefore the main method of control. In the past growers obtained derogation to apply copper compounds once the disease was found. These treatments helped to slow the progress of the disease. Copper products are now not approved for use though occasionally there has been some emergency authorisation. The potato haulm is usually flailed off once blight becomes serious. Removing the haulm when there is less than 25% leaf infection can reduce yields. The haulm is best burnt off with a flame weeder straight after flailing to reduce the risk of blight spores being washed down to the tubers. Rounded rather than steep edged ridges can also help reduce the number of blight spores washed through the ridges. Flame burning will also clean the ridges of many weeds and ease harvesting a few weeks later. Only disease-free tubers should be stored.

Potato prices have been as much as double the conventional prices, but yields are very variable. Average marketable yields of organic maincrop potatoes are about 25 t/ha but can vary between 10 and 40 t/ha.

10.8.3 Field vegetables, e.g., cabbage

Organic vegetables are grown on a range of farm types. There are a few specialist organic vegetable holdings, although more generally they are included in the rotation on mixed farms where there is a supply of compost and nitrogen supplying legumes. Some small-scale producers concentrate on local markets and vegetable box schemes. Other growers market their produce through cooperatives or pack houses, which supply the supermarkets. Marketing needs to be organised before growing the crops.

Meeting the quality requirements for organic vegetables can be much more difficult than for conventionally grown crops, both in relation to size, maturity and damage from pest and disease. Storage losses in some crops such as white and red cabbages can also be higher than for conventional crops. The procedures used for growing including irrigation and harvesting are similar to those for conventionally grown crops except for rotation, crop protection and fertilisers use (See Chapter 17).

With cabbages for example, rotations are no more than 1 year in four as this helps to reduce the incidence of some diseases. Cabbages require high fertility so are usually grown after a fertility building crop such as a grass clover ley. Organic manures either as farmyard manure or green waste are usually applied. The target should be to aim for a neutral pH so this is normally the time in a rotation where lime is applied if necessary, this will also help to reduce the risk of club root.

There is a large range of types of cabbages from summer cabbages to autumn harvested and stored red and white cabbages to winter hardy cabbages such as January King and Savoys. Varieties are often chosen not just for yield, time of harvesting, disease resistance and competitiveness against weeds, but also for taste. Currently there is only a limited number and range of varieties where organic seed is available. Growers must seek derogation from their certification bodies if they want to use untreated non-organic seed.

Organically raised transplants are mostly used so that there is plenty of time for seedbed cultivations and weed control prior to planting. Care must be taken with plant population; this should be adjusted depending on the fertility of the site so that marketable yield can be maximised. Autumn and winter cabbages are usually planted at 40–50 cm spacing within the rows. Date of planting of transplants will vary depending on type of cabbage and target harvesting period. Red cabbages for example are usually planted by the middle of June whereas Savoy cabbages may be transplanted up to the middle of July.

In crop weed control is best carried out using a harrow comb weeder and inter-row cultivator. Harrow combing (up to three times) should take place at each flush of newly emerged weeds and later followed by inter-row cultivations. The first weeding can take place just a week or two after planting as long as the transplants are well rooted. Little hand weeding is normally required.

If the crop is grown in an area of high incidence of cabbage root fly, then approval can be sought for use of a biopesticides, for example spinosad can be applied to transplants before planting. Cabbage caterpillars can be effectively controlled using biological agents such as *Bacillus thuringiensis*. Some small growers may use crop covers. Aphids can very occasionally be a problem though many farms try and attract predators by for example planting floriferous buffer strips. It is also possible to use an organic potassium soap spray though this is easily washed off leaves and is not persistent.

Organic cabbages can be affected by a range of diseases though other than liming against club root there are no effective methods of control other than using resistant varieties and clean seed. Foliar disease at harvest can increase harvesting costs due to more peeling back of leaves which can also lead to a higher amount of grade out.

Time of harvest will depend on type of cabbage, variety, sowing and transplanting date, weather conditions during growing season and soil fertility. Marketable yields for organically grown cabbages are commonly lower than those for standard grown crops.

10.9 Other systems

Another agroecological farming system which is practised by a few growers in the UK, but is more popular in other parts of Europe, is Biodynamic Farming. This is similar in many ways to the standard organic systems. It was started in the 1920s following the lectures of Rudolf Steiner. It is a whole farm system involving organic ideals as well as the spiritual approach. Several natural remedies are made from herbal and mineral preparations and are applied to improve the health of the crop, soil, and compost. There are 9 of these natural fermented remedies that can be bought or made on the farm. Studying the moon and planets is sometimes used when deciding on dates of planting or harvesting crops. Biodynamic produce is sold under the Demeter label.

10.10 Organic farming and the environment

Organic farmers are encouraged to work with natural systems, to protect sensitive habitats and maintain and increase biodiversity.

Many organic farms have mixed rotations and manage their hedgerows by only cutting every 2 or 3 years rather than annually, this management can all have an impact on biodiversity found on the farm from plants to birds. In many research projects comparing organic land with conventionally farmed land there are examples where there has been a recorded increase in plant diversity including increases in rare arable weeds. Several studies have also shown increases in insect species including predator beetles, some spiders and bees. Due probably to increases in floral diversity butterfly species and numbers have also been shown to increase. Many of the certification schemes include aspects of ecology/conservation in their certification process; For example, farms certified with Organic Farmers and Growers need to draw up and maintain a biodiversity and conservation plan for their holding. Existing features are included plus any plans to enhance the farm biodiversity.

10.11 Key points

- Organic farming is an example of a sustainable or agroecological farming system that is regulated and certified.
- As organic farms do not use manufactured nitrogen fertilisers, energy requirements are subsequently reduced compared with non-organic systems.
- The use of manures and mixed crop rotations including grass and clover leys are important for increasing the soil organic matter on organic farms.
- Yields are usually lower from organically grown crops though returns can be higher than from conventional crops due to premiums paid.
- Crop protection relies mainly on husbandry to help reduce disease pressure and weed and pest populations.
- Good crop management is essential to reduce the risk of crop failure.
- This farming system delivers many known environmental benefits including increased biodiversity.

Sources of further information and advice

Applied Plant Research, 2006. Practical Weed Control. Wageningen.
Balfour, E., 2006. The Living Soil. Soil Association Ltd.
Barker, A.V., 2021. Science and Technology of Organic Farming. CRC Press.
Barton, G.A., 2018. The Global History of Organic Farming. Oxford University Press.
Blake, F., 1994. Organic Farming and Growing. The Crowood Press.
Briggs, S., 2007. Organic Cereal and Pulse Production: A Complete Guide. The Crowood Press Ltd.
Davies, G., Lennartsson, M., 2006. Organic Vegetable Production: A Complete Guide. The Crowood Press Ltd.
Davies, G., Sumption, P., Rosenfeld, A., 2010. Pest and Disease Management for Organic Farmers, Growers and Smallholders: A Complete Guide. The Crowood Press Ltd.
Davies, G., Turner, B., Bond, B., 2008. Weed Management for Organic Farmers, Growers and Smallholders: A Complete Guide. The Crowood Press Ltd.
Lampkin, N., 2002. Organic Farming. Old Pond Publishing Ltd.
Lampkin, N., Measures, M., Padel, S., 2017. Organic Farm Management Handbook. Organic Research Centre.
Newton, J., 2004. Profitable Organic Farming, second ed. John Wiley and Sons ltd.
Sattler, F., Wistinghausen, E., 2014. Growing Biodynamic Crops: Sowing, Cultivation and Rotation. Floris Books.
Younie, D., Taylor, B.R., Welch, J.P., Wilkinson, J.M., 2002. Organic Cereals and Pulses. Chalcombe Publishing.
Younie, D., 2012. Grassland Management for Organic Farmers. The Crowood Press Ltd.

Websites

www.agricology.co.uk.
www.biodynamic.org.uk.
www.fibl.org.
www.gardenorganic.org.uk.
www.gov.uk/nitrate-vulnerable-zones.
www.ifoam.bio.
www.naturalengland.gov.uk.
www.organic-europe.net.
www.ofgorganic.org.
www.orgfoodfed.com.
www.organic-plus.net.
www.organic-world.net.
www.orgprints.org.
www.organicresearchcentre.com.
www.organicresearch.net.
www.soilassociation.org.
www.sustainablefoodtrust.org.

Plant breeding and seed production

11

11.1 Introduction

The fundamental objective of agricultural plant breeders is to provide crop plant varieties with improved yield and quality for the growing conditions experienced on farms. Yield is a product of many traits, including physiological traits such as reduced straw length in cereals to increase harvest index, and reduced susceptibility to lodging, disease and pests. Quality refers to suitability for the end use such as milling or malting in cereals, determined by the amount and type of starch, protein and oil in seeds, physical size, shape and structure of seeds, and resistance of the plant to lodging, pests and diseases that reduce quality. Thus, breeding for yield and quality requires selection for scores of traits under the control of hundreds, if not thousands of genes.

11.1.1 Principles of genetics

The biological functional unit of heredity is a gene, which is a segment of DNA (deoxyribonucleic acid), some viruses have genes made from ribonucleic acid (RNA). The original understanding was that one gene codes for one protein, although it is now known that the mechanism is more complex in some circumstances. DNA molecules are wrapped around protein structures to form chromosomes. Each chromosome may contain many genes. The position of each gene on the chromosome is known as the locus (plural = loci). The chromosomes are found in the cell nucleus, although there is also some DNA in mitochondria and chloroplasts, possibly a relic of when these organelles were independent organisms way back in their evolutionary history.

DNA is a polymer with a double helix structure with the two strands joined by bonds between the nucleotide bases adenine (A), thymine (T), cytosine (C) and guanine (G). The base pairing rule is that A always pairs with T and C with G. A DNA strand is used as a template using the base pair rules (although RNA uses uracil U in place of T) to produce messenger RNA (mRNA) during the process of transcription. The mRNA template is transferred to a ribosome where it is used as a template to assemble amino acids into the protein, known as translation. This is facilitated by transfer RNAs (tRNA) which bind to a specific amino acid, which then binds with the mRNA according to the base pairing rules. The sequence of bases in the DNA of a gene determines the sequence of amino acids in the protein produced, and consequently the shape and functionality of the protein.

Sometimes mistakes occur during DNA replication which may change the amino acid that is coded for, or even introduce a 'stop codon' which ends the amino acid chain. The result may be a protein which folds into a different shape and may have

Lockhart and Wiseman's Crop Husbandry Including Grassland. https://doi.org/10.1016/B978-0-323-85702-4.00008-X

altered functionality. Changes in the DNA code sequence are known as mutations. Different forms of a gene at a locus are called alleles.

Often mutations are deleterious, but sometimes the mutation may give the organism an advantage so that it is more likely to contribute to subsequent generations, known as fitness. Alleles that confer a fitness advantage will become more frequent in the population through the process of natural selection.

The genotype is all the genes possessed by an individual plant. The phenotype is the observable characteristics of the plant and is a product of the genotype and the environment. Some phenotypic traits are under the control of a single gene, so that the inheritance of the trait follows the laws of classical genetics, as proposed by Gregor Mendel. For example, most barley has covered grains with the hull glued to the grain, but some barley threshes free and is called naked barley. All naked barley varieties have the same deletion mutation of a DNA sequence at the *NUD* locus, disabling a transcription factor needed to express a gene involved in producing the glue that cements the lemma and palea to the grain. There are two alleles at the *NUD* locus, *Nud*, the dominant wild-type, and *nud*, the recessive loss-of-function mutant. The phenotypic trait of covered versus naked grain segregates according to classical Mendelian genetics, regardless of environment: plants with two copies of *Nud* (homozygous dominant) or one of each *Nud* and *nud* (heterozygous) have covered grains; plants with two copies of *nud* (homozygous recessive) have naked grain.

Quantitative or complex traits are under the control of the cumulative effects of several or many genes (polygenic) and do not follow simple classical genetics, but instead show a continuous range of variation, and are also often influenced by environmental factors. An example of a quantitative trait is crop height: a cross between a tall and a short wheat variety will produce plants in the F2 generation with a continuous range of heights; and environmental factors such as sowing date, seed rate and crop nutrition will interact with genotype to influence height.

11.1.2 A brief history of plant breeding

Most of the crops grown in the United Kingdom originated in other parts of the world, where they evolved by natural selection. A pioneer in the understanding of crop origins was the Russian geneticist and agronomist Nikolai Vavilov who proposed his centres of origin or centres of diversity in 1924. Einkorn wheat, emmer wheat, barley, field bean, pea, chickpea, lentil, and flax all originated in an area of south-west Asia sometimes known as the Fertile Crescent, and so the first farmers had sources of starch, protein, oil, and fibre. Oats and rye were secondary crops, growing as weeds in wheat and barley before being domesticated as agriculture moved north and west into cooler, wetter climates in Europe.

The crop wild ancestors in the Fertile Crescent evolved an annual life cycle as a way of coping with the hot dry summers, surviving as a seed and germinating with the autumn rains, to produce seed in the late spring and continue the cycle. There is a possible fitness advantage in larger seeds that produce more vigorous seedlings, and certainly a high proportion of the world's largest-seeded grasses are found in and around the Fertile Crescent. Domestication is a permanent change in the genetics of an organism. The key change in cereals was a mutation resulting in a tough, rather than fragile and shattering rachis, which meant that seeds did not shed before harvest. Alleles for a tough rachis would quickly disappear in the wild as the plant is unable to disperse its seeds without human intervention to harvest and resow.

Agriculture spread into Europe over the course of several thousand years, and during that time locally adapted populations known as landraces developed by human and natural selection. A landrace population is genetically diverse, though tends to have some uniformity in phenotype, for example all the plants are around the same height and flower and mature at around the same time, because significant deviation from the population mean, e.g., a very short or late maturing plant, would have a fitness disadvantage. The knowledge of how to select seeds and the traits to look for was passed on from one generation of farmers to the next along with the seeds themselves.

For the first 10,000 years of agricultural history, landraces were the only type of crop variety. In the 19th Century plant breeders found that they could select the best individual plants from the diverse landraces and improve yields and standardise quality for the large mills, bakeries and breweries that were appearing with industrialisation. Examples of pure line varieties, from single plant selections, are Chevallier spring barley, Squareheads Master wheat and the Potato Oat, so named because it was found in a potato field.

There is no diversity within pure line varieties to select from for further improvement, and even within a diverse population, it may be that no individual plant has the combination of all the desired traits. Hybridisation or crossing two plants with different genotypes was developed in the early 20th Century to address these limitations. Gregor Mendel (1822−84), a monk based in what is now the Czech Republic, developed the laws of inheritance around the same time (1850 and 1860s) as Charles Darwin was working on the Theory of Evolution. Unfortunately, the true significance of Mendel's work was not recognised at the time, but Mendel and Darwin's work was rediscovered around 1900 and combined to form the new science of genetics, especially by William Bateson, who coined the term genetics. Rowland Biffen, professor of agricultural botany at the University of Cambridge and later the first director of the Plant Breeding Institute (PBI) wrote a paper in the first volume of the Journal of *Agricultural Science* in 1905, titled 'Mendel's laws of inheritance and wheat breeding', and bred the first British wheat varieties bred using the science of genetics, Little Joss in 1908 and Yeoman in 1916.

Commercial plant breeding in the UK perhaps stagnated after the Second World War with the wheat area dominated by imported varieties especially Cappelle-Desprez. The Plant Varieties and Seeds Act in 1964 allowed UK breeders to receive royalty payments following grant of Plant Breeders Rights. This stimulated a surge in plant breeding activity at PBI and commercial breeders, with varieties such as Maris Huntsman wheat increasing yields. PBI was the leading crop breeder in the UK for much of the 20th Century and was a public institution until it was privatised in 1987. In the 1960s and 70s PBI varieties had the prefix Maris, after the road it was based on, and some of these varieties are still grown today such as Maris Piper and Maris Peer potatoes.

The Green Revolution in the 1950s and 1960s began with the work of Norman Borlaug in Mexico using semi dwarf wheat bred from a Japanese variety Norin 10. Dwarfing genes allow a higher harvest index (grain as a proportion or percentage of the total above ground dry weight) and greater inputs of nitrogen fertilisers to achieve an optimum crop canopy size without lodging. Semi dwarf wheat varieties were widely adopted in the UK from the late 1970s onwards.

The discovery of the double helix structure of DNA by Watson, Crick, Franklin, and Wilkins in the 1950s began the science of molecular genetics, allowing breeders to understand and directly alter the genotype, rather than indirectly through observation of phenotype. Molecular markers, genome mapping, sequencing, genetic modification, and gene editing are all tools available to modern plant breeders. Increased computing power, essential for sequencing and statistical analysis, has been fundamental in the development of modern breeding techniques.

11.2 Plant breeding methods

11.2.1 Conventional plant breeding

The breeder begins by carefully choosing the parents for the cross. The objective for making the cross may be to introduce a trait, such as a new resistance R gene, often using intermediate material produced by pre-breeding programmes, or to improve yield. Crossing the 'best with the best' can improve yield by transgressive segregation, where some of the progeny have a greater yield than either parent due to new combinations of alleles. A good example of the power of transgressive segregation is UK spring barley, where impressive yield increases over the last 20 years have been achieved by crossing within a quite limited pool of genetic material.

Wheat, barley, oats, and peas are all examples of species that almost entirely self-fertilise, so the breeder must prevent 'selfing' to make the cross. This is done by removing the anthers while they are still green and before the pollen is mature. The emasculated ears are covered with a pollen-proof bag and a day or two later, when the stigma are mature and receptive, pollen from the male parent is transferred onto the stigma to pollinate. The first-generation seeds and plants produced by this cross are the F1 (F1 = first filial generation). At all gene loci where the parents carried different alleles the F1 plants are 100% heterozygous, having one copy of each allele. The F1 plants are allowed to self-pollinate to produce the F2 seed and plants, which are 50% heterozygous and start to segregate. With each generation of 'selfing' the percentage of heterozygous gene loci halves (F1 = 100%, F2 = 50%, F3 = 25%, F4 = 12.5% etc.), and so it becomes more likely that a plant will 'grow true' and show the same phenotype as its parent.

The system the breeder uses to handle the segregating material after the cross has two objectives: to progress the material through the generations of selfing to achieve genetically stable Recombinant Inbred Lines (RILS); and to select superior lines as candidates for variety registration.

Breeders usually aim for 2000–3000 F2 plants, though the size of the F2 population depends on how many gene loci are segregating; where the parents have different alleles at many loci, a large F2 population is needed to recover recombinant plants with the desired alleles at all loci. In the pedigree system, used through the 20th Century, the F2 plants are planted in a grid pattern and selected F2 plants form the foundations for families. Subsequent generations are grown out as ear rows, with further selection between and within families. Once the lines are sufficiently uniform in phenotype and there is little variation to perform further selection within lines, then the lines can be tested using replicated plots for yield, disease resistance and lodging. By careful observation and record keeping each line can be traced back to the F2. An advantage of the pedigree system is it allows classical Mendelian genetic information to be captured by observation of the ratios in which phenotypes segregate. This was especially important before the development of modern molecular techniques.

Disadvantages of the pedigree system are the resources needed in skilled labour and field space, and the time taken to produce a variety, typically around 12 years. The resources required can be reduced by growing the F2, F3 and F4 generations as populations without selection within them, called bulks, and starting selection in the F5, a successful strategy sometimes used in developing countries, and sound if the F2 bulks are large enough to retain sufficient diversity from the cross. Growing in bulks allows some natural selection in the conditions in which the resulting variety will be grown, so can produce resilient varieties, especially in farmer-participatory breeding programmes, which use a mother (the breeding station) and baby (the satellite trials on farms) system.

For commercial breeders, reducing the time for variety development from the cross to variety release is essential to keep up with competitors. The time for F2 to F4 generations can be shortened using single seed descent (SSD), where plants are grown in small module trays in glasshouses to produce a single ear per plant, only one seed replanted, and development accelerated using artificial lighting. Immature seeds may be harvested, and dormancy broken using gibberellins, cutting out the time otherwise needed for ripening. 'Speed Breeding' protocols for accelerated SSD which allow up to six generations in a year have been developed for many crops at the University of Queensland. In 2021 a vernalisation protocol of exposing seeds on the soil surface to 22 h light, 2 h dark at 10°C was added to the speed breeding technique, allowing up to five generations a year of winter wheat and winter barley.

Shuttle breeding was developed by Norman Borlaug, by growing spring wheat over the winter in the Yaqui Valley of northern Mexico and resowing the seeds for the summer season at high altitude near Mexico City, hence growing two generations in a year. An accidental, but useful, by-product of this strategy was that Borlaug's Green Revolution wheats were day-length-neutral and did not need a specific day-length to trigger flowering, so could be used in many areas around the World. Today, European breeders shuttle breed spring crops using New Zealand during the Northern Hemisphere winter.

The double haploid (DH) technique allows the breeder to achieve 100% homozygous lines within one generation. In wheat, the cross is made as normal and the F1 plants emasculated and pollinated using maize pollen. Plant hormone treatment stimulates the early stages of embryo development, but the wheat and maize chromosomes cannot combine so that the embryo is haploid. The endosperm does not form but the embryo is rescued by dissecting it away from the grain and culturing it on a tissue culture media containing nutrients and plant growth substances. Then colchicine is applied to double the chromosomes, resulting in the double haploid plant. Anther culture (androgenesis) from the F1 followed by colchicine-mediated chromosome doubling may also be used to create double haploids.

11.2.2 Molecular markers

Molecular markers allow the breeder to select by genotype directly, rather than by phenotype. A marker is a sequence of DNA that differs between plants with different genotypes and can be tested for using molecular probes that bind with specific sequences. A common type of marker are SNPs (Single Nucleotide Polymorphism, often referred to as snips), where different forms (polymorphisms) differ from each other by a single nucleotide base pair. Most of the DNA in an organism is not genes coding for proteins, but semi-random sequences called introns, and most markers are found within the introns. Random mutations in the introns can accumulate over thousands of years because they are usually neutral in their effect on fitness and inheritance. A 'perfect marker' is a section of code within the gene of interest. Markers that are physically closer on the genome to a gene of interest are more likely to segregate with the gene when crossovers and recombination occur during meiosis, known as genetic linkage. The result is a linkage map that helps breeders identify parents for the cross and follow the alleles through the segregating generations.

For quantitative traits such as yield, height, or drought tolerance, under polygenic control, breeders undertake Quantitative Trait Locus (QTL) analysis. QTL analysis starts with a cross to produce a mapping population and statistical analysis shows the probability of regions of the genome, identified by the position of the markers, being linked with the trait of interest. GWAS (genome-wide association study) uses similar techniques to QTL but starts with a randomly mated population, rather than a cross, such as a screen of gene bank accessions. Transcriptomics uses next generation sequencing to sequence the genome and then identify the active genes that are being transcribed, by reverse transcribing the mRNA.

11.2.3 Hybrid varieties

In contrast to conventional varieties, where a variety is only released after enough generations of self-pollination to achieve homozygous recombinant inbred lines, a hybrid variety is released as F1, the first generation after the cross. Hybrid varieties often benefit from hybrid vigour or heterosis, as recessive alleles from one parent, which are often deleterious, are supressed by the dominant alleles from the other parent. Hybrids are also phenotypically uniform, which is important for fresh produce, where uniform fruits and vegetables are required. Farmers are not able to legally save and resow seed of F1 hybrids, instead must purchase new seed every year, hence hybrids potentially provide a greater income stream for breeders to reinvest. Saving seed from an F1 would be biologically inadvisable because self-pollination would result in segregation in the F2 and a loss of uniformity. Also, hybrid vigour is lost in the F2 and subsequent generations.

To produce F1 hybrid seed, one parent must be prevented from self-pollinating so that it can be cross pollinated by the other parent. For high value seeds produced in relatively small quantities such as tomatoes and cucumbers, it is commercially viable to manually remove the anthers and cross pollinate in the same way as the crossing process when breeding a conventional variety. Maize, with its separate male (tassels) and female organs located at different heights on the plant can be mechanically detasselled; in the early 20th Century this was done by workers, often students; now specially designed mowers detassel the rows intended as female, leaving other rows as pollinators. Other field crops require use of chemical hybridising agents or genetics in the form of cytoplasmic male sterility (CMS) systems. CMS plants are unable to produce viable pollen themselves so cross pollination rather than self is guaranteed. Usually, the pollinator line contains a restorer gene that allows the F1 plants to be self-fertile restored hybrids (RH), as used in hybrid oilseed rape. Most maize hybrids are produced using CMS to avoid the need for mechanical detasseling.

11.2.4 Genetic modification

Genetic modification (GM) techniques allow the breeder to directly insert genes into a genome from a donor species that could not breed naturally, even from a bacteria, fungus, or animal into a plant. This is possible due to the universal nature of the DNA genetic code. Common methods of achieving this is via *Agrobacterium*-mediated transformation, or by biolistics, which are tiny gold or tungsten pellets coated with DNA shot into the plant cells to be transformed. *Agrobacterium tumefaciens* is a bacterial pathogen that causes crown gall disease on fruit trees. *Agrobacterium* produce Ti (tumour inducing) plasmids, which are circles of DNA that transfer a section of DNA known as T DNA containing genes for gall formation and opine (a source of energy, carbon and nitrogen made by the plant, but only used by the bacteria) biosynthesis into the plant genome. Molecular scissors called restriction endonucleases are used by scientists to cut out the gall and opine genes and replace them with a T DNA construct of a promotor to ensure expression, the genes the breeder wants to introduce, and often an antibiotic resistance marker to allow selection of transformed cells. The transformed cells are then grown on tissue culture medium to regenerate plants. The modified plants are known as transgenic; where the inserted gene comes from a plant that could cross naturally, but has been inserted using GM techniques, this is sometimes called cisgenic, for example late-blight resistance genes from wild potato into a cultivated potato.

Worldwide, only four crops make up much of the GM crop area: maize, soya, cotton, and canola (oilseed rape); and the most common traits are herbicide tolerance and insecticide production, especially Bt crops that have genes from the bacteria

Bacillus thuringiensis for insecticide Cry proteins. In all countries where GM crops were allowed, the GM varieties quickly reached near 100% of the seed market share for that crop. Tolerance to the herbicide glyphosate simplified weed management by replacing several applications of selective herbicides with one or two applications of glyphosate, which is also safer for the environment than some of the herbicides such as atrazine that it replaced. Likewise, Bt crops that produce their own insecticide reduced the need for insecticide sprays. From the mid 1990s onwards non-GM varieties in USA quickly became a niche for the organic and labelled GM-free markets.

In the EU, only 1 GM variety has ever been grown commercially, a Bt maize MON 810, resistant to the European corn borer and Mediterranean corn borer, approved in 1998 before the current GM regulations came into force. GM crops are not banned in the EU, but the regulatory process is so long and expensive that effectively a ban is the result. Some individual EU member states do have national bans in place.

Opposition to GM crops in the UK and EU comes from several quarters on environmental grounds, potential human health risks, seed sovereignty and rights of farmers, and anti-capitalism, anti-globalisation, and anti-corporate politics. Many of the initial concerns over human health have proved to be unfounded, with no health problems being proven to be linked to GM crops in over 25 years of commercial use. There are concerns sometimes raised over the safety of glyphosate, though it has been found safe by regulatory bodies around the world. Initially herbicide tolerant and Bt GM crops achieved impressive reductions in the need for pesticide spraying. Over the years resistance to glyphosate caused by high selection pressure of repeated use has meant growers have had to start to increase again their use of alternative herbicides with different modes of action.

One of the major concerns was that after a series of acquisitions and mergers a small number of agrochemical and biotechnology companies have a monopoly of both the seed and pesticide market. Plant Breeders' Rights for conventional varieties have two key exemptions from the rules that apply to other intellectual property (IP): firstly once a variety is on the National List (NL) (actually once it is in year two of NL testing) it is public domain and any other breeder may use it in crosses to produce a new variety; secondly, farmers can save seed produced on their own holding (but not sell, trade or barter seed) for their own use as long as the royalty is paid. In the case of GM crops both exemptions are removed. Patenting of the modified traits prevents their use by other breeders without a licencing agreement, potentially restricting competition and locking out challenger companies. On the other hand, a greater guarantee on their investment afforded by patenting could encourage investment; a situation illustrated by the greater genetic gains in GM maize in the USA compared with non-GM wheat. The loss of the ability to farm save seed is seen by some as a loss of farmers' rights, but this is already the case with hybrid varieties.

The controversy around GM crops is showing little sign of abating within the EU, and opinion polls show continuing public scepticism. Perhaps the development of crops such as potatoes resistant to late blight and purple tomatoes with cancer-fighting anthocyanins will inspire more public enthusiasm than previous traits, such as glyphosate tolerance.

11.2.5 New breeding technologies

Mutation breeding, where plant breeders induce mutations using gamma radiation or chemical mutagens such as EMS (ethyl methane sulphonate), has been used since the 1950s. The mutations occur at random, and most will be harmful to the plant, but occasionally useful variations arise. Golden Promise spring barley released in the 1960s, and still grown today for specialist malts, was produced by gamma ray mutation of an older variety Maythorpe, and the semi-dwarf mutation with short stiff straw created is still used in current UK spring barley varieties.

The gene editing techniques ZFNs (Zinc Finger Nucleases), TALENs (Transcription Activator-Like Effector Nucleases) and CRISPR-Cas9 (Clustered Regulatory Interspaced Short Palindromic Repeats) use nuclease enzymes to cut DNA strands and create mutations at precise locations, rather than the random effect of radiation or EMS. Of these New Breeding Technologies (NBTs), CRISPR is generating the greatest interest and excitement. CRISPR repurposes a bacterial immune system defence against bacteriophages (viruses that infect bacteria). A single guide RNA (sgRNA) recognises a target sequence in the phage DNA by binding with it using the Watson-Crick base pairing rules, and so guides the Cas-9 nuclease to its target, where it cuts the DNA strand to destroy the phage. By replacing the sgRNA sequence with a sequence from a plant gene, breeders can guide Cas-9 to make cuts in the gene of interest. The plant cell repairs the cut but sometimes mistakes occur, resulting in a mutation in the gene of interest. The breeder can also supply a repair DNA template, that the cell's DNA repair machinery incorporates into the strand when it repairs the cut made by Cas-9.

Most scientists see gene editing with CRISPR as a more precise form of mutation breeding, which has a long history of safe use, and so gene edited crops should be regulated as conventional plant varieties. Several countries including Argentina, Australia, Brazil, and USA regulate gene edited crops as conventional crop varieties, providing that only small edits have been made, without introducing sequences from other organisms, and that the mutations could have occurred naturally. In 2018 the European Court of Justice (ECJ) ruled that gene editing counts as genetic modification and that any crop produced by CRISPR would have to go through the GM crop registration process in the EU, effectively banning gene edited crops. In 2021, following the UK leaving the EU, the UK Government announced that it would change the rules to allow the use of gene edited crops. Opponents of GM and gene editing maintain that NBTs are GM by a different name and should be regulated as GM crops (i.e., banned). There are concerns that the edited traits could be patented, as are GM, and as such restrict farmers' rights to save seed.

11.2.6 Pre-breeding and plant genetic resources

Some breeding activities, for example crops to cope with predicted climate in the future, are too commercially risky or have a pay-off too far in the future to be of commercial interest to breeding companies. Pre-breeding is the identification of useful genes in unadapted material, unsuitable for crossing directly into a breeding programme, which may be landraces, old varieties or wild relatives, known as the donor, and transfer into well-adapted genetic background or elite breeding lines known as the recipient. The intermediate material produced in the pre-breeding programme is then made available to commercial breeders, along with any information on genetic markers. An example of pre breeding in wheat is the use of synthetic hexaploids: the original hybridisation between tetraploid emmer or durum wheat and diploid goat grass to produce bread wheat (*Triticum aestivum*) only occurred once; synthetic hexaploids repeat the hybridisation using different strains of goat grass to introduce new genes in to the wheat gene pool. Pre-breeding research is often publicly funded through research grants and is undertaken in research institutions and universities, often in partnership with consortia of commercial breeders. In the UK the genetic improvement networks such as WGIN (Wheat Genetic Improvement Network) and VeGIN (Vegetable Genetic Improvement Network) are excellent examples of public-private collaboration in this way.

As the science of plant breeding advanced in the 20th Century, some scientists realised that the success of the new improved varieties, leading to replacement of landraces and traditional varieties with high yielding varieties, could jeopardise the very genetic diversity that plant breeding relies upon. Nikolai Vavilov was one of the first scientists to develop *ex situ* conservation of Plant Genetic Resources (PGR), where varieties, landraces and crop wild relatives are collected, catalogued, and stored in gene banks. Vavilov's gene bank in Petersburg (formally Leningrad) was the scene of great bravery and tragedy during the Second World War, where several scientists died of starvation during the Siege of Leningrad rather than eat the precious seeds in the vaults. Today, the Svalbard Global Seed Vault, sometimes called the Doomsday Vault, acts as a back-up for the world's crop collections where duplicate samples are stored in case of loss of individual collections. The key UK collection is held at the Germplasm Resource Unit at the John Innes Centre in Norwich and includes collections going back to the early days of PBI, and to the botanist AE Watkins, who used the connections of the Board of Trade officials to source wheat landraces from around the World in the 1930s.

Advances in high throughput molecular genetics means that the slowest and therefore most restricting step in the use of plant genetic resources is now phenotyping. Automated systems for phenotyping using arrays of cameras and lasers for measurement are a recent innovation, either a conveyor belt system with the plants in pots in a glasshouse, such as the National Plant Phenomics Centre at Aberystwyth University, or a moving gantry system in the field such as the Field Scanalyzer at Rothamsted Research.

11.3 Plant breeding targets

11.3.1 Breeding for yield

The main driver of yield improvement of many crops during the 20th Century was the increase in harvest index. In wheat, the harvest index of the pre-Victorian landraces is around 0.3 (30% of the final above ground dry weight is grain), and this was steadily increased to 0.45 by the early 1970s by selection for shorter straw: a landrace wheat such as Red Lammas or Hen Gymro will grow in excess of 1.5 m; compared to Cappelle-Desprez or Maris Huntsman with a height of just over 1 m. The next breakthrough in UK wheat yield came with the introduction of semi-dwarf wheat varieties in the late 1970s, the first being Hobbit with a harvest index of around 0.5 and a height of around 0.8 m.

In wheat, the most widely used dwarfing genes are *Rht1 (Rht-B1b)* and *Rht 2* (*Rht-D1b*) which both confer insensitivity to the plant growth substance gibberellin. *Rht 2* is the most used in the UK, simply because it happened to be the dwarfing gene in the South American material used by PBI in the 1960s and 1970s when they were crossing the Green Revolution wheat into a Cappelle-Desprez background; *Rht 1* has been in some continental varieties grown in the UK, Soissons in the 1990s for example. In Mediterranean countries *Rht 8* which is independent of gibberellin is preferred because it does not reduce coleoptile growth, which is important to allow deeper sowing in dry conditions. *Rht 8* is closely linked to *Phd-D1a*, an allele for photoperiod-insensitivity, which causes early flowering and maturity, a very useful trait in Mediterranean climates, but not useful in Northern Europe.

Dwarfing genes are also very widely used in barley, including the *sdw1/denso* locus, which has several dwarfing alleles, and the *ari-e* mutant from Golden Promise. In oats dwarfing genes have been less successful as the dwarfing trait was found to be linked with low specific weight, such as the varieties Buffalo and Hendon in the early 2000s. In oilseed rape a few semi-dwarf varieties are available that have the *bzh* allele, originally produced by mutation breeding, but they have lower biomass and yield, so that they are only used in specific situations such as exposed locations.

There is little scope to further increase harvest index by using more extreme dwarfing genes because these reduce biomass, and hence yield. Increase in yield must come from increases in biomass while maintaining the harvest index. In wheat since the 1980s, this has been through aggregation of marginal gains, such as erect leaves that increase light penetration through the canopy, allowing lower leaves to contribute. In winter oilseed rape and winter barley, hybrid varieties increase biomass and greater vigour helps in more marginal conditions. It is hoped that in the near future that hybrid wheat can deliver similar benefits.

11.3.2 Breeding for quality

Producing wheat varieties with good baking quality that could be grown in the UK was one of the first challenges addressed by breeders in the early 20th Century. From the 1870s onwards, millers and bakers preferred imported hard wheat, especially from Canada, due to its very strong elastic gluten and high rates of water absorption. Rowland Biffen at PBI realised that these qualities were heritable and crossed a soft-grain weak-gluten Squarehead-type wheat Browick with Red Fife, a Canadian hard red spring wheat to produce Yeoman in 1916. The baking quality of Red Fife and Yeoman is associated with the high molecular weight glutenin sub-unit 1A, still used in some varieties today. In the early days, selection for strong gluten was done by the breeder chewing a sample of grains to develop the chewing gum-like gluten structure. In the second half of the 20th Century breeders were able to use techniques such as electrophoresis to identify the high molecular weight glutenin sub-units that confer baking quality and follow them using classical Mendelian genetics. Now breeders have molecular markers, which are very useful where the phenotype is difficult to identify by field observation.

In malting barley breeders have increased hot water extract and decreased β-glucan, and over the last 20 years non-GN (Glycosidic Nitrile) producing varieties have been developed for distilling. GN reacts in the presence of a copper catalyst, such as a copper whisky still, to produce ethyl carbamate, a carcinogen, and so now only non-GN producing varieties are accepted for distilling. Food uses of barley have been neglected but there is now progress on naked barley that threshes free from the hull and has a higher content of β-glucan soluble fibre. β-glucan has several health benefits including reducing cholesterol and moderating blood sugar levels, reducing the risk of type-2 diabetes. Oats are being bred for higher kernel content and increased β-glucan. Since the 1990s, naked oat varieties, with papery hulls allowing the grains to thresh freely have been released.

Plant breeders have greatly improved the usefulness of rapeseed oil: wild and older varieties of brassicas have high levels of the undesirable compounds erucic acid and glucosinolate. Single low oilseed rape was produced by Canadian plant breeders (Canola = Canadian Oil Low Acid) and now all food use varieties are double low with less than 2% erucic acid, and less than 25 μmol/g glucosinolate. HOLL (High Oleic Low Linolenic) oilseed rape varieties are a more recent innovation, which produce oil that is stable for high temperature frying without hydrogenation. HEAR (High Erucic Acid Rape) varieties have been bred for industrial use.

Field beans with low levels of the anti-nutritional compounds vicine and convicine, making them more useful for monogastric diets, have been released within the last 5 years. Recent research using wrinkled peas (fresh or garden pea varieties left to maturity and combine harvested dry) shows that these have a high content of resistant starch that moderates blood sugar and reduces the risk of type-2 diabetes, and now there is breeding work on wrinkled pea varieties specifically for combining dry. Phaseolus or New World beans are widely eaten in the UK as baked beans, kidney beans, butter beans etc. but not grown in the UK, but there are now renewed breeding efforts to produce a British baked bean.

GM and gene editing techniques are now being used to improve the health benefits of a number of crops: purple tomatoes modified with genes from snapdragon with increased antioxidants; *Camelina* in GM and gene edited versions that produce long-chain omega-3 fatty acids; CRISPR gene-edited wheat with very low free asparagine so that carcinogenic acrylamides are not produced during toasting.

All the above are examples of the role of plant breeders producing varieties with the quality traits needed to allow farmers to access new markets and provide safe and healthy food.

11.3.3 Breeding for disease and pest resistance and tolerance

Genetic resistance and tolerance to diseases and pests has long been an aim of plant breeders. In recent years there has been more urgency in the need to offer genetic solutions to diseases and pests following the loss of plant protection products due to tightening legislation, and the continuing evolution of disease, pest and weed populations for resistance to pesticides. Genetic resistance is a key cultural control in the Integrated Pest Management (IPM) toolbox.

Plant genetic resistance to disease may be based on major or minor genes. Major gene or gene-for-gene resistance works by the product of the resistance (R) gene recognising the product of an avirulence (Avr) gene in the pathogen and initiating the plant defences known as the hypersensitive response. Mutations in the pathogen Avr gene may allow strains of the pathogen to evade detection and cause disease in plants supposed to be resistant. When an R gene is widely deployed, for example when large areas of one variety or related varieties with the same source of resistance are grown, then there is a strong selection pressure in favour of the pathogen races able to overcome the resistance. This is referred to as resistance breaking, though it is important to note that when a variety 'breaks down' it is due to changes in frequencies of Avr alleles in the pathogen population, rather than a change in the genetics of the plants.

Minor gene resistance, field resistance or non-R-gene resistance may be achieved by a variety of mechanisms which do not rely on recognition of specific races of the pathogen. An important example of non-race specific resistance is the *mlo* allele used in spring barley to provide durable resistance to powdery mildew (*Blumeria graminis*). The dominant *Mlo* allele found in wild type plants codes for transmembrane transporter protein that the pathogen uses as a weak point to enter plant cells, while the recessive mutant *mlo* allele produces an altered form of the protein that the pathogen is unable to use.

Often minor genes have an additive effect and do not provide total resistance, but rather slow the growth of the pathogen. The need to keep many genes together makes minor gene resistance harder to handle in breeding programmes than R genes.

The ban on neonicotinoid insecticides has hastened the development of varieties with resistance to virus diseases that were previously managed by insecticide seed treatments to control the aphid virus vector. Oilseed rape is an excellent example of the speed of progress: Amalie was added to the 2016/17 Recommended List as a specific recommendation with resistance to TuYV (Turnip Yellows Virus), it was the first variety with this trait and its yield was behind the fully recommended varieties; on the 2022/23 AHDB Recommended List there was a choice of 21 varieties with TuYV resistance with no yield penalty compared to non-resistant varieties, and all the new varieties added were resistant. A few varieties of winter barley have tolerance to BYDV (Barley Yellow Dwarf Virus), and the first UK winter wheat variety with Resistance to BYDV was added to Recommended List in 2020. Sugar beet with resistance to virus yellows is expected to be available very soon.

The most outstanding success in breeding for pest resistance in the last two decades has been resistance to Orange Wheat Blossom Midge (OWBM), starting with Brompton which had a specific recommendation, followed by Robigus which had full UK recommendation and was very widely used as a parent. Currently around two thirds of the winter wheat varieties on the RL have resistance to OWBM.

Breeding to aid weed management has received some attention with oilseed rape, and now sugar beet varieties, selected for tolerance to certain ALS herbicides, allowing selective control of closely related weeds and volunteers, such as weed beet. Breeding for increased competition with weeds is not a target for mainstream breeders, although differences between released varieties have been observed.

11.3.4 Breeding forage crops

In the early 20th Century it was recognised that the UK was too dependent on imports of animal feed, especially press cake from oilseeds, and that plant breeding had a role to play. The Welsh Plant Breeding station at Aberystwyth under its first director, Sir George Stapledon, developed techniques to breed forage grasses, especially ryegrass. As a cross-pollinating species, ryegrass is bred using recurrent selection, requiring a different approach to that used in self-pollinating crops such as wheat. In the 1960s tetraploid ryegrasses, produced using colchicine-mediated chromosome doubling, were introduced. Breeding targets for clover are to increase persistence and a long-term aim is reduce the compounds responsible for bloating in livestock.

A key development over the last 30 years are the Aber HSG (High Sugar Grass) varieties of ryegrass. As the name suggests, these varieties have high concentrations of water-soluble carbohydrate (WSC) which release energy much quicker than the cellulose, hemi-cellulose and lignin found in plant cell wall, providing energy to the rumen microbes so that they can use protein more efficiently. This results in better animal performance in producing meat and milk and less nitrogen being excreted.

Festulolium is a hybrid between ryegrass (*Lolium* spp.) and fescue (*Festuca* spp.), gaining greater tolerance of drought and waterlogging from fescue and the yield and palatability of ryegrass. AberNiche was the first Festulolium variety released by IBERS in 2011, and further work is ongoing to provide more resilient forage as a response to climate change. Another benefit of Festolulium's better rooting is the production of a more porous soil structure, able to increase infiltration and reduce surface run-off, hence mitigating flood risk from intense rainfall.

11.3.5 Breeding potatoes and sugar beet

The potato sector stands out from other field crops in that large parts of the market are dominated by older varieties, although there is also significant innovation as well, especially in processing varieties. The top 10 varieties account for 40% of the potato area, and the most widely grown is Maris Piper first introduced in 1966. Maris Piper was from a programme of back-crossing to introduce resistance to one of the species (*Globodera rostochiensis*) of Potato Cyst Nematode (PCN) from landrace potatoes from the Andes, but its continuing success has been due to its eating and processing qualities. Maris Piper, and many other varieties, are not resistant to *Globodera pallida* PCN, so that this species is now dominant, an excellent example of how widespread growing of varieties with the same type of resistance causes selection pressure on the pathogen population. Producing a high marketable yield of evenly sized and shaped tubers with the required skin finish and dry matter content is more important than total yield. The major disease threat to potatoes is late blight (*Phytophthora infestans*) and most varieties are susceptible. The Sarpo varieties bred by the Sarvari Research Trust have impressive late blight resistance from material originally in Vavilov's collection, but have not been accepted by supermarkets and processors, although have a following in the garden and allotment market and organic box-schemes. Using GM techniques to transfer resistance genes from wild and landrace potatoes while preserving the eating and processing qualities of the recipient variety has been successfully field-tested but needs consumer acceptance to be commercial.

Perhaps the most important breakthrough for sugar beet in the 20th Century was the breeding of monogerm varieties. The multigerm ancestors of sugar beet produce flowers in fused clusters and several seedlings germinate from each seed ball, requiring thinning. Monogerm seed allows precision planting and equal space for each beet to grow. Improved bolting resistance allowing earlier sowing and better canopy retention in the autumn have increased crop biomass. There is a very fast turn-over of sugar beet varieties; of the 17 varieties on the BBRO recommended list for 2022, the oldest is from 2017 and 10 are from 2020 or 2021. Current priorities are resistance to virus yellows, following the ban on neonicotinoid seed treatments, ease of mechanical harvesting and reduction of root-tip breakage that leads to storage losses.

11.3.6 Breeding fresh produce

Breeders of fruits and vegetables face many demands for consistent size, shape, colour, skin finish, robustness for storage and transport, and eating quality. By breeding innovative varieties, they help growers compete for supermarket shelf space. Many vegetable varieties are F1 hybrids because of the uniformity F1s provide. Some varieties are marketed in a closed loop with a small group of growers contracted to a particular wholesaler or supermarket.

11.4 Seed quality

11.4.1 Variety registration and seed certification

The variety registration and seed certification system ensure that farmers have access to healthy seed of superior varieties that meet the needs of end users, and that breeders receive payment to invest in variety development. For example, the use of Certified Seed and fungicide seed treatments have made the formally very damaging disease of wheat, bunt, now quite rare. For seed to be legally marketed it must be of a variety on the National List, having passed tests for DUS (Distinct, Uniform and Stable) and VCU (Value for Cultivation and Use). DUS is currently assessed using a set of phenotype descriptors such as shape of the ear, glumes, and grains, although there is debate as to whether molecular markers should be adopted in DUS testing. The aim of DUS is to ensure that the variety is distinct, rather than a renaming of an existing variety, and that the variety is uniform and stable so the variety will grow true to type. VCU is a set of minimum standards for yield, quality for the variety's intended end use, and agronomic traits such as disease resistance. Tests for entry on to the National List and granting of Plant Breeders Rights take 2 years and the system is managed by the Animal and Plant Health Agency (APHA).

Within the European Union when a variety is added to the National List of any member state, it is included in the EU Common Catalogue and as such can be legally marketed in any EU member state. At the time of writing, varieties that were included in the Common Catalogue before the UK left the transition period on December 31, 2020 carry across and can be marketed in the UK, but from January 1, 2021 new varieties must go through the UK National List system.

A further tier of testing that varieties may be entered for is the Recommended List system. The Recommended List was established in the 1940s to give buyers confidence in the quality of British wheat for milling and baking. Currently the Recommended List is managed by AHDB, leading a project consortium of the British Society of Plant Breeders (BSPB), Maltsters' Association of Great Britain (MAGB) and UK Flour Millers (UKFM—formally known as NABIM, National Association of British and Irish Millers). The AHDB Recommended List covers winter and spring wheat, barley and oats, winter oilseed rape and has associated Descriptive Lists for spring oilseed rape, spring linseed and winter rye and triticale. The Processors and Growers Research Organisation (PGRO) in 2020 changed the Recommended Lists of combining peas and field beans to a Descriptive List, in recognition that the requirements of end users were too diverse to recommend varieties across whole market sectors. AHDB Potatoes (formally British Potato Council) currently produces the potato variety database. The sugar beet recommended list is produced by BBRO (British Beet Research Organisation). Recommended Grass and Clover Lists (RGCL) are funded by plant breeders through the BSPB and AHDB and HCC, representing levy boards in the ruminant sector.

11.4.2 Seed production

Seed quality in terms of viability, physical quality, freedom from seed-borne disease and contamination with seeds of weeds or other crop varieties is ensured by the seed certification process. Only certified seed can be legally traded; farmers may save seed from their own crops in some circumstances, but this may only be used as seed on the holding on which it was produced. The seed certification authority for England and Wales is the Animal and Plant Health Agency (APHA); in Scotland the Science Advice for Scottish Agriculture (SASA); and in Northern Ireland the Department of Agriculture and Rural Affairs (DAERA). NIAB runs the seed certification scheme in England and Wales on behalf of APHA.

The categories of certified seed for most crops are as follows, with each category normally representing one growing season in the multiplication process: Breeders Seed, Pre-Basic Seed, Basic Seed, Certified Seed first Generation (C1), and Certified

Seed second Generation (C2). Seed of the generations up to and including Basic Seed are used in further seed production and may only be produced with permission of the breeder. Farmers that grow commercial crops for seed use Basic seed to produce C1, or C1 seed to produce C2. C1 and C2 are the generations usually purchased by farmers for commercial crop production. No further commercial seed production is allowed after C2 (except for linseed and flax which may be C3), although farmers may save seeds of non-hybrids for as many generations as they wish. Most oilseed rape is traded on FOSFA (Federation of Oils, Seeds and Fats Associations) contracts that exclude produce of farm saved seed of hybrids and limit conventional farm saved seed to one generation due to the potential to outcross and lose the double low quality. For cereals and some other crops there are Higher Voluntary Standards (HVS), which exceed the minimum legal standards.

The official labels on seed bags are important documents containing information of the variety and seed lot reference number. Colour coding of the labels identifies the generation as follows: Pre-Basic Seed—white with purple stripe; Basic Seed—white; C1 seed—blue; C2 seed—red.

Fields used for seed crops must be carefully chosen to avoid contamination by volunteers from previous crops or by cross pollination by neighbouring crops. Fields that are heavily infested with weeds such as wild oats should also be avoided. For all cereals except rye, Pre-Basic and Basic seed must be produced in fields that have not produced any cereals for 2 years, and for C1 and C2 seed the minimum break is 1 year without any cereals and at least 2 years free from the same species. For rye, 1 year clear of rye or triticale is required. Pea or field bean seed crops require 2 years previously free from peas or beans, including cereal-legume mixed crops. For oilseed rape the minimum is no seed-bearing cruciferous crop in the previous 2 years, but it is good practice to have no cruciferous crops in the previous 5 years.

The minimum distance of isolation from other crops of the same species depends on whether the species is self or cross-pollinating. Wheat and conventional barley are self-pollinating, and the pollen is heavy, only moving about 1 m, and a 2 m buffer is all that is required. The exception is hybrid barley, which requires 100 m for Pre-basic and Basic, and 50 m for Certified Seed. Triticale has higher rates of out-crossing than wheat so requires 50 m for Pre-basic and Basic and 20 m for C1 and C2. Rye is out-crossing and requires 300 m for Pre-basic and Basic and 250 m for Certified Seed; the requirements for hybrid rye are 100 m for Pre-basic and Basic when using male sterility, and 500 m for Certified Seed. Pea seed crops require a 2 m gap or physical barrier to recognise the crop boundary, but are self-pollinating with no isolation requirement. Field beans are cross pollinating and smaller area seed crops require a larger isolation, because a smaller area produces less volume of pollen to dilute pollen from neighbouring fields: the minimum isolation for a Basic seed crop of under 2 ha is 200 m, and 100 m if the area is greater than 2 ha. Oilseed rape is approximately 30% out-crossing and requires 200 m (or 500 m for hybrids) isolation for Pre-basic and Basic and 100 m (300 m for hybrids) for Certified Seed. Sugar and fodder beet are outcrossing, requiring minimum isolation of up to 1000 m, depending on the variety.

Seed crops are inspected in the field by official or licenced crop inspectors trained to identify the specific characteristics of the variety, as set out in the DUS report produced when the variety was tested for the National List. For most species the best time to carry out crop inspections is around flowering. The number of inspections required depends on the crop species. The inspector will record number of plants in a sample area that do not conform to the official description of that variety, also the presence of certain weeds such as wild oats, and any relevant diseases, such as loose smut in wheat and barley. As an example, the maximum number of wild oat plants permitted in a seed crop of wheat is 7 plants/ha for Pre-basic and Basic, 50 plants/ha for CI or 7 plants/ha for C1 HVS. Care must be taken to avoid lodging in seed crops, because pre-harvest sprouting will ruin seed quality. For HVS the crop must not be more than one third lodged at the time of inspection, and at the minimum standard a crop will be rejected if the level of lodging makes the inspection impossible. Seed grain must be dried carefully, as high temperatures will reduce or prevent germination.

An official control plot of a representative sample of the seed lot used to sow the seed crop is grown by NIAB, to authenticate the variety and assess varietal purity. Once harvested and processed (cleaned etc.) the seed must be sampled by a licenced seed sampler and tested by a licenced seed testing station or an official seed testing station, a list of which can be found on the APHA website. The sample is tested for analytical purity, other seed content and germination. Standards for analytical purity and seeds of other species are stricter for the HVS. For cereals the minimum germination of any category of Certified Seed is 85%, except for naked oats and naked barley where the standard is 75%, because the exposed embryo of these varieties is very vulnerable to damage, making achieving 85% very difficult. The current rules and regulations for seed production can be found in the handbooks on the APHA website.

In the UK potatoes are always grown from seed tubers rather than true seed, and as such are vegetatively propagated clones, and special care must be taken to avoid virus accumulation. The Seed Potato Classification Scheme (SPCS) is administered by APHA in England and Wales and SASA in Scotland and imposes stringent control of all stages of the seed potato production chain to ensure seed health. At the start of the process, virus-free plants are produced from tissue culture (TC) of meristem cells and grown in aphid-free controlled glasshouses. There are three categories of seed potatoes: Pre-Basic, Basic, and Certified. Pre-Basic starts with Pre-Basic TC, followed by a maximum of four generations for Pre-Basic FG (Field Grown), known as Pre-Basic 1–4. Basic seed is graded S, SE, or E according to how many field generations the parent seed has been used for, health of the crop at field inspection and health of the tubers after harvest. Certified seed is graded A or B. A maximum of nine field generations is allowed in the production of potato seed. Pre-Basic and Basic seed is used to produce seed potatoes, Certified seed to produce ware potatoes. Land used to grow seed potatoes must be free of potato wart disease, potato cyst nematode (PCN), *Rhizomania*, ring rot, brown rot, or any other quarantine disease. The land must not have grown potatoes for 7 years for pre-basic, 5 years for basic and 4 years for certified seed.

11.4.3 Farm-saved seed

Saving seed grown on a holding for replanting the next season on the same holding offers farmers a number of advantages including saving costs. Farmed Saved Seed

(FSS) use must be declared to British Society of Plant Breeders (BSPB), and payment per tonne made when the seed is processed by a mobile seed processor registered with BSPB, or by the farmer declaring the area grown and paying per hectare. Guidance, a list of eligible varieties, and rates for FSS payments can be found on the BSPB website. Typically, plant breeders' royalty payments for FSS are around half that paid on Certified Seed, although some varieties are now only available on the Breeders' Intellectual Property Office (BIPO) scheme which charges the same rate per area for Certified and farm saved seed, as well as requiring farmers to sign a Grower Agreement and be subject to auditing and crop inspections to ensure compliance. Farm saved seed may not be sold, traded, bartered, or otherwise transferred to another business for replanting, and F1 hybrid varieties may not be farm saved.

Saving seed allows farmers to buy a smaller quantity, perhaps one or two tonnes, of a new variety to trial, and then grow a larger area next season if the variety does well. Farm saved seed is often processed on-farm by a mobile seed processor that cleans the seed and applies a fungicide seed treatment. An advantage of farm saving seed is that it can be cleaned very hard, to remove smaller seeds and which therefore improves vigour. The cost savings of farm saved seed are very helpful when the farmer uses higher seed rates and later drilling as cultural controls for black-grass. Seed may be sown untreated, but it is wise to first send samples to a laboratory to test germination and for seed-borne diseases such as loose smut and bunt in wheat. When seed is to be tested it is important to get a representative sample of the seed lot. Sowing untreated seed saves costs if the seed is free from disease, and untreated seed often emerges quicker because some fungicide seed treatments may inhibit the action of gibberellin. If a grower does their own germination test, it is better to sow seeds in trays of soil at outdoor temperatures, rather than on kitchen paper in the airing cupboard, as a test of vigour.

11.5 Breeding for organic, agroecology and regenerative agriculture

The market for organic seed in the UK is too small to interest commercial breeders to specifically breed for organic farming. Organic farmers and research organisations such as the Organic Research Centre (ORC) trial already released varieties to see which are most suited to organic growing. There are a small number of varieties bred in countries with a larger organic sector, such as Austria, marketed as part of the EU Common Catalogue. Of course, organic farmers benefit along with conventional farmers from improvements in disease and pest resistance but weed suppression and tolerance to weeds is the most important trait in many organic crops. In wheat, breeding for higher harvest index and shorter straw has reduced the ability to compete with weeds, and older tall varieties may perform better in organic systems. There is also evidence that the root length density, measured as length of roots per volume of soil, has declined, reducing the ability to scavenge nutrients and compete with weed roots. Perhaps selecting varieties for high yield under conditions of high NPK fertiliser has inadvertently rewarded partitioning to the shoots rather than the roots. This would be consistent with the 'trade-off' theory of evolutionary biology. Unfortunately, for organic farmers, there is only one tall wheat variety, Maris Widgeon, still maintained on the National List, and therefore having seed legally available.

A very different system of breeding is to use Composite Cross Populations (CCPs), a technique first developed by Harry Harlan in the USA in the 1920s with spring barley. Early 20th Century breeders found that they could improve yields by making pure line selections from landraces, but these often then 'broke down' to disease. Harlan recognised that the cause was lack of genetic diversity in the pure lines. A CCP is made by making many crosses and then combining the F2s into a composite population. In the UK, the late Professor Martin Wolfe led the development of the Wakelyns ORC wheat population, also known as Wakelyns YQ, at the ORC with the John Innes Centre. Wakelyns YQ was made by crossing 20 parents in every possible combination, a total of 190 crosses. Despite most of the parents being semi-dwarf varieties, Wakelyns YQ is taller than commercial varieties, and had a much greater ability to suppress weeds in on-farm experiments under organic conditions. In a CCP, plants compete as individuals with their neighbours, so that natural selection favours more vigorous and competitive plants. CCPs may also be able to respond to changing climate due to their genetic diversity. The diversity of resistance genes, major and minor, found in a CCP, where every plant may have a different combination, makes it very difficult for any pathogen to accumulate enough mutations to evade resistance.

CCPs are designed to be diverse not uniform, and to evolve rather than to be stable, so do not fit DUS criteria, and could not be entered onto the National List, so seed cannot be certified and traded. A temporary derogation from 2014 to 2021 was granted by the EU to allow marketing of CCP seed, the success of which resulted in marketing of Organic Heterogenous Material being made legal in the EU as part of the new EU Organic Regulation 484/2018, coming into effect in 2022. Wakelyns YQ is not accepted in any group under the UK Flour Millers (formally NABIM) system but is used by several bakeries and mills operating as local supply chains.

A growing number of farmers are adopting Conservation Agriculture (CA), sometimes combining the CA practices of no-till, continuous soil cover by crops, cover crops or residues, and crop rotation with further techniques such as herbal leys and mob grazing. Direct-drilled crops do not benefit from the mineralisation of nitrogen that would occur due to tillage, so varieties need good early vigour. Like the situation with organic, varieties are not bred under CA systems, but varieties most suited to CA are identified by trials after the variety has been released.

11.6 Key points

- Plant breeders apply the science of genetics to provide farmers with varieties with improved yield and the quality traits needed to access markets.
- Varieties with resistance to diseases and pests play a vital role in Integrated Pest Management (IPM). Evolution of the pathogen and pest populations poses a challenge to plant breeders to find new sources of resistance.

- Genetic diversity is the vital resource that breeders work with. For example, landraces and crop wild relatives are a source of new disease resistance genes. Collections of plant genetic resources are held in gene banks.
- Breeders have made use of advances in molecular genetics and computing and incorporated techniques such as DNA sequencing and marker assisted breeding into their programmes.
- Genetic modification (GM) and gene editing (GE) are powerful techniques that allow breeders to introduce genes from other organisms or to produce new variations by making targeted mutations in genes. These techniques offer great potential but currently face scepticism from the public.

Sources of further information and advice

Acquaah, G., 2012. Principles of Plant Genetics and Breeding. Wiley-Blackwell, Chichester.

Agriculture and Horticulture Development Board (AHDB), latest ed. AHDB Recommended List for Cereals and Oilseeds. Available at: https://ahdb.org.uk/knowledge-library/recommended-lists-for-cereals-and-oilseeds-rl.

Animal and Plant Health Agency (APHA), 2016. Technical Standards Supplements. Available at: https://www.gov.uk/government/organisations/animal-and-plant-health-agency.

Animal and Plant Health Agency (APHA), 2017. Instructions to Licensed and Official Crop Inspectors in England and Wales. Available at: https://www.gov.uk/government/organisations/animal-and-plant-health-agency.

Animal and Plant Health Agency (APHA), 2017. Instructions to Licensed and Official Seed Samplers in England and Wales. Available at: https://www.gov.uk/government/organisations/animal-and-plant-health-agency.

Animal and Plant Health Agency (APHA), 2020. Explanatory Guide to the Seed Potato Classification Scheme and Approved Stock Scheme 2020/21. Available at: https://www.gov.uk/government/organisations/animal-and-plant-health-agency.

Brown, J., Caligari, P., Campos, H., 2014. Plant Breeding. Wiley-Blackwell, Chichester.

Denison, R.F., 2012. Darwinian Agriculture: How Understanding Evolution Can Improve Agriculture. Princeton University Press, Princeton, NJ.

George, R.A.T., 2011. Agricultural Seed Production. CABI Publishing, Wallingford.

Hesser, L., 2009. The Man Who Fed the World. Durban House, Dallas.

Kingsbury, N., 2009. Hybrid: The History and Science of Plant Breeding. University of Chicago Press, Chicago.

Lammerts van Beuren, E.T., Myers, J.R. (Eds.), 2012. Organic Crop Breeding. Wiley-Blackwell, Chichester.

Lynas, M., 2018. Seeds of Science. Bloomsbury-Sigma, London.

Pringle, P., 2009. The Murder of Nikolai Vavilov. JR Books, London.

Slater, A., Scott, N.W., Fowler, M.R., 2008. Plant Biotechnology. Oxford University Press, Oxford.

World cropping systems

12

12.1 Introduction

World wide growers produce a multitude of different crops, which superficially can seem very different from each other: cassava is markedly different to potatoes, and rice is different to wheat, as is oil palm from oilseed rape. However, whilst the crop species differ the crops that are grown are unified by the products that are harvested. These are namely the macronutrients of carbohydrates, proteins and fats.

There are around a quarter million species of flowering seed-bearing plants classified as the angiosperms, and as well as being the most numerous of the types of vascular plant they are also by far the most important agriculturally, and of these only around 150 species are actively cultivated. Unsurprisingly, of the few species that are grown, they share the characteristic of producing high yields of macronutrients for human consumption, and lend themselves to domestication and agriculture. Therefore, cropping centres around crops that contribute to human nutrition and consists of crops high in carbohydrates, proteins and or oil.

Growers across the world are unified in growing crops which are most viable to their business, and grow what there will be in demand, balanced against cost of production and value of the produce. The specific crops grown will be the those best suited to the local environmental conditions that prevail.

Table 12.1 lists the top 10 most grown crops by land area (as land use is the best overall measure of a crops significance, given that land is limited, any land devoted to a particular crop is important). This list is unsurprisingly dominated by cereals which are staples for the world's population, and six of the top 10 crops grown are cereals, with the remaining four crops that make up the top 10 most grown are soybean, rapeseed, beans and groundnuts. These crops are grown because of their high protein (and oil) content which is also crucial in a balanced diet.

The aim of this chapter, is to broadly quantify and assess major cropping globally and highlight why the crops which dominate global agricultural production do so. It should be noted, that whilst key crops will be the chapter's focus, any crop that is grown by a grower will be important for their business and whilst not globally significant, all crops are important, and have their role in food (and non-food) production. In order to explain global cropping, it is necessary to quantify global demand, and then chronicle why these crops are specifically grown instead of alternatives.

Lockhart and Wiseman's Crop Husbandry Including Grassland. https://doi.org/10.1016/B978-0-323-85702-4.00015-7

Table 12.1 Global land use – food crops. 2020 FAO. Over a third of crop land use is devoted to just three crops: Wheat, Rice and Maize. Note that seed cotton also has 32 million hectares grown in 2020, the most significant non-food crop by land use. Over half of crop land is attributed to just 10 crops. Based upon estimate of 1.67 billion hectares of cropland (FAO).

Crop	Land use (million ha)	Approximate % global agricultural land use	Cumulative land use %
Wheat	219	13.1	13
Maize	202	12.1	25
Rice	164	9.8	35
Soybeans	127	7.6	43
Barley	51	3.1	46
Sorghum	40	2.4	48
Rapeseed	35	2.1	50
Beans	35	2.1	52
Millet	32	1.9	54
Groundnuts	32	1.9	56

12.2 Global demand

The primary demand for agriculture globally is for food production, with around 2000 k calories of energy required per person per day (but specific individual demand is dependent on age, gender, weight, activity, and if pregnant). Given that the human population is over seven billion then it follows that the global demand for food is considerable and will continue to grow with an increasing global population.

Together with the need for the daily energy requirement to be met, there is also a daily protein requirement (for growth and repair). Whilst protein contains energy (around 4 k calories per gram, which is the same as for carbohydrates) typically an average adult requires around 50 g of protein per day. Unlike for carbohydrates, however, the quality of the protein is also an important consideration, and is required to contain the compliment of essential amino acids (i.e., those that cannot be synthesised by the body and need to be part of a diet). To meet this additional requirement meats are beneficial in that they contain all the essential amino acids, whilst vegetable proteins tend to be deficient in at least a few of them. However, sourcing proteins from different vegetable proteins that complement each other means that vegetarian diets can meet the specific requirements for essential amino acids. For example cereals are deficient in the essential amino acid lysine, whilst soybeans are deficient in methionine, but with a diet containing both, the deficiencies are overcome (Table 12.2).

12.3 Efficiency of production

Yield is undoubtedly a crucial factor in determining which crops are grown. However, gross yields are deceptive, and may mask the global significance of a crop. This is

Table 12.2 Energy and macronutrient content (per 100 g) for selected most widely grown crops. Crops tend to be grown to primarily meet human dietary requirements, reflected in them either having a high carbohydrate content or a high protein content. Fats (oils) are also required, but less than proteins specifically, and crops high in oils (of the top 10 most widely grown crops) tend also to be high in protein (values taken from McCance and Widdowson, 2015 for soybeans, beans and peanuts, and for the grains from INRAE-CIRAD-AFZ feed tables).

Food	Energy/100 g (kC)	Carbohydrates/ 100 g	Protein/ 100 g	Fat/ 100 g
Wheat grain	378	60.0	11.0	1.4
Maize grain	384	63.8	7.6	3.6
Rice grain (hulls included)	373	56.6	7.5	2.2
Soybeans (dried, raw)	370	15.8	35.9	18.6
Beans (red kidney, dried)	266	44.1	22.1	1.4
Groundnuts (peanuts, plain)	564	12.5	25.8	46.0

because the water content contributes to gross yields, and crops with the highest water contents inevitably will also be the highest yielding. Table 12.3, below, shows the gross yields of selected crops to highlight the significance of yield. Whilst for all crops it is desirable for them to be grown with a higher yield (after gross margins and quality factored in), a comparison between crop gross production serves little value. By virtue, primarily of average yields being the highest, sugar cane has the highest gross production of any crop globally.

Table 12.3 Average yields of selected crops 2020 (FAO), ranked highest to lowest, including the highest yielding crop and the lowest and their gross production annually.

Crop	Gross yield (tonnes)	Gross production (millions of tonnes, vanilla excepted)
Sugar cane	70.6	1870
Sugar beet	57.0	253
Cucumber	40.4	91
Bananas	23.0	120
Potatoes	21.8	359
Oil palm fruit	14.6	418
Cassava	10.8	303
Maize	5.8	1162
Soybean	2.8	353
Vanilla	0.08 (80 kg)	7600 tonnes

Whilst it is established that water content singularly contributes most to the yield of many crops, macronutrient composition is also a key factor: crops with a higher carbohydrate content generally have higher yields than crops with a higher protein or oil content. This is because, anabolically, carbohydrates require less energy to synthesise than proteins (despite having similar calorific values), and because fats and oils are more energy dense that carbohydrates, they also require more energy, and hence they are more metabolically expensive to synthesise than carbohydrates.

With notable exceptions (i.e., fresh produce) a key determinant of global cropping is how much energy and how much protein is yielded per unit area of land. This follows, since land use is limited, and land is required to be as productive as possible all things being equal (as farmers will grow crops for profit, and the gross margins will compromise growing for the ultimate highest yield, which is usually uneconomic to do so). Table 12.3 estimates how much energy and protein on average are harvested for the 10 most grown crops together with some other crops for comparison. It can be seen that the most widely grown crops unsurprisingly tend to have the highest yields, either of energy or protein harvested, and assists with explaining their dominance globally. Whilst important to the grower, and also supplemental to a balanced and varied diet, fresh produce like lettuce have low yields of energy and protein, and are grown far less than the crops which constitute the basis of global diets. It is notable, however, that sugar beet and sugar cane are very productive in the amount of energy that can be harvested per hectare. On this basis, these crops are beneficial, although simple refined sugar tends to make up additional calories in a diet, rather than a replacement for complex starch carbohydrates found in cereals, and root crops including potatoes and cassava. Energy harvested per hectare of oil palm is also amongst the highest. Furthermore, both sugar cane and oil palm require less inputs relative to output than the other crops, in part because cultivations costs (and energy inputs) are lower. The ratio of energy harvested to the energy input is an important consideration, and partially explains why sugar cane is cheaper to produce than sugar beet, despite sugar beet having a higher energy yield per hectare on average than sugar cane.

Some caution is required when comparing different crops, as although beans yield poorly, and subsequently have a lower protein yield than cereals per hectare they are still amongst the most widely grown crops globally. This is attributed to beans and other legumes providing a complementary high-quality vegetable protein in diets, coupled with legumes value agronomically with improving soil fertility and therefore they are of great value in a mainly cereal rotation.

Crops tend to be high in either energy or protein, but not both. Beans are the exception, with their low global yields, means that whilst having a high protein content, the amount of protein per hectare is low, however, the quality of the protein is high. Crops including lettuce comprise important supplements and variation to a diet, and this is why they are grown, but do not yield either high amounts of energy or protein, and is one reason why lettuce is grown less than other crops, see Table 12.4.

Table 12.4 Estimated energy and protein per hectare of selected crops (2020 FAO values used for yields within the calculations).

Crop	Energy/ha (GC)	Protein/ha (tonnes)
Wheat	11.2	0.39
Maize	20.2	0.43
Rice	17.2	0.35
Soybeans	11.6	1.00
Groundnuts	10.1	0.46
Beans (dry, red kidney)	2.1	0.18
Sugar cane (@12% sugar)	32.3	0
Sugar beet (@18% sugar)	42.0	0
Oil palm (mesocarp only)	25.8	0
Lettuce	2.5	0.27

12.4 Global trends

With an increasing world population, it is inevitable that there will be an increased demand for food. Food production has so far kept pace, with the increase in production being achieved either by increasing the amount of crops grown (i.e., an increase in land use) or by increases in yield. However, both the amount of additional land available for crop production is limited (as well as tending to be more marginal and less productive land) and yield increases have a ceiling which is close to being reached. Hence, it will be increasingly challenging to meet global demand in the future.

The demand for vegetable oil, which dramatically increased after the second world war, and the increase in demand for processed foods, continues to increase. This is not just driven by keeping up with global population increases, but also can be attributed in part, to increase in demand for biodiesel as a renewable liquid fuel substitute for diesel fossil fuel. Similarly, there is also an increased demand for carbohydrates for non-food uses, especially for the production of bioethanol as a road transport biofuel (see Chapter 16).

The production of wheat, maize and rice (the three most grown cereals, globally) was just over 200 million tonnes of each in 1961 (FAO) and by 2020 had increased to around 750 million tonnes for wheat and rice and as much as 1.2 billion tonnes of maize annually. The increase in production of all three crops is attributed both to an increase in land use and also to an increase in yield. The increased production reflects growing demand in line with world population growth.

12.5 Climatic requirements

Climatic conditions vary around the world, which influences cropping and the required agronomy of the crops. Together with light, photosynthesis depends upon temperature, and the availability of sufficient water and nutrients which vary geographically and temporary.

Broadly, the world can be split into five climatic zones. These are:

1) Polar Zones: no cropping
2) Cold-Temperate Zones: suitable for forage only, short growing season
3) Cool-Temperate Zones: long growing season, suited to many crops
4) Warm-Temperate Zones: Subtropical, hot summers, cool winters
5) Tropical Zones: high temperatures year long, with usually a wet and dry season.

Of these zones, the cool-temperate zone is the most important agriculturally with around 30% of the world's potential agricultural land, and also tend to have fertile soils which retain their nutrients well (due to having a high cation exchange capacity). In contrast, the cation exchange capacity (CEC) (the capacity of the soil to retain nutrients) is very low in tropical soils, which mean that they lose nutrients very rapidly. This is not a problem for natural habitats like the rainforest, as they are characterised by closed nutrient cycling, as perennial plants die, their nutrients are released and are quickly taken up for plant growth by the surrounding vegetation. This explains why rainforests are amongst the most productive environments in the world for plant growth, but this diminishes when used agriculturally with annual cropping permitting nutrients to be lost, and fertility reduces.

12.6 Land use

Whilst there is around five billion hectares of land used agriculturally, around two thirds of this is for grazing, with only a third, or about 1.67 billion hectares used for cropping (of which 90% is devoted to annual crop production, and 10% permanent crops (like oil palm and fruit trees). Initially, this may seem encouraging, the potential for more land to be utilised for crop production in the future. However, grassland is often of marginal value for crop production, and also hill and mountain sides can only readily be exploited for agricultural production through livestock grazing.

Bearing in mind that soils formation is dependent on five factors, the underlying parent material, the climate, fauna and flora, topography and time. This ultimately explains the potential for land use, and that the potential to improve is limited. In mild climates the soil production rates are estimated at just 1 cm of soil formed between 200 and 400 years (i.e., up to almost half a millenia timeframe) and in the wet tropics in order to make a soil fertile can take 3000 years. This highlights how crucial it is to conserve soils, especially since soil loss is a far more rapid process, and soils are being lost far more quickly than they can be created. Hence, it is essential that losses of agricultural land from erosion and other causes is prevented. Furthermore, the amount of the land with the most potential agriculturally unexploited is minimal, together with the impact of soil degradation (i.e., those with compromised soil health) which reduces crop yield by 10% globally.

The amount of arable land per capita as well as the productivity of the arable land is an important consideration when assessing the potential to produce food of a country.

Table 12.5 Arable land per capita of selected countries. Arable land values and population taken from CIA World factbook.

Country	Arable land (million hectares)	Population (millions)	Arable land/ha/ populace
USA	165	335	0.49
UK	6.1	67	0.09
Kenya	5.7	55	0.10
India	157	1339	0.12
Brazil	72	213	0.34
World	1670	7772	0.21

Hence, the US dominates exports given its abundance of productive arable land per capita and can generate a surplus. In contrast, many developing countries are characterised by having low amounts of arable land and less productive and consequently less able to produce a surplus. Table 12.5, highlights the amount of arable land per capita of selected countries. Generally, Higher Income Countries (HICs) have more arable land per capita than those from developing countries.

As published by the FAO, cropland area per capita between 2007 and 2016 was smallest in Asia (0.13 ha per capita), followed by Africa (0.22 ha per capita), the Americas and Europe (0.40 ha per capita), and Oceania (1.21 ha per capita).

Given that the amount of land available for crop production is finite, with the majority of the most suitable land already farmed decades or indeed centuries ago, then it is inevitable that the amount of global cropland has dropped from 0.45 ha per capita in 1961 to under half that (i.e., 0.21 ha per capita) in 2016.

12.7 Threats to global production - losses to yield attainment

Yields are generally higher in cool-temperate zones than in other zones. This can be attributed to a number of factors, which includes more fertile soil, and diminished pest pressures compared with the sub-tropics and tropical regions (whereby the year-round high temperatures, and humid conditions allow pests to thrive). The significance of crop losses is substantial, with around a third of potential losses coming from weed competition and pathogens of around a fifth.

Losses can be either abiotic (i.e., light, water, temperature and nutrients) or biotic (i.e., weeds, animal pests and pathogens). The combination of both abiotic and biotic factors explains why the loss of potential yield attainment is usually higher in the tropics and sub-tropics than it is in temperate regions and wheat yields in the UK are over twice the global average for example. Abiotically, water and nutrients are likely to be limiting factors to crop growth in the tropics. Globally, water is the most limiting resource, and this is unsurprising given that even crops with high water use efficiency, like sorghum, require at least 150 L of water to produce 1 kg of dry matter.

Of the abiotic factors, there is limited potential to affect the light and temperature (global warming excepted), but for water limitation crops can be irrigated and fertiliser added for nutrients, and as such there is potential to reduce crop losses and maximise yield attainment. Of the biotic factors improved crop protection can serve to reduce yield losses, although pests are so called for a reason, and it is not a simple case of better agronomy to eradicate the losses incurred, and there is always a need to base inputs on economic thresholds for control.

12.8 Post-harvest losses

As well as pre-harvest losses, the amount of food produced is reduced from post-harvest losses, whereas these are greatest in fresh produce (i.e., crops with a high water content) they are also significant in all crops. Due to generally worse storage conditions and the high ambient temperatures in the tropics, losses are highest in these regions compared with temperate climates. Losses to harvested yield can be as high as 50% in developing countries.

12.9 Most important crops grown globally

Cropland use is dominated by food production, with an exception of cotton, which is grown primarily for non-food uses, and has 32 million hectares of land devoted to its production in 2020, which makes cotton the joint ninth most widely grown crop globally. As they also take up a significant use of land, cash crops, will also be discussed in this chapter.

12.9.1 Cereals (crops grown principally for carbohydrates and as an energy source)

Cereals dominate global agriculture for providing an energy source for humans daily dietary needs. Other crops are also grown primarily for their carbohydrate (i.e., energy) content, which include the root crops potatoes (with over 16 million hectares grown in 2020) and cassava (with 28 million hectares grown in 2020), making them the 17th and 13th most grown food crops globally. Root crops have deceptively high gross yields, which means by production tonnages they rank far higher than by land use for amount produced annually. However, whilst they are productive crops, they differ from the cereals in that they have a very high water content (around 80% for potatoes), and also are more difficult to cultivate (being a root crop), grow (requiring more water), transport (being bulky) and store (due to their high water content). All these factors help explain why they are less significant than cereals as a carbohydrate feed source.

Please refer to Chapter 13: Cereals for details on the growth stages common to all cereals together with in depth agronomy of wheat and barley. This section, aims to complement that chapter, and will explain why specific cereals dominate different regions globally.

Currently, between wheat, maize and rice there are over 2.5 billion tonnes of cereal grains produced each year (with around half of this coming from maize). Whilst much of this will be produced for animal feed, their vast production does highlight how much food is produced annually: given that the yearly energy requirements for a human would be met by around just 200 kg of grain (that is one tonne of cereal grain meets the energy needs for five people annually), from these crops alone, this is sufficient for 12.5 billion people. This highlights how important and useful cereals are.

Cereals dominate global agriculture with over a third of global cropland devoted to three crops, wheat, rice and maize. This is unsurprising given the many benefits of growing them. Whilst they differ from one another they can be considered largely interchangeable in that they are grown primarily to provide energy in the form of carbohydrates.

Overall, there is very little overlap in the growing regions of these crops (and other cereals too) which highlights that local growers will grow whichever of the crops will be most suited to their local conditions. They make up the staple diet for the majority of the planet's human population, and on average, supply the majority of a person's daily energy requirements.

Unlike fruit and vegetables, cereal grains have a low water content. Generally, this makes them easier to grow, partly because they are not as water demanding as fruits and vegetables at key stages of their growth. Another advantage is that, being so dry, grains are energy dense, together with their low water content making them both easy to transport and to store. Different cereals are grown in different regions of the world. In temperate regions wheat dominates, whilst in the humid tropics, particularly the south and southeast of Asia, rice is grown the most. In semi-arid regions sorghum and millet prevail.

Adaptations of cereals. Considering the conditions in which they grow, it is unsurprising that wheat is frost-hardy, sorghum and millet drought resistant and rice resistant to water logging, given that the majority of rice is grown in paddy conditions (i.e., within standing water). In addition, maize, sorghum and millet are most suited to hot and sunny conditions. All these crops exhibit a mixture of structural, physiological and biochemical adaptations which allow them to thrive in the specific conditions in which they are grown.

In the case of sorghum and millet that are adapted to growing in hot, dry semi-arid conditions and are amongst the most drought resistant. Indeed, where possible, growers would have a preference to grow maize, which whilst it has some drought resistance, has a greater water requirement than either sorghum or millet. Therefore, with decreasing rainfall, and increased aridity, maize production is displaced by the more drought resistant sorghum, which in turn is replaced by millet in even more arid regions, and of the three crops, millet has the lowest water requirement and maize the highest.

These crops are adapted to drought conditions, the more drought resistant sorghum and millet have fewer stomata, which are sunken and below the surface of the leaves, which serves to reduce water loss through transpiration. They also have thickened cuticles which also assist with reducing water loss. In the semi-arid regions, lack of water and extreme heat are the main abiotic stress factors that plants face. Excessive heat has the potential to denature a plant's proteins and ultimately kill the plant. Sorghum and millet can tolerate high temperatures better than other cereals. This is because such xerophytic plants (i.e., those adapted to arid growing conditions) are capable of synthesising heat shock protein (HSPs) when the temperature rises. Temperatures in these areas can rise extremely quickly so it is important that these HSP's are produced rapidly so preventing denaturing of crucial enzymes.

Another major feature maize, sorghum, and millet have for surviving in hot conditions is a modified form of photosynthesis that is common to many tropical plants. This is called C4 photosynthesis and are so named simply because the first products of photosynthesis are molecules with four carbon atoms rather than three carbon atoms (in C3 plants, including wheat and barley). This is advantageous because C4 plants do not photorespire, which C3 plants do. Photorespiration is the uptake of oxygen and release of carbon dioxide during photosynthesis, which is a problem for a plant to photosynthesise and convert carbon dioxide and water into useful carbohydrate because photorespiration reduces the net rate of photosynthesis. Photorespiration occurs in C3 plants because the key photosynthetic enzyme, rubisco, has an affinity not only for carbon dioxide, but also for oxygen, especially when the carbon dioxide concentration within the leaf is low and the temperature is high. The C4 plants use a different enzyme (PEP carboxylase) in photosynthesis, which has an exceptionally high affinity for carbon dioxide, even when present at low concentrations. This means that in the right conditions C4 plants are more productive than their C3 counterparts.

12.9.1.1 Wheat

By area, more wheat is grown than any other crop, which highlights its importance both as a staple for human consumption and as an animal feed. Wheat dominates cereal production in the cool temperate zones, and unsurprisingly very large countries produce the most, with China, India and the Russian Federation being the top three producers globally producing over 40% of the world's wheat.

It is noted that the global yield is relatively low, at around 3.5 t/ha lower than for maize and rice. This reflects that both maize and rice are grown in warmer and sunnier climates, and provided that other factors are not limiting (e.g., water and nutrients) then they are able to photosynthesise more, and this translates into a greater yield potential.

The United Kingdom only contributes just over 1% of the wheat produced globally, but does so on relatively little land, and has amongst the best yields globally of wheat. Unusually, the US has yields below the world average, and this is primarily due to a shorter growing season. However, due to the extensive area of production, then the US is the fourth largest producer of wheat globally. In contrast, Western Europe has the highest wheat yields globally, but the amount of land available is limited (Table 12.6).

Table 12.6 Top 10 global wheat producing countries (ranked by production, millions of tonnes) 2020, along with land use (million hectares) and yields (t/ha), and includes the United Kingdom and total world production, for comparison (FAO, 2020).

Country	Production (MT)	Area (million ha)	Yield (t/ha)
China	134	23.8	5.7
India	107	31.4	3.4
Russian Federation	86	28.9	3.0
USA	49	14.9	3.3
Canada	35	10.0	3.5
France	30	4.5	6.7
Pakistan	25	8.8	2.9
Ukraine	25	6.5	3.8
Germany	22	2.8	7.8
Turkey	21	6.9	3.0
United Kingdom	10	1.4	7.0
World	761	219	3.5

12.9.1.2 Maize

Maize lends itself to a rotation with soybeans, alternating between the two. This simple rotation is effective in that the two crops are from different families and also benefits from the nitrogen fixing of the leguminous soybean.

The majority of maize is grown for animal feed. In the United States, maize production centres on the mid-western states, with Iowa the largest producer, along with its neighbouring states of Nebraska, Minnesota and Illinois. The abundance of corn means that both the pig industry and bioethanol industry are also in these areas, as they both ultilise maize in their production.

It is notable that of the cereals, maize has the highest average yield globally, and whilst grown on slightly less land than wheat the overall amount produced is about a third more, and one of only two crops which have a crude production of over one billion tonnes per year (sugar cane is the other) (Table 12.7).

12.9.1.3 Rice

Unlike wheat and maize, rice is grown primarily for human consumption rather than for animal feed, and currently around half the world's population have rice as their staple. Consequently, whilst wheat and maize are grown more extensively, rice can arguably be considered the single most important crop in the world.

Wetland paddy rice grows in regions that are characterised by heavy, silty soils, with very poor drainage, high temperatures and heavy seasonal rainfall. Consequently, the majority of the world's rice is produced in waterlogged paddy fields.

Rice, however, is not a true aquatic plant since it can be grown in non-waterlogged conditions (although the yields tend to be lower) and will die if it is totally and continually submerged. Initially, it may appear odd that rice is grown in this way, and is

Table 12.7 Top 10 global grain maize producing countries (ranked by production, millions of tonnes) 2020, along with land use (million hectares) and yields (t/ha), and total world production, for comparison (FAO, 2020).

Country	Production (MT)	Area (million ha)	Yield (t/ha)
USA	360	33.4	10.8
China, mainland	261	41.3	6.3
Brazil	104	18.3	5.7
Argentina	58	7.7	7.5
Ukraine	30	5.4	5.6
India	30	9.9	3.1
Mexico	27	7.2	3.8
Indonesia	23	3.9	5.7
South Africa	15	2.6	5.8
Russian Federation	14	2.7	5.1
United Kingdom	N/A		
World	1162	202	5.7

especially true when the requirement for a complex system of irrigation is needed to maintain the water level so that the bases of the rice stems remain submerged. A major advantage, however, is that less fertiliser is required because cyanobacteria in the water fix nitrogen. Rice is amongst the least drought tolerant of crops and is the most water demanding cereal: keeping it well watered ensures good growth, whilst for all other major cereals the seedbed must be well aerated and free-draining.

Rice is adapted to thriving in waterlogged conditions by adaptations to its root system, which are able to respire aerobically despite the anaerobic conditions that prevail in waterlogged soils. Both the stems and the roots have tissues with air spaces and channels, called aerenchyma, and these allow oxygen to penetrate. Although rice is far from unique in having aerenchyma, as other cereals can produce it in response to anaerobic conditions, however it is better developed in rice. Furthermore, if the aerenchyma fails to supply enough oxygen, and the roots begin to respire anaerobically, rice has a greater tolerance of these conditions than other cereals.

Unsurprisingly, rice is grown where there is an abundance of water – and this lends itself to tropical regions with a distinct monsoon season and a distinct dry season (when the crop is harvested). The south of India is dedicated to rice growing, whilst in the drier north in the Punjab, wheat dominates cereal production. The temperatures remain relatively constant throughout the year, so crops can be grown year long, however, they are restricted by the amount of rainfall that will determine whether multiple crops can be grown each year. Whilst countries that produce rice are limited by the distribution of rainfall, in contrast temperate regions like the United Kingdom generally are more limited by temperature which varies throughout the year, whilst in contrast the amount of rainfall does not vary that much from month to month (e.g., in the United Kingdom can typically expect around 50 mm of rainfall each month) (Table 12.8).

Table 12.8 Top 10 global paddy rice producing countries (ranked by production, millions of tonnes) 2020, along with land use (million hectares) and yields (t/ha) (FAO, 2020).

Country	Production (MT)	Area (million ha)	Yield (t/ha)
China, mainland	212	30.1	7.0
India	178	45.0	4.0
Bangladesh	55	11.4	4.8
Indonesia	55	10.7	5.1
Vietnam	43	7.2	5.9
Thailand	30	10.4	2.9
Myanmar	25	6.7	3.8
Philippines	19	4.7	4.1
Brazil	11	1.7	6.6
Cambodia	11	2.9	3.8
World	757	164.2	4.6

12.9.1.4 Barley

The fourth most grown cereal, barley is distinct from the three more abundantly grown cereals of wheat, maize and rice in that it does not constitute a major part of the human diet, and whilst grown for food and drink produce is grown mainly for animal feed. It can be grown in temperate regions, and has a similar distribution to that of wheat. It does not replace wheat, however, as maize or rice may do in a human diet, and of those three crops there is generally little overlap in where they are grown. Please also refer to section 13.4 for husbandry of barley in the United Kingdom.

Table 12.9 shows the 10 countries which produce the most barley and unlike the other crops of the top 10 grown globally, the United Kingdom features as a major producer. Europe dominates barley production, with the United Kingdom having amongst the best yields of barley globally, although the mainland European country of Germany has higher yields as its climate is better suited to growing barley.

Table 12.9 Top 10 global barley producing countries (ranked by production, millions of tonnes) 2020, along with land use (million hectares) and yields (t/ha) (FAO, 2020).

Country	Production (MT)	Area (million ha)	Yield (t/ha)
Russian Federation	20.9	8.3	2.5
Spain	11.5	2.7	4.1
Germany	10.8	1.7	6.5
Canada	10.7	2.8	3.8
France	10.3	2.0	5.2
Australia	10.1	5.0	2.0
Turkey	8.3	3.1	2.7
United Kingdom	8.1	1.4	5.8
Ukraine	7.6	2.4	3.2
Argentina	4.5	1.1	4.1
World	157.0	51.6	3.04

Table 12.10 Top 10 global grain sorghum producing countries (ranked by production, millions of tonnes) 2020, along with land use (million hectares) and yields (t/ha) (FAO, 2020).

Country	Production (MT)	Area (million ha)	Yield (t/ha)
United States of America	9.5	2.1	4.6
Nigeria	6.4	5.2	1.2
Ethiopia	5.1	1.8	2.8
India	4.8	5.5	0.9
Mexico	4.7	1.5	3.2
China, mainland	3.6	0.7	4.9
Brazil	2.8	0.9	3.2
Sudan	2.5	5.8	0.4
Niger	2.1	3.7	0.6
Burkina Faso	1.8	1.9	1.0
World	58.7	40.2	2.8

12.9.1.5 *Sorghum*

The sorghum belt of the USA is in the southern states, where the water is limited and is grown primarily for animal feed. Sorghum is typically grown where there is not enough water to grow maize. Inevitably the yields of sorghum are lower than they are for maize, given that it grows in less favourable conditions (Table 12.10).

12.9.1.6 *Millet*

Unlike maize, the majority of millet is grown for human consumption and global production is primarily within Africa and Asia. Typically, if water is less limiting, farmers will favour growing Maize, however, as the environment becomes increasingly arid then sorghum will prevail.

Whilst there are 'only' 32 million hectares grown of millet globally, it is notable that the average global yield is only an average of 0.95 t/ha (2020), and hence only a fraction of the production of maize and sorghum (at around 30 million tonnes produced per year). Almost all the global production of millet comes from either Asia (48.8% in 2020) or Africa (47.1% in 2020). It is however, important, not to underplay the importance of millet (and indeed any crop that is grown) as clearly, they will be crucial for the specific growers that farm them. Unlike maize, sorghum, wheat and barley, along with rice, the millet that is produced will be grown to feed humans rather than livestock, and furthermore, despite the low yields (which are around a sixth of maize, globally), represent maximum productivity for the areas in which it is grown, and unsurprising the yields are so low given the limiting resources associated with where it is grown.

12.9.1.7 Rye

Rye is primarily grown in Europe and is amongst the most frost resilient cereals, and hence lends itself to production in countries with a harsh winter. In 2020, rye was grown on only 4.4 million hectares of land globally. Refer to section 13.6 for details of rye production within the United Kingdom.

12.10 Pulses: crops grown primarily for protein

Whilst the amount of energy from carbohydrates and protein is the same, crops grown for protein tend to have a lower yield than crops with a higher carbohydrate content. For example the yield of a high protein wheat for bread making would be expected to be lower yielding than a wheat grown for feed, which has a lower protein content. Hence, to meet daily energy requirements, carbohydrates dominate, and are less expensive to grow than crops with a higher protein content at similar yields. Cereals have comparable protein/hectare as many pulses, but they have higher yields, and the protein composition is less.

As there is a nutritional need to have protein in the human diet of around 50g per person daily, then it also follows that produce high in protein will be in demand. Whilst animal produce is often a good source of protein (and complete in essential amino acids) the cost of production is generally high, which is inevitable given the requirement for livestock to eat plant produce, and loss of energy through trophic levels (because livestock convert a fraction of energy consumed into growth). In contrast, plant-based sources of protein are cheaper to produce, and the principle protein source for human nutrition globally. Good sources of plant protein come from legumes, and principally pulses, which are legumes that are harvested as dry grains. Whilst the composition of harvested pulses can vary widely, over 20% protein is common (kidney beans are around 22%, and red lentils around 24% protein).

Soybeans (see Chapter 14) and groundnuts (peanuts) are excluded from being defined as pulses as they characteristically also have a high oil content, which pulses do not (Table 12.11).

Table 12.11 World pulse production (ranked by production, millions of tonnes) 2020 along with land use (million hectares) and yields (t/ha). Soybeans are also a legume, but classified as a vegetable oil crop primarily. There are many different varieties of pulses (FAO, 2020).

Crop	Production (MT)	Area (million ha)	Yield (t/ha)
Pulses	73	81	0.9
Dry beans	23	30	0.75
Chickpeas	13	14	0.93
Cowpeas	6	11	0.55
Lentils	5	4	1.25
Broad beans (dry)	3	2	1.5
Peas (dry)	11	6	1.8
[Soybeans]	256	112	2.3

Table 12.12 Top 10 global soybean producing countries (ranked by production, millions of tonnes) 2020, along with land use (million hectares) and yields (t/ha) (FAO, 2020).

Country	Production (MT)	Area (million ha)	Yield (t/ha)
Brazil	122	37.2	3.3
United States of America	113	33.3	3.4
Argentina	49	16.7	2.9
China, mainland	20	9.9	2.0
India	11	12.1	0.9
Paraguay	11	3.6	3.0
Canada	6	2.0	3.1
Russian Federation	4	2.7	1.6
Bolivia (Plurinational state of)	3	3.6	2.1
Ukraine	3	1.4	2.0
World	353	127.0	2.8

12.10.1 Soybeans

Soybeans, have amongst the highest protein content, at about 35%. Coupled with soybeans valuable high fat content (around 20%) it is unsurprising that this crop is the most widely grown after the three main cereal staples (wheat, maize and rice) globally. As a legume soybean is beneficial in improving soil fertility, and a valuable crop to have in a rotation, especially considering the dominance of cereals grown globally, which inevitably will compromise the crop rotation. The table above lists the top global producers of soybean. Production is dominated by Brazil in South America and the United States in North America (Table 12.12).

12.11 Oil seeds: crops grown primarily for oil

As maize grain has a relatively high oil content for a cereal of around 5% which combined with its vast production makes it a significant potential producer of vegetable oil globally. However, other crops are grown primarily for their high oil content whereas maize is grown primarily for its carbohydrate content and ease of transport and storage. With the exceptions of oil palm and coconut, the major vegetable oils also have a high protein component too. Hence, these crops will be grown for their protein content as well as for their oil content.

There has been a burgeoning demand for vegetable oils since the 1950s which has been driven by the increase in processed food. Note, that this demand has been primarily driven by demand for human consumption primarily rather than for non-food uses (e.g., biodiesel production) which is also significant.

Whilst soybeans are grown for their high protein content (which is almost double that of its oil content), because it is high in oil and its extensive cultivation, soy vegetable oil is the second most consumed globally (60 million metric tonnes in 2021),

Table 12.13 World vegetable oil crop production (ranked by production, millions of tonnes) 2020, along with land use (million hectares) and yields (t/ha) (FAO, 2020).

Crop	Production (MT)	Area (million ha)	Yield (t/ha)
Soybeans	353	127	2.8
Rapeseed	72	35.5	2.0
Groundnut	54	31.6	1.7
Oil palm	418	28.7	14.6
Sunflower	50	28	1.8
Coconuts	62	11.6	5.3

following on from palm oil (at around 74 million metric tonnes from the mesocarp, and another eight million metric tonnes from the kernel).

Table 12.13 shows the gross production of soybeans and the five others most widely grown oil crops. The yields of oil palm stand out, with yields almost a magnitude higher than that of groundnuts. Whereas soybeans, rapeseed and groundnuts all have relatively high protein contents as well as high oil content, oil palm as the name suggests is grown primarily for its oil content, with a small protein component from within its kernel. As the majority of the oil is derived from the fleshy mesocarp of the palm, by definition, it is the most grown fruit globally (although the fruit is refined to produce oil, rather than eaten directly as would be the case of typical fruits like bananas).

Oil Palm as a perennial, does not require cultivation each year, and the costs of production of oil palm are less than for annual crops. Typically, oil palms only need replanting every 20–25 years, when harvesting becomes difficult due to increased height and diminished yields. This factor, along with its high yields, makes it inevitable that it is grown in such abundance, in order to meet the global demand for vegetable oil. It is grown either side of the equator, and requires high rainfall, production is dominated by Indonesia, Malaysia, Nigeria and Thailand.

Growing oil palm is increasingly contentious, given the clearing of natural habitats that has occurred to increase its expansion. Whilst many countries have stopped clearing land to grow oil palm, it does not detract from the significance of this crop as it has such a high yield of vegetable oil (which is also increasing in demand), and is inexpensive to grow. It does, however, highlight the use of non-waste vegetable oils for non-food uses, for example for the production of biodiesel, palm oil has a lower commodity price than other oils, and so lends itself to use for biodiesel.

12.12 Crops grown primarily for export: cash crops

Semantically, other than crops grown for subsistence (which still have a monetary value) all crops are grown for financial reasons or as a 'cash crop'. However, the term 'cash-crop' usually relates to crops that are grown primarily for export. By this definition, cereals cannot be considered cash crops given that despite the millions of tonnes exported annually this is still a relatively small proportion of the amount grown globally. Cash crops also tend to have limited nutritional value, and hence tea, coffee and sugar (from sugar cane) are all cash crops.

12.12.1 Tea, coffee and cocoa

Many developing countries are reliant upon exporting cash crops to generate foreign revenues, and by land use (sugar cane excepted) tea, coffee and cocoa dominate cash cropping. There are around 11 million hectares of coffee grown, 12 million hectares of cocoa and five million hectares of tea (and also around 13 million hectares of non-food rubber plantations), so a significant amount of land is devoted to their production. However, when the global demand is considered, and is met by this land use, then comparatively little land is attributed to their production (i.e., around 2% of crop land use). Whilst, beneficial to the countries to export these commodities it is worth noting that tariffs of the raw produce tend to be low or tariff free into developed importing countries, who benefit from not economically being able to produce these commodities and are not in competition. Furthermore, tariffs on higher value processed products are usually substantially higher, which limits high value products being produced and exported as an alternative to the base commodity.

A major problem for producers is that there is considerable competition between less developed countries (which have the correct climatic conditions to grow these cash crops) which suppresses the value of these commodities. Furthermore, the final product can be considered price inelastic, as the retail price of tea and coffee is little affected by dramatic changes in the wholesale value of the raw unprocessed commodity.

12.12.2 Vanilla

As a specific example, under 20 countries grow vanilla, dominated by Madagascar which has almost 80% of the World's vanilla cropping. However, this is still under 75,000 ha (and under 100,000 ha grown globally), with Indonesia second, accounting for around 15% of land used for vanilla production. This specific example highlights that the global demand for this cash crop is minimal in terms of the land requirement to meet the demand. It is a significant crop for major producers, as vanilla is an extremely high value crop, with an average yield of around 80 kg per hectare, and a total global production of under 8000 tonnes per year.

12.12.3 Sugar cane and sugar beet production

Sugar is unique in that the commodity can be produced from two distinctly different species, the temperate root crop sugar beet (see Chapter 15) or from tropical sugar cane, which is from a different plant family, but the final refined product is identical irrespective of which crop it is derived from.

Table 12.14 Land use (millions of hectares) of sugar cane and sugar beet, in last 5 decades.

Year	Sugar cane	Sugar beet
1970	11.0	7.6
1980	13.3	8.9
1990	17.1	8.7
2000	19.3	6.0
2010	24.6	4.7
2020	24.5	4.4

Globally, sugar cane land use (and production) is increasing, Whilst for sugar beet it continues to reduce (Table 12.4). Annually around 1.9 billion tonnes of sugar cane is harvested which makes its the highest of any crop globally (compared with maize, with around 1.2 billion tonnes annually). This is deceptive as sugar cane has a far higher water content than cereals but is still a very productive crop on a dry weight basis.

Just as the root crop sugar beet is harvested annually, so is sugar cane. However, a key difference is that sugar cane will typically be only replanted every 3 years. The cost of production of sugar cane is less than for sugar beet and it is unsurprising that globally it dominates crops grown for sugar. Unlike many tropical crops, sugar cane has a high yield, and whilst the amount of sugar per hectare is less than that expected from sugar beet, the difference is not as marked as it is for other crop products.

Globally, there is a reduction in the amount of sugar beet being produced, but the overall amount of sugar produced globally is still increasing. This is explained by the increasing demand for refined sugar, and is best met by sugar cane production rather than by sugar beet, given that it has lower production costs (and is also more efficient to grow, requiring less energy input than sugar beet) (Table 12.14).

Brazil has twice as much land devoted to sugar cane production than its nearest rival, with over 10 million hectares in 2020 (and with an average yield of over 75 tonnes/ha) compared with India devoting around five million hectares of sugar cane (crude yield of 80 t/ha). However, despite this, Brazil produces less raw sugar than India, because over half of the sugar cane is grown to produce bioethanol directly.

The average global yield in 2020 of sugar beet was around 57 t/ha and for sugar cane was 71 t/ha. However, as sugar cane has typically around only two thirds the sugar content of sugar beet per hectare the amount of sugar produced is broadly comparable. For both sugar crops the sugar content will alter through the season, and harvest is optimised to process the crops when the content is highest, but this is not always possible. The sugar content of both crops can also differ year on year, and the specific growing conditions of the year will be a factor in determining sugar content.

12.13 Future challenges for global crop production

There will always be huge demand for crop products globally, and it is essential that sufficient can be grown to meet this ongoing demand. Hence, whilst it may seem comforting that the future of crop production is guaranteed, there are huge challenges to be faced to ensure that safe, sustainable and plentiful food can be grown. Growers globally will need to work toward improving both the agricultural efficiency of food production at the same time as increasing productivity. This is a huge challenge, especially given that energy efficiency tends to diminish with increasing intensity of production. As the amount of useful agricultural land for crop production has plateaued, any increase in production is likely to need to be met by increasing yields rather than increasing land use.

This is further compounded because culturally people are linked to the produce that they have traditionally grown, and as such difficult to change behaviours. For example Asians are likely to favour rice as their staple even if maize is abundant and cheaper, and Europeans will favour wheat over alternative cereals.

Whilst both the amount of land devoted to cropping and their yields have increased, which has mitigated the increased demand with the growing human population, the rates of increase have slowed. This means that feeding the growing population is becoming increasingly difficult.

Whilst the yields can improve with increased crop protection and fertiliser inputs, the rate of increase suffers from diminishing returns, and as such, achieving the highest yields will not be the most energy efficient means of production. To further confound the challenge, the need to increase production comes at a time when there is an urgent need to farm in a more sustainable way, and mitigate for climate change. Therefore, increasingly a challenge is to increase production yet also to do so in a sustainable way.

Somewhat paradoxically, whilst intensive agriculture can sustain the world's population in that it can produce the highest yields yet it is not sustainable given the energy required in production, often from non-renewable sources. Low input agriculture is sustainable but cannot sustain the world's population as the yields are too low.

To overcome the need to increase production, it will be necessary to increase the overall average yields rather than the maximum yield (e.g., if world average wheat yields increased from 3.5 t/ha to 4 t/ha this has a markedly greater impact on global production than achieving the theoretical maximum from a relatively small area, and with inefficient use of crop inputs). In addition, increasing urban agriculture may offset some of the problems incurred from limited land availability in the future, and increasingly the global shift to regenerative agriculture is to be promoted.

12.14 Key points

- Crops are grown primarily to meet human nutritional requirements, and cereals dominate arable land use, due to their ease of growth, transportation and storage and high carbohydrate (energy) yields.
- Crops are also grown primarily as a protein source, with soybean the most significant crop grown for protein.

- Farmers globally are united in growing whichever crop is best suited to their specific climatic growing conditions, and favour different crops accordingly to produce carbohydrates, proteins and vegetable oils.
- Whilst global demand dictates which crops are ultimately grown, farmers' priority will be to grow whichever crop is most viable economically, but these are dominated by crops for food although cash crops (typically of low nutritional value) also take a significant proportion of crop land globally.
- It is unsurprising that sugar cane and oil palm are grown so widely, given that they are highly productive in the amount of energy that can be harvested per hectare.
- There is limited potential to increase productivity through increasing land used for cropping, most will have to come from increasing yields, and/or reducing the proportion of crops grown for animal feed or non-food uses (e.g., biofuels).
- With an increasing global human population, coupled with threats to food production, it is increasingly challenging to meet future demands for food, especially in a sustainable way.

Sources of further information and advice

Jeffery, S., Verheijen, F.G.A., 2020. A new soil health policy paradigm: pay for practice not performance! Environmental Science & Policy 112, 371–373 (Elsevier).

McCance, R.A., Widdowson, E.M., 2015. McCance and Widdowson's the Composition of Foods, Seventh Summary Edition. Institute of Food Research.

Oerke, E.C., Dehne, H.W., 2004. Safeguarding production – losses in major crops and the role of crop protection. Crop Protection 23, 275–285 (Elsevier).

Oerke, E.C., 2006. Crop losses to pests. Journal of Agricultural Science 144, 31–43 (Cambridge University Press).

Savary, S., Willocquet, L., Pethybridge, S.J., Esker, P., McRoberts, N., Nelson, A., 2019. The global burden of pathogens and pests on major food crops. Nature Ecology & Evolution 3, 430–439.

Scholes, R.J., Montanarella, L., Brainich, E., Barger, N., ten Brink, B., Cantele, M., Erasmus, B., Fisher, J., Gardner, T., Holland, T.,G., Kohler, F., 2018. IPBES: summary for policymakers of the assessment report on land degradation and restoration. The Intergovernmental Science-Policy Platform on Biodiversity and Ecosystem Services, United Nations.

Tivy, J., 2017. Agricultural Ecology. Taylor & Francis Group, 297 pages.

Wakatsuki, T., Rasyidin, A., 1992. Rates of weathering and soil formation. Geoderma 52, 251–263. Elsevier.

Useful websites

FAOSTAT. Fao.org/faostat/en/~data/QCL.

INRAE-CIRAD-AFZ feed tables http://feedtables.com.

The Word Factbook cia.gov/the-world-factbook.

World Health Organisation who.int.

Cereals

13

13.1 Introduction

The cereals sector is the third biggest agricultural sector in Europe in terms of output value after the vegetable/horticultural plant sector and the dairy sector. In the UK, cereals account for around 50% of the cropped area with the area of wheat ranging from 1.5 − 2 million hectares and the area of barley ranging from 1.0 − 1.5 million hectares annually depending on weather and soil conditions at planting. These crops are the cornerstone of many arable rotations. In Southern Europe, grain maize is also an important cereal and oats and rye also contribute to the total area of cereals across Europe.

General principles such as identification, growth and development, components of yield and harvesting and drying processes which are common to all types of cereals will be described at the start of this chapter before the individual species, their markets and agronomy are described in greater detail.

13.1.1 Cereal identification

In the vegetative stage, different cereal plants can be identified by their leaves and the presence or absence of auricles, as well as from the size of their auricles (Fig. 13.1). At the seedling stage all cereal leaves twist clockwise except oats which tend to twist anti-clockwise.

The different cereal crops are more easily recognised by their grains or flowering heads later in their growth cycle.

The flowering head of cereals is usually referred to as the ear or spike. In wheat, barley and rye ears the spikelets that contain the grain are attached to the main stem (rachis) without a stalk (Figs. 13.2−13.5).

In wheat the spikelets alternate on opposite sides of the rachis. 1−5 grains develop in each spikelet (Fig. 13.2). A few varieties have long awns.

In six row barley, all three flowers on each spikelet are fertile (Fig. 13.3). Awns are attached to the grains.

Two row barley heads hang down when ripe: each grain has a long awn; the small infertile flowers are found on each side of the grains (Fig. 13.4).

In rye the grain is easily seen in the spikelets (Fig. 13.5).

In oats the spikelets are stalked, giving an open ear or panicle (Fig. 13.6).

Lockhart and Wiseman's Crop Husbandry Including Grassland. https://doi.org/10.1016/B978-0-323-85702-4.00016-9

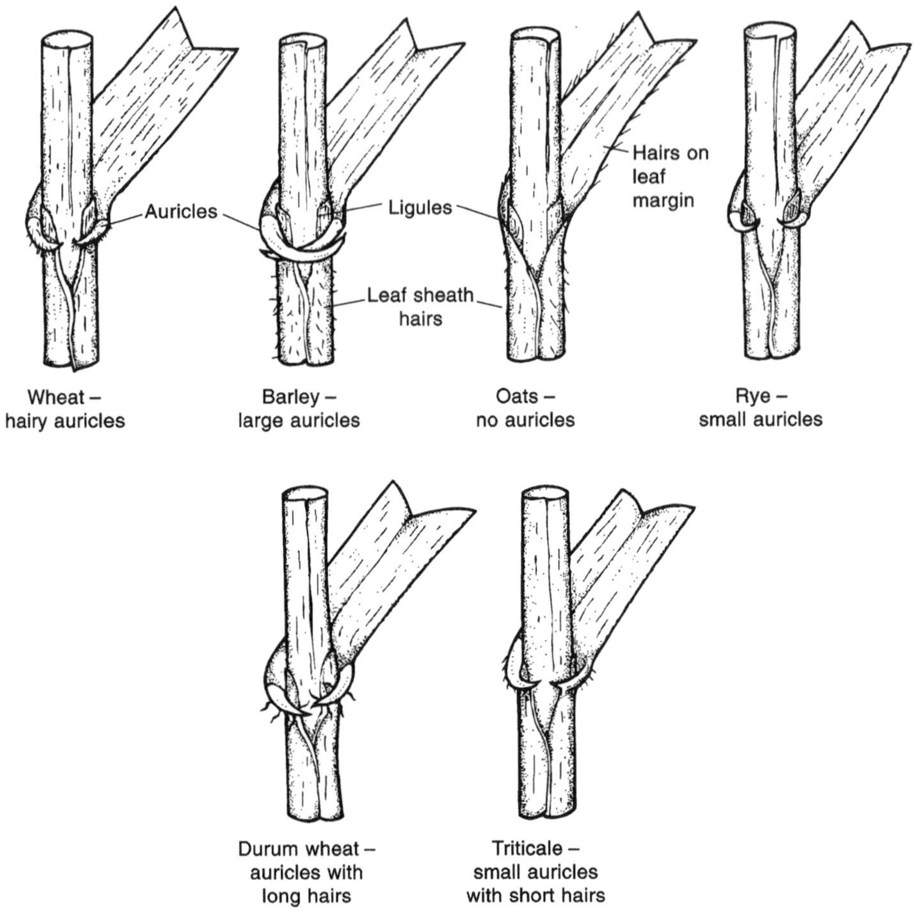

Figure 13.1 Using vegetative features to identify cereals.

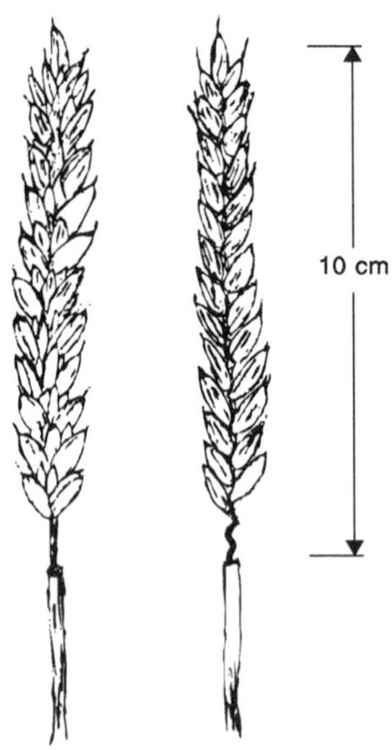

Figure 13.2 Wheat ear.

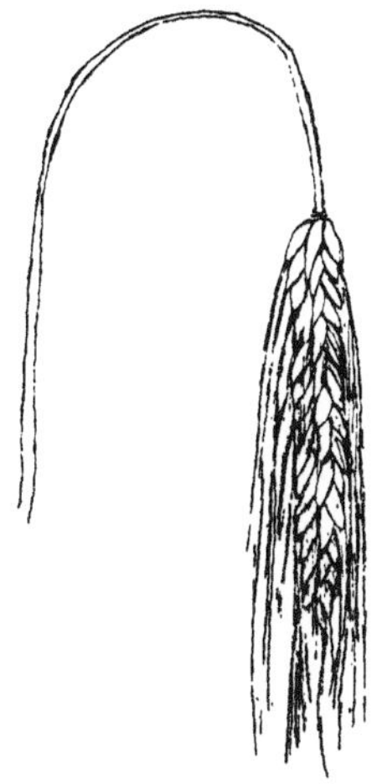

Figure 13.3 Six row barley ear.

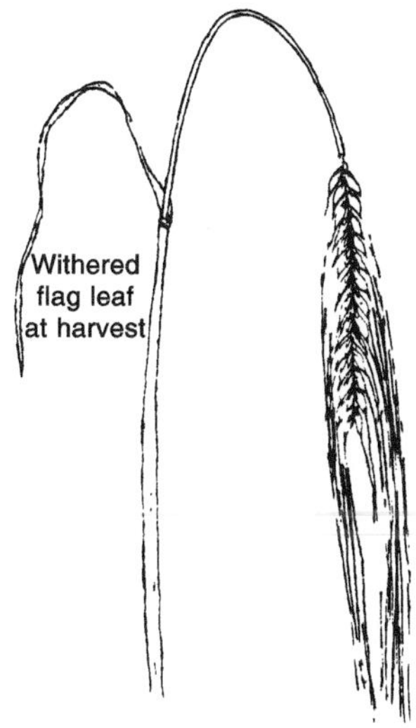

Figure 13.4 Two row barley ear.

Wheat, rye and maize grains consist of the seed enclosed in a fruit coat (the pericarp) and are referred to as 'naked' caryopses (kernels). In barley and oats, the kernels are enclosed in husks formed by the fusing of the glumes (palea and lemma) and are referred to as 'covered' caryopses (Fig. 13.7). However, note that naked oat and barley varieties are also now commercially available.

Figure 13.5 Rye ear.

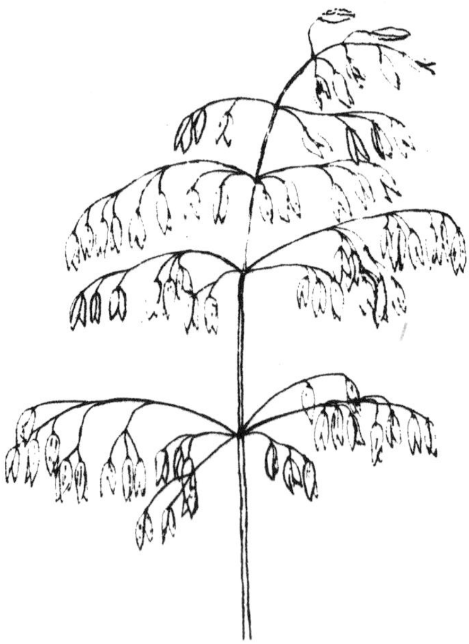

Figure 13.6 Oat panicle.

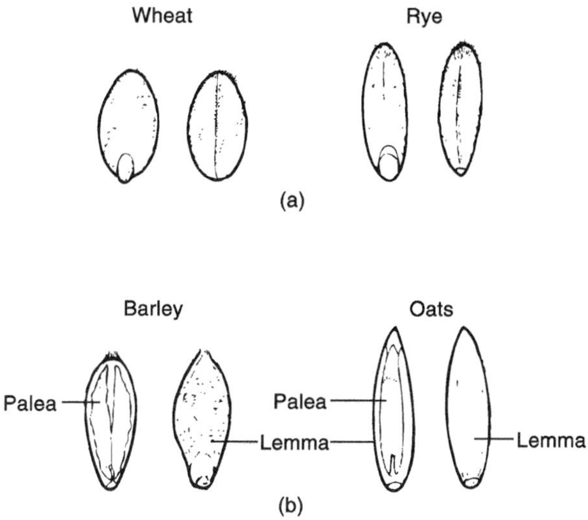

Figure 13.7 Kernels: (a) 'naked' kernels, (b) 'covered' kernels.

13.1.2 Cereal growth and development

Throughout the growing season, the crop changes in form (development) and also accumulates dry matter (growth). Development is influenced by sowing date, variety, temperature, the requirement for vernalisation and day length and cannot be influenced by management decisions during the season. Growth is also influenced by these factors but can also be manipulated by management decisions made during the growing season e.g., nutrition and weed, pest and disease management.

Development of cereals can be considered in the following broad phases. For most of the cultivated species of cereals both winter and spring sown varieties are available which will influence the time of year at which the phases of growth described below will occur.

- *Foundation.* This is the period from sowing to the start of stem extension during which yield bearing shoots (tillers) and primary roots are forming and the canopy is incomplete (Fig. 13.8). In an autumn sown crop this period occurs over the winter months when light levels are low. In a spring sown crop it will occur over a shorter period in early spring when light levels are increasing and it will therefore be shorter in duration.
- *Construction.* This is the period from the formation of the first node (start of stem extension) to flowering (anthesis) when the leaves that will contribute directly to yield are forming and the fertile florets are also forming (Fig. 13.9). Rapid growth occurs during this period maximising canopy size and the interception of solar radiation during critical periods when the potential yield is being determined.
- *Production.* This is the period after flowering when grains fill and ripen and the canopy starts to senesce. Maintenance of the green canopy during this period is important to maximise grain fill. In some cereal species significant contributions to grain fill come from the redistribution of water-soluble carbohydrates stored in stems especially during periods of stress e.g., drought.

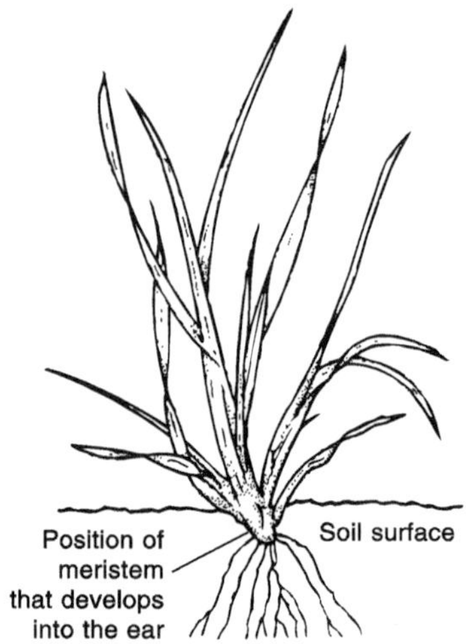

Figure 13.8 Leafy winter plant with four tillers.

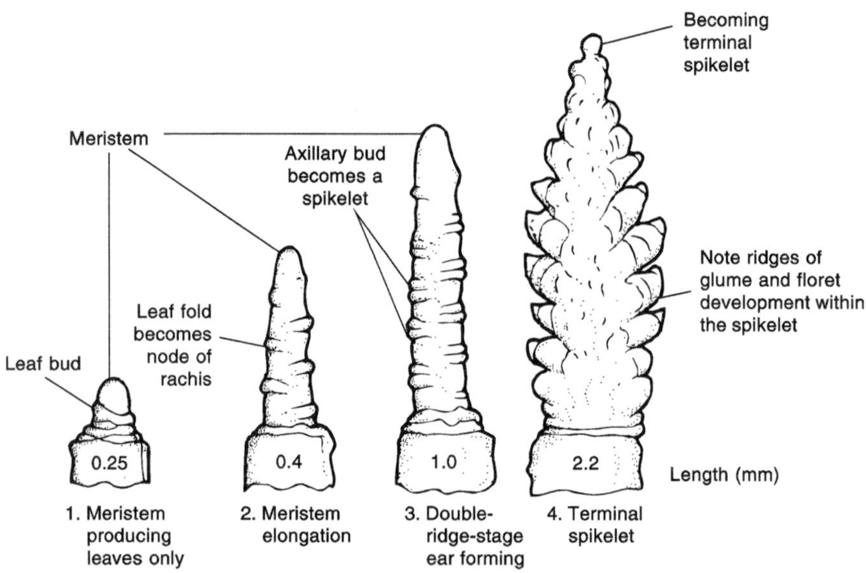

Figure 13.9 Simplified diagram of an ear forming from the tiller apex of winter wheat (not to scale). (Visible under a microscope or possibly with a good pocket lens.) The apical meristem changes from leaf production to ear formation following cold treatment (vernalisation) and increasing day length (photoperiod).

In practice, key agronomic decisions such as nutrient and crop protection product applications are made at very specific growth stages and a decimal growth stage key is commonly used which breaks the above phases down into much greater detail (Table 13.1). The key describes 10 primary stages of growth (subdivided into

Table 13.1 The decimal code for growth stages (GS) of small cereal grains.

Code 0 Germination. Subdivided – 00 dry seed to 09 when first leaf reaches tip of coleoptile.
Code 1 seedling growth
10 1st leaf through coleoptile
11 1st leaf unfolded
12 2 leaves unfolded
13 3 leaves unfolded
14 4 leaves unfolded
15 5 leaves unfolded
16 6 leaves unfolded
17 7 leaves unfolded
18 8 leaves unfolded
19 9 or more leaves unfolded
Code 2 tillering
20 main shoot only
21 main shoot and 1 tiller
22 main shoot and 2 tillers
23 main shoot and 3 tillers
24 main shoot and 4 tillers
25 main shoot and 5 tillers
26 main shoot and 6 tillers
27 main shoot and 7 tillers
28 main shoot and 8 tillers
29 main shoot and 9 or more tillers
Code 3 stem elongation
30 ear at 1 cm (pseudostem erect) – visible only if the growing point is dissected
31 1st node detectable
32 2nd node detectable
33 3rd node detectable
34 4th node detectable
35 5th node detectable
37 flag leaf just visible
39 flag leaf ligule collar just visible
Code 4 40–49 booting stages in development of ear in leaf sheath to when awns just visible – GS 49.
Code 5 50–59 stages in the emergence of the inflorescence or ear. Ear fully emerged – GS 59
Code 6 60–69 stages of anthesis (flowering). Flowering complete – GS 69
Code 7 Milk development stages in grain. Late milk – GS 77
Code 8 Dough development stages in grain. Hard dough – GS 87
Code 9 Ripening stages in grain. Grain hard and not dented by thumb nail – GS 92

secondary stages) from sowing through to the vegetative leaf and tillering stages, stem elongation, ear emergence, flowering to grain filling and ripening.

It is very important to be able to identify growth stages accurately in the field to make timely decisions which maximise the efficacy of inputs and indeed many pesticides have latest timing restrictions in terms of growth stages. The main stem should always be used when checking the growth stage. The main stem can be split with a knife to determine when the switch from tillering (Fig. 13.8) to stem extension is occurring (growth stage 30 – Fig. 13.10) and the same technique can be used throughout stem extension to determine the number of nodes present. However, note that the correct timing for fungicides corresponds with the emergence of the key yield forming leaves e.g., final leaf 3 and final leaf 1 (flag leaf) (Fig. 13.11). These are not always fully aligned with decimal growth stages so the careful dissection of the furled leaves in the stem is a more accurate method for determining fungicide timings.

13.1.3 Components of yield in cereals

The yield of a cereal crop is determined by the contributions made by the three components:

1. number of ears per unit area,
2. number of grains per ear,
3. weight (size) of the individual grains.

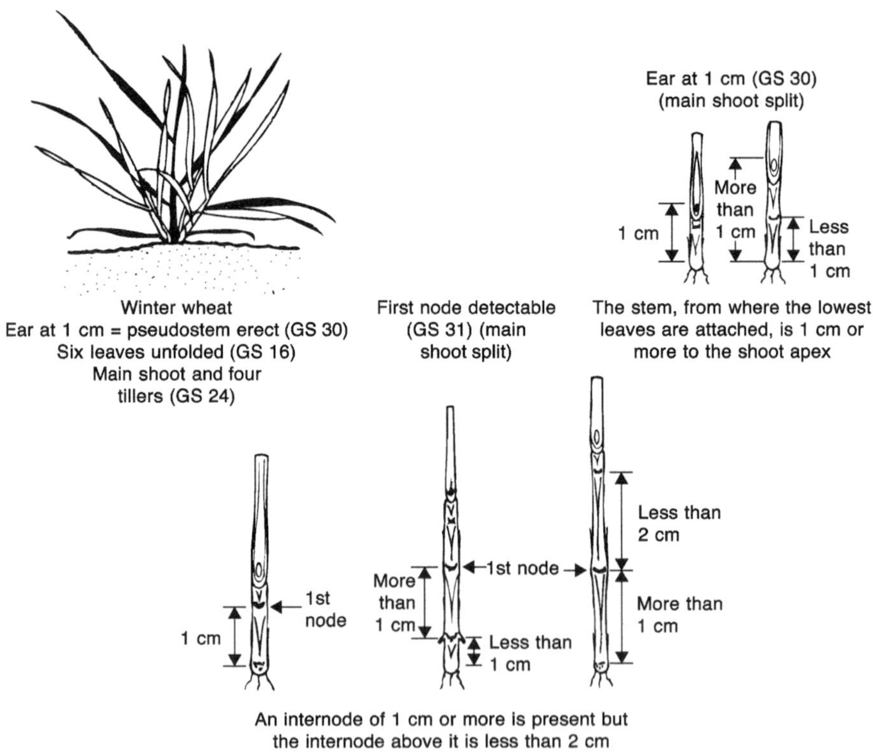

Figure 13.10 Early stages of stem elongation in wheat.

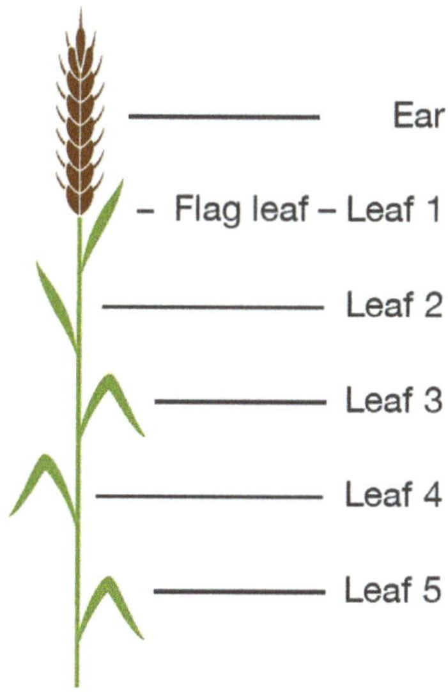

Figure 13.11 Numbering of leaves on a cereal plant.

Ear number is influenced by plant population and growth during tillering up to flag leaf emergence. Growth between the flag leaf emerging and flowering affects numbers of grains per ear and grain size is affected by conditions post flowering.

These components are inter-related. By increasing the number of ears per unit area (e.g., by denser plant populations or more tillering), the number of grains per ear may be reduced and also the size of the grains. The flexibility for one component to compensate for another depends on the species of cereal. Benchmark values for the three components of yield for winter wheat and barley are shown in Table 13.2 below. Grain yield represents about half of the above ground dry matter produced, usually expressed as a percentage, this value is termed the harvest index.

13.1.4 Grain quality in cereals

There are a range of markets for cereals, including milling for human consumption, animal feed, malting, distilling, seed, export and industrial uses including production

Table 13.2 Benchmark values for yield components in high yielding feed varieties of winter wheat and winter barley (2 row).

Component of yield	Winter wheat	Winter barley (2 row)
Ears/m^2	460	775
Grains/ear	48	25
Individual grain weight (mg)	50	46

of bioethanol. Grain quality requirements will be affected by the proposed market and will be described in more detail in relation to the individual cereal species later in this chapter. However, some standard tests when selling grain relate to all cereal species and include the following:

- *Moisture content.* This is very important when storing grain. For long term storage it should be about 14%, too high a moisture content will be penalised or not accepted. Grain which is overheated when being dried, or in storage, can be spoiled for seed, malting or milling. Overheating can also damage the germination ability as well as the protein quality.
- *Sample appearance and purity.* Good quality grain is clean and attractive in appearance, free from mould growth, pests and bad odours. Sample purity, freedom from contamination with other cereals or weeds will affect the potential value of grain, as will the amount of shrivelled or broken grain. Careful setting of the combine will help to minimise broken grains and produce a cleaner sample and some grain drying systems also have grain-cleaning facilities.
- *Specific weight (bushel weight).* Specific weight is a measure of grain density reported in kilograms per hectolitre (kg hl^{-1}). A high specific weight is preferred for all end uses and thresholds will apply for the different markets. Of the cereal species, wheat has the highest specific weight and oats have the lowest. Specific weight is to an extent a varietal trait but can also be affected by agronomy (e.g., time of sowing, nutrition, pest and disease control) as well as by weather conditions during grain fill and harvest.

Most merchants now only buy grain that comes from farms that have been assured by a registered assurance scheme. These schemes include strict protocols on crop production and grain storage.

13.1.5 Harvesting

Threshing is the separation of the grains from the ears and straw. In wheat and rye the chaff is easily removed from the grain. In barley, only the awns are removed from the grain; the husk remains firmly attached to the kernel. In oats, each grain kernel is surrounded by a husk that is fairly easily removed by a rolling process as in the production of oatmeal; the chaff enclosing the grains in each spikelet threshes off. Varieties of naked barley and oats have now been bred which thresh free from their husks.

The majority of cereal crops are harvested using a combine harvester. Only a few crops are traditionally harvested with a binder where the straw is required for thatching, and some crops are foraged for whole crop silage (see chapter 18).

Good weed control, to minimise green material at harvest and avoidance of lodging, is also very important; combining is easier, and costs are less if no drying of grain is required. The combine capacity should be adequate to harvest the crops as they ripen. Delay can result in poor quality grain and shedding losses. Timing of harvesting will be affected by area and crops grown as well as quality requirements, combine and drying capacity, labour, use of pre-harvest desiccants and weather conditions.

Combines can be fitted with straw spreaders or choppers, which are helpful in ensuring the straw is spread evenly for incorporation.

Cleaning combines to remove weeds, e.g., wild oats, black-grass, barren brome, is very important before moving to a clean field. It is essential that the combine is clean before moving into a seed crop.

13.1.6 Grain-drying

In some years, when all the cereals are combined at below 15% moisture content, no drying will be necessary, only possibly cooling. When drying is required a range of options are available. While using high-temperature driers, it is important not to overheat the grain as this may result in loss of quality premiums. Seed, malting and milling samples need to be dried at relatively low temperatures as the seed must remain viable. Feed samples can be dried at higher temperatures and therefore the output of the dryer will be increased. Maximum drying temperatures will depend on the type and size of dryer being used and even the individual make of machine.

Continuous-flow drier. The principle is that hot air removes the excess moisture and ambient air then cools the grain to 10–15 °C for safe storage. However, in very warm weather this may not be possible and night time air may have to be used to cool the heap after drying. Wet grain (over 22% moisture content) may require two passes through the drier.

Batch driers. The drying principle is similar to that of the continuous-flow drier, but the grain is held in batches in special containers during the drying process increasing efficiency of the drying process.

Ventilated silos or bins. Cold/ambient or slightly heated air is blown through the grain in the silo or bin. This can be a slow process, especially in damp weather. The bins can range in size up to 100's of tonnes and may be grouped together in a circle, around a central reception pit, or in straight lines with a reception pit and elevators at one end and conveyors running along the top and bottom of the bins to aid moving grain in and out of the system.

Bulk storage/on floor drying. A large volume of cold/ambient or slightly heated air is blown through the grain (possibly over a few days) to remove excess moisture. The air may enter the grain in several ways:

(a) from a fan, through a main duct (roughly 2m high by 1m wide) and then into lateral ducts at about a metre apart placed on, or formed in, the floor;
(b) from a single duct positioned in the centre and at the base of a large heap;
(c) through a perforated floor (often called a 'pallet' or 'drive over' floor constructed from timber or precast concrete panels;
(d) using individual spear type perforated circular steel pipes with a small fan on top, positioned at regular intervals throughout the grain heap.

Floor drying is very popular because it can be carried out in a general purpose building. It is a cheap method and requires very little labour when filling. A common problem with floor drying is to use heat when drying very wet grain as this over-dries the lower layers around the ducts, and the moisture settles out in the cooler upper layers where it forms a crust which impedes the air flow.

Grain stirrers can be installed which overcome the problems of uneven drying and 'crust' forming. They move automatically through the heap of grain on a steel framework and thoroughly mix the grain as it dries. The relative humidity (RH) of the air being blown through the grain determines the final moisture content of the grain. Using heat should be saved for the later stages of drying i.e., down to about 14% moisture content.

Mobile grain driers Most are of the recirculating batch type, but portable versions of well-established static batch or continuous flow machines are also in use.

Membership of co-operatives. Another option which farmers have is to dry and store grain with a co-operative group. This store will provide drying facilities as well as gravity separators for improving grain quality.

13.1.7 Grain storage

13.1.7.1 Dry grain

Grain can be stored in bulk for up to 1 month at 16%–17% moisture content, but for longer term storage it should be dried down to 14% moisture content. Only fully ripe grain in a dry period is likely to be harvested in the UK at 14% moisture. In a wet harvest period, the moisture content may be over 20% and the grain will have to be dried for storage.

Damp grain will heat up due to the growth of moulds, mites and respiration of the grain. Moulds, beetles and weevils may damage grain that is stored at a high temperature, e.g., grain not cooled properly after drying or grain from the combine on a very hot day. Ideally, the grain should be cooled to 15 °C within 2 weeks to prevent saw-tooth beetles breeding. For long-term storage the grain should be cooled to below 5 °C to prevent mites and grain weevil breeding.

Before storing grain, it is essential to clean and dry the store thoroughly as most problems arise from pests already present in the fabric of the store (primary pests) rather than those that are introduced with the harvested grain. Once thoroughly cleaned, bait bags and floor traps should be used to monitor for the presence of pests. A small range of approved insecticides is available to treat the fabric of the store and fumigants can also be used for inaccessible parts of the store.

The temperature of stored grain should be checked regularly. Any rise in temperature can indicate an insect infestation. A decreasing range of approved residual insecticides is available to treat pest infestations in stored grain. Any such treatments require movement of the grain over a conveyor for accurate application.

Rodents and birds must also be kept out of the store. When grain is sold, any chemical treatments must be declared. Any kind of grain stored for human consumption must be kept under conditions that satisfy the Food Safety Act 1990. This means that the farmer must exercise 'due diligence' in keeping stores rodent- and bird-proof; these are aspects covered in the various assurance schemes. Other commodities, such as fertiliser, must not be stored in the same building.

13.1.7.2 Moist grain

Most grain is stored dry. There are, however, alternative methods of preserving moist grain that can be simpler and cheaper than more conventional systems if grain is destined for animal feed on the farm where it was grown.

The cereal is combined three to 4 weeks earlier than normal when the grain dry matter is at around 55%–75%. The grain is then crushed in a machine called a 'crimper' and ensiled in an airtight clamp to ferment. A preservative or organic additive is normally used to reduce clamp spoilage. Yield of crimped grain is usually higher than grain harvested at the normal time.

13.1.8 Cereal straw

Straw is an increasingly valuable commodity reflected in long term price increases over and above short-term fluctuations in response to seasonal weather patterns. Markets today include renewable energy systems, for both domestic heating and power generation, overwinter carrot production and mushroom production as well as the traditional markets for livestock feed and bedding. Simultaneously, inorganic nutrient prices have risen and increasing attention has been given to soil health and the benefits of incorporating straw to increase organic matter levels in arable soils.

Barley and oat straw is predominantly baled for livestock bedding and feed. The feed value for wheat straw is lower than for barley and oats. The use of wheat straw will depend on the area where it is grown and the proximity of livestock enterprises or other end users such as power generation or field vegetables. A decision on whether to sell or incorporate it should consider practicalities such as the effectiveness of the straw chopper on the combine and contractor availability as well as a calculation of the equivalent market value of replacing the nutrients it contains. Information on the nutrient content of the different cereal straw types is available in the latest version of The AHDB Nutrient Management Guide (RB209). Reciprocal muck-for-straw deals can be of benefit to both arable and livestock businesses.

13.2 Wheat

Wheat has been cultivated for more than 10,000 years, arriving in the UK around 5000 years ago and by the turn of the 19th century, wheat was the UK's most significant crop grown for human consumption. However, this wheat was very different to the wheat varieties we are familiar with today, reaching 160 cm tall, and with great genetic diversity. These 'landrace' varieties were created by generations of natural selection and farmers saving diverse seed year after year. Over time, the landrace would become adapted to the specific soil and climate of the region, as the genotypes that do best in those environments became more prevalent. Around the mid-1800s, the first plant breeders realised that if they saved the best ears out of the landrace they got single varieties that yielded more. During the 20th century genetics became better understood and it was discovered that although imported Canadian grains did not produce a good yield in our climate, they still produced high protein grain, and the newly formed Plant Breeding Institute set about capturing this trait by crossing it with some of the British varieties and modern breeding programmes started. Indeed, over 80% of the wheat used by millers in the UK is now home grown due to improvements in wheat quality and milling technology. Interestingly, there is currently also some renewed interest in 'heritage' wheat varieties such as Red Lammas, and 'heritage' grains such as Einkorn,

Emmer and Spelt, using some of their genetics to bring in desirable traits and for their nutritional value.

Yields increased rapidly in wheat compared with the other cereals over the period from the mid 1960's to the mid 1990's (Introduction, Fig 0.3). In the UK, for example, the average wheat yield is now 8.4 t/ha, compared with 6.3 t/ha for barley. The relatively cool, damp climate and long growing season is particularly conducive to autumn planted cereals hence these are among the highest yields in the world. The initial increases in yield were largely due to improvements in varieties and husbandry techniques. Fertiliser use, particularly nitrogen and application of crop protection chemicals for the control of pests, diseases and weeds were also important in improving yields. Now, with a number of years of less predictable weather conditions and increases in pesticide resistance, yield increases are relatively small despite the continuous introduction of new varieties.

Currently, nearly half of the cereal area in Europe is wheat. France, Germany and the UK are the major producers. In the UK, 1.5 – 2 million ha are grown annually (Intoduction, Fig 0.1) and it is the most widely grown arable crop.

Both winter and spring sown varieties of wheat are available. Winter wheat normally yields much better than spring wheat, but the latter is usually of very good milling and baking quality and input costs are lower. The area of spring wheat often increases when poor planting conditions in the autumn prevent the planned area of winter wheat from being sown. However, there is also a longer-term trend with the area of spring wheat increasing due to the need to include more spring crops in rotations for weed, pest and disease management purposes.

13.2.1 Markets and quality

The main markets are for human consumption (milling for flour, starch for sweeteners and distilling), animal feed, export and bioethanol production.

The endosperm is the starchy inner portion of the wheat kernel which is milled into flour. When milled, the endosperm fractures along the cell walls, and separates from the bran layers. The ease of separating the endosperm from the bran layers is determined by the wheat's characteristics. Hard wheats (used for breadmaking) will allow easy separation, while the endosperm of soft wheats (used for biscuit flours) does not separate as easily from the bran layers.

Wheat varieties are classified by the UK Flour Millers organisation into their suitability for the different markets. Group 1 varieties are hard endosperm varieties which exhibit consistent milling and baking qualities, attracting the highest premium over feed wheat providing they meet the required specification. Group 2 varieties are those with bread-making potential but are less consistent or desirable in the required traits compared to group 1 varieties and this is reflected in a lower premium over feed wheat. Group 3 varieties are soft endosperm and are used for biscuits and cakes. Group 4 comprises of both hard and soft endosperm varieties which are suited mainly to animal feed or starch and bioethanol production. The feed price is generally the lowest price in the market.

13.2.1.1 Milling wheat

Approximately six million tonnes of wheat a year are processed into flour in the UK. Wheat flour provides 20% of the energy and protein consumed by the UK population. Between 75% and 85% of this demand is met by the home-grown crop, the exact percentage being dependent on the quality of the domestic crop each year. Imported wheat, generally from Germany, Canada, France and the USA makes up for any shortfall in quality requirements.

All milling wheat should satisfy the following general standards:

- Be free of pest infestation, discoloured grains, objectionable smells, ergot and other injurious materials.
- Not overheated during drying or storage.
- Moisture content of 15% or less.
- Maximum impurities less than 2% by weight.
- Pesticide residues within limits prescribed by legislation. A 'passport' describing pesticide application and vehicle cleanliness must accompany all loads taken to the mill.
- Specific weight – usually at least 76 kg/hl required.
- Mycotoxins - before harvest some fungi present on grain can produce mycotoxins, e.g., *Fusarium* spp. which can have negative health impacts if consumed at high levels. Toxicity of the mycotoxins varies with fungal species. There are now legal limits for the *Fusarium* mycotoxins deoxynivalenol (DON) and zearalenone (ZON) in unprocessed grain, flour and finished products as shown in Table 13.3 below. Wet conditions during flowering and wet, delayed harvests tend to increase the amount of mycotoxins present.

Specific tests for bread making:

- Hagberg Falling number (HFN) – a measure of the activity of the enzyme alpha amylase which breaks down the grain's starchy endosperm into sugars during natural sprouting processes. A small amount of alpha-amylase enzyme is desirable for changing some starch to sugars for feeding the yeast in bread making. However, excessive levels of alpha-amylase resulting in high levels of sugar will produce a sticky crumb and a collapsed loaf, which cannot easily be sliced. The HFN is the time in seconds for a plunger to fall through a mixture of wheat flour and water, plus 60 seconds for heating time before the plunger is dropped. A high HFN (>250) occurs when the plunger falls slowly because the mixture is thick with starch due to low levels of alpha-amylase. A low HFN (<250) occurs when starch is being converted to sugars by high enzyme activity. If grain has started germinating or sprouting before harvest (see Fig. 13.12), alpha-amylase activity will be high and consequently the HFN will be low.

Table 13.3 Legal mycotoxin limits for wheat for human consumption.

End-use	DON (ppb)	ZON (ppb)
Unprocessed wheat	1250	100
Flour	750	75
Finished products	500	50
Infant food	200	20

(A)

(B)

Figure 13.12 (a) Lodged wheat crop (b) Wheat grains sprouting in the ear due to lodging.

- *Protein content.* An indicator for gluten level which gives strength to the dough. A high protein content (>13%) is required for bread making (Table 13.4).
- *Protein quality.* The quality of the protein in the grain affects the baking characteristics (when the protein in wheat flour is mixed with water it produces gluten). Protein quality can be assessed by several tests, Zeleny index, SDS sedimentation, gluten washing index or the Chopin Alveograph test. Tests undertaken will depend on the target market. In the Zeleny test white flour is mixed with lactic acid and isopropyl alcohol and the resulting sedimentation is measured, the greater the amount of sediment the better the quality. The SDS test is similar but faster. Another method is to wash the gluten out of the ground grain and assess it for colour, elasticity and toughness (strength). For the export milling market the UK uses the Chopin Alveograph method. In this test the flour is assessed for baking strength. Strong flour for bread making will have a high strength (W value > 200). Weak flour is required for biscuit making with a W value < 100. In this test air is pumped into a sample of milled grain made into dough and the amount of elasticity (L value) and dough resistance to pressure (*P* value) are recorded. For bread making, the flour needs to produce a

Table 13.4 Summary of quality standards for milling wheat.

Quality criteria	Milling wheat
HFN	>250
Protein content	>13%
Specific weight	>76 kg/hl

strong elastic dough with a P/L ratio of 0.5–0.9. For biscuit making, the flour needs to form an extensible dough with a P/L ratio of 0.2–0.4. If grain is overheated during drying (more than 60 °C) then the protein can be spoiled (denatured) and not be suitable for bread making.

Grain close to these standards may be accepted with price deductions in years where there is under-supply.

13.2.1.2 Animal feed

Both hard and soft endosperm wheat varieties can be used for animal feed. The specification for feed wheat is less stringent than that for milling wheat with a specific weight of 72 kg/hl, a moisture content of below 15% and a maximum admixture of 2%. Guidance, rather than legal limits apply for mycotoxins in the case of feed wheat as shown below in Table 13.5.

13.2.1.3 Export

Export provides a potential market for growers located within an 80-mile radius of a port. Exports range from 0.5 – 3 million tonnes per year depending on domestic production levels. Approximately 75% of these exports are to EU countries. AHDB Exports have produced two brands to aid export of UK wheat; ukp is premium semi-hard wheat for bread making and uks is soft wheat for a variety of markets. These classifications are indicated on the AHDB recommended list and this classification system also means that international buyers do not have to have detailed knowledge of UK varieties. Grain is analysed using the Chopin Alveograph Test when selling into this market as it gives a good indication of end use.

Table 13.5 Guidance mycotoxin limits for wheat for animal feed.

End-use	DON (ppb)	ZON (ppb)
Feed grains	8000	2000
Complete feedstuffs for pigs	900	250
Complete feedstuffs for piglets and gilts	900	100
Complete feedstuffs for calves, lambs and kids	2000	500

13.2.1.4 Bioethanol

Standard petrol in the UK has recently been mandated to contain 10% renewable ethanol as part of the UK's obligation to reduce carbon dioxide emissions. This potentially increases the demand for domestic production of bioethanol from wheat after production plants were 'mothballed' a few years ago when the use of domestic wheat could not compete with imported bioethanol. Soft wheat types will give higher ethanol yields as it is derived from starch. Animal feed is produced as a by-product of processing for bioethanol.

13.2.2 Wheat agronomy

13.2.2.1 Site selection

Wheat is a deep-rooted plant that grows well on heavy soils and in the drier eastern and southern parts of the country. Winter wheat will withstand most frosty conditions, but it can be killed by waterlogged soils. The soil pH should be above 6.0.

13.2.2.2 Rotation

The most profitable position in the rotation is as a first wheat following a break crop such as oilseed rape, peas, beans, a grass ley or root crops. Subsequent second and third wheats suffer yield penalties due to soil and trash-borne diseases such as take-all (*Gaeumannomyces graminis*) and eyespot (*Oculimacula acuformis* and *Oculimacula yallundae*). In some situations, wheat is grown continuously and yields will increase again from the lowest yielding second and third wheat as the populations of antagonists to take-all build up and 'take-all decline' occurs but will not match those of a first wheat crop. Increasing problems with herbicide resistant grass weeds and the need to diversify rotations as part of disease and pest management strategies make continuous wheat growing less viable.

13.2.2.3 Varieties

Once a market is decided upon detailed variety information is available in the AHDB Recommended Lists updated annually, available for both winter and spring varieties. The preference of local millers should be considered if growing a milling variety. Yield, quality and agronomic characteristics can be compared using the recommended list.

Varieties are rated from 0 (weak) to 9 (good) for their resistance to a number of foliar diseases and disease pressures will differ in relation to geographical location. Both fungicide treated and untreated yields are presented which can also help inform decisions as it will influence what fungicides need to be applied, what rates are required and therefore cost of the fungicide programme that will be needed. Other agronomic considerations will include resistance to lodging, particularly for early sown crops and very fertile or exposed sites. Resistance is also available to orange wheat blossom midge in the majority of varieties. There is also a recently introduced variety with resistance to barley yellow dwarf virus.

Other considerations may include whether it is a first or second wheat in the rotation and whether it exhibits any yield advantage on particular sowing dates or soil types, the data for which is available.

13.2.2.4 Seed treatment

A range of fungicide seed treatments (often termed single-purpose dressings) are routinely used to control seed-borne diseases such as bunt, loose smut and seedling blights. There is also a fungicide seed treatment that can be co-applied with a single purpose treatment in a second wheat situation to reduce the impact of take-all. An insecticide seed treatment is currently available for the suppression of wireworm and wheat bulb fly for use in specific situations where those pests pose a threat. On deficient sites, readily available manganese can also be applied as a seed treatment.

There is increasing interest in phosphite and nutritional seed treatments as well as some biological treatments containing endophytes aimed at enhancing seedling growth.

13.2.2.5 Seedbed preparation

Wheat is tolerant of a relatively coarse seedbed compared to smaller seeded crops and a coarse seedbed also helps prevent 'soil-capping' in a mild, wet winter. However, it should be noted that very coarse, unconsolidated seedbeds will favour the activity of slugs. When residual herbicides are used a finer seedbed will help to optimise their activity. While ploughing and combination drilling are still commonly used, growers are again looking at increasing the amount of minimum tillage and no-till establishment techniques on suitable soil types.

Ideally, the crop should be sown at a uniform depth of 2.5–4 cm, depending on the moisture level in the soil. Row width is commonly 15–18 cm.

13.2.2.6 Time of sowing

Winter wheat can be sown from September to March. Early sowings from the middle of September can result in the highest yields but may require additional fungicide and growth regulator inputs. Delayed autumn sowing is important for second wheats, which should be sown from mid-October to reduce the risk of take-all. Delaying drilling is also a useful weed management tool at sites with a heavy grass weed burden to allow time for some stale seedbed cultivations to control the first flush of the seedling weeds before the crop is sown.

For early drilling slow-developing varieties should be chosen and for later drilling faster-developing varieties should be used. An indication of suitability for early and late drilling is available on the AHDB recommended list which also shows the 'latest safe sowing date' which relates to the vernalisation requirement of different varieties. Where sowing is delayed seed rate should be increased.

Spring wheat can be sown from late autumn to April the following year.

13.2.2.7 Seed rates and plant population

Yield is determined by the number of fertile tillers (ears)/m^2 multiplied by the number of grains per ear multiplied by the individual grain weight. These components are able to compensate for one another to an extent e.g., a low tiller population is likely to produce more grains per ear, so although plant population impacts on the components of yield it does not necessarily compromise overall yield above a low threshold. Benchmarking has shown that 400–500 fertile shoots/ears per m^2 are required to maximise wheat yield. If sown in mid- September this may be achieved by sowing 150 plants m^2 which each produce a main-stem and 3 fertile tillers but if sown in November 300 plants m^2 would be needed with each plant producing just one tiller as the potential tillering period is reduced. For every 2 weeks that drilling is delayed, an extra 50 plants/m^2 are typically needed to compensate for reduced tillering potential.

The number of seeds required to be sown to achieve the target plant population will depend on seedbed conditions, the weather and time of sowing. As a guide, good sowing conditions at the optimum drilling date should give an establishment of about 85%, moderate conditions about 65%–75%, and poor conditions as low as 50%. Individual seed weight, which is commonly measured as 1000 grain weight (TGW) in wheat varies from around 40-55g and must also be included in the seed rate calculation below.

$$\text{Seed rate (kg / ha)} = \frac{\text{target plant population}/\text{m}^2 * \text{TGW}}{\%\ \text{establishment expected}}$$

Spring wheat has less time to produce tillers and as a consequence is established at higher plant populations of 400–500 plants/m^2.

Caution should be exercised in using very low seed rates where there is a high grass weed burden as increasing the competitiveness of the crop provides a useful addition to an integrated weed management strategy. However, very high seed rates may be associated with an increased lodging and disease risk.

Variable seed rate technology is also now available and can be used to even out in-field establishment variations caused by factors such as different soil types, topography, pest/weed pressure and seed-bed characteristics.

13.2.2.8 Crop nutrition

Phosphate and potash recommendations for wheat are shown in Tables 13.6 and 13.7. Recommendations differ depending on whether straw is incorporated (Table 13.6) or removed (Table 13.7). These are based on winter wheat yields of 8 t/ha and spring wheat yields of 6 t/ha with straw yield estimated to be 50% of grain yield and should be adjusted using the current AHDB Nutrient Management Guide (RB209) if expected yields are larger or smaller than this. At Index 2, phosphate and potash can be applied when convenient during the year, but at Index 0 and 1, they should be applied and worked into the seedbed.

Table 13.6 Phosphate and potash recommendations (kg/ha) for wheat – straw ploughed in/incorporated.

	Soil P or K index			
	0	**1**	**2**	**3 or higher**
Winter wheat				
Phosphate Potash	110 105	80 75	50 45 (2−) 20 (2+)	0 0
Spring wheat				
Phosphate Potash	110 95	80 65	50 35 (2−) 0 (2+)	0

Table 13.7 Phosphate and potash recommendations (kg/ha) for wheat – straw removed.

	Soil P or K index			
	0	**1**	**2**	**3 or higher**
Winter wheat				
Phosphate Potash	115 145	85 115	55 85 (2−) 55 (2+)	0 0
Spring wheat				
Phosphate Potash	110 130	80 100	50 70 (2−) 40 (2+)	0 0

These recommendations have been based on AHDB Nutrient Management Guide (RB209) Section 4 updated in January 2021 by the Agriculture and Horticulture Development Board.

Nitrogen rates for wheat, as shown in Table 13.8, depend on the Soil nitrogen supply (SNS) index and any use of organic manures which must be accounted for. Where past experience of growing wheat indicates that yields above 8 t/ha can be realistically expected, the recommended rate may be increased by 10 kg N/ha for each 0.5 t/ha additional yield, up to a maximum of 13 t/ha. Conversely, for low-yielding crops, the recommended rate can be reduced by 10 kg N/ha for each 0.5 t/ha reduction in expected yield.

Nitrogen is applied in the spring just prior to or during the stem extension phase when the crop is growing rapidly. For winter wheat where the total requirement is less than 120 kg/ha it can be applied as a single dressing by early stem extension but not before early April otherwise it risks being lost through runoff or leaching. Where the total requirement is greater than 120 kg N/ha, 40 kg N/ha can be applied early between mid-February and mid-March to second wheat crops at risk of

Table 13.8 Nitrogen recommendations (kg/ha) for wheat.

	SNS[a] index						
Soil category	**0**	**1**	**2**	**3**	**4**	**5**	**6**
Winter wheat 8 t/ha (sown up until the end of January)							
Light sand	180	150	120	90	60	0–60	0–40
Shallow Medium	280[a]	240	210	180	140	80	0–40
Medium	280[a]	220	190	160	120	60	0–40
Deep clay	250[a]	220	190	160	120	60	0–40
Deep silty	240[a]	210	170	130	100	40	0–40
Organic	–	–	–	120	80	40–80	0–40
Peat	–	–	–	–	–	0–40	
Spring wheat							
Light sand	160	130	100	70	40	0–40	0
All other mineral soils	210[a]	180	150	120	70	40	0–40
Organic				120	70	40	0–40
Peat						0–40	

These recommendations have been based on AHDB Nutrient Management Guide (RB209) Section 4 updated in January 2021 by the Agriculture and Horticulture Development Board.
[a]These N recommendations exceed the N max limit that applies within NVZs. Note the N max limit is calculated for the whole of the area of a crop type grown on farm and not for individual fields. For more details, see gov.uk/nitrate-vulnerable-zones.

take-all or very backward crops to promote tillering. This early application is unlikely to be of benefit to very well-tillered crops which may be prone to lodging and higher disease levels. The balance of the application should be applied in one or two dressings during early stem extension. Where more than 120 kg N/ha remains to be applied, half should be applied at the start of stem extension (not before April) and half at least 2 weeks later (not after early May).

For spring wheat drilled before March, nitrogen should be applied at early stem extension. For rates above 70 kg N/ha, the first 40 kg N/ha should be applied to the seedbed, except on light sand soils. On light sandy soils 40 kg N/ha should be applied at the three-leaf stage. For late-drilled crops, all the nitrogen can be applied in the seedbed, except on light sandy soils, where amounts more than 70 kg N/ha should be split, with 40 kg N/ha in the seedbed and the remainder by the three-leaf stage.

Note that these recommendations are based on a break-even ratio of 5.0 (cost of fertiliser N in £/kg divided by value of grain in £/kg), if this ratio changes significantly then rates should be adjusted using the information available in the AHDB Nutrient Management Guide. Advice is also available on using information on grain protein concentration to monitor whether rates are appropriate and refine accordingly.

Wheat intended for breadmaking where protein levels above 13% are required commonly receives an extra 40 kg/ha of nitrogen. This is most effective when applied as foliar urea at the milky ripe stage. If applied too early there is a risk that it will increase yield and as a consequence actually dilute protein levels. When applying at this growth stage in warm weather conditions, care should be taken to ensure that it does not scorch the flag leaf of the crop by applying early or late in the day in cooler conditions.

On sandy soils and in regions with high rainfall, wheat crops are likely to respond to sulphur and 20–50 kg/ha of a sulphate containing fertiliser should be applied in spring. Tissue and grain analysis can be used to confirm a deficiency. It is particularly important in wheat grown for milling where sulphur deficiency is linked to the formation of acrylamide, a processing contaminant.

The most common micronutrient deficiencies in wheat are copper, manganese and zinc which can be diagnosed using visual symptoms. Manganese is the most prevalent, causing pale, floppy younger leaves streaked with white/grey. Manganese is routinely applied, sometimes multiple times during a growing season as a foliar treatment on sandy, organic and peaty and alkaline soil types. The problem will be exacerbated by unconsolidated seedbeds and low soil temperatures.

13.2.2.9 Pests

In the autumn, slugs can be a serious problem in trashy, cloddy seedbeds especially when wheat is sown after oilseed rape in the rotation. The slugs can hollow out the seed and shred the young seedling leaves. Slug pellets should be applied after monitoring has been carried out if there is a high risk. Early-drilled wheat can be affected by aphid-transmitted barley yellow dwarf virus (BYDV). A fall in temperature as drilling is delayed will reduce aphid migration into the crop and therefore reduce the risk of infection. The optimum time for control is when the second generation of aphids spread the virus within the crop and this can be calculated using a T-sum method although control may be limited as insecticide resistance is a widespread issue in aphids. Other pest problems occur in specific rotational positions e.g., frit fly (*Oscinella frit*), leatherjackets (*Tipula paludosa*) or wireworms (*Agriotes* spp) after grass or wheat bulb fly (*Delia coarctata*) after row crops and are described in more detail in chapter 7. During grain fill, aphids and wheat blossom midges can cause direct damage to developing grains. If, after monitoring, threshold numbers are present, treatment is usually advisable but the use of insecticides is generally undesirable at this time of year when many crops and wild plants are flowering. The majority of varieties on the current AHDB Recommended list are resistant to attack by orange wheat blossom midge (*Sitodiplosis mosellana*).

13.2.2.10 Diseases

Yield and quality of wheat can be affected by stem-base, foliar and ear diseases. Of the stem base/root diseases take-all and eyespot are the most common. Take-all disease can be a serious problem when growing wheat continuously, particularly on light soils until the 'take-all barrier' has been passed. In second and third wheat rotational positions drilling should be delayed if possible and an appropriate seed treatment should be considered.

The main foliar diseases in wheat are *Septoria* leaf blotch (*Zymoseptoria tritici)* formerly and most often known as *Septoria tritici*, yellow rust (*Puccina striiformis*), brown rust (*Puccinia triticina*); leaf and glume blotch (*Parastagonospora nodorum)* and powdery mildew (*Blumeria. graminis* f. sp *tritici*).

The main ear disease of concern is Fusarium head blight (*Fusarium gramnearum)* which is exacerbated by wet conditions during flowering. Rotation, varietal resistance, cultivations and drilling date are important tools in managing these diseases but fungicide programmes are also very widely used. Fungicide programmes should combine a range of active ingredients with different modes of action offering both protectant and eradicant activity to reduce the risk of resistance occurring.

In winter wheat a T0 fungicide timing in late March/early April (growth stage 30 – start of stem extension) may sometimes be required if high amounts of septoria tritici or yellow rust inoculum are seen prior to stem extension. This is followed in mid-late April by the TI timing (growth stage 32 – second node detectable), which protects final leaf 3 and is also the best time to control eyespot. In wheat the T2 (growth stage 39 – flag leaf ligule visible) timing in mid-late May gives the largest economic response as it protects the flag leaf (final leaf 1). Finally, a T3 timing (growth stage 63–65 - anthesis) is used to protect the ear against fusarium head blight especially on milling varieties with statutory mycotoxin limits (see Table 13.3) and will also 'top up' disease control on the flag leaf. Spring wheat programmes are likely to be based around the T1 and T2 timings only.

13.2.2.11 Weeds

In winter wheat, autumn-germinating grass weeds such as black-grass (*Alopecurus myosuroides*), the bromes (*Bromus* spp*)* and ryegrass (*Lolium* spp) are most difficult to control. The use of 'cultural' controls specific to the species present are very important due to the limited range of effective herbicides now available. Robust herbicide programmes for these weeds will commonly start with the use of pre-emergence residual products containing more than one active ingredient which will be followed up again later in the autumn once the crop has emerged.

The control of annual meadow grass (*Poa annua*) and broad-leaved weeds in the autumn is generally more straight forward and a post-emergence herbicide application from when the crop is at the 2-3 leaf stage will usually be sufficient.

Spring herbicide treatments may be required to treat any weeds that evaded treatment in the autumn or any that have germinated subsequently. Individual products will specify the latest growth stage at which they can be applied but in general all weed control should be completed by growth stage 39 of the crop.

13.2.2.12 Plant growth regulators

Lodging is the term used when the crop 'falls over', and is predicted to affect up to 20% of winter wheat crops every 3–4 years. It can occur at the stem base, when it is known as 'root lodging' or slightly higher up the stem when it is known as 'stem lodging'. It usually occurs during grain filling when the ears are becoming heavier and is more likely to occur during windy and wet weather conditions. The earlier that lodging occurs during grain filling, the greater the yield loss which can be up to 75%, with quality also reduced. Drying costs of a lodged crop may also be increased. Lodging risk is affected by the lodging resistance score of the variety (rated 1-9 on the AHDB Recommended list, where nine indicates a high level of resistance to lodging), levels of soil and applied N, plant population, the exposure of the site and weather conditions.

Plant growth regulators (PGR's) are used on most winter wheat crops during the period of stem extension, to shorten the internodes and strengthen the stem walls, to mitigate against the risk of lodging occurring. Those used early in stem extension inhibit the synthesis of giberellins in the plant, shortening and strengthening the lower internodes and those used later in the programme shorten the upper internodes by blocking the transport of auxins.

13.2.2.13 Typical crop protection programme for winter wheat

Despite the range of pests, diseases and weeds listed which may affect a crop (see also Chapters 5–7), an individual crop is unlikely to face all of these problems in practice. In addition, the control of more than one problem is often carried out simultaneously by tank mixing treatments to reduce the number of 'passes' the sprayer makes through the crop. A typical crop protection programme for a winter wheat crop is shown in Table 13.9.

13.2.2.14 Harvesting

Glyphosate can be applied as a harvest aid to crops at GS 87, when grain moisture is at 30% (grains hold a thumbnail indentation) to reduce immature tillers, often found in tramlines or to eliminate perennial weeds. It can also help to protect grain quality by enabling the crop to be harvested slightly earlier than if unsprayed. The crop can be harvested 7 days after application. It should not be used in crops that are being grown for seed production.

Winter wheat ripens before spring wheat. The crop is normally harvested in August and early September.

13.2.2.15 Yield

Wheat yields have plateaued over the last 10 years and the current 5-year average for winter wheat is 8.4 t/ha. However, there is wide variation between years influenced by establishment conditions in the autumn and growing conditions in the spring (Introduction, Fig 0.3).

13.3 Durum wheat

Durum wheat is a tetraploid species of wheat (*Triticum turgidum* spp. *durum*) compared with hexaploid common or bread wheat (*Triticum aestivum*). It is widely

Table 13.9 Typical crop protection programme for winter wheat.

Crop growth stage	Indicative date	Treatment type	Target
GS 00	Sept/Oct	Herbicide (residual)	Pre-emergence grass weed control
GS 00-13	Sept/Oct	Molluscicide	Slugs
GS 12-13	Oct/Nov	Herbicide (residual + contact)	Post emergence broad leaved weeds and grass weeds[a]
		Insecticide	Aphid vectors of BYDV
GS 29/30	Late Mar-early Apr	PGR	Shortening/strengthening lower internodes
		Fungicide	Protection against foliar and stem-based diseases
		Herbicide (contact)	'Tidy up' treatment for autumn germinating weeds
GS 32	Mid-late April	PGR	Shortening/strengthening lower internodes
		Fungicide	Protection of leaf 3 against foliar diseases
		Herbicide (contact)	Treatment of spring germinating weeds
GS 39	Mid May	PGR	Shortening upper internodes if required[b]
		Fungicide	Protection of flag leaf and leaf 2 against foliar diseases
GS 65	Early June	Fungicide	'Top up' foliar disease control and control of ear diseases
GS 87	Mid-late July	Non-selective herbicide	Pre-harvest desiccation and perennial weed control

[a]In situations where high populations of black-grass, ryegrass or bromes are present this is likely to be a follow up treatment to the residual herbicide applied at GS 00.
[b]It is likely that only two out of the potential three PGR timings would be required.

grown in Canada, North America and southern Europe, with only a very small area being grown in the UK. If properly grown, durum wheat has an inherently high protein and hard, amber-coloured vitreous endosperm. On milling it breaks into fairly uniform large fragments called semolina, which is made into a range of high quality pasta products by extruding a stiff dough (semolina mixed with warm water) through dies of various shapes. This high quality market (up to a 50% premium) requires grain which is a light amber colour and meets minimum standards, i.e., 13% protein, Hagberg 280, vitreous grains 70%, specific weight 78 kg/hl, and less than 2% impurities, free of mould and sprouted grains. Some durum wheat is used for breakfast cereals. It is essential to have a contract with a company and to follow the company's management protocol.

13.3.1 Durum wheat agronomy

13.3.1.1 Soils and climate

Durum wheat will grow on a wide range of arable soils, but is best suited to medium textured soil. It tolerates dry conditions better than ordinary wheat. As it is a Mediterranean crop it tends to yield better in the drier eastern wheat growing areas of England. The problem when growing this crop is achieving the required quality and it yields poorly compared with conventional wheat.

13.3.1.2 Seedbeds and sowing

Seedbed requirements and methods of sowing are similar to those for wheat crops.

13.3.1.3 Time of sowing

As durum wheat does not require vernalisation, it can be sown in either October or March. Autumn-sown crops are susceptible to frost damage.

13.3.1.4 Varieties

Varieties all have awns on the ear and are therefore described as 'bearded'.

13.3.1.5 Seed rates

The target should be 300 plants/m^2.

13.3.1.6 Fertilisers

Recommendations as for milling wheat.

13.3.1.7 Crop protection

As for wheat. However, note that as a minor crop there are only a limited number of products that are approved for use in durum wheat, mainly as Extensions of Authorisation for Minor Use (EAMUs).

13.3.1.8 Harvesting

An autumn-sown crop is usually combined in August. Spring-sown Durum wheat is normally harvested in September. If possible, it should be given priority (about 20%–22% moisture content for combining) to obtain good vitreous grains and a high Hagberg number. If it is left too late, the endosperm is white and there is a risk of sprouting. Grain should be stored at 15% moisture content.

13.3.1.9 Yield

Durum is not as high yielding as conventional wheat, averaging about 20% less. The range is 3–7 t/ha.

13.4 Barley

After wheat, barley is the next most important cereal crop grown in Europe. The Russian Federation is the largest producer globally but about a third of the world trade in barley is from Europe; France, Germany and Spain are the main producers followed by the UK. The UK barley area has increased over the last decade with approximately 1.2−1.5 million ha of barley now being grown with the balance now being in favour of spring barley. This is partly in response to autumn planting conditions over the last few years but also reflects a general move to include more spring crops in the rotation as part of integrated weed management strategies.

13.4.1 Markets and quality

The main markets are for animal feed and malting or distilling.

13.4.1.1 Animal feed

Barley is a palatable feed with high levels of energy and starch. It is commonly fed to pigs, poultry and fish (ground), cows (as a supplement to forage) and also used in intensive cereal beef systems (rolled). Much of this will be home produced and fed but specifications for feed barley when sold are usually based on 15% moisture and a specific weight of 63 kg/hl Some feed barley is also exported, predominantly to the Middle East and North Africa.

13.4.1.2 Malting

The UK has the third largest malting industry in the world and over a third of barley sold in the UK goes for malting/distilling. The Malting Barley Committee (MBC) of the MAGB (Maltsters' Association of Great Britain) carries out malting, brewing and distilling tests on new varieties to assess their suitability for these markets, producing an approved list each year. The British Beer and Pub Association (BBPA) and the MAGB also subject plant protection products to a testing regime to check for any adverse effects on malt or beer quality before they are accepted for use on malting barley in the UK. The results of these tests are also publicised to growers.

Germination is critical for maltsters, with a minimum 98% viability being the standard. Next, meeting the exacting specifications for nitrogen laid down by brewer and distiller customers is critical. Malt for brewing requires a grain N range from 1.65% to 1.85% while distilling generally requires a grain N below 1.65%. A premium, which varies from year to year, is paid for good malting barley. Some malting barley is also exported.

Malt is also used in breakfast cereals and bakery products.

In the malting process the grain is soaked (steeped), and then the grain is allowed to sprout. During this germination period the starchy endosperm is converted to sugars. Once germinated, the grain is dried in a kiln at high temperatures and finally the rootlets (culms) removed to leave the sweet malt grains. In the brewing process the malt is crushed and then soaked in warm water in mash tuns. The sugars then dissolve in the liquor before being drained off. The remainder of the malt (brewer's grains) is used as cattle food. The sugary liquor (wort) is boiled with hops to give it a bitter flavour and keeping quality, and to destroy the enzymes. The strained wort is then fermented with yeast which converts the sugars to alcohol and so produces beer. To make whisky, hops are not added; the fermented wort is distilled to produce a more concentrated alcoholic liquid (malt whisky).

13.4.2 Barley agronomy

13.4.2.1 Site selection

Barley can be grown on arable land throughout the UK, provided the pH of the soil is about 6.5; it is more affected by a low pH than wheat. Barley grows better than other cereals on thin chalk and limestone soils. However, it will grow on a wide range of soils provided they are well drained. On organic and very fertile soils, it may lodge, especially in a wet season, and it is more difficult to achieve malting quality.

13.4.2.2 Rotation

In rotations it is a popular alternative for second wheat crops in areas where there is demand for grain for feed or malting. As an early harvested crop, winter barley is commonly grown before winter oilseed rape, allowing more timely establishment of the oilseed rape than when sowing after wheat. The increase in the spring barley area over the last few years is due to the need to include more spring crops in rotations and the fact that the agronomy of spring barley is well understood and it yields reliably compared to alternative spring sown crops.

13.4.2.3 Varieties

Once a market is decided upon detailed variety information is available in the AHDB Recommended Lists updated annually, available for both winter and spring varieties. The recommended list indicates all varieties that are approved by the MBC but the preference of local maltsters should also be considered if growing a malting variety. The majority of MBC approved varieties are spring sown as they are usually more likely to meet the required N specification.

If growing winter barley for feed, then both six row and two row (conventional) varieties are available (Figs. 13.3 and 13.4). Nearly all six row varieties are hybrids, bred using male and female parents to produce a first-generation hybrid seed which now accounts for one third of the feed market. Six row hybrids offer vigorous growth which helps suppress grass weeds, gives high yields and produces a rooting advantage. Hybrid varieties are earlier maturing than most conventional two row varieties, which makes them a useful entry for winter OSR. Hybrid barleys have longer straw than conventional varieties requiring robust PGR programmes but have generally good disease resistance. Historically, the hybrid varieties suffered with low specific weights but this is no longer the case. The seed from hybrid varieties cannot be farm saved as it eventually reverts to its parentage.

Yield and agronomic characteristics can also be compared using the recommended list. Varieties are rated for their resistance to a number of foliar diseases and disease pressures will differ in relation to geographical location. Both fungicide treated and untreated yields are presented in AHDB recommended lists which can also help inform decisions as it will influence what fungicides need to be applied, what rates are required and therefore cost of the fungicide programme that will be needed. Other agronomic considerations will include resistance to brackling and lodging. Brackling is the buckling of the stem at a node higher up the stem than the point at which stem base lodging occurs. It occurs as the straw is drying out before the crop ripens and is ready to harvest, resulting in losses during harvest as the lowest hanging heads are cut off and left on the ground.

13.4.2.4 Seed treatment

A range of fungicide seed treatments (often termed single-purpose dressings) are routinely used to control seed-borne diseases such as leaf stripe (*Pyrenophora graminea*), loose smut (*Ustilago nuda*) and seedling blights in barley. On deficient sites, readily available manganese can also be applied as a seed treatment.

There is increasing interest in phosphite and nutritional seed treatments as well as some biological treatments containing endophytes aimed at enhancing early growth.

13.4.2.5 Seedbed preparation

Barley is tolerant of a relatively coarse seedbed compared to smaller seeded crops and a coarse seedbed also helps prevent 'soil-capping'. However, it should be noted that very coarse, unconsolidated seedbeds will favour the activity of slugs in autumn sown crops in particular. When residual herbicides are used a finer seedbed will help to optimise their activity. While ploughing and combination drilling are still commonly used, growers are again looking at increasing the amount of minimum tillage and no-till establishment techniques on suitable soil types. One consideration when min-till techniques are being used to establish barley as a second cereal after wheat is the difficulty in removing volunteer wheat from a growing barley crop.

Fields destined for spring barley can be ploughed over winter, allowing 'one-pass' establishment methods in spring, thus conserving moisture. Spring barley also facilitates the use of cover cropping over winter.

Ideally, the crop should be sown at a uniform depth of 2.5–4 cm, depending on the moisture level in the soil. Row width is commonly 15–18 cm.

13.4.2.6 Time of sowing

The window for sowing winter barley is narrower than for winter wheat. A low vernalisation requirement means barley is less suited to very early sowing, while yield declines faster when it is sown after mid-November. The optimum window is from mid-September to early November. For the six row hybrids the optimal drilling window is mid-September to the end of October. Crops that are too advanced can be susceptible to frost damage as well as higher levels of foliar disease and BYDV.

Spring barley can be sown from January to early April. A good seedbed is more important than early drilling. Yields are usually reduced from drilling after the end of March.

13.4.2.7 Seed rates and plant population

Barley is not able to compensate for a low seed rate later in its growth cycle by producing more grains per ear to the same extent as wheat. The ears are limited to the production of either one (2-row) or 3 (6 row) grains per spikelet whereas wheat can respond to the level of assimilate supply during ear development and increase the number of grains per spikelet. Therefore, achieving the optimum number of ears/m^2 is more critical than in wheat. The target population in spring should be 300 plants/m^2. For hybrid 6-row varieties the target is reduced to 180 plants/m^2, excessively high populations will potentially result in decreased specific weight.

The number of seeds required to be sown to achieve the target plant population will depend on seedbed conditions, the weather and time of sowing. As a guide, good sowing conditions at the optimum drilling date should give an establishment of about 85%, moderate conditions about 65%–75%, and poor conditions as low as 50%. Thousand grain weight (TGW) in barley varies from around 45-60g and must also be included in the seed rate calculation below.

$$\text{Seed rate (kg/ha)} = \frac{\text{target plant population/m}^2 * \text{TGW}}{\text{\% establishment expected}}$$

13.4.2.8 Crop nutrition

Phosphate and potash recommendations for barley are shown in Tables 13.10 and 13.11. Recommendations differ depending on whether straw is incorporated

Table 13.10 Phosphate and potash recommendations (kg/ha) for barley – straw ploughed in/incorporated.

	Soil P or K index			
	0	**1**	**2**	**3 or higher**
Winter barley				
Phosphate	110	80	50	0
Potash	95	65	35 (2–) 0 (2+)	0
Spring barley				
Phosphate	105	75	45	0
Potash	90	60	30 (2–) 0 (2+)	0

Table 13.11 Phosphate and potash recommendations (kg/ha) for barley – straw removed.

	Soil P or K index			
	0	**1**	**2**	**3 or higher**
Winter barley				
Phosphate	115	85	55	0
Potash	130	100	70 (2–) 40 (2+)	0
Spring barley				
Phosphate	105	75	45	0
Potash	125	95	65 (2–) 35 (2+)	0

These recommendations have been based on AHDB Nutrient Management Guide (RB209) Section 4 updated in January 2021 by the Agriculture and Horticulture Development Board.

(Table 13.10) or removed (Table 13.11). These are based on winter barley yields of 6.5 t/ha and spring barley yields of 5.5 t/ha with straw yield estimated to be 50% of grain yield and should be adjusted using the current AHDB Nutrient Management Guide (RB209) if expected yields are larger or smaller than this. At Index 2, phosphate and potash can be applied when convenient during the year, but at Index 0 and 1, they should be applied and worked into the seedbed.

Nitrogen fertiliser recommendations for barley are lower than for wheat. If growing crops for malting where a low grain nitrogen is required then less N is applied than on a crop grown for feed.

Nitrogen rates for barley, as shown in Tables 13.12 and 13.13, depend on the SNS index and any use of organic manures which must be accounted for. Where past experience of growing winter barley indicates that yields above 6.5 t/ha can be realistically expected, the recommended rate may be increased by 10 kg N/ha for each 0.5 t/ha additional yield, up to a maximum of 11 t/ha. For spring barley where yields above 5.5 t/ha can realistically be expected, the recommended rate may be increased by 10 kg/ha for each 0.5 t/ha additional yield, up to a maximum of 9 t/ha. Conversely, for low-yielding crops, the recommended rate can be reduced by 10 kg N/ha for each 0.5 t/ha reduction in expected yield.

For winter barley nitrogen is applied in the spring just prior to or during the stem extension phase when the crop is growing rapidly. For winter barley where the total requirement is less than 100 kg/ha it can be applied as a single dressing by early stem extension. Where the total requirement is greater than 100–200 kg/ha, half can be applied during late tillering, between mid-February and early-March and half at early stem extension (GS30-31). Where the recommended rate is above 200 kg/ha it should be split three ways between late tillering (40%), GS30-31 (40%) and GS32 (20%).

Table 13.12 Nitrogen recommendations (kg/ha) for winter barley.

	SNS[a] index						
Soil category	**0**	**1**	**2**	**3**	**4**	**5**	**6**
Feed barley							
Light sand	170	140	110	80	60	0–40	0
Shallow Medium	220[a]	190	150	120	60	20–60	0–20
Medium and deep clay	190[a]	170	140	110	60	20–60	0–20
Deep fertile silty	170	150	120	80	40	0–30	0
Organic	–	–	–	110	60	0–40	0
Peaty	–	–	–	–	–	0–40	
Malting barley							
Light sand	130	90	60	0–40	0	0	0
All other mineral soils	150	120	90	50	0–40	0	0
Organic	–	–	–	50	0–40	0	0
Peat	–	–	–	–	–	0	

[a]These N recommendations exceed the N max limit that applies within NVZs. Note the N max limit is calculated for the whole of the area of a crop type grown on farm and not for individual fields. For more details, see gov.uk/nitrate-vulnerable-zones.

Table 13.13 Nitrogen recommendations (kg/ha) for spring barley.

	SNS[a] index						
Soil category	**0**	**1**	**2**	**3**	**4**	**5**	**6**
Feed barley							
Light sand	140	110	70	50	0–40	0	0
All other mineral soils	160[a]	140	110	70	30	0–30	0
Organic	–	–	–	70	30	0–30	0
Peat	–	–	–	–	–	0–30	
Malting barley (1.8% grain N)							
Light sand	110	80	40	0–40	0	0	0
All other mineral soils	130	110	70	40	0–30	0	0
Organic	–	–	–	40	0–30	0	0
Peat	–	–	–	–	–	0	

These recommendations have been based on AHDB Nutrient Management Guide (RB209) Section 4 updated in January 2021 by the Agriculture and Horticulture Development Board.
[a]These N recommendations exceed the N max limit that applies within NVZs. Note the N max limit is calculated for the whole of the area of a crop type grown on farm and not for individual fields. For more details, see gov.uk/nitrate-vulnerable-zones.

For spring barley crops drilled before March, nitrogen should be applied at early stem extension. For amounts greater than 70 kg N/ha, the first 40 kg N/ha of the total can be applied in the seedbed, except on light sandy soils. On light sandy soils

40 kg N/ha should be applied at the three-leaf stage. For late-drilled crops, all the nitrogen can be applied in the seedbed, except on light sandy soils, where amounts more than 70 kg N/ha should be split, with 40 kg N/ha in the seedbed and the remainder by the three-leaf stage.

In the case of both winter and spring crops grown for malting the tables are based on a target grain N of 1.8%. Where the target specification is lower than this, N rates from the malting sections of the tables should be reduced further based on the detailed guidance in Section 4 of the AHDB Nutrient Management Guide. In the case of winter barley grown for malting, all N should be applied by GS30-31 and for spring malting barley all N should be applied by the end of March.

Note that these recommendations are based on a break-even ratio of 5.0 (cost of fertiliser N in £/kg divided by value of grain in £/kg), if this ratio changes significantly then rates should be adjusted using the information available in the AHDB Nutrient Management Guide.

The most common micronutrient deficiencies in barley are copper, manganese and zinc which can be diagnosed using visual symptoms. Manganese is the most prevalent, causing pale, floppy younger leaves streaked with white/grey. Manganese is routinely applied, sometimes multiple times during a growing season as a foliar treatment on sandy, organic and peaty and alkaline soil types. The problem will be exacerbated by unconsolidated seedbeds and low soil temperatures.

13.4.2.9 Pests

In the autumn, slugs can be a problem, particularly in trashy, cloddy seedbeds although less so than in wheat because barley does not usually follow OSR in the rotation. The slugs typically will not hollow out the seed as they do in wheat as the husk provides some protection but they will shred the young seedling leaves. Slug pellets should be applied after monitoring has been carried out if there is a high risk. Early-drilled winter barley can be affected by aphid-transmitted barley yellow dwarf virus (BYDV). A fall in temperature as drilling is delayed will reduce aphid migration into the crop and therefore reduce the risk of infection. The optimum time for control is when the second generation of aphids spread the virus within the crop and this can be calculated using a T-sum method although control may be limited as insecticide resistance is a widespread issue in aphids.

13.4.2.10 Diseases

The seed-borne diseases such as leaf stripe, *Fusarium*, and loose smut are largely controlled by the use of certified seed and seed treatments. The mosaic viruses (barley yellow mosaic virus (BaYMV) and barley mild mosaic virus (BaMMV)) can be a problem where winter barley has been grown too frequently. The only means of control is rotation and use of resistant varieties.

The main foliar diseases in both winter and spring barley are rhynchosporium (*Rhynchosporium secalis*), powdery mildew (*Blumeria graminis* f. sp. *hordei*), brown rust (*Puccinia hordei*), net blotch (*Pyrenophora teres*) and Ramularia leaf spot (*Ramularia collo-cygni*). Rotation, varietal resistance, cultivations and drilling date are important tools in managing these diseases but fungicide programmes are also very widely used. Fungicide programmes should combine a range of active ingredients with different modes of action offering both protectant and eradicant activity to reduce the risk of resistance occurring. In winter barley leaves 2 and 3 contribute more to yield than in wheat and therefore, it is the T1 (GS 30−32 – start of stem extension to second node detectable) fungicide timing in winter barley that gives the greatest yield response compared with the T2 (GS39 – flag leaf ligule just visible) application. Occasionally a pre-T1 treatment is required if there is early infection. The T3 (GS 59 – ear fully emerged) treatment gives the lowest response in barley and is only necessary if the T2 treatment was inadequate or not applied. In spring barley, the T2 timing gives the largest yield response compared with the T1 timing at GS 26−31. Normal practice in barley is for at least a two spray programme at T1 and T2. For effective control and to reduce the risk of resistance, fungicides with different modes of action should be used.

13.4.2.11 Weeds

Autumn herbicide treatment is preferred in winter barley for both annual broad-leaved and grass weeds, using approved herbicides. However, there are very limited active ingredients available for the control of some grass weeds in winter barley and serious consideration should be given as to whether the crop should be grown where resistant black-grass in particular is present. There is evidence to suggest that the vigorous canopy growth of the six row hybrids is better able to suppress grass weeds than conventional varieties. In contrast, spring barley offers a very good opportunity to control these difficult autumn germinating grass weeds before drilling using stale seedbed techniques.

The control of annual meadow grass and broad-leaved weeds in the autumn is generally more straight forward and a post-emergence herbicide application from when the crop is at the 2-3 leaf stage will usually be sufficient. Spring herbicide treatments may be required to treat any weeds that evaded treatment in the autumn or any that have germinated subsequently. Individual products will specify the latest growth stage at which they can be applied but in general all weed control should be completed by growth stage 39.

In spring barley where it is being used as part of an integrated strategy to control black-grass then a pre-emergence residual herbicide may be used to 'mop up' any black-grass that has evaded the stale seedbed strategy. In the absence of black-grass, spring barley will often only require a broad-leaved weed herbicide plus possibly a wild oat herbicide.

13.4.2.12 Plant growth regulators

Barley can be susceptible to lodging. Stiff-strawed varieties should be chosen. To control lodging in winter barley, one or two spray PGR programmes are commonly used. A three spray PGR programme is recommended on the hybrid varieties.

The highest risk of lodging in spring barley is on thick crops growing in fertile soil, especially if there are wet and windy weather conditions after flowering.

13.4.2.13 Typical crop protection programme for barley

Despite the range of pests, diseases and weeds listed which may affect a crop (see also Chapters 5–7), an individual crop is unlikely to face all of these problems in practice. In addition, the control of more than one problem is often carried out simultaneously by tank mixing treatments to reduce the number of 'passes' the sprayer makes through the crop. A typical crop protection programme for a winter barley crop is shown in Table 13.14. In spring barley, a similar programme is likely to start with a herbicide during the tillering stages in March/April, depending on when the crop was sown, except in cases where grass weeds are present and a pre-emergence herbicide may be used. The rest of the programme will be broadly similar.

Table 13.14 Typical crop protection programme for winter barley.

Crop growth stage	Indicative date	Treatment type	Target
GS 00	Sept/Oct	Herbicide (residual)	Pre-emergence grass weed control
GS 00-13	Sept/Oct	Molluscicide	Slugs
GS 12-13	Oct/Nov	Herbicide (residual + contact)	Post emergence broad leaved weeds and grass weeds*
		Insecticide	Aphid vectors of BYDV
GS 29/30	Late Mar-early Apr	PGR	Shortening/strengthening lower internodes
		Herbicide (contact)	'Tidy up' treatment for autumn germinating weeds
GS 31	Mid-April	PGR	Shortening/strengthening lower internodes
		Fungicide	Protection of upper leaves against foliar diseases
		Herbicide (contact)	Treatment of spring germinating weeds
GS 39	Mid May	PGR	Shortening upper internodes if required
		Fungicide	Protection of flag leaf and leaf 2 against foliar diseases

13.4.2.14 Harvesting

Winter barley is ready for combining from mid-July to early August. Spring barley combining normally starts between mid-August and early September. In a crop containing late tillers, harvesting should start when most of the crop is ready.

13.4.2.15 Yield

Yields of barley can be more variable than yields of winter wheat. Spring barley grain yields are usually about 1 t/ha lower than those for winter barley. The 5 year average winter barley yield for the UK is 6.8 t/ha and the 5 year average spring barley yield is 5.6 t/ha.

13.5 Oats

Oats are the UK's third largest home grown cereal and the area of oats has risen significantly over recent years, fuelled by demand for human consumption and most recently by demand in this sector for innovative products such as 'plant-based' milk. They require less inputs than wheat and barley and can fit well in a 'second cereal' position in the rotation. Both winter and spring sown types are available, the spring sown varieties fulfilling the need for more spring crops in rotations.

The best quality oats are used for bread making, oatcakes, porridge and breakfast foods. Oats for human consumption should have a high specific weight of 50 kg/hl, a moisture content of between 12% and 15%, a high kernel content of 75%–78% and less than 6% screenings through a 2 mm sieve. Non-food consumption of oats includes use in cosmetics and shampoos. A small amount of UK oats are currently exported, predominantly into the EU but they are unlikely to be able to compete post-Brexit so this market may decline.

As an animal feed, oats are particularly good for horses and are also valuable for cattle and sheep. However, they are not very suitable for pigs because of their high husk (fibre) content.

Naked oats are varieties that naturally thresh free from the husk during harvest with improved energy density and digestibility. They can be used as a whole grain, for flaking, malting and in animal feeds including pigs and poultry.

The majority of oat straw is baled and used for bedding/feed. Quality of the straw is usually better than for wheat.

13.5.1 Oat agronomy

13.5.1.1 Site selection

Oats will grow on most types of soil and can perform well on the lighter soil types as they have scavenging roots. Grass weed management in the crop is a particular challenge, as there are very few effective herbicides approved. As a result, fields with significant black-grass, ryegrass, brome or wild oat populations should be avoided.

13.5.1.2 Rotation

Oats can be grown as a first cereal but are commonly used in a 'second cereal' position in the rotation as they are not affected by take-all and are often a lower input crop than other cereals.

13.5.1.3 Varieties

Detailed variety information is available from the AHDB Recommended Lists updated annually, available for both winter and spring varieties. The preference of local millers should also be considered. Yield and agronomic characteristics can also be compared using the recommended list. Varieties are rated for their resistance to crown rust (*Puccinia coronata)* and powdery mildew (*Blumeria graminis* f.sp. *avenae*) and disease pressures will differ in relation to geographical location. Both fungicide treated and untreated yields are presented which can also help inform decisions as it will influence how robust or costly a fungicide programme will be needed. Other agronomic considerations will include resistance to lodging.

13.5.1.4 Seed treatments

A range of fungicide seed treatments (often termed single-purpose dressings) are routinely used to control loose smut and seedling blights.

13.5.1.5 Seedbed preparation

Similar to the requirements of wheat and barley described earlier in this chapter.

13.5.1.6 Time of sowing

The optimum period for sowing winter oats is from mid-Sept to mid-Oct but they can be sown into November. Spring oats should be sown from late February until early April but can also be sown from November onwards if conditions allow, enabling the crop to produce a deeper rooting structure to sustain it during a dry spring.

13.5.1.7 Seed rates and plant population

The target plant population is 250 plants/m^2 for winter oats and 300 plants/m^2 for spring oats. Much like wheat, oats have the potential to compensate for low plant population through increased tillering and the development of more grains per panicle.

The number of seeds required to be sown to achieve the target plant population will depend on seedbed conditions, the weather and time of sowing. The same calculation should be used as for seed rates in wheat and barley. Seed rates are lower for naked oat varieties.

13.5.1.8 Crop nutrition

Phosphate and potash recommendations for oats are shown in Tables 13.15 and 13.16. Recommendations differ depending on whether straw is incorporated or removed.

Table 13.15 Phosphate and potash recommendations (kg/ha) for oats (winter and spring) – straw incorporated.

	Soil P or K index			
	0	**1**	**2**	**3 or higher**
Phosphate	110	80	50	0
Potash	95	65	35 (2–) 0 (2+)	0

Table 13.16 Phosphate and potash recommendations (kg/ha) for oats (winter and spring) – straw removed.

	Soil P or K index			
	0	**1**	**2**	**3 or higher**
Phosphate	115	85	55	0
Potash	160	130	100 (2–) 70 (2+)	0

These recommendations have been based on AHDB Nutrient Management Guide (RB209) Section 4 updated in January 2021 by the Agriculture and Horticulture Development Board.

These are based on yields of 6.0 t/ha and should be adjusted using the current AHDB Nutrient Management Guide (RB209) if expected yields are larger or smaller than this. At Index 2, phosphate and potash can be applied when convenient during the year, but at Index 0 and 1, they should be applied and worked into the seedbed.

Nitrogen fertiliser recommendations for oats are lower than for wheat or barley, although they have been increased in recent years as breeding has reduced straw height, enabling more N to be applied without increasing the risk of lodging.

Nitrogen rates for oats, as shown in Table 13.17, depend on the SNS index and any use of organic manures which must be accounted for.

For winter oats nitrogen is applied in the spring just prior to or during the stem extension phase when the crop is growing rapidly. For winter oats where the total requirement is less than 100 kg/ha it can be applied as a single dressing by early stem extension. Where the total requirement is greater than 100 kg/ha, 40 kg/ha can be applied during late tillering, between mid-February and early-March and the remainder at early stem extension (GS30-31). If the remaining N is 100 kg N/ha or more, then it should be applied in two dressings, half at early stem extension (not before late March) and half at least 2 weeks later. These recommendations assume appropriate measures are taken to control lodging. Amounts should be reduced by 40 kg N/ha for oats if lodging risk is high. The timings for spring oats are the same as those for spring barley.

Table 13.17 Nitrogen recommendations (kg/ha) for oats.

Soil category	SNS index 0	1	2	3	4	5	6
Winter oats							
Light sand	150	110	80	20–60	0–40	0	0
All other mineral soils	190	160	130	100	70	0–40	0
Organic	–	–	–	100	70	0–40	0
Peat	–	–	–	–	–	0–40	
Spring oats							
Light sand	90	60	30	0–30	0	0	0
All other mineral soils	140	110	70	40	0–30	0	0
Organic				40	0–30	0	0
Peat						0	

These recommendations have been based on AHDB Nutrient Management Guide (RB209) Section 4 updated in January 2021 by the Agriculture and Horticulture Development Board.

Note that these recommendations are based on a break-even ratio of 5.0 (cost of fertiliser N in £/kg divided by value of grain in £/kg), if this ratio changes significantly then rates should be adjusted using the information available in the AHDB Nutrient Management Guide.

13.5.1.9 Micronutrients

Oats are particularly susceptible to manganese deficiency, causing pale, floppy younger leaves streaked with white/grey and should be treated with manganese if symptoms are identified. The problem will be exacerbated by unconsolidated seedbeds and low soil temperatures.

13.5.1.10 Pests

The main pest problem is that of aphids transmitting BYDV. Oats are the most susceptible cereal crop. A fall in temperature as drilling is delayed will reduce aphid migration into the crop and therefore reduce the risk of infection. The optimum time for control is when the second generation of aphids spread the virus within the crop and this can be calculated using a T-sum method although control may be limited as insecticide resistance is a widespread issue in aphids.

13.5.1.11 Diseases

Powdery mildew and crown rust are the major foliar diseases of oats. Winter oats generally require a two-spray programme of a fungicide at GS 31 against mildew and at GS 39 against crown rust and mildew.

13.5.1.12 Weeds

Grass weeds are very difficult to control in the winter oat crop. Few chemicals are recommended for grass weed control and therefore it is preferable to avoid growing winter oats where there is a black-grass or wild oat problem. However, despite the lack of approved grass weed herbicides, spring oats can actually be used as part of a grass weed control strategy due to their spring sowing date and vigorous growth. Broad leaved weeds are less problematic in oats with a good range of approved herbicides. Oats grow rapidly in spring and caution should be exercised in using multi-way tank mixes of herbicides and PGRs on the soft, lush growth.

13.5.1.13 Growth regulators

Oats can be very susceptible to lodging, although recently introduced varieties show greater resistance, enabling higher N rates to be used with the aim of achieving higher yields. Two spray PGR programmes at GS 30 and GS31/32 are commonly used to reduce the risk. Grain from lodged crops will suffer from a reduced specific weight.

13.5.1.14 Harvesting

Winter oats are usually harvested just before winter wheat. Likewise, spring oats are normally harvested before spring wheat. Oats, like barley, are liable to shedding when ripe. The grain is not very dense and requires 40% more storage space than wheat.

13.5.1.15 Yield

Yields are variable, especially for spring oats in response to soil moisture conditions in the spring. A good yield for a winter oat crop would be 8 t/ha and for a spring oat crop 7 t/ha.

13.6 Rye

Rye is a relatively minor cereal grain crop in the UK, grown for human consumption and animal feed. The human consumption market remains relatively static with around 2000ha required for milling for the crispbread market and small quantities required for flour milling and distilling. In terms of animal feed, the pig feed market in particular has increased as benefits to health and welfare from the inclusion of rye in the diet are being seen. The plentiful, long, tough straw it produces can be good for thatching and bedding.

13.6.1 Rye agronomy

13.6.1.1 Site selection

Rye is a relatively low input winter sown crop which will grow on poor, light, acid soils and in dry regions where the yield of the other cereals may be severely compromised. It is mainly grown in such conditions for grain because, on good soils, although the output may be higher, it does not yield as well as other cereals and markets are more limited. It is extremely frost-hardy and withstands much colder conditions than the other cereals.

13.6.1.2 Rotation

Rye can replace other cereals in a rotation, especially where the fertility is low. It can be grown continuously on poor soils with occasional breaks. It is more resistant to take-all than wheat or barley and so it can be grown as a third or fourth cereal crop.

13.6.1.3 Varieties

Detailed variety information is available from a descriptive list, updated annually by the AHDB. The majority of the varieties now grown are hybrids.

13.6.1.4 Seed treatments

A range of fungicide seed treatments (often termed single-purpose dressings) are routinely used to control seedling blights.

13.6.1.5 Seedbeds preparation

The seedbed requirements and methods of sowing are similar to those for wheat described earlier in this chapter.

13.6.1.6 Time of sowing

The optimum time to sow is during September, although it can be sown throughout October.

13.6.1.7 Seed rate

In September 175–200 seeds/m^2 should be sown, increasing to 220–250 seeds/m^2 for October sowings.

13.6.1.8 Crop nutrition

Where straw is retained phosphate and potash requirements are the same as those for oats (see Table 13.14). However, where straw is removed, due to its large volume, phosphate and potash requirements are higher than for other cereals.

Nitrogen recommendations on mineral soils at index 0 are 150 kg/ha. However, these recommendations were based on non-hybrid varieties yielding 6 t/ha, whereas modern hybrid varieties are much higher yielding. However, evidence of higher yields must be provided as justification to increase recommendations on farm.

13.6.1.9 Crop protection

A range of grass and broad-leaved herbicides are approved, often as Extensions of Authorisations for Minor Use (EAMU's) due to the small size of the market. The competitive nature of the crop can help to out-compete weeds in the spring.

The main foliar disease problem of concern is brown rust in the hybrid varieties for which a robust T1 and T2 fungicide should be applied. Powdery mildew may also be a concern and can be treated at these timings.

Because rye is tall and often weak-strawed, a robust two spray growth regulator programme is required.

13.6.1.10 Harvesting

The crop is normally harvested after winter barley and before winter wheat. It is combined when the grain is hard and dry and the straw is turning from a greyish to a white colour. The grain sprouts very readily in a wet harvest season, so should be combined as soon as possible once it is ready.

13.6.1.11 Yield

Average yields are difficult to obtain due to the small area grown and the fact that in national statistics it is combined with triticale and mixed corn. The modern hybrids are capable of producing yields up to 12 t/ha but 7.5 t/ha would seem to be a realistic figure for on farm yields.

13.7 Triticale

Triticale is produced by crossing tetraploid durum wheat with diploid rye and treating seedlings of the sterile Fl plants so that their chromosome number is doubled and they become reasonably fertile. It was developed to combine the grain quality of wheat with the low input requirements and hardiness of rye. It will grow on more marginal soils where the yield of the other cereals may be severely compromised. It typically sells into both the ruminant and non-ruminant animal feed market at a discount to wheat in the UK where the area grown is small. In other European countries where winter hardiness is a more critical requirement it is more popular.

13.7.1 Triticale agronomy

13.7.1.1 Site selection

Triticale is probably best suited to marginal and lighter soil where it can outperform other cereals. It is winter hardy and tolerates drought well. It will tolerate relatively acidic soils down to a pH of 5.5.

13.7.1.2 Rotation

It is particularly well suited to a second cereal position in the rotation as it is unaffected by take-all but can frequently outperform other cereals in a first cereal position, its uptake in this situation being limited by available markets.

13.7.1.3 Varieties

Detailed variety information is available from a descriptive list, updated annually by the AHDB. Straw requirement, yield and quality should be considered when selecting a variety.

13.7.1.4 Seedbed preparation

The seedbed requirements and methods of sowing are similar to those for wheat.

13.7.1.5 Seed rate

A seed rate of 270—330 seeds/m^2 is required depending on sowing date and establishment conditions.

13.7.1.6 Time of sowing

The crop can be sown from mid-September to October. Winter triticale has a requirement for vernalisation so it should not be sown after the middle of February.

13.7.1.7 Crop nutrition

Nitrogen requirements are similar to wheat but in practice due to the lodging risk rates are reduced to a maximum of 150 kg/ha.

13.7.1.8 Crop protection

Triticale is a very competitive crop, though herbicides are still required. There are more restrictions on the use of some grass weed herbicides in triticale than with the wheat crop. Several broad-leaved weed herbicides are recommended and some have EAMUs for use in triticale.

The most problematic foliar disease is yellow rust, to which many varieties are quite susceptible. Fungicide programmes based around triazoles with good activity on yellow rust are required.

The crop is on average 20 cm taller than wheat and lodging is a risk, particularly on less droughty soil. A robust PGR programme is required.

13.7.1.9 Harvesting

The crop is harvested in August, as for wheat. Drying and storage advice is also the same as for wheat.

13.7.1.10 Yield

Average yields are difficult to obtain due to the small area grown and the fact that in national statistics it is combined with rye and mixed corn. A realistic figure for on farm yields would seem to be around 9 t/ha.

13.8 Maize for grain

Maize is a tall annual grass plant originating from the sub-tropics which unlike most other UK crops photosynthesises using a C4 mechanism making it more tolerant of hot, dry conditions. In the UK the crop is widely ensiled for use as forage or as feedstock for anaerobic digestion as described in detail in chapter 18. However, in many areas of the world it is more commonly grown for grain, indeed in Europe grain maize is actually the third most important cereal after wheat and barley, growing well in the Mediterranean climate of southern France and Italy.

In the UK there are two markets for grain maize. The small specialised grain maize market demands varieties with yellow kernels which are dried and sold for animal feed and the larger crimped grain market, where moist maize grains (approx. 65% dry matter) are treated with a preservative to create a digestible ready to use high energy animal feed.

In the UK grain maize production is largely confined to southern England where there is greater likelihood of receiving the required number of heat units to reach >65% dry matter. In more marginal areas the crop can be grown under plastic so that it can be sown and harvested earlier. It is commonly not ready to combine until October or November and if it is not being crimped then it usually requires quite a lot of drying. As well as earliness to reach maturity, varieties with good standing power are needed as well as good cob cover as this reduces their susceptibility to fusarium infection later in the season.

Currently the UK imports around 1–2.5 million tonnes of maize for animal feed annually but the area that is home-grown is likely to remain limited until more cold-tolerant varieties are produced.

13.9 Key points

- Cereals are the most widely grown arable crop in Europe and account for approximately 50% of the cropped area in the UK.
- A 'first' wheat crop that follows a break crop in a rotation is the most profitable crop in the arable rotation for many growers. Barley is well suited to a 'second' cereal position following a first wheat.
- Both winter and spring types of wheat and barley are available, winter types yielding more than spring types but requiring more inputs to achieve the higher yields.

- A range of markets exist for which there are specific quality requirements, some of which are related to the management of the crop during the growing period but some of which are determined by weather conditions.
- It is very important to understand the specific requirements for the intended market and tailor variety choice and other agronomic decisions accordingly.
- A programmed approach related to specific growth stages is usually used to manage fertiliser and crop protection inputs to the crop.
- The agronomy of cereals is well understood resulting in high yields but some problems in controlling weeds, pests and diseases have arisen due to the dominance of winter sown cereals in arable rotations.

Sources of further information and advice

Agriculture and Horticulture Development Board (AHDB), latest ed.. AHDB Recommended list for Cereals and Oilseeds. Available at: https://ahdb.org.uk/knowledge-library/recommended-lists-for-cereals-and-oilseeds-rl.

Agriculture and Horticulture Development Board (AHDB), 2021. Grain Storage Guide. Available at: https://ahdb.org.uk/knowledge-library/grain-storage-guide.

Agriculture and Horticulture Development Board (AHDB), 2021. Managing Weeds in Arable Rotations. Available at: https://ahdb.org.uk/knowledge-library/how-to-manage-weeds-in-arable-rotations-a-guide.

Agriculture and Horticulture Development Board (AHDB), 2021. Nutrient Management Guide (RB209) Section 4. Available at: https://ahdb.org.uk/rb209.

Agriculture and Horticulture Development Board (AHDB), 2021. Wheat Growth Guide. Available at: https://ahdb.org.uk/knowledge-library/wheat-growth-guide.

Agriculture and Horticulture Development Board (AHDB), 2019. Wheat and Barley Disease Management Guide. Available at: https://ahdb.org.uk/cereal-dmg.

Agriculture and Horticulture Development Board (AHDB), 2018. Barley Growth Guide. Available at: https://ahdb.org.uk/knowledge-library/barley-growth-guide.

Gooding, M.J., Shewry, P., 2022. Wheat: Environment, Food and Health. Wiley.

Wieser, H., Koehler, P., Scherf, K., 2020. Wheat: An Exceptional Crop. Woodhead Publishing and AACC International Press.

Rosentrater, K., 2021. Storage of Cereal Grains and Their Products. Woodhead Publishing.

Wrigley, C., Batey, I.L., Miskelly, D., 2016. Cereal Grains: Assessing and Managing Quality. Elsevier Science and Technology.

Useful websites

Agriculture and Horticulture Development Board. https://ahdb.org.uk/.

The Maltsters Association of Great Britain (MAGB) https://www.ukmalt.com/.

UK Flour Millers organisation. https://www.ukflourmillers.org/.

Seed suppliers also often provide useful information and comparisons of crop types and their management as well as more specific information on varieties.

Oilseeds and pulses

14

14.1 Introduction

Oilseed and pulse crops are integral to many cereal based rotations as non-cereal crops that can be combine harvested for their oil or protein containing seeds i.e., combinable break crops. Generally, each individual species or in some cases group of similar species, should not be grown more often than 1 year in 4 or 5 in the rotation.

The range of crops, often with autumn- and spring-sown varieties available, offers an option to suit most soil types, pH and crop rotations. Seasonal weather may adversely influence crop establishment, harvesting and quality.

On a global basis a large range of crops produce edible oil with palm, soya, oilseed rape (canola) and sunflower dominating. The high protein meal that results from the crushing process to extract the oil from these crops is used in intensive livestock systems. The pricing of both oilseed and protein crops in the UK are influenced by global supply and demand of these four major oilseed crops. This is because despite the fact that not all of them are grown on any scale in the UK, for many end-uses oil and high protein meal from different oilseeds can substitute for one another.

14.2 Oilseed rape

Oilseed rape (*Brassica napus*) is a plant from the Brassicaceae family, recognised by its bright yellow flowers in April/May and grown for its oil-rich seeds (42%–46% oil), which can be used as vegetable oil for human consumption or in biodiesel. The processing of the seed also produces a high protein meal used as animal feed and competing with imported soya bean. It is the third largest source of vegetable oil and the second largest source of protein meal globally.

The market is volatile as it is influenced by the fact that for many end-uses oil and meal from different oilseeds can substitute for one another and therefore it is affected by changes in global oilseed and protein production. Despite this volatility and some agronomic challenges, oilseed rape remains the most widely grown and profitable combinable break crop in the UK.

14.2.1 Site selection

The crop will grow in a wide range of soil and climatic conditions provided the land is well drained, pH over 6, and the soil and subsoil structure is good.

Lockhart and Wiseman's Crop Husbandry Including Grassland. https://doi.org/10.1016/B978-0-323-85702-4.00005-4

14.2.2 Rotation

Ideally, oilseed rape and other *Brassica* crops should not be grown more than 1 year in 5, in order to avoid a build-up of diseases such as Clubroot (*Plasmodiophora brassicae)*, Sclerotinia (*Sclerotinia sclerotiorum*) and Verticillium Stem Stripe (*Verticillium longisporum)* and also of pests.

14.2.3 Varieties

Winter and spring oilseed rape varieties are available but the majority (95% in 2020) of the oilseed rape area is sown to winter varieties because of their more favourable gross margins. However, the area of spring oilseed rape is increasing due to some of the specific establishment challenges that winter oilseed rape now faces and to the rotational benefits of spring cropping.

Both hybrid and conventional varieties are available with hybrids perceived to offer greater vigour in the autumn which may be important when establishment conditions are challenging, although seed costs are higher. Most breeding programmes are focused on hybrid varieties and these are likely to dominate the seed market in the future.

Most oilseed rape varieties grown in the UK are classified as 'double-low' or '00' – being low in erucic acid content (less than 2% of total fatty acid) and low in glucosinolates (less than 18 μmol/g of seed), making the seeds suitable for both human and animal consumption. A smaller area is sown to varieties that are high (approx 50%) in Erucic acid (HEAR), the oil from which is used for industrial purposes such as high temperature lubricants, slip agents, printing ink and cosmetics. It is recommended that HEAR is not grown within 5m of any double low oilseed rape and the grower should not be growing oilseed rape more than 1 year in three to avoid cross contamination. Another relatively small area is sown to HOLL varieties which have a high oleic low linolenic (HOLL) fatty acid oil profile. These varieties produce healthy, low-transfat and stable frying oil favoured by the food manufacturers and food service companies. Similarly, to HEAR varieties these must be grown on land that has had a 4–5 year break from growing oilseed rape. Both HEAR and HOLL are grown on contract and attract a premium over double low oilseed rape. Yields of HEAR and HOLL varieties have historically been slightly lower than double low varieties but they are now more comparable.

Other specific characteristics of varieties that may be considered by growers are Turnip Yellows Virus (TuYv) resistance and Clubroot resistance.

One further category of varieties worthy of note are those with Clearfields technology, which provides tolerance to certain imidazolinone herbicides, allowing for the control of weeds from the brassica family.

Varieties and their characteristics are detailed in the AHDB *Recommended Lists for Cereals and Oilseeds* updated annually. The winter oilseed rape lists are regionalised between North and East/West and varieties are compared for their yield, gross output, maturity, disease and lodging characteristics.

14.2.4 Seed treatment

Options are limited to an insecticide with some activity against flea beetles and some biological and biostimulant products.

14.2.5 Seedbed preparation

Good establishment is critical with oilseed rape which is particularly vulnerable to pest damage during its early growth stages. The winter crop is sown in August/September when both time and moisture are at a premium which has led to over 50% of the crop being established using either a minimum tillage or single pass system that sows the seed, often places seedbed fertiliser by the seed and consolidates the tilth. Another 25% is established by direct drilling. These techniques can be very successful providing that moisture is available at the time of sowing and that the soil is well structured as yield will be affected by compaction if it impedes the growth of the tap root. These techniques commonly place the seed in bands up to 30 cm apart, ideally 2–3 cm deep. The remaining 25% of the area of oilseed rape crop is established using either a plough and combination drill, a precision drill or an auto-casting system.

14.2.6 Time of sowing

Winter crops are sown from early August to early September. Where winter barley is included in the rotation it can be used to provide an earlier entry than winter wheat for the subsequent oilseed rape crop. Spring sown crops should be sown from mid-February through to the end of April.

14.2.7 Seed rates and plant population

The ultimate aim is to produce the optimum number of pods and seeds per square metre so that the available light is used efficiently for seed fill. It should be noted that there is a yield penalty associated with excessively thick canopies. For winter varieties the aim should be to achieve 25–40 plants/m^2 in the spring by sowing 30–60 seeds/m^2 based on 70%–80% establishment. Spring varieties should be sown at 80–90 seeds/m^2 to give the required higher plant populations of 70 plants/m^2. Seed size varies considerably and should always be taken into account in seed rate calculations.

14.2.8 Crop development

Growth stages for oilseed rape are shown in Table 14.1. The most critical growth stages in terms of the success of the crop are germination and emergence because the crop is small seeded and its seedlings relatively weak. In the winter oilseed rape crop the leaf development phase continues through the autumn until mid Feb-early March when stem extension starts as temperatures increase. Flowering occurs in April/May and the crop is open-pollinated. Seed number is a critical yield component

Table 14.1 Principal growth and secondary growth stage key (BBCH system) for oilseed rape.

GS0	***Germination and Emergence***
00	Seed dry
01	Seed imbibition starts
03	Seed imbibition complete
05	Radical (embryonic root) emerged from seed
07	Cotyledons emerged from seed
09	Cotyledons emerged through soil surface
GS1	***Leaf Development***
10	Cotyledons completely unfolded
11	First leaf unfolded
12	Second leaf unfolded
19	Nine or more leaves unfolded
GS2	***Side shoot formation***
20	No side shoots
21	First side shoot detectable
22	Second side shoot detectable
29	Nine or more side shoots detectable
GS3	***Stem elongation/extension***
30	No internodes (rosette stage)
31	One visibly extended internode
32	Two visibly extended internodes
39	Nine or more visibly extended internodes
GS4	***Does not apply to oilseed rape***
GS5	***Inflorescence/flower bud emergence***
50	Flower buds present and enclosed by leaves
51	Flower buds visible from above ('green bud' stage)
52	Flower buds free and level with the youngest leaves
53	Flower buds raised above the youngest leaves
55	Individual flower buds (main inflorescence) visible but closed
57	Individual flower buds (secondary inflorescences) visible but closed
59	Flower buds closed with first petals visible ('green bud' stage)
GS6	***Flowering***
60	First flowers open
61	10% of flowers on main raceme open
62	20% of flowers on main raceme open
65	50% flowers on main raceme open, older petals falling (full flower stage)
67	Flowering declines, most petals fallen
69	End of flowering
GS7	***Pod/seed (fruit) development***
71	10% of pods have reached final size
72	20% of pods have reached final size
79	90% of pods have reached final size
GS8	***Pod/seed (fruit) ripening***
80	Ripening starts (seeds green and filling pod cavity)
81	10% of pods ripe, with seeds dark and hard
82	20% of pods ripe, with seeds dark and hard
89	90% of pods ripe, with seeds dark and hard
GS9	***Senescence***
90	Plant death starts
97	Plant dead and dry
99	Harvested product

Source: AHDB Oilseed Rape Growth Guide, 2020.

which is determined by photosynthesis during a 3 to 4 week period after mid-flowering. Seed filling lasts for around 40 days after which ripening and senescence occurs with harvest generally taking place in late July.

Table 14.2 Nutrient recommendations (kg/ha) for oilseed rape.

	SNS[a], P or K index						
	0	**1**	**2**	**3**	**4**	**5**	**6**
Winter oilseed rape 3.5 t/ha							
Nitrogen (N)							
Mineral Soils	220	190	160	120	80	40–80	0–40
Organic Soils	–	–	–	120	80	40–80	0–40
Peaty Soils	–	–	–	–	–	40–80	
Phosphate (P_2O_5)	110	80	50	0	0	0	0
Potash (K_2O)	100	70	40 (2–) 20 (2+)	0	0	0	0
Spring oilseed rape 2 t/ha							
Nitrogen (N)							
Light Sand Soils	120	80	50	0–40	0	0	0
Mineral Soils	150	120	80	50	0–40	0	0
Organic Soils	–	–	–	50	0–40	0	0
Peaty Soils	–	–	–	–	–	0–40	
Phosphate (P_2O_5)	90	60	30	0	0	0	0
Potash (K_2O)	80	50	20 (2–) 0 (2+)	0	0	0	0

These recommendations have been based on AHDB Nutrient Management Guide (RB209) Section 4 updated in January 2021 by the Agriculture and Horticulture Development Board.
[a]Soil Nitrogen Supply Index.

14.2.9 Fertilisers

Phosphate and potash recommendations for oilseed rape are shown in Table 14.2. These are based on winter oilseed rape yields of 3.5 t/ha and spring oilseed rape yields of 2 t/ha and should be adjusted using the current Nutrient Management Guide (RB209) if expected yields are larger or smaller than this. At Index 2, phosphate and potash can be applied when convenient during the year, but at Index 0 and 1, they should be applied and worked into the seedbed.

An application of 30 kg/ha of nitrogen can be placed with the seed or applied to the seedbed for winter oilseed rape crops to encourage autumn growth on soils with SNS index 0–2. Crops sown after early September are unlikely to respond.

Spring nitrogen rates for winter oilseed rape, as shown in Table 14.2, depend on the SNS index and any use of organic manures which must be accounted for. The recommendations are for crops yielding 3.5 t/ha. Where previous experience of growing the crop indicates that yields above 3.5 t/ha can be realistically expected, the recommended rate may be increased by up to 30 kg N/ha per 0.5 t/ha of yield, up to an expected yield of 5.0 t/ha. This adjustment should be used with caution because applying too much early nitrogen to crops with large canopies can increase lodging and ultimately reduce yield.

Estimates of Green Area Index (GAI) which indicates the amount of nitrogen already in the crop (about 50 kg/ha per unit of GAI) and Soil Mineral Nitrogen (SMN) in the top 90 cm of soil in February just before the start of spring growth are increasingly used as a tool to manage nitrogen timing and adjust rates in the spring. For crops with a small GAI in mid-February then the nitrogen should be applied in two equal splits, half at the start of stem extension (end of Feb-early March) and half at green bud (mid March - early April). For those with a large GAI then the first dose can be reduced or missed altogether.

For spring oilseed rape crops all of the nitrogen should be applied to the seedbed except on light sand soils where the total rate is more than 80 kg N/ha, the dressing should be split, with 50 kg N/ha in the seedbed and the remainder by early May.

Foliar N applied as a solution of urea at 40 kg N/ha after the end of flowering has also been proven to increase gross output, even where the optimum amount of nitrogen has been applied to soil. The economic benefit of this treatment is dependent upon the cost of the nitrogen product and the price of oilseed rape.

The yield of most oilseed rape crops on mineral soils will now respond to sulphur because of the reduction in atmospheric deposition, so 50–80 kg/ha of $S0_3$ should be applied at the start of stem extension. This has been shown to reduce the amount of chlorophyll that remains in the seed at harvest and is a problem during processing. Samples with over 4% green seed may be rejected.

At Mg indices 0 and 1, between 50 and 100 kg/ha of MgO (30–60 kg Mg/ha) should be applied every three or 4 years.

The most common micronutrient deficiencies in oilseed rape are manganese, boron and molybdenum. Visual symptoms may be seen but co-formulated products containing these micronutrients are often applied prophylactically on soil types where deficiencies are likely e.g., on light soils with a high pH.

The current nutrient recommendations for oilseed rape are summarised in Table 14.2.

14.2.10 Crop protection

14.2.10.1 Pests

Slugs may be a serious problem of winter oilseed rape at the cotyledon and first true leaf stages, particularly in trashy or cloddy seedbeds, slow growing conditions or when late established. Slug trapping should be used to indicate numbers and pellets should be applied where there is a high risk.

Cabbage stem flea beetle (*Psylloides chrysocephala*) has proved a particular challenge to establishment in recent years, since the revocation of neonicotinoid seed dressings and the development of resistance to the commonly used pyrethroid insecticides. The adults damage the cotyledons and first true leaves during establishment and the larvae invade the stem later in the autumn and can cause significant crop damage and indeed failure. With a lack of approved and effective insecticides integrated pest management strategies are particularly important with any measures that can be taken to promote strong early growth such as early sowing and fertiliser placement being the most successful. Other more novel methods such as trap cropping or companion cropping with alternative hosts and defoliation of the crop by grazing or mowing over winter are also showing some promise.

Unless a Turnip Yellows virus resistant variety is being grown the peach-potato aphid (*Myzus persicae*) which is the vector of this disease, may need an insecticide spray in October/November, but options are limited as this aphid species is resistant to several groups of insecticides.

In the spring, at the green/yellow bud stage the pollen beetle (*Meligethes aeneus*) can be a problem, causing flower buds to abort resulting in blind stalks. Recognised economic damage thresholds should be used with thin crops tolerating greater numbers of the pest than thick crops. Resistance to pyrethroid insecticides is endemic so approved alternatives must be used.

During and after flowering cabbage seed weevil (*Ceutorhynchus obsrtictus*), bladder pod midge (*Dasineura brassicae*) and mealy cabbage aphid (*Brevicoryne brassicae*) can also infest the crop and may require treatment with approved insecticides.

Pigeons can also cause serious damage in some situations, especially in areas with a high pigeon population, near woods, where only a small area of rape is grown and where there is little else for the pigeons to eat in the winter. The greatest damage is caused by grazing in late winter and early spring when the new shoots and buds are developing. This leads to uneven flowering and ripening and creates problems for timing of spraying and harvesting, as well as loss of yield and a lower oil percentage. A variety of different scaring devices such as bangers, kites, balloons and scarecrows are available and when used they should be moved around from time to time. Shooting is also helpful, particularly when the weather is severe and in the more vulnerable periods. A full crop is always likely to be less damaged than a thin patchy crop and direct drilling into a long cereal stubble can put them off landing in the field during early autumn.

14.2.10.2 Diseases

As the frequency of oilseed rape in rotations has generally increased over the last 50 years the soil-borne disease of all cruciferous plants, Clubroot (*Plasmodiophora brassicae*) has become more prevalent. It causes the development of galls on the roots which reduce nutrient and water uptake by the plant, reducing yield and in severe cases causing complete crop failure. The pathogen can survive for up to 15 years in the soil with a half-life of 3 to 4 years. Lengthening rotations, minimising movement of infected soil and organic materials, maintaining drainage, controlling cruciferous weeds and maintaining soil pH above 7 are important factors in reducing the spread of the disease. Resistant varieties are also available which mitigate some of the associated yield loss but should be used with caution as they may increase the selection pressure on the pathogen for resistant strains to emerge.

Phoma leaf spot caused by the pathogens *Leptosphaeria maculans* and *Leptosphaeria biglobosa* can occur in the autumn, appearing as beige lesions with black pycnidia on the upper leaf surface, developing into stem canker in the spring if not controlled. Varietal resistance can be used to help manage the disease but fungicide applications in the autumn are frequently also required when 10%–20% of plants are affected. A follow up application 4 weeks later is advised if infection is still actively occurring. A forecasting system is available to help predict the date on which thresholds are likely to be reached at a given location based on weather data.

Light leaf spot (*Pyrenopeziza brassicae*) can also be a problem in both late autumn and spring time and varietal resistance is available to help manage the risk. Yield loss in untreated crops can be up to 1 t/ha. Weather conditions in Northern and Western regions are more favourable for disease development so particular attention should be paid to varietal resistance in these regions. A localised forecasting tool is available on the AHDB website. Initially, when infection spreads from previous crop residues, light green lesions occur on the leaf surrounded by white spore masses. When the disease spreads to stems and lateral branches, elongated fawn lesions, surrounded by black speckling, appear. Fungicides are frequently used in late autumn in high risk situations or in spring when symptoms are seen.

Sclerotinia (*Sclerotinia sclerotiorum*) occurs during flowering. White stem lesions form with black sclerotia within the stem cavity. Infection can cut off the supply of nutrients and water to the canopy leading to premature senescence and yield loss. Close rotations and wet weather during flowering increase the risk. Varietal resistance is not available. In high risk situations a first fungicide should be applied at the early flowering stage and followed up 4 weeks later if the risk persists.

Verticillium stem stripe (*Verticillium longisporum*) is a sporadic but persistent soil-borne disease characterised by brown/grey vertical stripes on stems and branches in early July. The use of clean seed and longer rotations will help to reduce the risk.

14.2.10.3 Weeds

Winter oilseed rape develops into a large canopy with the potential to suppress weeds but canopy closure in the autumn is often slow and it is not competitive in the early stages so good weed control is crucial. Grass weeds, cranesbill (*Geranium* spp), poppies (*Papaver rhoeas*), charlock (*Sinapis arvensis*), runch (*Raphanus raphanistrum*), hedge mustard (*Sisymbrium officinale*) and cleavers (*Galium aparine*) are among the most challenging. Weeds can reduce the yield and also contaminate the sample affecting the marketability of the crop. Historically, most of the herbicides available for broad leaved weed and annual meadow-grass control have been those with residual activity aimed at the pre-emergence or very early post-emergence timing. However, with the threat to establishment posed by cabbage stem flea beetle, understandably, many growers have been reluctant to invest too much in herbicides before the crop is well established. The residual properties of these herbicides and their autumn timing have also led to many of them being found in water resulting in some revocations and some restrictions on use. More recently, an improved range of products for post-emergence use in the autumn and early spring have been introduced.

In minimum tillage situations where high populations of volunteer cereals are likely to be present and competing with the crop at an early stage then graminicides should be used. For resistant grass weeds, programmes are based around graminicides and residual herbicides that are applied later in the autumn.

The brassica weeds charlock, runch and hedge mustard have always been particularly challenging to control in oilseed rape. Historically, various herbicides have held extensions of authorisation for minor use for control of these weeds but have not been highly effective. The introduction of Clearfields technology, which provides hybrid varieties with tolerance to certain imidazolinone herbicides has provided a better opportunity to control them. However, the system should be implemented with some caution as subsequent oilseed rape volunteers will be tolerant to some herbicides used in other crops in the rotation.

A reduced range of herbicides is available for use in spring oilseed rape where weed control really relies on good, rapid establishment and canopy closure as residual herbicides are less reliable in spring soil conditions.

14.2.10.4 Typical crop protection programme for winter oilseed rape

Despite the range of pests, diseases and weeds listed which may affect a crop (see also Chapters 5–7), an individual crop is unlikely to face all of these problems in practice. In addition, the control of more than one problem is often carried out simultaneously by tank mixing treatments to reduce the number of 'passes' the sprayer makes through the crop. A typical crop protection programme for a winter oilseed rape crop is shown in Table 14.3.

Table 14.3 Typical crop protection programme for winter oilseed rape.

Crop growth stage	Indicative date	Treatment type	Target
GS 9-12	Sept/Oct	Molluscicide	Slugs
GS 12-18	Sept/Oct	Herbicide	Broad leaved weeds
		Graminicide	Volunteer cereals/grass weeds
GS 12-18	Oct/Nov	Fungicide	Phoma
GS 16-19	Nov/Dec	Herbicide	Follow up treatment if black-grass (*Alopecurus myosuroides*) present
GS 31-39	Mar	Fungicide	Light leaf spot
GS 59	Apr	Insecticide	Pollen beetle
GS 60-65	Apr	Fungicide	Sclerotinia
GS 85-89	Jul	Non-selective herbicide	Desiccation

Note that while Cabbage stem flea beetle is a major pest, insecticides for its control are not included here as they are largely ineffective due to widespread resistance.

14.2.11 Growth regulation

Growth regulators in oilseed rape are used to reduce lodging and increase rooting, but also alter the structure of the flowering canopy, making it more open and more efficient at capturing sunlight, thus increasing seed set. If canopies have a GAI above one at green bud stage or above two at yellow bud stage then it is likely that they will benefit from a growth regulator application. Many of the products used for growth regulation in oilseed rape also have fungicide activity so product choice will also be based on what diseases are present and the balance required between disease control and growth regulation in a particular situation.

14.2.12 Harvesting

The indeterminate growth of oilseed rape results in uneven ripening between the early- and late-formed branches which can make it a challenge to harvest. Total herbicides such as glyphosate are routinely used to desiccate oilseed rape pre-harvest as a harvest aid and also to mitigate against the risk of shedding losses if harvesting of the crop is delayed. Pod sealers comprised of polymer blends can also be added to the application to reduce the shedding risk. Pod shatter resistance is also a focus of current breeding programmes due to the significant losses and volunteer problems it can incur. Desiccants are applied 14–21 days pre-harvest when the seed is below 30% moisture. Timing is important as if applied too early the seed filling period will be reduced and there is a risk of immature seed in the sample. It should also be borne in mind that application of glyphosate is usually recommended to crops that are actively growing whereas in this situation they are being applied to a crop that is senescing so their activity can be slow and incomplete. It should not be applied to crops grown for seed or those that are severely lodged.

An alternative is swathing, whereby the crop is cut when the seeds in the top third are green and green/brown, those in the middle third are green/brown and those in the bottom third are dark brown/black and laid down in swaths on a tall stubble (20–30 cm) to complete the ripening process. Swathing suits exposed sites and upright or leaning crops, but not lodged or leaning short crops and is usually carried out by contractors.

Crops that are ripening evenly and are free from weeds and lodging do not routinely require desiccation. Driving through the crop at this late stage can cause some seed losses if the crop is knocked down in the tramlines so direct combining should be considered where possible. This is the lowest cost option and can result in the highest yields but harvest may be delayed by a few days compared to desiccated or swathed crops.

14.2.13 Drying and storage

The contract price is usually based on 9% moisture content and 2% admix. Seed should be dried to slightly below 9% moisture for storage but care is needed as, if dried to below 6% moisture seed can be rejected.

14.2.14 Yield

The 5-year average yield for oilseed rape is 3.5 t/ha, well below the calculated potential UK yield estimated at 6.5 t/ha. There is significant variation between years with best yields occurring when good conditions are seen at establishment followed by a sunny, dry April so the optimum number of seeds are set and a cool wet May and June to prolong flowering. In terms of agronomic factors, cabbage stem flea beetle damage, poor selection and management of varieties, inadequate nitrogen and sulphur nutrition and increases in cropping frequency, leading to greater amounts of soilborne disease are associated with lower yields. For spring oilseed rape the average is 2 t/ha.

Oil content is also important as most contracts will pay a bonus per 1% of oil over 40% with deductions below 40%.

14.3 Linseed

Linseed is an oilseed crop which has fluctuated in its popularity over recent decades partly in response to changes in support payments and partly in response to demand. Both winter and spring types are available, the winter sown area is increasing as an alternative to winter oilseed rape but the spring sown area still dominates with growers using it as an opportunity to control grass weeds or diversify rotations. Compared to other combinable break crop options it has relatively low input costs and leaves behind a good soil structure. One of its perceived disadvantages is that it is difficult to harvest which has sometimes been the case but the development of new varieties with reduced stem fibre levels has produced varieties that are easier to cut.

The seed contains about 40% oil and both brown and yellow seeded varieties are available. The oil produced from brown seeded linseed is mainly used for industrial oils and oil from yellow varieties is mainly but not exclusively used for human consumption and trades at a premium over brown. The seed of either type can also be used whole for human consumption. It is the nutritional benefits for human consumption and a lack of supply from other regions that have fuelled the recent increase in demand. The meal that remains after crushing is a high protein animal feed which is also high in alpha-linoleic acid (ALA). Livestock fed on linseed-based feedstuffs produce meat and milk that are higher in Omega 3.

There is no futures market for linseed and therefore it is generally advised to grow on a buy back contract to reduce exposure to the liquidity of the market.

14.3.1 Site selection

Linseed can be grown in any of the arable areas of the UK on most soil types, although, moisture retentive soils are preferable because of the small root system of the plant. The pH should be about 6.5.

14.3.2 Rotation

Because of the comparatively small area grown, pests and diseases are not generally a serious problem. Nevertheless, linseed should not be grown more often than 1 year in 5 to reduce the risk of the soil-borne pathogen Fusarium wilt (*Fusarium oxysporum f.* sp. *Lini*). It is not related to any other break crop so it can be grown in all rotations. It leaves a good seedbed for a following crop of wheat with a fine tilth and less slugs than oilseed rape.

14.3.3 Varieties

Once a choice has been made between a winter or spring sown crop other important characteristics are earliness to harvest, standing ability and yield. Current varieties of spring linseed are detailed in the AHDB Descriptive list for Spring Linseed updated annually.

14.3.4 Seed treatment

No chemical seed treatments are currently available but some biological seed treatments have an extension of authorisation of minor use and have been shown to decrease seed-borne diseases such as Botrytis and Fusaruim.

14.3.5 Seedbed preparation

As a small-seeded crop, linseed requires a relatively fine, firm seedbed. Minimum tillage and direct drilling techniques can be used to conserve moisture.

14.3.6 Time of sowing

The winter crop should be sown in September and the spring sown crop should be sown in March or April when soil temperatures are above 7°C.

14.3.7 Seed rates and plant population

The seed should be sown 15–25 mm deep into a fine, firm and moist seedbed. For winter crops 450 seeds/m^2 are required to achieve a plant population of around 400 seeds/m^2. For spring crops 650–800 seeds/m^2 should be sown to achieve a plant population of 600 plants/m^2. Thousand seed weight is variable and must be taken into account when calculating seed rates.

14.3.8 Crop development

After germination a pair of cotyledons appear first followed by a succession of true leaves. One main stem develops but basal branching may also occur. Buds are formed at the top of the plant.

Most varieties have pale blue (a few have white) flowers which appear in early July. The plant flowers early in the morning, but only for a few hours and the flowering period for the crop may last several weeks. It is normally self-pollinated, after which the ovary starts to swell forming the seed head, known as a ‘boll’ or capsule in which the seeds develop, maturing over a period of 30–40 days.

14.3.9 Fertilisers

Fertiliser requirements for linseed are relatively low, reducing input costs compared to other combinable break crops. The phosphate and potash recommendations shown below in Table 14.4 are based on yields of 1.5 t/ha and can be adjusted using the current AHDB Nutrient Management Guide if larger or smaller yields are likely. At index 0 and 1 the phosphate and potash should be worked into the seedbed but at index 2 it can be applied at any time of year.

All of the nitrogen requirement can be applied to the seedbed except on light sand soils requiring more than 80 kg/ha where 50 kg/ha should be applied to the seedbed with the remainder by early May. For winter linseed up to 125 kg/ha is recommended, as a 40:60 split with the first application during March and a later application just before flowering.

Table 14.4 Nutrient recommendations (kg/ha) for spring linseed.

	SNS[a], P or K index						
	0	**1**	**2**	**3**	**4**	**5**	**6**
Nitrogen (N)							
Light Sand Soils	80	50	0–40	0	0	0	0
Mineral Soils	100	80	50	0–40	0	0	0
Organic Soils	–	–	–	0–40	0	0	0
Peaty Soils	–	–	–	–	–	0–40	
Phosphate (P_2O_5)	90	60	30	0	0	0	0
Potash (K_2O)	80	50	20 (2–) 0 (2+)	0	0	0	0

These recommendations have been based on AHDB Nutrient Management Guide (RB209) Section 4 updated in January 2021 by the Agriculture and Horticulture Development Board.
[a]Soil Nitrogen Supply Index.

14.3.10 Crop protection

14.3.10.1 Pests

Spring linseed can be affected by Flax flea beetles causing shot-holing damage to the cotyledons and first true leaves. If the pest is present and damage is seen, repeat applications of approved pyrethroid insecticides may be required to protect new growth as it appears. Winter linseed is not affected.

14.3.10.2 Diseases

Kabatiella (*Kabatiella lini*) may affect winter sown crops in the autumn, causing lesions on the cotyledons and early true leaves. Septoria (*Septoria linicola*) also known as pasmo causes dark brown leaf lesions and is favoured by wet conditions in the spring. Botrytis infects when damp weather occurs during and after flowering and causes a grey mould on the surface of bolls and leaves which can cause premature senescence. A range of triazole fungicides can be used for the control of these diseases and also for their growth-regulatory effect.

14.3.10.3 Weeds

Linseed competes poorly with weeds. A range of pre- and post-emergence herbicides can be used, many of them approved as extensions of authorisation for minor use. Several graminicides can also be used to control non-resistant grass weeds. Caution should be exercised over the use of tank mixes and sequences as the crop is quite sensitive to herbicides. Any weed problem not controlled during the growing season can potentially be controlled by a pre-harvest non-selective herbicide.

14.3.11 Harvesting

Winter linseed should be ready for harvest from mid-July to mid-August and spring linseed from mid-August to mid-September. The crop should be cut at the first opportunity when ready as leaving too long will result in the stem fibres separating from the woody stem (retting), making it more difficult to harvest. Linseed is ready for harvesting when the top of the plant is dry, the stems are yellow-brown and the leaves have fallen from the base. The stem can still be green. The seeds should ideally be between 7% and 11% moisture and rattle in the bolls. Linseed straw is very tough and wiry and so the combine knife must be kept sharp and it will harvest more easily on a dry, sunny day when relative humidity is below 65%.

The crop can be desiccated before combining to control weeds and condition the stems to aid harvest. The straw is exempt from the burning ban and can be burnt in the swath or baled and used in on-farm boilers to provide heat or chopped depending on the weather and condition of the chopper.

14.3.12 Drying and storage

Linseed should be dried to about 9% for safe storage. For on-floor drying systems it should be stored at a maximum depth of 1 m.

14.3.13 Yield

Linseed typically has a yield range of 1.7–2.5 t/ha.

14.4 Sunflowers

The commercial sunflower crop is restricted to a handful of growers in the south-eastern part of England because of the required length of growing season and their need for heat to reach maturity. Modern varieties are hybrid, semi-dwarf varieties with a requirement from drilling to harvest of 1300−1400 accumulated day degrees (over a base of 6 °C) which usually equates to around 120 days depending on summer temperatures. The decorticated (dehusked) seed contains over 40% of valuable, high quality edible oils but with no commercial crushing facilities in the UK, production is mainly used for bird seed. Recently small areas have also been grown for 'pick your own' ventures and photo opportunities!

A well-drained soil with a pH of over 6.5 is essential. A rotation of 1 year in 5 or 6 is recommended to reduce the risk of infection from Sclerotinia (*Sclerotinia sclerotiorum*). A fine, firm and moist seedbed is required and 25−50 kg/ha of nitrogen together with 40−60 kg/ha of both phosphate and potash should be applied to the seedbed as a compound. Too much nitrogen can cause lodging, can favour disease development, delay maturity and lower seed oil content. Molybdenum, copper and manganese deficiencies can be a problem in very acidic soils and the crop is also sensitive to boron deficiency.

The crop is sown in late April/early May when soil temperatures in the top 10 cm reach 8 °C. Sowing rate depends on seed size and the aim should be to achieve 8−11 plants/m^2 using either a precision or air drill at a depth of 2−5 cm.

Slugs can be a problem during establishment, as well as pigeons and rabbits.

Weed control options are limited due to the small area of the crop grown but some pre-emergence residual herbicides have extensions of authorisation for minor use. Mechanical weed control is also a useful option.

The main disease threat is Botrytis which causes head rot and is common in cool, damp (15−25°C) conditions.

The crop is ready for harvesting in mid-September/October when seed has reached 30% moisture content or less. At this stage the back of the head is deep yellow and the bracts have begun to turn brown. A conventional cereal combine can be used with similar settings as for harvesting beans. Seed is dried on a drying floor at a maximum depth of 1 m using cold air until the moisture content reaches 15%. After cleaning, seed should be dried to 9% using heat. Yields are typically in the range of 1.5−2.5 t/ha.

14.5 Soya beans

Soya bean is a nitrogen fixing legume crop, however, it is classified as an oilseed rather than a pulse due to its high oil content and its primary use as a source of vegetable oil.

A global increase in the consumption of livestock products has driven a rise in demand for protein-based animal feeds. Globally, the most common vegetable protein comes from soya bean meal, a highly digestible by-product of oil extraction which represents 55% of the global production of oilseeds. Despite this demand, historically the soya bean varieties introduced to the UK were late to mature and produced pods too close to the ground making them notoriously difficult to harvest. The gross margin was also not comparable to other combinable break crops. In recent years the price of soya has risen significantly as domestic demand, especially for non-GM (identity preserved) protein-based animal feed has increased. The varieties have also improved in their earliness and ease of harvest and the agronomy of the crop has become better understood making soya a viable, low input, alternative spring sown combinable break crop in the south of England (south of York).

The crop is suitable for most soil types with a pH between 6 and 8 and is sown in late April/early May facilitating the use of stale seedbeds for grass weed control. It requires a fine, firm seedbed and should be sown 25–40 mm deep, preferably into moisture at a seed rate calculated to achieve around 50–60 plants/m^2. The seed is provided pre-inoculated with rhizobia bacteria. 25 kg/ha of nitrogen and a maintenance dressing of phosphate and potash should be applied to the seedbed.

Pest and disease problems are minimal with the relatively small area grown but pigeons can be a significant issue during the establishment phase. Bean seed fly (*Delia* sp.) may also require treatment with approved pyrethroid insecticides if present during establishment. An approved fungicide application for Sclerotinia may be required where there is a history of the disease in the rotation. For weed control an approved pre-emergence residual herbicide should be applied and a post-emergence follow-up is also likely to be required.

The crop is ready to direct cut in Sept/Oct and yields are in the range of 1.25–2.5 t/ha.

14.6 Combinable pulses

More than 40 species of grain legumes belonging to the Leguminosae family are grown throughout the world and unlike other crops do not need nitrogen fertiliser because the *Rhizobium* bacteria found in nodules on the roots are able to fix atmospheric nitrogen. They are excellent break crops providing a yield advantage in the following crops which benefit from the residual nitrogen. The nutritional value of the grain legume seeds is of major importance in both human and animal nutrition providing protein, starch and fibre. The grains of these crops are traded as pulses but this term excludes leguminous oilseeds such as soya beans.

In the UK, the main pulses grown are field beans and combining peas with some interest in lupins. The grain from these crops is primarily used for animal feed but premium markets are available for human consumption and for pigeon feeding (mainly maple peas and small seeded 'tic' beans).

These crops have been hampered by inconsistency in supply and demand and a lack of investment so progress has somewhat lagged behind that of cereals and oilseed rape.

Yields of spring sown crops are very dependent on moisture availability during early summer and therefore gross margins are variable. However, the current interest in diversifying rotations, sustainability and soil health may encourage the production of these crops.

14.6.1 Field beans

Typically, around 30% of the 150–200,000 ha of field beans grown in the UK are exported to North Africa and the Middle East for human consumption where beans are a staple component of the diet. The UK competes with France and Australia for this market. This market requires a bean with a pale colour and good skin quality which can be affected by variety, pest problems and weather conditions around harvest. The remaining 70% is used domestically for animal feed, some for on-farm use but the majority for compounding for ruminant feed, competing on price with soya and rapeseed meal but with the advantage of a guarantee of being GM free. Use in non-ruminant feeds and also for aquaculture both domestically and for export are growing markets.

Both winter and spring sown varieties are available.

14.6.1.1 Site selection

Winter field beans do well on the heavier soils provided they have a good structure and are well drained. Spring beans are more suited to medium soils but may then suffer from drought.

14.6.1.2 Rotation

Field beans are a good break from cereals and are often followed by winter wheat which benefits from the residual nitrogen. To reduce the risk of build-up of persistent soil-borne diseases such as footrots caused by *Fusarium solani, Fusarium culmorum* and *Didymella pinodella* and pests such as nematodes, field beans and related crops should not be grown within 5 years of themselves or each other. For spring beans, in addition, stem rot (*Sclerotinia sclerotiorum*) may be a problem and host crops such as linseed and oilseed rape should therefore have the same rotational restrictions.

14.6.1.3 Varieties

These are detailed on the Processors and Growers Research Organisation (PGRO) Website. Beans are classified as winter or spring beans and also by hilum colour or as tic beans. Pale hilum types are suitable for the human consumption export market. Winter varieties are generally large seeded while spring varieties are smaller. Tic beans are smaller and rounded and are used in the pigeon feed market. Downy mildew resistance may also be of interest when selecting a variety.

14.6.1.4 Seed treatment

Certified seed is advised but home saved seed is commonly used to reduce costs. No fungicide seed treatments are currently available for field beans. When using home-saved seed it should be tested for germination and seed-borne leaf and pod spot (*Ascochyta fabae*) infection. It is advised that seed for commercial production (C2) should not be used if it contains more than 1% leaf and pod spot infection. It is also recommended that all field bean seed is tested for stem and bulb nematode (*Ditylenchus* spp) and should not be used for seed if nematodes are present. PGRO offer a testing service.

14.6.1.5 Seedbed preparation

Beans will tolerate relatively cloddy seedbed conditions in comparison to other smaller seeded crops but yields will benefit from good soil structure and the removal of any compacted layers at depth. Winter beans were historically sown at a time of year when soil moisture levels were high and were therefore often broadcast or drilled onto the soil surface and then covered by shallow ploughing. However, both seed distribution and emergence can be very uneven and so most winter and spring beans are now drilled. Direct drilling or minimum tillage drilling methods for establishing both winter and spring beans are increasingly popular. Beans must be covered by a minimum of 3 cm soil and can be sown up to a depth of 12.5 cm.

14.6.1.6 Time of sowing

Winter beans should not be sown before the second week of October as crops which are too forward are more prone to disease and to the effects of severe winter weather. Sowing from mid-October to early November is usually the optimum time but sowing can continue right through December and even January. Spring beans are best drilled as early as possible from late February onwards provided soil conditions are satisfactory. Later sowing delays harvest and may subject beans to summer drought stress at flowering.

14.6.1.7 Seed rates

Winter beans require an established plant population of 18–20 plants/m^2 post-winter and the chosen seed rate should allow for 15%–20% field loss. For spring beans the required population is 50–55 plants/m^2 (reduce to 35–45 for very fertile sites) with a field loss of up to 5%. Dense populations tend to increase disease problems whereas thin crops are less competitive to weeds. Thousand seed weight should also be taken into account when calculating seed rates.

14.6.1.8 Crop development

Field beans are an indeterminate crop, thus producing flowers, pods and seed over a period of time throughout crop development while vegetative growth is continuing (Table 14.5).

Table 14.5 Principal growth and secondary growth stage key (BBCH system) for field beans.

GS0	***Germination***
00	Dry seed
07	Shoot emerged from seed (plumule apparent)
08	Shoot growing toward soil surface
09	Emergence: Shoot breaks through soil surface
GS1	***Leaf Development***
10	Pair of scale leaves visible
11	First pair of leaves unfolded
12	Two pairs of leaves unfolded
19	Nine or more pairs of leaves unfolded
GS5	***Inflorescence/flower bud emergence***
50	Flower buds present still enclosed by leaves
51	First flower buds visible outside leaves
55	First individual flower buds visible outside leaves but still closed
59	First petals visible, many individual flower buds, still closed
GS6	***Flowering and pod set***
60	First flowers open
63	Flowers open on three nodes per plant
65	Flowers open on five nodes per plant
66	Full flowering, flowers open on 6 or more nodes per plant
67	First pod set, flowering declining
69	End of flowering
GS7	***Development of fruit***
70	First pods have reached final length
75	Fifty percent of pods have reached final length
79	Nearly all pods have reached final length
GS8	***Ripening***
80	Beginning of ripening: Seed green, filling pod cavity
85	50% of pods ripe, seeds dry and hard
89	Fully ripe: nearly all pods dark, seeds dry and hard (desiccation stage)
GS9	***Senescence***
93	Stems begin to darken
95	Fifty percent of stems brown or black
97	Plant dead and dry
99	Harvested product

Source: Pulse Growers Research Organisation.

14.6.1.9 Fertilisers

The fertiliser requirements of beans are small and no nitrogen is required. Magnesium, phosphate and potash requirements are shown in Table 14.6 which assumes a bean yield of 3.5 t/ha. Where yields are likely to be larger or smaller than this then rates should be adjusted. At Index 0 and 1 the required fertiliser should be incorporated into the seedbed.

Table 14.6 Nutrient recommendations (kg/ha) for field beans.

	P, K or Mg index						
	0	**1**	**2**	**3**	**4**	**5**	**6**
Phosphate (P_2O_5)	100	70	40	0	0	0	0
Potash (K_2O)	100	70	40 (2−) 20 (2+)	0	0	0	0
Magnesium	100	50	0	0	0	0	0

These recommendations have been based on Nutrient Management Guide (RB209) Section 4 updated in January 2021 by the Agriculture and Horticulture Development Board.

The micronutrients manganese and boron may also be required on deficient soil types.

14.6.1.10 *Crop protection*

14.6.1.10.1 Pests

Slugs may be a serious problem requiring an application of approved slug pellets where the winter crop is late sown into a cloddy seedbed. Rooks and pigeons may also damage the crop and bird scarers or other methods of control may be required.

Pea and bean weevil (*Sitona lineatus*) can cause yield loss due to larval feeding on the root nodules and is more likely to be a problem in spring beans. Insecticides can be applied to prevent egg laying when adult damage in the form of 'U' shaped notches on the leaf margins is seen at the early true leaf stages of establishment.

Stem nematode (*Dictylenchus dispaci*) can cause severe yield loss and complete crop failure and can live in the soil for up to 10 years. The use of clean, nematode free seed and long rotations will help to reduce the risk.

Bean seed beetle or Bruchid beetle (*Bruchid rufamanus*) is a particular problem in crops for human consumption due to contamination and the holes in the beans caused by larval feeding. Egg laying adults can be controlled by a pyrethroid insecticide at first pod set when temperatures reach 20°C for two consecutive days and repeated 7 days later.

Dense colonies of Black bean aphids (*Aphis fabae*) can cause direct feeding damage and transmit viruses more commonly in spring beans than winter beans. Approved insecticides may be required during and after flowering but caution should be exercised in using insecticides in flowering crops due to the potential impact on pollinators.

14.6.1.10.2 Diseases

Downy mildew (*Peronospora viciae*) can be a problem in cool, damp conditions in spring beans causing a greyish-brown felty growth on the underside of the leaf. Varietal resistance and fungicide treatments are available.

Leaf and pod spot (*Ascochyta fabae*) should be minimised by the use of clean seed but fungicides will give some control if required.

Chocolate Spot (*Botrytis fabae)* produces brown leaf lesions which enlarge and coalesce in cool, damp conditions, it is mainly a problem in winter beans. Rust (*Uromyces fabae*) causes red-brown pustules on upper leaf surfaces which can be damaging if they appear during flowering or early pod set. Later infections are less damaging. Fungicide applications at early flowering and 4 weeks later are routinely used to control Chocolate Spot and Rust.

14.6.1.10.3 Weeds

Initially, field beans are susceptible to weed competition, although later they are a very competitive crop. Tall, climbing weeds can also cause lodging. The majority of approved herbicides are residuals for pre-emergence use with very limited post-emergence options, apart from graminicides for volunteer cereal and grass weed control.

Glyphosate can be used pre-harvest for perennial weed control (except for seed crops) when the crop is at 30% moisture. The pods will be dry and black although stems may still be green at this stage.

14.6.1.10.4 Typical crop protection programme for spring field beans

Despite the range of pests, diseases and weeds listed which may affect a crop (see also Chapters 5–7), an individual crop is unlikely to face all of these problems in practice. In addition, the control of more than one problem is often carried out simultaneously by tank mixing treatments to reduce the number of 'passes' the sprayer makes through the crop. A typical crop protection programme for a spring sown field bean crop is shown in Table 14.7. Spring sown crops dominate the combinable pulse acreage and hence are chosen for inclusion in the table below.

14.6.1.11 Harvesting

Winter beans are usually ready for harvesting in August or September; spring beans from late August onwards. They ripen unevenly, lower pods first, and although the leaves wither and fall off early, combining has to be delayed until the stems lose their green colour. Loss of shed seed when combining may be reduced by combining in dull

Table 14.7 Typical crop protection programme for spring sown field beans.

Crop growth stage	Indicative date	Treatment type	Target
GS 00	Feb/Mar	Herbicide	Broad leaved weeds and some grass weeds
GS 11-12	Mar/Apr	Insecticide	Pea and bean weevil
GS 60-65	June/July	Fungicide	Foliar diseases
		Insecticide	Black bean aphid

weather or in the morning or evening. The crop is at its easiest to combine when the seed is at 18% moisture content. Beans for seed and specialist markets such as human consumption or pigeons must be carefully harvested and handled.

14.6.1.12 Drying and storage

Field bean seeds are large and must be dried carefully to ensure the moisture is removed from the centre of the seed and to avoid cracking. In continuous dryers, seed of over 20% moisture should be dried in two stages. On-floor dryers may be successfully used but if the ducts are widely spaced some beans near the floor may not dry properly as the air escapes easily. For long-term storage, beans should be dried to 14% moisture content and the marketing standard is normally 14% moisture and 2% impurities or a combination of the two up to 16%.

14.6.1.13 Yield

Indicative yields in t/ha are shown in Table 14.8.

14.6.2 Dry harvested peas

Dried (combining) peas are a useful break crop that can be used for human consumption or as a high protein component of pet and livestock feeds (compounding or micronising) and are a good source of residual nitrogen for the following cereal crop.

14.6.2.1 Types of peas

14.6.2.1.1 White flowered

- *Marrowfat.* This is the main type used for human consumption, the seeds are large and wrinkled with a blue-green seed coat. A good colour and freedom from waste and stains are required quality factors so timeliness of harvest is important.
- *White peas (yellow).* These have round, smooth seeds with a white/yellow seed coat. They are mainly used for animal feed but are also sold for use in soups and as split peas.
- *Small blues (green).* These have small, round, smooth, seeds with a blue-green seed coat. There is limited use for canning as small processed peas.
- *Large blues (green).* These have large, round, smooth seeds with a blue-green seed coat. Samples with good colour may also be used for micronising to be used for pet food or in packets for human consumption.

Table 14.8 Field beans indicative yields (t/ha).

	Average	High
Winter beans	3.5	5.0
Spring beans	3.25	4.5

14.6.2.1.2 Coloured flowered

- *Maple peas.* These have purple flowers and small round brown speckled seeds. These varieties are grown as whole-crop forage or for the seed to be sold for pigeons.

14.6.2.2 Site selection

Peas grow best on well-drained loams and lighter soils. The pH should be about 6–6.5; if it is too high, manganese deficiency is likely and may reduce the quality of the harvested sample. Lodging may be a problem where fertility is high and so suitable varieties should be chosen.

Compared to the other pulses, peas are more tolerant of drought stress and high rainfall may make combining difficult and result in stained produce which may be less marketable.

14.6.2.3 Rotation

Peas should not be grown more often than 1 year in 5 to avoid the build-up of foot rots caused by *Fusarium solani*, *Aphanomyces euteiches* and *Didymella pinodella* which can survive as resting spores for long periods in soil. It is also recommended to leave a similar interval between peas and other pulse crops to avoid problems such as the Pea cyst nematode (*Heterodera gottingiana*).

14.6.2.4 Varieties

The main market for combining peas is for animal feed and high yield is important. For the specialist markets such as human consumption, pigeons and micronising for pet food, choice of an appropriate variety is the main consideration and the premium will need to compensate for the lower yield. Disease resistance, standing ability and ease of combining are also important. Varieties are detailed in descriptive lists on the PGRO website.

14.6.2.5 Seed treatment

Certified seed is advised but home saved seed is commonly used to reduce costs. No fungicide seed treatments are currently available for combining peas but some nutritional and biostimulant seed treatments can be used to encourage healthy early growth and boost the plants own natural defences. When using home-saved seed it should be tested for germination and seed-borne leaf and pod spot (*Ascochyta fabae*) infection. PGRO offer a testing service.

14.6.2.6 Seedbed preparation

Ploughing in the autumn and allowing weathering to occur will facilitate the preparation of a seedbed with a minimum number of cultivations because peas are sensitive to compaction. On very light land, spring ploughing can work well. Peas can also be established successfully by minimum tillage or direct drilling techniques. Recommended row width is 20 cm.

14.6.2.7 Time of sowing

Peas can be sown after mid-February and those sown early can achieve higher yields and earlier maturity but sowing into suitable seedbed conditions is the most important consideration.

14.6.2.8 Seed rates

Seed rates vary with the type of pea being sown. The marrowfat types require an established population of 60–70 plants/m^2 to achieve optimum yields whereas the white and blue types require an established population of 80–90 plants/m^2 to achieve optimum yields.

The seed rate calculation should allow for 15%–20% field losses and should also take in to account 1000 seed weight. Seed rates may be reduced slightly on very fertile soils and increased on very drought-prone sites.

14.6.2.9 Crop development

Table 14.9 Growth stages in combining peas.

Code	Definition	Description
Germination and emergence		
000	Dry seed	
001	Imbibed seed	
002	Radicle apparent	
003	plumule and radicle apparent	
004	Emergence	
Vegetative stage refers to main stem and recorded node. Two small-scale leaves appear first and the nodes where these occur are not recorded.		
101	First node	(Leaf fully unfolded, with one pair of leaflets, no tendrils present)
102	Second node	(Leaf fully unfolded, with one pair of leaflets, simple tendril)
103	Third node	(Leaf fully unfolded, with one pair of leaflets, complex tendril)
10×	*x* node	(Leaf fully unfolded, with more than one pair of leaflets, complex tendril found on later nodes)
	Last recorded node	(Any number of nodes on the main stem with fully unfolded leaves according to cultivar)

Continued

Table 14.9 Growth stages in combining peas.—cont'd.

Code	Definition	Description
Reproductive stage refers to main stem, and first flowers or pods apparent.		
201	Enclosed buds	Small flower buds enclosed in terminal shoot
202	Visible buds	Flower buds visible outside terminal shoot
203	First open flower	
204	Pod set	A small immature pod
205	Fat pod	
206	Pod swell	Pods swollen, but still with small immature seeds
207	Pod fill	Green seeds fill the pod cavity
208	Green wrinkled pod	
209	Yellow wrinkled pod	Seed 'rubbery'
210	Dry seed	Pods dry and brown, seed dry and hard
Senescence stage refers to lower, middle and upper pods on whole plant.		
301	Desiccant application stage. Lower pods dry and brown, seed dry, middle pods yellow and wrinkled, seed 'rubbery', upper pods green and wrinkled	
302	Pre-harvest stage. Lower and middle pods dry and brown, seed dry, upper pods yellow and wrinkled, seed 'rubbery'	
303	Dry harvest stage. All pods dry and brown, seed dry	

These definitions and codes refer to the main stem of an individual plant. Definitions and codes for nodal development refer to all cultivars; only nodes where a stipule and leaf stalk develop are recorded; descriptions in brackets refer to conventional leaved cultivars only.
Source: Pulse Growers Research Organisation.

14.6.2.10 Fertilisers

The fertiliser requirements of peas are small and no nitrogen is required. Magnesium, phosphate and potash requirements are the same as for beans as shown in Table 14.6 which assumes a yield for peas of 4 t/ha. Where yields are likely to be larger or smaller than this then rates should be adjusted. At Index 0 and 1 the required fertiliser should be incorporated into the seedbed. Peas may suffer from sulphur deficiency on sensitive soil types. Where deficiency is possible, 25 kg SO_3/ha should be applied.

If manganese deficiency occurs (sometimes called marsh spot), symptoms being yellowing of leaves between veins which remain green and a brown spot in the centre of the pea seed, manganese sulphate should be applied at full flower and again 10–14 days later. It is most likely to be a problem on organic soils or those with a high pH and should be routinely applied on soils with a pH above 7.

14.6.2.11 Crop protection

14.6.2.11.1 Pests

Pea and bean weevil (*Sitona lineatus*) adults cause characteristic leaf margin 'U' shaped notches but it is the damage caused by the larval stage eating the root nodules that can cause yield losses. It is more likely to be a severe problem in the earliest sown crops. Insecticides can be applied to prevent egg laying when adult damage is seen at the early true leaf stages of establishment. A monitoring system which detects adults when they begin migration into crops in spring is available to aid spray decision making.

Pea aphids (*Acyrthosiphon pisum)* if present in large numbers may result in a severe reduction in yield through flower abortion and growth distortion. They can also transmit viruses. A predictive model is available and an insecticide should be applied when 20% or more of plants are infested.

Pea moth (*Cydia nigricana*) larvae are an important pest which feed upon the developing seeds reducing the value of the crop particularly those for human consumption or seed. The timing of control with insecticides is a very narrow window between egg hatch and larvae entering the pods. Pheromone traps should be placed in the field, monitored very regularly and used in combination with a model run by PGRO to determine an accurate spray application date.

14.6.2.11.2 Diseases

Downy mildew (*Perenospora viciae*) persists in the soil from previous crops and infects seedlings, with secondary spread within the crop then occurring by windblown spores. Varietal resistance is important as there are currently no approved seed treatments or foliar fungicides for control.

The main disease threat later in the growth cycle is from leaf and pod spot diseases (*Mycosphaerella pinodes, Didymella pinodella and Ascochyta pisi*) which cause lesions on the leaves, pod and stem and are most likely when wet weather occurs during flowering. The use of disease-free seed and foliar fungicides during flowering and pod set is advised. Grey mould *(Botrytis* spp*)* is also a risk during petal fall in wet weather and can be controlled with fungicides applied during this period.

14.6.2.11.3 Weeds

Good weed control is essential in peas which are not a very competitive crop and as well as increasing yield will aid combining and drying of the grain. Broad leaved weed control is based around the use of pre-emergence residual herbicides as post-emergence options are very limited. Good activity relies on good seed bed conditions and soil moisture being present. A leaf wax test should be carried out to ensure crop safety prior to the use of some post-emergence broad leaved weed herbicides. Grass weeds can be removed relatively easily with the use of post-emergence graminicides.

14.6.2.12 Harvesting

The price of combining peas for human consumption is largely dependent on quality and great care should be taken at harvest. Wet weather pre-harvest can cause staining in a lodged crop but in contrast the sun can cause 'bleaching' which is also undesirable. Peas can be harvested when the moisture content of the seed is 18% moisture. If they are left in the field too long, or if they are over-dried, the crop may be unsuitable for human consumption, the percentage of 'non-soakers' increases and the seed may split and crack. Some growers will harvest at up to 20% moisture and dry them carefully down to 18% to reduce this risk. There is more flexibility over the timing of harvest with those being grown for animal feed which can be harvested at around 16% moisture to reduce drying costs but pod shatter and shelling out becomes a risk at very low moisture contents.

If the crop is very weedy or uneven in maturity, a pre-harvest desiccant can be applied to aid combining as long as it is permitted by the processor. Moisture content of the pea seed should have fallen to below 30% at application and harvesting should be possible 10–14 days later.

Most varieties of peas lodge to some extent before harvest, although modern semi-leafless and stiff-strawed types are less prone to this, combining is most easily carried out in the opposite direction or at an angle to the direction of lodging using efficient lifters.

14.6.2.13 Drying and storage

As with beans, peas must be dried and stored with care. The relatively large size of the seed makes drying more difficult and low temperature dryers are safer. Peas for human consumption are dried at a lower temperature than peas for animal feed and if moisture content is over 24% a lower temperature should be used and two passes may be more appropriate. For long-term storage, peas should be dried to 14% moisture content and the normal marketing standard is 14% moisture and 2% impurities or, a combination of the two up to 16%.

14.6.2.14 Yield

Yields for marrowfat varieties are in the range of 4 - 4.5 t/ha, while those for white and blue varieties will be in the range of 4.5 – 5 t/ha.

14.6.3 Lupins

Lupin seed has a protein content of 30% to 40% and therefore if the crop could be successfully grown in the UK it would be a very good non-GM protein source to compete with imported soya and be of particular interest to pig and poultry producers. Modern varieties have been bred for low alkaloid content (sweet lupins) and new, more determinate winter- and spring-sown varieties are available with a yield potential of 2–3 t/ha.

The main commercial interest is in two species:

- White lupin (*Lupinus albus*) with winter and spring varieties.
- Blue lupin (*Lupinus angustifolius*) with spring varieties.

Lupins require a soil pH below 7 and the seed should be inoculated with *Rhizobium* bacteria to encourage nodulation. Spring sown varieties dominate and should be sown from mid-March to mid-April.

The crop does not require nitrogen and on fertile soils has little requirement for phosphate and potash. Limited crop protection products are available with Extensions of Authorisation for Minor Uses (EAMUs) approval. It is usually ready to harvest in mid-late September.

14.7 Key points

- Oilseed and pulse crops are integral to many cereal based rotations as non-cereal i.e., 'break' crops that can be combine harvested for their oil or protein containing seeds.
- The pricing of both oilseed and pulse crops in the UK is influenced by global supply and demand for oilseeds and protein.
- The dominant combinable break crop in the UK is winter sown oilseed rape producing oil for human consumption and biodiesel.
- Linseed is a lower input alternative oilseed break crop to oilseed rape.
- Combinable pulses (field beans and combining peas) grown for animal feed and human consumption can provide a yield advantage in the following crops which benefit from the residual nitrogen but yield and profitability can be variable.
- Small areas of sunflowers, lupins and soya beans are grown in the UK but can present challenges in terms of the suitability of the varieties for the UK climate and reliability of yield.

Sources of further information and advice

Agriculture and Horticulture Development Board (AHDB), latest ed. AHDB Recommended list for Cereals and Oilseeds. Available at: https://ahdb.org.uk/knowledge-library/recommended-lists-for-cereals-and-oilseeds-rl.

Agriculture and Horticulture Development Board (AHDB), 2021. Nutrient Management Guide (RB209) Section 4. Available at: https://ahdb.org.uk/rb209.

Agriculture and Horticulture Development Board (AHDB), 2020. AHDB Oilseed Rape Growth Guide. Available at: https://ahdb.org.uk/knowledge-library/oilseed-rape-growth-guide.

Useful websites

Seed suppliers often provide useful information and comparisons of crop types and their management as well as more specific information on varieties.

PGRO, Information and advisory service for pulse growers and users. www.pgro.com.

Premium crops. www.premiumcrops.com.

Senova. https://www.winter-beans.co.uk.

Soya UK. https://www.soya-uk.com.

United Oilseeds. https://www.unitedoilseeds.co.uk/grow-your-crop.

Root crops

15

15.1 Introduction

Cash root crops are an important part of the cropping systems of a significant number of farmers throughout the main arable areas of the UK. On appropriate soil types they can be part of rotational diversity of cropping, acting as a break from combinable crops and, if grown well, can provide higher gross margins than cereals, oilseeds and legumes. Both sugar beet and potatoes require capital investment in specialist machinery, or the ability to use contractors, and most potato enterprises need investment in buildings suitable for long-term storage of the produce as well as grading/handling systems. There are well established markets and supply chains in place for the crop both in the UK and to a lesser extent for export for potatoes. Yields can be dramatically affected by the weather and are increasingly grown on a contract basis.

Root crops are usually spring sown with options suited to most soil types with harvest from early summer for new potatoes or throughout the winter for sugar beet.

On a global basis root crops are rarely traded on a large scale between continents due to the higher water content and lower dry matter compared to cereal crops. In these cases, national production tends to meet market demand. Sugar beet competes with cane sugar in certain markets including the UK where alternative sources of sugar are available. Sugar beet is a dual-purpose crop which can be used for sugar extraction or animal feed whereas potatoes are grown for human consumption or for seed.

15.2 Potatoes

Potatoes (*Solanum tuberosum*) is a plant from the Solanum family, with a starchy plant tuber which is one of the most important food crops, cooked and eaten as a vegetable. Globally potatoes are the fourth largest food crop after maize, wheat and rice. It is recognised by its leafy green canopy called the haulm and the harvested tubers that grow below ground. It is the potato tubers that are grown for specific markets for human consumption or seed production. High costs of production mean that potatoes are not usually grown for livestock feed.

15.2.1 Potato marketing and the supply chain

Historically in the UK growing more than 2 ha of potatoes would require a levy to be paid to the Agriculture and Horticulture Development Board. In 2021 potato growers

Lockhart and Wiseman's Crop Husbandry Including Grassland. https://doi.org/10.1016/B978-0-323-85702-4.00021-2

and buyers however voted overwhelmingly to abolish their sector's compulsory levy. The final year for collecting the levy was in the 2021/22 growing season.

In the UK, there are now only around 2000 growers compared to 80,000 in 1960. The area grown has also decreased to around 125,000 ha but total production has remained relatively constant due to improvements in yield. Approximately 80% of the crop is grown for human consumption known as 'ware' with an additional 10% for seed and 10% for stock feed. Within the human consumption market approximately 45% are sold fresh by the multiple retailers as 'prepacks', i.e., washed and packed in plastic bags with the other 55% used for processed potato products such as chips, crisps and ready meals. For both the pre-pack and processing market the majority of the crop is now grown under contract.

Growing a high yielding crop of potatoes is relatively straightforward but achieving the high quality that the market requires is where the skill and attention to detail by the grower is required. Subject to the specification of the buyer, good quality ware potato tubers require certain characteristics that will affect appearance, storability and cooking/eating qualities.

These include potatoes that

- Have not dehydrated.
- Are not damaged, frosted or diseased.
- Are free of greening (an indication that a toxic substance, solanine, is present).
- Are free from secondary growth irregularities such as misshapes, cracks, sprout growth.
- Have a bright clean skin finish for supermarket fresh potatoes.
- Are of good shape and size (40–80 mm depending on target market).
- Have clean skins with shallow eyes for easy peeling.
- Have not been damaged by pests such as wireworms, slugs and cutworms.
- Have flesh which does not turn brown or blacken before or after cooking.
- Do not break down when being boiled or fried.
- Free from any internal defects such as hollow heart, spraing or rust.

In addition, potatoes must have special qualities to be suitable for processing. For crisps, chips and dehydration the tubers must have a high dry matter (starch) content and a low reducing sugar content.

15.2.2 Site selection

The crop will grow in a wide range of soil and climatic conditions provided the land is well drained and the soil and subsoil structure are good. Potatoes grown on heavier soils risk damage to soil structure at harvest with potential crop losses in a wet autumn. Potatoes grown on lighter sandier soils will require irrigation in a dry season. Access will be required for large harvesting machinery so steep sloping fields and any obstacles such as pylons should be avoided if possible.

Early potatoes are most productive on light, well drained soils in areas free of late frosts, but some frost protection is possible with irrigation. These areas are usually coastal, where the sea acts as a storage heater and a buffer against cold temperatures. Covering with polythene sheeting or microfleece can hasten crop establishment. This is removed when the crops grow bigger. Anecdotal evidence suggests that darker wetter soils hold onto the temperature from the sun during the day.

Maincrops are best suited to deep, fertile, loam soils because high yields are very important, as this can offset low prices. Potatoes will grow in acid soils (under pH 5.5); common scab can cause problems on high pH soils in a dry season. Irrigation, if it is possible, should be considered in order to suppress the disease.

15.2.3 Rotation

Ideally, potato crops should not be grown more than 1 year in 5, in order to avoid a build-up of soil borne diseases such as Rhizoctonia (*Rhizoctonia solani*), and pests such as nematodes. Potatoes are usually used as a break crop in cereal heavy rotations. Seed crops cannot be grown on PCN infested land.

15.2.4 Varieties

There are many factors that need to be considered when selecting which potato variety to grow. Over time plant breeders have selected traits so that potato varieties can be selected according to the intended market, yield potential, pest and disease resistance, storage qualities, harvest date and season length. A grower will usually know the intended market before planting and this will be a key influence when selecting a variety.

In the United Kingdom, varieties are classified as first earlies, second earlies and maincrop. The difference in times of maturity between varieties is due to varieties being either determinate or indeterminate. Indeterminate varieties will carry on growing for longer and, in some environments, will need some form of management to stop leaf and stem growth, prior to harvesting. Determinate varieties are shorter season crops and will mature quicker. Early potatoes are usually determinate varieties as they need to mature earlier to reach a lucrative early season market. Early potatoes are usually grown in warmer climatic conditions that are frost free, for example Cornwall and Pembrokeshire. Yields are lower than for maincrops, but prices are often high, especially very early in the season. Early potato production can be difficult to predict as demand and supply are hard to match. Yields can be variable at this time and this can influence prices for potatoes that are not grown on contract. Second earlies are planted early in the season and can be either harvested early for sale or stored long term. Maincrop potatoes are planted in the spring and usually grown to maturity for long term storage, depending upon the contractual arrangements (Table 15.1).

Other specific characteristics of varieties that may be considered by growers are resistance to potato blight (*Phytophthora infestans*) and potato cyst nematode (*Globodera* sp.).

Varieties and their characteristics are detailed in the AHDB *Potato Variety Database*. This includes independent and commercial plant breeder information on a range of potato varieties.

The Seed Potato Classification Scheme (SPCS) provides classification of all potatoes produced and marketed in England and Wales, depending on the class of the

Table 15.1 The top 10 varieties grown by area in 2020 In the UK.

Rank	Variety	Area planted (Ha)	Maturity	Usage
1	Maris Piper	13,920	Maincrop	Retail fresh and processing
2	Markies	5870	Maincrop	Retail fresh and processing
3	Melody	4730	Maincrop	Retail fresh
4	Taurus	3860	Early maincrop	Processing - crisping
5	Sagitta	3740	Early maincrop	Multipurpose
6	Innovator	3580	Second early	Processing - French fries
7	Maris Peer	3540	Second early	Retail fresh - salads
8	Royal	3430	Early maincrop	Processing - crisping & French fries
9	Lady Rosetta	3310	Early maincrop	Processing - crisping
10	Nectar	2480	Early maincrop	Retail fresh

parent seed, and the health of the crop and tubers. The aim of the SPCS is to provide assurance that seed potatoes delivered to buyers and growers meet specified minimum health and quality standards. In England and Wales, the SPCS is administered on behalf of the Department for Environment, Food and Rural Affairs (Defra) by the Animal and Plant Health Agency (APHA).

Every EU member state operates certification schemes to ensure that all seed potatoes marketed within the EU are officially classified, and comply with regulations.

The SPCS classifies all potatoes that are marketed in England and Wales. The SPCS inspects the quality of seed potatoes to protect purchasers and to ensure that seed potatoes meet the Seed Potatoes (England) Regulations 2015 and Seed Potatoes (Wales) Regulations 2016.

The inspections, tests and controls required by the SPCS are carried out by Plant Health and Seeds Inspectors (PHSI) in order to determine a seed potato's health, trueness to variety and freedom from mixtures.

When classified, seed potatoes are given a Union grade which is a harmonised grade allowing marketing throughout the EU.

Seed potato crops must be entered for classification in one of three categories: (Table 15.2). For each grade there is a maximum field generation as well as different field tolerances to diseases including viruses.

Table 15.2 Seed potato classification.

Pre-basic	Basic grades	Certified grade
PBTC (tissue culture)	S	A
PB (field grown)	SE	B
	E	

Pre-basic and basic seed is mainly intended for the production of seed crops, while certified seed is mainly intended for ware production.

15.2.5 Seed treatment

Options are limited to a small number of plant protection products to control seed and soil borne diseases along with some biological and biostimulant products. They can be applied to clean tubers on a roller table or on the planter.

15.2.6 Crop advancement – chitting (breaking dormancy)

Chitting is the process by which potatoes are encouraged to break dormancy by exposure to warmer temperatures and artificial light.

Fewer than half of the potato crops in the United Kingdom are grown from chitted seed. It is a necessity with earlies to obtain high early market prices and chitting should be started in the winter before planting so that single sprouts develop on each tuber (an apical dominance effect). It is also desirable to chit maincrop potato seed as this can increase yield by 3–5 t/ha. Some growers prefer to plant cold stored seed (non-chitted) as delayed planting can lead to long and weak chits that can cause planting difficulties as they are easily knocked off.

Seed crops also benefit from sprouting as the crop bulks earlier and so the haulms can be destroyed early which can check aphid transmitted viral diseases. To obtain high yields from maincrops and seed crops, several sprouts should be encouraged to grow on each tuber. The physiological age of seed tubers can have a marked effect on how most varieties develop. Physiologically old seed is produced from early planted seed crops which are lifted early in warm conditions and, when in store, have heat units added to them by keeping the temperature at levels above 4 °C. Such seed produces earlier crops with fewer tubers, but the crop is more susceptible to drought. Physiologically young seed, produced and stored in cooler conditions, is better for maincrops which grow on to produce high yields.

The rate of growth of the sprouts is controlled by temperature and is usually quickest at about 16 °C, but varieties differ considerably. The strength of the sprout is controlled by light. Short, sturdy sprouts are formed in well-lit buildings such as glasshouses or barns fitted with warm-white fluorescent tubes. Storage buildings must be frost-proof and well ventilated to avoid high humidity and condensation. Chitting trays, each holding about 15 kg (60–80 per tonne), are still the most popular containers for sprouting, especially for earlies. Higher capacity white plastic trays, or wire, metal frames with nest crates and pallet boxes holding at least 500 kg, can speed up the loading of high output maincrop planters while reducing labour inputs.

15.2.7 Seedbed preparation

Potato cultivations will depend upon many factors including previous cropping, soil organic matter status, soil structure, soil texture, climate, time of planting and soil depth. The aim is to produce a deep seedbed of fine tilth that is free from stones and clods. Heavier silt/clay soils are likely to require additional cultivations when compared to lighter sandy soils. There is a wide range of machinery available to potato growers and there are many systems that are specific to local conditions. Digging a soil profile pit is a useful way to examine the soil structure; any areas of compaction may influence cultivation choice. Potatoes are known for their intensive machinery and cultivation requirements which makes them a relatively expensive crop to grow. The majority of potatoes are grown on a bed system where following subsoiling and ploughing (if required) large ridges are made using a bed former. From this point onwards, the soil where the potatoes will be planted into is kept separate from the wheeling's of the cultivations/planting that will follow. If soils are cloddy, especially for silty/clay soils then bed tilling (rotorvating) may be required which is a time and diesel consuming process. In dry planting conditions light soils can be prone to wind erosion whereas planting on sloping fields can lead to soil water erosion in wet seasons. These issues can be mitigated by planting across gentle slopes or using water saving measures such as a tied ridger that encourages infiltration rather than run-off. Increasing soil organic matter can help reduce wind and water erosion.

Some soils are very stony and clods can still remain, which is likely to cause harvesting problems, misshapen potato tubers and tuber bruising damage. The stones can be collected mechanically and removed, or mechanically separated between the ridges by a 'destoner'. Any stones and clods that have not been removed from the seedbed may cause damage to the potato crop on the harvester or in the trailer. Stone/clod separation is a slow and expensive field operation. The destoner passes the ridge or bed soil over a moving web and the stones and clods can be windrowed alongside the row. The spaces on the webs of the destoner should ideally match those of the harvester. This should mean that, at harvest, all the soil drops away leaving just potatoes on the harvester webs. This makes harvesting easier, causes less damage to the tubers and prevents stones in the soil from restricting growth.

While cultivating and planting potatoes has been seen to be heavily reliant on machinery, labour and fuel there are systems both old and new which aim to reduce the input costs. On light, stone free soils, potatoes can be planted directly into a flat field that has been cultivated to produce a fine seedbed. This system does not require deep bed forming cultivations. Single pass systems for cultivations are available which will break up clods, separate stones and form beds that can be planted directly into.

Systems developed in other countries now have a single pass system that cultivates the soil and plants the crop in one go. These systems will only be effective where the soil type is uniform across the whole field and where space is available to move and transport the machinery from field to field. Many fields in the UK are not uniform and the soil type may be very variable within the field.

15.2.8 Timing of cultivations and planting

The earliest crops are planted in January and will be protected by plastic film or fleece. The majority of crops are planted in March and April, with some planted in May in a wet spring.

15.2.9 Seed rates and seed spacing

The decision on the most profitable seed rates for potatoes is complicated by factors such as cost of the seed and expected price for the crop; variety and size of the seed must also be considered carefully. The figures given in Table 15.3 are for normal-sized seed (30–60 mm), but if healthy small seed (20–30 mm) is used the rate can be reduced by at least 25%.

In some countries it is common practice to cut larger seeds into pieces by hand or machine before planting. The cut pieces are more likely to rot than whole seed and this practice is not advised in the UK.

Only basic or certified seed can be bought or sold. Growers can plant uncertified home-grown seed, but this is risky if the crop is from poor stock and is not protected by suitable insecticides against aphids, which spread leaf roll and mosaic virus diseases. Some varieties have a high resistance to virus diseases and so are better suited to growing-on as home-grown seed. A reliable tuber-testing service should be used to check any doubtful stocks of seed before planting.

UK seed crops of potatoes are grown in Northern Ireland, Scotland or at a high altitude in England. Disease and aphid levels are lower in these areas.

The planting density of potatoes is very much dependent on the market that the crop is destined for. Smaller salad potatoes that will be less than 40 mm in size when sold will require many more stems per unit area than say potatoes for baking. Baking potatoes which are 65–85 mm in size require more space to bulk up therefore fewer stems. Seed potatoes are best planted when they are all of a similar size as this will encourage synchrony of emergence and harvest. Potato seed can be size graded into specific fractions commonly <35 mm, 35–45 mm, 45–55 mm. Some seed supplied for processing crops are simply labelled as small or large, to differentiate between the two at planting.

Row width for earlies and seed potatoes is 60–70 cm; for maincrop, the commonest width is 90 cm. There is less yield loss and fewer problems in producing good ridges, keeping them free of clods, and fewer 'greened' and blighted tubers with the wider rows. The spacing between the sets will depend on the seed rate and the size of the seed. Table 15.3 gives an indication of optimum seed rate and set numbers per hectare. This assumes that the seed size is 35–55 mm, the rows are 90 cm apart and is about twice the likely price for the ware crop.

Table 15.3 Yield and seed rate for potatoes.

	Yield of tubers (t/ha)	Seed rate (t/ha)	Time of planting	Time of harvesting
Earlies	10–30	4–5	Feb–Mar	Late May/July
Maincrop	35–65	3–4	April	Aug/Oct
Seed	25–40	4–5	April	Sept/Oct
Canning	10–20	5–7	Mar–June	June/Oct

On highly fertile soils, especially if irrigated, and where higher yields (over 70 t/ha) are expected, some of the tubers may be too large (over 80 mm). Maincrop potatoes are usually size graded into <45 mm, 45–65 mm and 65–85 mm. Anything over 85 mm is unmarketable as a baking potato due to its size.

In these conditions, the seed rate should be increased by 10% to give more tubers of desirable size (Table 15.4).

15.2.10 Planting

Nearly all potato crops in the United Kingdom are planted mechanically with two to eight row machines available. Hand planting is still practised on a handful of farms growing early varieties. The ridges must be well-formed to protect the new tubers from blight spores and light, which would cause greening. Some farmers who grow on lighter soils plant potatoes using the bed system. There are usually two to four rows in beds 150–180 cm wide. This is preferable for irrigation (there is less run-off), and it produces more tubers in the 40 mm size fraction for salad potatoes and less greening. Planting depth is usually around 15 cm from the top of the ridge.

15.2.11 Fertiliser

When grown in soil with a good structure, potatoes are capable of producing extensive root systems that are efficient in taking up water and nutrients, therefore, every effort should be made to ensure seedbeds are free of compaction. The value of the potato crop is dictated by the marketable yield, not the total yield and, in consequence, decisions about fertiliser rates should be considered together with factors such as site selection and seed rates.

Fertiliser decisions for potatoes can be difficult and need to relate to the variety, weather, soil type and intended market. Specific advice should be sought from a FACTS qualified advisor and experienced potato agronomists.

Potato crops benefit from organic manures such as farmyard manure and slurry which should be taken into account when calculating any inorganic fertiliser requirements. Manure can be applied in the spring prior to planting at approximately

Table 15.4 Optimum populations and seed rates.

Sets/50 kg	Tuber spacing 90 cm row	Sets/ha	t/ha
630	31 cm	36,000	2.86
650	25 cm	45,000	3.46
600	32 cm	35,000	2.96
700	30 cm	35,500	2.70
800	31 cm	36,000	2.25
750	33 cm	34,000	2.26
750 (*Cara*)	17 cm	65,000	4.30

30–50 t/ha of farmyard manure or slurry equivalent. Normally this is ploughed—in during the spring cultivations. Fertilisers are either broadcast during the spring cultivations or, more efficiently, placed in the ridges by the planter. Too much nitrogen applied to the seedbed usually delays tuber development which can reduce yields in some seasons. Too much nitrogen can also delay haulm senescence and skin set. This can lead to a harvest where damage can occur to un-set skins and to lower dry matters. The tuber size is increased by nitrogen and potash, while phosphate increases the number of tubers. Dry matter can be decreased by high levels of potash.

Nitrogen rates for potatoes are not straightforward. Growers must consider the soil nitrogen supply, the length of growing season and the longevity of the haulm – linked to maturity of the crop. The last factor is a varietal one and varieties fall into one of four categories:

Short haulm longevity (determinate varieties e.g., Estima).
Medium haulm longevity (partially determinate varieties e.g.,.Melody).
Long haulm longevity (indeterminate varieties e.g., Maris Piper).
Very long haulm longevity (Indeterminate varieties e.g., Markies).

Nitrogen can be placed in the seedbed and topped up when the crop has emerged and the tubers are beginning to form. There is a direct relationship between crop yield potential and total N uptake. Most of the N uptake occurs relatively early within the growing season. The fertiliser policy chosen for P, K and Mg depends on many factors that are crop, soil and farm specific. Potatoes will often respond to fresh phosphate (P) but the rate of response is less at high soil P Indices. Within a typical arable rotation, the potato crop is more likely to respond to P than most other crops, such as cereals or oilseed rape. Potassium (K); a 50 t/ha potato crop will remove approximately 300 kg K/ha (5.8 kg K for each tonne of tuber yield) and higher or lower yields will remove more or less K. Since this is always more than the amount needed to maximise yield, a policy based on applying K to replace removal is appropriate.

For specific information on fertiliser rates use the AHDB nutrient management guide section 5.

15.2.12 Crop development

Potatoes can take from 2 to 6 weeks to emerge depending on the soil temperature, with early planted crops planted into cold soils taking the longest. The emerging crop usually grows rapidly as the soil and air temperature begins to increase in the spring. Uniform and synchronous emergence is favoured in order to produce a uniform canopy capable of producing potatoes that are of a similar size to meet the market specification. Once emerged the canopy will soon cover all of the bare soil which will reduce any weed competition. Leaves rapidly develop and as the plants begin to meet along the rows the tubers will begin to form on the stolons below ground. This is known as tuber initiation. The leaf development phase continues until the canopy produces flowers which is usually in June for most maincrop varieties. Potatoes produce little pollen compared to cereal crops and pollination is not required for the formation of the harvestable crop. As potatoes are flowering the tubers that have initiated will begin to bulk up and increase in size. The rate and extent of bulking will be reliant on an adequate supply of irrigation/rainfall and sunshine. Harvest will begin when either the crop has met the market specification or contractual commitments need to be fulfilled. The early potato harvest can be from May onwards whereas maincrop varieties for storage will begin in September.

15.2.13 Crop protection – pests (see chapter 7)

Slugs may be a serious problem in potatoes when the tubers have formed and the crop remains in the soil prior to harvest. Heavier, cloddy seedbeds are most at risk from slug damage especially where soils are kept moist from rainfall or irrigation. Slug trapping should be used to indicate numbers and pellets should be applied when there is a high risk. Tubers damaged by slugs are unmarketable.

Potato cyst nematode (PCN) (*Globodera Sp.)* is one of the most damaging pests of potatoes in the UK. Estimated annual yield losses cost the sector as much as £26m and in 2016 a survey has shown that 48% of potato fields tested showed the presence of PCN. There are two different species of PCN. White PCN (*G. pallida*), with pale coloured females and golden PCN (*G. rostochiensis*), with golden coloured females, though these colours are only seen on recently emerged cysts. These cysts can remain dormant in the field for up to 20 years until another *Solanum* crop is planted. Once roots of the potato crop start to grow the eggs within the cysts hatch and the newly released nematodes invade the root. This root feeding damage reduces crop growth and the subsequent yield of potato tubers. Increasing the number of years between potato crops and selecting resistant varieties can help reduce the effects of PCN. Chemical control options are available in the form of granular and liquid nematicides applied and incorporated into the soil. Left uncontrolled PCN will multiply in numbers resulting in an increased risk to any subsequent potato crops. PCN can also be reduced by using trap and biofumigation crops such as nightshade and mustard cover crops.

Free-Living Nematodes (FLN) are considered a significant constraint to profitable potato production most prominently in processing varieties. FLN cause damage directly by feeding on potato roots, reducing yields and quality, and indirectly by infecting roots and transmitting tobacco rattle virus (TRV) leading to necrotic lesions in the tuber flesh called spraing. Relatively low levels of TRV infections can render entire crops unsaleable, both for the fresh and the processing industries. The problem is further compounded by beneficial and pathogenic FLN species co-existing making accurate species identification by laboratory analysis important.

Aphids, particularly those that are virus vectors may need controlling in seed crops. Aphids pick up and spread the virus quickly. The Peach-potato aphid, *Myzus persicae,* is considered the most efficient vector of PVY. Symptoms include chlorotic mosaics in the haulm with possible leaf deformation, reduced leaflet size and overall reduction in vigour. Susceptible varieties could also produce smaller tubers. Aphid numbers can be assessed using monitoring and forecasting services with chemical control possible using insecticides although insecticide resistance is a problem in aphid populations. Seed crops are usually grown in higher altitude regions where temperatures are lower reducing the risk of virus transmission by aphids.

Wireworms (*Agriotes* spp) are the soil-dwelling larvae of click beetles and can damage potato crops in a number of ways. They feed for up to 5 years before pupating and cause damage by burrowing into roots and tubers, leading to significantly reduced marketability, yield loss or seedling death. There are no specific chemical control options but risk can be reduced by avoiding planting potatoes in fields that have been in long term grass pasture and leys. Some PCN nematicides will have incidental activity to help reduce wireworm numbers. Pheromone traps can be used to monitor the adult click beetles or soil traps to assess the numbers of wireworm larvae. Accurate sampling though is difficult and the numbers trapped do not always relate to the level of crop damage.

15.2.14 Weed control

Potatoes are a very competitive crop once they meet across the rows at full canopy, however, early weed emergence can reduce yields due to competition for light, water, space and nutrients. Weeds can also affect potato quality and ease of harvesting.

Weeds, by tradition, have been controlled in good weather by cultivations, including ridging after planting. However, cultivations may be impossible in wet seasons and there can be a considerable loss of valuable moisture and damage to crop roots in dry seasons. Re-ridging of potatoes can control weeds culturally but in most potato crops, weeds are controlled by herbicides; often a mixture of contact and residual products.

15.2.14.1 Pre-planting

Perennial weeds should, if possible, be controlled by using a non selective herbicide applied pre-harvest in the previous combinable crop (the weeds must be green and actively growing), or by a stubble treatment in the autumn.

15.2.14.2 Pre-emergence of potato shoots

Weed control in potatoes is best achieved following planting of the crop but before emergence. Once the crop has emerged, chemical control options are greatly reduced.

Since the loss of fast acting contact herbicides growers are using a range of residual herbicides soon after planting. These residual herbicides work best when soil moisture is present in order to maximise uptake into the emerging weeds. Product choice will depend on the weeds likely in that location, soil type and variety of potatoes grown. Some herbicides have variety and soil type restrictions. Always check the product label to ensure compatibility with variety and soil type.

15.2.14.3 Post-emergence of potato shoots

There are very few products which can be used post-emergence in potatoes. These should only be relied upon to control weeds in a season where residual activity has been poor or in a cold spring where weeds may be emerging at the same time as the crop. Some post-emergence herbicides will check the crop and show signs of chlorosis for a few days from which crops usually recover. Several graminicides are approved for grass weed control where populations are high following cereals.

15.2.15 Haulm (leaves and stems) destruction

Chemicals are normally applied to destroy the haulm; this encourages skin set, prevents further bulking and reduces the spread of blight to the tubers. Mechanical defoliation of the haulm is increasingly practised as the choices of fast acting defoliation chemicals is further restricted. Mechanical defoliation is done by using specialised flail machines that match the shape of the potato beds but can be risky due to compaction or removing soil from the ridge. Potatoes that are destined for storage or the supermarket shelf (apart from new potatoes) will require a set skin to extend shelf life and preserve quality. Skin set may take 3 weeks from haulm destruction depending upon the maturity of the crop canopy.

Volunteer potatoes from tubers and seed are an increasing problem in other crops and these should be controlled using chemical or cultural methods such as cultivations.

Prior to haulm destruction potatoes that are being grown for long term storage (not for seed) may have a chemical application to the green canopy of a product that will supress sprouting (re-growth) in potato stores. This chemical is translocated from the leaves and into the potato tubers.

15.2.16 Disease control (see also Chapter 6)

15.2.16.1 Blight - *Phytophthora infestans*

The name Phytophthora comes from the Greek words for plant decay and infestans from the Latin verb for infesting. This is the worst fungal disease which attacks the potato crop. It can seriously reduce yield by killing the foliage early; during periods of heavy rain the spores of the fungus can be washed into the soil and onto the tubers, so causing them to rot in the ground or during storage. The disease spreads rapidly in warm, humid conditions which, when they occur, are known as 'Blight Periods'. These are recorded by the Meteorological Office, and information can be obtained by farmers by mobile text or through the Internet. Growers can then respond by an application of fungicides; a wide range is available. Most fungicides have protectant and systemic action and are effective if sprayed regularly with the leaves being well covered. Spray intervals range from 7 to 14 days depending on material used and pressure of disease. The use of irrigation in the summer usually triggers a Blight Period and a spray programme should be started soon after the crop has reached the rosette stage (fully emerged crop not meeting along the row).

To reduce the risk of spores spreading from the leaves to the tubers, the haulm should be destroyed if blight has not been controlled effectively. Chemical fungicide applications can be made to specifically target blight infections to the tubers. It is also important to think about preventative measures to stop the initial spread of blight. Potato dumps and volunteers in neighbouring fields are important sources. Reject and waste potatoes should be used as livestock feed or in an anaerobic digestor. Volunteer potatoes should be controlled in other crops before they become a source of infection.

Other diseases of potatoes are mainly seed-borne and can be controlled by treating seed with fungicides. They include gangrene, silver scurf, dry rot, skin spot and stem canker. Control methods for disease can be in the form of foliar sprays, varietal resistance, seed treatments and in furrow soil applied products.

15.2.17 Physiological disorders

In addition to a reduction in yield and quality caused by pests and diseases there are a wide range of physiological disorders that can arise during the growing season. These can be specific to variety, climate, soil conditions and management practices. The following are potential physiological potato disorders.

- Little potato
- Growth cracks
- Secondary growth
- Mechanical damage
- Bruising
- Internal rust spot
- Hollow heart
- Vascular browning
- Frost damage
- Chemical damage (Table 15.5)

Table 15.5 Typical crop protection programme for maincrop potatoes.

Crop developmental stage	Indicative date	Treatment type	Target/purpose
Seed treatment	Pre-planting	Fungicide spray on seed potatoes	Seed and soil borne diseases.
At planting	March/April	Fungicide powder or spray on planter	Seed and soil borne diseases.
At planting	March/April	Soil nematicide granule or spray	PCN, FLN, wireworm
Pre-emergence of crop	March/April	Contact/Residual herbicide spray	Grass and broad leaved weeds
Post-emergence of crop	April	Contact herbicide spray	Remaining weeds
Rosette stage	April ongoing	Fungicide spray (weekly while crop is green)	Protection from blight.
Full canopy	June	Aphicide spray	Aphids
Pre-harvest	September	Sprout suppression spray	Reduce in store sprouting.
Pre-harvest	September	Desiccation spray	To achieve skin set for storage
Pre-harevest	September	Fungicide spray on seed potatoes	Tuber blight
Storage	October–June	Sprout suppression by gas/fog	Reduce in store sprouting.

15.2.18 Irrigation

In dry seasons, very profitable returns can be obtained from irrigating the potato crop. Quality may be improved by the tubers being more uniform in size and having less common scab, but the dry matter percentage may be lowered if an excessive amount of water is used late in the crop's life. It is important to aim to stop irrigation 4 to 5 weeks before the intended lifting date. If water is applied to potatoes which have almost died due to drought, secondary growth may develop as misshapes and cracks; this will obviously affect marketability.

It is vital that irrigation water is used economically and effectively. Restrictions on licences for water abstraction are likely to increase in the future and the cost of water will rise. Accurate irrigation scheduling is vital and over-watering should be avoided. Many growers are investigating the use of trickle irrigation systems and methods of preventing run-off from conventional irrigators (see also section 8.2.6).

Early potatoes may be protected from frost damage by keeping the soil moist on those warm sunny days which are followed by radiation frosts at night. In more severe cases, they can be protected by spraying with about 2–3 mm of water per hour during the frosty period; the latent heat given out, prevents damage to the leaves. Irrigation may also be used to assist in breaking down clods when preparing spring seedbeds in a dry time and to aid harvesting. Dry soils will fall through the harvester quickly where some moisture will allow a cushion of soil to carry the potatoes up the harvester. This reduces bruising damage.

15.2.19 Harvesting

Early potatoes are harvested when the crop is still growing. To make lifting and later cultivations easier, the tops should be destroyed with a flail-type machine. Early potatoes must be treated gently because the skins are very soft. Maincrops, especially if they are to be stored, should not be harvested until the tuber skins have set. This is usually about 3 to 4 weeks after the tops have senesced. Crops may have senesced due to their determinacy or by using herbicides to desiccate. Many crops are now being defoliated using mechanical flailing operations due to the reduction of desiccation products.

A few early crops are picked by hand, but most maincrops are lifted mechanically. One-row, two-row, four row and bed harvesters of various designs and capabilities are available to lift and load the crop into trailers or into boxes and bags. These machines can be trailed or self-propelled using tracks to reduce soil compaction. Harvesters can be manned or unmanned to remove stones, clods or poor quality potatoes while in the field. Very high rates of working are possible with some machines in stone and clod-free conditions, when it is reasonably dry.

Damage to tubers by harvesting operations can result in serious losses due to bad design and/or faulty operation of the machines. Damage can be extremely serious in very dry soil conditions when the tubers are somewhat dehydrated and more susceptible to damage, and the clods are very hard. Machinery and handling equipment is now designed to limit the number and intensity of impacts that the potato suffers during the harvesting, storing and grading process.

15.2.20 Storage

A high proportion of the maincrop has to be stored. Storage may only be for a short period for some of the crop, but in other cases the potatoes may have to be stored until May or June the following year when home-grown early potatoes come to the market. Potatoes that are harvested in wet conditions that have been damaged mechanically or have been exposed to pest and disease pressure will not store well long term. Dry, soil and disease free with no mechanical or pest damaged potatoes are suitable for long term storage. Skins must be set to retain moisture and crops should be harvested when soils are not waterlogged, with soil, weeds and stones remaining in the field.

Most crops are now stored indoors, either in on floor bulk stores or in buildings designed to hold large 1000 or 1500 kg boxes which are stacked on top of each other from floor to ceiling. The pallet boxes are expensive (£80–100 each), but ventilation within the box is usually adequate and the tubers retain a good appearance. Care is required when filling bulk stores to ensure an even distribution over the floor and to avoid any soil which can prevent air circulation. If the stored potatoes are more than 2 m deep, some form of duct ventilation will be required to prevent overheating. In more sophisticated buildings, very accurate temperature control is possible by using refrigeration and recirculation ventilation and most growers of contracted maincrop potatoes have invested in this type of store. Ventilation and allowing the store to warm up to 15 °C for 7–10 days after filling, helps to dry the tubers and to heal wounds. Potatoes also need to be cured for 7–10 days before going into a boxed refrigerated store.

For late storage (by whatever method), sprout growth has to be prevented in the spring. This can be done by keeping the store cool (4–7°C) by cold air ventilation and refrigeration. This method is not satisfactory for crisping and chipping potatoes, because some of the tuber starch is converted to sugar which can burn during the cooking process. These burnt sugars can contain acrylamides which are potentially harmful to human health. The fry colour therefore is an important quality measure for processing potatoes. Sprout suppressants can be used on processing potatoes which are applied as a gas/fog when the crop is being stored. Examples of sprout suppressants are ethylene, spearmint oil, orange oil and by the application of products to the growing crop close to harvest.

15.2.21 Grading

Grading of potatoes is the process whereby the crop can be separated in to specific sizes, stones and soil can be removed and any defects can be removed. This process can start in the field, when the crop is going into the potato store, coming out of the potato store or a combination of all three. Potatoes from box stores can be sold straight to the packer or processor in the box that they have been stored in. Potatoes stored in a bulk store are loaded into a vegetable bulker for transport to the processing site. Bulk potatoes are almost always destined for processing rather than a retail shelf. Maincrop potatoes are usually size graded into <45 mm, 45–65 mm and 65–85 mm. Anything over 85 mm is unmarketable as a baking potato due to its size. Processing potatoes for french fries are required to have a tuber size around 75 mm to meet the specifications of the retailer.

15.2.22 *Yield*

Average yields for potato crops are shown in Table 15.3. Yields will vary greatly depending on length of growing season, solar radiation, water, pests, disease, harvesting conditions, harvest time and variety.

15.3 Sugar beet

Sugar beet (*Beta vulgaris*) is part of the Amaranthaceae family of plants and is grown commercially in many temperate parts of the world for its high extractable sucrose content. Sugar beet is a biennial crop which means that in the first year of its growth a large root develops, followed by a flowering stage in year two. However, commercial sugar beet growers treat the crop as an 'annual' harvesting it before its second year of growth.

The early 1900s saw the beginning of the UK sugar beet industry as we know it today and the crop has now been a major player within UK agriculture for over 100 years. With a maritime climate that suits sugar beet, it has historically been grown in a wide geographical area across the UK. But, as a crop that is 80% water, which means expensive haulage costs, the traditional heartlands for beet growing have developed close to the sugar beet processing factories. Today these factories, owned and operated by the monopoly processor British Sugar, are based toward the eastern counties of the UK, and are situated at Newark, Bury St Edmunds, Cantley and Wissington.

Approximately 3000 growers, contracted to British Sugar, produce the beet that supplies these sites, predominantly on farms no more than 50 miles from a factory. The area of beet grown varies slightly each year, but has been in the region of 100,000ha for some time. Depending on the weather in the growing season, this area can produce as much as 8 million tonnes of beet from which 1.4 million tonnes of white sugar is extracted for both domestic consumption, industrial use and exports.

Home-grown UK beet sugar supplies about half of the UK's demand for sugar, the balance being imported from both cane and beet growing countries in the EU, Africa and South America. The dominant UK retail beet sugar brand is Silver Spoon with imported cane sugar marketed by Tate and Lyle. The majority of home-grown beet sugar, approximately 85%, is destined for the food and drinks manufacturers, with a small tonnage exported to the continent.

The price growers receive for their beet is negotiated by British Sugar and the NFU before the crop is planted, allowing growers to fine tune their crop area ahead of the growing season. The price fell dramatically in the early years of the 21st century to a level where farmers were questioning their ability to grow the crop profitably. However, steady and consistent yield improvements over the past 30 years has meant that for many growers it is still a profitable crop in the rotation.

The contract price the growers receive reflects the sugar content of the beet; beet with a sugar content over 16% attracts a higher price than beet with sugars below 16%. Average sugars are in the region of 17%–18% but can be as low as 14% in a difficult growing season (in a diseased crop for example) and as high as 21% with good summer sunshine and regular rainfall. Sugar beet growers are paid in 'adjusted tonnes' on a sliding scale reflecting the sugar content of the beet at the point of delivery with 16% being the reference point. So, for example 100 tonnes of beet delivered at 16% sugar, the grower will be paid for 100 'adjusted tonnes'. But 100 tonnes delivered at 14% sugar equates to only 87.5 'adjusted tonnes' while 100 tonnes at 18% sugar means that the grower will be paid for 112.5 'adjusted tonnes'. The incentive, clearly, is to produce beet with a high sugar content. The sugar scale is agreed between NFU and British Sugar and details can be found in the Inter Professional Agreement.

The UK sugar market has for many years been tightly controlled by a series of quotas and tariffs which severely restricted imports and exports from the country. However, changes to these regulations has meant that from 2017, overseas sugar producers have, for the first time, access to the UK market, and UK growers access to overseas markets. Although this is a major change to sugar trading, and potentially allows global sugar producers to further develop their market reach by importing sugar to the UK at costs that cannot be matched by domestic sugar beet growers, in practice there is little change to date. This is mainly due to the large global sugar producers focussing on domestic bioethanol production rather than expanding their overseas exports. However, if this changes and cheap sugar is imported into the UK, the impact on the home-grown beet sugar industry would be swift and dramatic as the UK processor would not be able to compete, leading to the collapse of sugar beet as a viable break crop within the UK.

British Sugar is the sole processor of sugar beet in the UK and in addition to extracting and refining beet sugar has successfully developed new and novel revenue streams from wastes associated with the beet crop. Bioethanol, liquid carbon dioxide, animal feed, topsoil and the liming material marketed as LimeX are examples of co-products developed from what was previously considered 'waste'. Medicinal cannabis is now being grown in an 18ha glasshouse adjacent to the Wissington factory site utilising waste heat and carbon dioxide pumped from the factory to provide ideal growing condition for the specialist crop.

Other UK markets for home grown sugar beet include anaerobic digesters (AD) and stockfeed. Historically, fodder beet has been preferred to sugar beet by many livestock producers, but with the growth of on-farm AD plants requiring a high energy feedstock such as sugar beet, an increased area of sugar beet has been grown across the UK. This has meant that sugar beet can now be offered as a feed option to many growers who previously only had access to fodder beet.

In the 1960s and 70s sugar beet could be harvested in a manner that allowed the leafy above-ground material to be collected and removed from the field to provide a useful additional source of animal feed from the crop. At that time, these sugar beet 'tops' provided additional feed of between 15 and 20 tonnes (fresh weight) per hectare. Although seen as providing useful additional feed, the practice has become redundant with all modern harvesters now removing all leaf material with a flail at the time of harvesting. This material is then incorporated into the seedbed to provide useful nutrients to the following crop.

15.3.1 Soils and climate

Sugar beet will grow on most of the UK's soil types, however difficulties in establishing an appropriate seedbed in the spring, and harvesting during the winter months means that the crop is usually associated with lighter, free draining soils. Stoney conditions can be problematic, with larger stones having to be removed with cleaner-loaders at harvest, which adds cost to the operation.

For high sugar yields, the beet crop needs to be established early in the spring to allow full canopy development by the longest day, June 21st. With a crop that is drilled on 50 cm rows, mid-June is the time for the beet to be 'touching in the rows'. This allows for maximum interception of sunlight by the leaves, and the efficient production of sugar. The quality and quantity of sunshine over the summer months can dramatically affect the photosynthetic efficiency of the crop, and the resultant sugar production; dull overcast conditions in July and August can lead to disappointing sugars at or below 16% at the start of the processing campaign.

With a warm spring, timely drilling can lead to the early development of the crop canopy with resultant high sugar content beet with above average yields; conversely, a cold spell of weather post drilling can lead to bolting, and the crop producing a seed-head, with low yields and poor sugars. Managing this drilling window is critical to establish a successful beet crop.

15.3.2 Place in rotation

As a spring planted break crop, sugar beet can play an important role in contributing to a diverse rotation. Historically there were contractual obligations that limited how frequently beet appeared in a rotation, but now that decision is left to the individual grower. Good agronomic practice means that the rotation should be no closer than 1:3, preferably longer. Also, brassica species in the rotation should be avoided as these can help build beet cyst nematode populations.

Sugar beet frequently follows a cereal crop. This allows a 'window' for subsoiling, use of over-winter cover crops, stale seedbeds, winter ploughing and grass weed control (particularly black-grass) before the beet is drilled in the spring. Winter weathering of ploughed ground can be a useful way of effectively 'breaking-down' heavier soil types to produce an appropriate seedbed. This requires careful management to ensure that these bare soils are not excessively eroded and lost, and that environmental regulations are adhered to.

Sugar beet is very sensitive to agrochemical inputs; careful use of residual herbicides in previous crops is required to ensure that there is no lasting impact on the germinating seed or establishment of the subsequent beet crop.

15.3.3 Seedbed preparation

Given the wide range of soil types that beet is successfully grown on, there is no 'blueprint' for the production of a sugar beet seedbed. The principles have always been to maintain moisture within the seedbed, ensure that all compaction has been removed prior to drilling, and produce a level consolidated seedbed. Although the traditional plough followed by power-harrow combination is less common now as growers look toward faster, more economical and less intensive cultivations, ploughing ahead of beet retains its popularity. Spring ploughing with a furrow press can produce perfectly acceptable seedbeds on the lighter soil types; heavier soils require late autumn ploughing and winter weathering to help produce a seedbed. On these soil types, a single one-pass cultivation before drilling in the spring should aim to produce a level consolidated seedbed that allows good soil-to-seed contact, which will, later in the season, reduce harvest losses.

On lighter soil types susceptible to soil erosion, a plough and furrow press combination can leave the seedbed with a mini-ridge surface structure that can help prevent wind erosion. Previous crop residue can also help prevent soil erosion by creating a mulch on the soil surface, again helping to consolidate the soil surface and prevent a soil 'blow'.

Strip tillage, and other reduced cultivation practices have been trialled for establishing sugar beet, but for the majority of beet growers, the plough still features in the rotation, commonly just prior to the beet crop, as the need to provide a deep loosened soil is key for a successful beet harvest. Non inversion techniques may help prevent weed beet germinating, but the development of herbicide tolerant (ALS) sugar beet varieties has given beet growers a more reliable tool for establishing a beet crop on fields with a high weed beet burden. For more information on ALS tolerant beet, check the latest BBRO Recommended List and the relevant seed breeder.

15.3.4 Varieties

The British Beet Research Organisation (BBRO) in partnership with the British Society of Plant Breeders (BSPB) conduct a trials programme to ensure that growers gain access to the best genetics available to the UK market. A Recommended List is produced each year allowing beet growers to choose varieties appropriate to their own farm and local conditions. Varieties are ranked according to yield, bolting, establishment and disease resistance (further information is provided on the annual Recommended List of Sugar Beet Varieties).

Sugar beet seed breeders are viewed with envy by other sectors of UK agriculture as they consistently deliver new varieties with improved yields, traits and characteristics. Beet yields (sugar yield, adjusted yield, root yield) have improved consistently over the past 30 years and much of this success has been driven by the introduction of new high-yielding varieties.

The threat from the soil borne disease Rhizomania in the 1980s was swiftly countered with the introduction of new varieties tolerant to the disease, and has led to the situation today where all recommended varieties are now resistant to Rhizomania. Similarly, where the new aggressive strain of Rhizomania (AYPR) has been identified, varieties have been introduced that will successfully grow in the presence of the disease. Other traits, such as tolerance to beet cyst nematodes, have been developed and further work to produce varieties tolerant to Virus Yellows is ongoing.

Beet seed is supplied to growers by Germains, a sister company to British Sugar, or direct from the seed breeders.

15.3.5 Manures and fertilisers

Sugar beet crops require adequate and timely supplies of a range of nutrients to achieve maximum yields, particularly during the early months of their growth. For fertiliser recommendations the AHDB Nutrient Management Guide (RB209) should be consulted to ensure the correct quantities of nutrients are applied in an appropriate fashion.

15.3.5.1 Nitrogen

A key nutrient, and essential for rapid canopy development early in the season. Beet crops require a total of 220–250 kgN/ha to produce maximum yields.

Nitrogen recommendations for beet should take account of;

- residual N left in the field by the previous crop, or mineralised from soil organic matter during the autumn
- N that is mineralised from soil organic matter during the growth of the crop
- N if applied in FYM and other manures/organic material.

An 'average' sandy loam with about 2% organic matter would provide approximately 120 kgN/ha (60 kgN/ha from the previous cereal crop and 60 kgN/ha mineralised during the season) which means that the crop requires a further 120 kgN/ha to support maximum yield. This is usually provided from inorganic fertilisers such as ammonium nitrate, and less commonly, urea.

To ensure maximum utilisation of the nitrogen and to ensure that little is lost to leaching or volatilisation, one third of the total nitrogen required is applied at, or soon after drilling, with the remainder being applied when the beet has emerged and is at the cotyledon growth stage and is visible 'in the rows'. Nitrogen applied later than this risks scorching the crop with the prills trapped in the leaf canopy.

Although beet yields have increased dramatically over the past 30 years, the N requirement remains the same. Repeated experimental work by BBRO has demonstrated that 120 kg N/ha remains the optimum nitrogen input, with additional N simply increasing beet impurity levels but not yield. Simply put, nitrogen recommendations do not vary with yield. Table 15.6 below gives more detail.

Table 15.6 Nitrogen recommendations (kg/ha) for sugar beet.

	Soil index				
	0 and 1	**2**	**3**	**4**	**5**
	kgN/ha				
Mineral Soils	120	100	80		0
Organic Soils	—	—	—	0–40	0
Peaty Soils	—	—	—	—	—

These recommendations have been based on AHDB Nutrient Management Guide (RB209) Section 4 updated in January 2021 by the Agriculture and Horticulture Development Board.

15.3.5.2 Phosphate, potassium and magnesium

The timing of applications of these nutrients is important to ensure that soil damage from heavy fertiliser spreaders and applicators is kept to a minimum.

Also, fertiliser applied too close to drilling can cause damage to young beet plants and must be avoided. To avoid this, the recommendation is for all inorganic fertilisers to be applied at least a fortnight before drilling, and incorporated into the topsoil. In practice this means that for many soil types, excluding the lighter soil types, beet fertiliser is ploughed-down in the autumn and winter. Clearly on lighter soils prone to leaching this is inappropriate, and here the fertiliser would be applied in January/February.

Phosphate and potash deficiency symptoms in the growing crop are quite rare; however, magnesium deficiency is surprisingly common and easily diagnosed. Magnesium deficiency is characterised by the margins of the older leaves turning yellow with the edges becoming dead and brown. These symptoms are best avoided by maintaining good background levels of Mg in the soil by using magnesium based fertilisers such as Kieserite (25% MgO), Kainit (5% MgO) and calcined magnesite (80% MgO) or by using magnesian limestone (15% MgO) when liming.

15.3.5.3 Sodium

Sugar beet is a maritime crop and growing in a salt-laden environment has developed a tolerance to sodium. For many years beet growers have taken advantage of this trait and supplied the crop with sodium (agricultural salt) as a cheap alternative to potash, and its recommendations are in Table 15.7. Excessive sodium within a soil is known to affect soil structure, but the recommendations are not at a level that has any adverse effect on structure and stability.

Table 15.7 Phosphate, potash, magnesium and sodium for sugar beet.

	P, K or Mg index				
	0	**1**	**2**	**3**	**4 and higher**
	Kg/ha				
Phosphate (P_2O_5)	110	80	50	0	0
Potash (K_2O)	160	130	100	0	0
Magnesium (MgO)	150	75	0	0	0
Sodium (Na_2O)	200	200	100	0	0

These recommendations have been based on AHDB Nutrient Management Guide (RB209) Section 4 updated in January 2021 by the Agriculture and Horticulture Development Board.

15.3.5.4 Sulphur

High yielding sugar beet crops may use as much as 100 kg S/ha, which can mean that lighter (and shallow) soils can struggle to provide sufficient sulphur for the crop. Regular applications of FYM and other organic material can provide sufficient sulphur, but additional sulphur may be necessary if deficiency symptoms have been noticed in other crops such as barley and oilseed rape. If a deficiency of sulphur is likely, an application of 25–50 kg SO_3/ha would be appropriate.

15.3.5.5 Manganese and boron

A large area of the UK's sugar beet is grown on soils with a high pH, or those that have recently received lime. These conditions can lead to the 'lock-up' and reduced availability to the crop of the micronutrients, manganese and boron. Fortunately, the early signs of deficiency are easily identified, and an appropriate foliar application will quickly address the problem. Many growers who regularly apply lime will routinely look to use a foliar manganese and boron feed, applied as soon as there is sufficient leaf area in the crop.

Manganese deficiency is described as 'speckled yellows' where there are sunken yellow/buff coloured spots between the leaf veins. Early boron deficiency is characterised by a 'laddering' on the stem, followed by a shrivelling and blackening of the heart leaves.

15.3.5.6 Lime and pH

Sugar beet requires a soil pH in the range 6.5–7.5. Acid conditions, particularly below pH 6.0, can severely depress crop growth and yields.

Regular soil testing is required to determine the pH, and if there is a deficiency an appropriate liming material should be applied. As many of the commercially available liming products are relatively insoluble and slow-acting, it is advisable to apply lime 18 months in advance of the beet crop to ensure the material is thoroughly mixed and available through the soil profile.

15.3.6 Seeds and sowing

Raw sugar beet seed is processed and covered with a clay coat to produce a high quality uniform pellet that is consistently 3.5–4.75 mm in size and has a minimum of 90% laboratory germination. Pelleting also allows the seed to be treated with a range of insecticides and fungicides for control of common soil-borne pests and diseases (the seed is also colour-coded to indicate the treatment(s) that has been applied). For up to date information on this check with BBRO or the seed breeders.

An early part of the treatment of the raw sugar beet seed is a priming process, to help a rapid and uniform germination under field conditions.

15.3.6.1 Seed rate

Sugar beet seed is packaged and dispatched to growers in boxes (units) of 100,000 seeds. The recommendation is to establish 100,000 plants/ha after field losses, and to do this on a standard row width of 50 cm and seed spacing of 17 cm requires 1.2 units/ha to be drilled. For drilling at different seed spacing and row widths, consult the BBRO seed rate tables.

15.3.6.2 Timing of sowing

If field conditions allow it, beet should be drilled by the end of March. This allows the crop the best opportunity to develop a full canopy by mid-June and optimise light interception and sugar production. Delayed drilling beyond April 4th can lead to yield losses of over 4 adjusted t/ha per week. Conversely, early drilling risks triggering bolting, which results in the crop flowering in year 1 rather than year 2, so growers have to be sure of their local weather conditions before starting, as cold conditions post-drilling can cause plants with a low vernalisation requirement to flower in the first year.

15.3.6.3 Precision drilling

The UK beet crop is planted with precision drilling equipment that allows for the seed to be spaced accurately and evenly. Most beet is drilled on either 50 cm or 45 cm row widths with seed spacing between 15 and 17 cm. As previously mentioned, the target plant population at establishment (when the beet has reached the six true leaf stage) is 100,000/ha. Lower numbers may adversely impact on yield, and a crop with less than 40,000/ha established may require re-drilling. Uniformity across the field is important to maximise growth during the season, but is also a key factor to ensure that harvest losses are kept to a minimum later in the year.

Beet needs to be drilled into a well prepared, but not over-worked, seedbed. The key to success here is to maintain soil moisture and not lose it through excessive cultivations. The aim is to have a minimum of 30% of soil particles less than 3 mm around the seed to improve availability of moisture to the seed, with larger aggregates and a more open soil structure below the seedbed. For a successful start beet must be drilled into moisture, anywhere between 2 and 5 cm.

15.3.6.4 Prevention of wind damage

Beet grown on lighter/organic soils can be at risk of wind damage ('wind-blows') particularly early in the season. Small soil particles can be carried by strong winds and cause severe damage to the young crop, sand-blasting and completely defoliating and killing it in the worst instance. On sites historically prone to a 'blow' there are a number of measures a grower can take to help the situation;

- A spring cereal cover crop. Commonly this is spring barley. This is planted about 2–3 weeks before the beet crop, with the sugar beet being drilled into the newly emerged barley. The young barley roots and leaves can help prevent the soil particles from being whipped up by strong spring winds, and protects the beet. Management and destruction of the barley

using herbicides is key here; killing the cover crop too early risks the beet being damaged by a subsequent wind blow, whereas a late destruction of the barley can mean that there is a yield penalty as the beet is effectively in competition with the barley.
- Crop residue/surface mulch. Shallow incorporation of a crop residue such as straw can provide a surface mulch that the beet can successfully be drilled into. This mulch effectively 'holds' the soil together and helps prevent the worst of the wind damage and soil erosion.
- A plough and furrow press can do a similar job, producing a 'rough' soil surface that is not so prone to the damaging effect of the wind. Strip tillage too offers potential to prevent the worst effects of a wind blow as only approximately 50% of the area is cultivated leaving the remainder to act as a wind-break.

15.3.7 Weed control

Sugar beet is very sensitive to competition but it is particularly the early emerging weeds in a beet crop that have the most potential to reduce crop yields and sugars. For example, 1 tall weed plant per m^2 can reduce crop yield by between 11% and 15%, so it is key to control the larger spring germinating weeds to ensure they do not out-compete the beet.

Regular crop monitoring to identify weeds and their growth stage is an essential part of an integrated management programme to ensure efficacy of chosen control options. Sugar beet is a spring planted break crop. This means that there are plenty of overwinter opportunities for weed control using, for example, stale seedbeds, cover crops and cultivations. Herbicides for control of perennial grass and broad leaved weeds can also be applied during this period to ensure that the beet is drilled into as weed free an environment as possible.

A sugar beet herbicide programme can include a mixture of pre-drilling, pre-emergence and post emergence herbicides. Sugar beet herbicides can be contact or residual in nature, and occasionally exhibit both characteristics. The choice of the active ingredients and the timing is very much down to a determination of the developmental stage of the crop, the weed (type, number and growth stage) and overall weather conditions. Annual broad-leaved weeds need to be controlled when they are small; preferably at the cotyledon or first true leaf stage. Any larger than this may mean a high herbicide dose rate that can 'check' crop growth. Graminicides are available for a range of grass weeds in a beet crop, but herbicide resistance issues makes this control route increasingly unsustainable and unreliable.

Many growers will choose to use a pre-emergence programme as this allows an early application of herbicide to the crop and serves to 'sensitise' the weeds, making them easier to kill with subsequent treatments. Others, however, choose not to apply a pre-emergence herbicide preferring to wait until the weeds emerge, and then tailoring a weed control programme specifically to the site.

Recent developments in weed control include commercially available herbicide tolerant sugar beet varieties (sold with the corresponding herbicide). Although expensive, they allow sites with high weed beet populations back into the rotation, as the beet herbicide kills the weed beet, but leaves the sugar beet crop unaffected. The protocols for growing herbicide tolerant beet include strict stewardship measures to ensure that the sites are regularly and comprehensively monitored to ensure that bolters from the tolerant varieties are removed and destroyed.

For details of all currently available beet herbicides, consult the latest BBRO herbicide tables.

As a row crop, sugar beet lends itself to inter-row hoeing to control small weeds and weed beet. Modern hoes, using GPS guidance, can cover up to 80% of the area between the beet rows, and when used in the developing beet crop are an effective weed control tool particularly when used early in the season.

Later in the season, if there is a height differential between the crop canopy and the tall weeds, the option to cut or weed-wipe becomes available. However, these last two weed-control systems are very much last-ditch methods used when all earlier options have failed.

15.3.8 Disease & pest control

(See also Chapter 6).

15.3.8.1 Disease control

Virus Yellows. From the mid 1990s neonicotinoid seed treatments were excellent for controlling the aphids that transmitted the virus yellow complex in sugar beet crops. This complex is made up of three different viruses; beet yellows virus, beet mild yellowing virus and beet chlorosis virus, with the disease symptoms a yellowing of the crop that starts in small patches and subsequently spreads.

The withdrawal of the neonicotinoids in 2019, coupled with the absence of approved alternative aphicide treatments led to a resurgence of the disease in subsequent years. Virus yellows can reduce beet yields by as much as 50%, so for the crop to be commercially viable there have to be effective control measures available. Aphids, particularly the Peach Potato Aphid (*Myzus persicae*), have developed resistance to many of the commonly available beet insecticides. This means that the seed breeders, who were very successful breeding varieties tolerant to virus that causes rhizomania, will be the likely route for control of virus yellows in the future, producing new beet varieties that are resistant to the virus yellows complex.

Cercospora (Cercospora beticola) and Ramularia (Ramularia beticola). These two leaf-spotting diseases may appear in the crop late summer/early autumn. Uncontrolled they can completely defoliate the crop, but a well-timed robust fungicide programme can check the disease progress. Strains resistant to the fungicide's active ingredients are developing which is undermining their efficacy and allowing the disease to spread. The latest BBRO fungicide tables will provide more information here about the current appropriate products available to use.

Rust (Uromyces betae) and powdery mildew (Erysiphe betae) can also be problematic, but the fungicide programme used for the leaf-spotting diseases also provides good control of rust and powdery mildew. Also, the varieties on the Recommended List have a disease score for resistance to these diseases allowing growers to choose more resistant varieties for difficult sites.

Downy mildew (Peronaspora farinose). The heart leaves become thickened and distorted, with purple/grey downy spores on the underside of the leaf. Rarely has an impact on yield, and there are no fungicidal control measures available.

Rhizomania (Beet Necrotic Yellow Vein Virus BNYVV) once a severe threat to the crop, is now effectively controlled by using tolerant varieties. The aggressive strain of rhizomania, AYPR, is also controlled via this route. See the current BBRO Recommended List for more information on the current AYPR tolerant varieties.

15.3.8.2 Pest control (see Chapter 7)

Springtails (Onychiurus armatus), symphylids (Myriapoda symphylidae) and millipedes (Blaniulus spp*)* make up the Soil Pest Complex and are most damaging in organic and silty soils. They attack the roots of young seedlings, and left untreated can cause significant crop damage. Control is with insecticide treated seed; check the BBRO guide for up to date information on available seed treatments.

Wireworms (Agriotes spp.*) and leatherjackets (Tipula palidosa).* These soil pests can be a particular concern if there is grass in the rotation. Their grazing of the roots and stems of the beet plants is particularly problematic the second or third year after a grass ley. Only cultural control methods are available.

Beet Cyst Nematode (Heterodera schachtii) was historically controlled by using a wide rotation (1 year of beet in a 20 year cycle). Seed breeders have, though, introduced new varieties that are BCN tolerant which has allowed beet back on to fields that were considered unsuitable for beet because of the high BCN presence.

Free Living Nematodes (Trichodorus spp. *Longidorus* spp.*)* causing Docking Disorder rely on cultural control methods such as biofumigation with brassica cover crops. They are particularly active in wet sandy soils, and can cause significant yield loss by damaging the roots. The developments of garlic-based nematicides are indicating promising levels of control.

Flea beetle (Chaetocnema concinna). The feeding damage on young plants is the familiar 'shot-hole' symptom commonly seen in crops of oilseed rape. Unlike OSR there is no damage from the larval stage of this pest.

Leaf miner (Diptera anthomyidae). Multiple generations through the early spring and summer can lead to a complete defoliation of the crop. As the leaf miner larvae grows and develops it burrows between the upper and lower leaf surfaces, causing significant damage and leaf death. Protected inside the leaf, it is difficult for the active ingredient in an insecticide to effectively target the pest, so commonly the pest remains untreated and is allowed to complete its life cycle.

Black bean aphid (Aphis fabae) feed in large numbers on a single plant. Although this looks bad, with the plant completely smothered in aphids, the plant invariably recovers well. An insecticide treatment here would be counter-productive as it would kill many of the beneficial insect species that were feeding on the aphids (Table 15.8).

Table 15.8 Typical crop protection programme for sugar beet.

Crop growth stage	Indicative date	Treatment type	Target
Pre-drilling	Nov–March	Herbicide	Perennial weeds
Pre-emergence	March/April	Seed treatment	Soil pest complex
		Herbicide	Annual broad leaved weeds
Post-emergence	March–June	Herbicide	Annual broad-leaved weeds
	April–June	Graminicide	Volunteer cereals/grass weeds
		Insecticide	Flea beetle, leaf miner, aphids
	July/October	Fungicide	Rust, ramularia, cecospora, powdery mildew

15.3.9 Irrigation

A fully developed sugar beet root is 80% water so to maximise yield and sugar potential it clearly needs access to sufficient water during the growing season. Irrigation during a dry year and on lighter soil types can significantly improve growth and therefore sugar yields, but as beet is often considered the 'low-value' crop in the rotation, frequently other high value crops, such as potatoes, have their water needs prioritised over beet.

The greatest benefit from irrigation is achieved by applications in June and July. Irrigation in August and September generally does not produce sufficient additional sugar yield to cover the cost of the water.

During a very dry spring it may be cost effective to irrigate to encourage the beet to germinate. While this may be successful, the water droplet size from a rain-gun can result in the soil surface 'running together' and when this dries it forms a particularly hard 'cap' that prevents the beet emerging. The answer here is either not to irrigate at all, or, once started, keep the soil moist until the beet has emerged.

At the other end of the season, irrigation can be useful to soften the ground ahead of harvest to keep root breakages to a minimum, thus reducing 'harvest losses', a key strategy for increasing harvested yield.

15.3.10 Weed beet

Weed beet is any unwanted beet that appears in a beet crop within and between rows of drilled beet. Unlike true beet, which is a biennial, weed beet behaves as an annual and produces seed in a single year. It occurs from several sources. Commercial seed if drilled in cold conditions in early March can produce significant numbers of bolters. These are beet plants that produce a seed head in year one, rather than year two groundkeepers. Small pieces of beet can grow from cleaner loader sites that establish to produce bolters.

A single weed beet can produce up to 2000 seeds with a dormancy range of 1 to 20 years, giving it a very effective survival strategy.

Uncontrolled, a weed beet population can quickly expand to levels where it becomes difficult to successfully grow a commercial crop.

The steps needed to keep fields weed beet free;

- Choosing varieties which are less likely to bolt (see the BBRO Recommended List)
- Drilling beet in late March rather than early March to avoid any cold periods that may induce bolting
- Tractor hoeing is an option early in the season which can effectively control the bolters that have germinated between the beet rows. Most hoes will not control the bolters between the beet plants in the row.
- Hand-pulling is labour intensive, but a highly effective method of control suitable for low weed beet populations.
- Cutting and topping is swift and reasonably effective but only removes the seed heads that are above the canopy and allows the lower shoots to develop viable weed beet seeds.
- Weed wiping can be used as the height differential between the beet canopy and the weed beet allows for this operation. Here a small amount of a non-selective herbicide such as glyphosate is applied to the weed beet or bolter by means of a wick or pad or roller.

Herbicide tolerant technology. CONVISO Smart is a system from Bayer and KWS where herbicide tolerant varieties are used in conjunction with dedicated herbicides based on ALS inhibitors. Although very effective in allowing a beet crop to be grown on land contaminated by high populations of weed beet, strict stewardship protocols are required to ensure that the herbicide tolerant varieties are not allowed to flower, produce seeds and appear in subsequent beet crops.

15.3.11 Harvesting

Sugar beet can be harvested ('lifted') from early September once the root has developed sufficient size to successfully pass through a beet harvester. Depending on the weather conditions, the beet can continue to grow during the autumn and winter, in some cases increasing in yield by as much as 50% between September to January. This late season beet growth is helped by;

- Modern high yielding beet varieties.
- Efficient six row harvesters that leave little unharvested or broken beet in the field.
- Improved beet agronomy that leads to a uniform, easy to harvest field.
- Improved fungicide programmes that keep the canopy disease free and able to intercept autumn and winter sunshine, and build sugar yield.
- Physiological 'greening' effect from the triazole/strobilurin fungicides programme keeping the canopy active for longer.
- More mild, 'open' autumn and winter weather conditions that allow for late season growth.

Crops with the poorest yield potential are harvested first, with the better crops harvested later. The beet harvesting and processing period, known as the 'campaign' is commonly 150–160 days, but in good years with higher yielding crops may stretch beyond 200 days. Growers are contracted to supply beet to the processor (British

Sugar) throughout this period which means that there may well be beet lifted during the winter months when field conditions are far from ideal. With an increased focus on managing soils sustainably, this is causing some growers to question their long-term commitment to the crop.

The industry operates a 'just-in-time' harvest and delivery programme between September and November, and looks to lift and store the beet for delivery after November. The aim here is to minimise stored beet in the early part of the campaign to allow the crop to capitalise on late season growth. Clearly, once beet has been harvested it can no longer increase in yield, and a long period in store can lead to significant sugar loss as the beet continues to respire. Longer term storage, from November onwards allows fields to be cleared and the next crop drilled without suffering too much of a yield penalty. Most beet crops are harvested by January, although on very light soil types hand-to-mouth lifting continues throughout the campaign.

In the UK, the sugar beet harvesting programme is dominated by 6-row beet harvesters. Although the larger farms may own their own machine, many choose to use contractors who offer the full service of harvesting and transport of the beet off the field back to the clamp or loading site. Modern well set-up harvesters in the hands of appropriately trained operators can achieve yield losses below 1 tonne/ha. Conversely, in wet or very dry conditions, or where there is high weed infestation, small beet or a gappy crop, beet breakages can exceed 8 tonnes/ha, and dramatically reduce the harvested yield.

Frost too can be a problem during the later months of the campaign. A strong healthy, disease-free canopy can protect against a mild frost (0°C to -3°C) but lower temperatures can cause the beet severe damage and make it un-processable. This potential for crop damage drives many growers to lift their beet when conditions are favourable, and store in long-term clamps rather than gambling on risking frost damage later in the season.

15.3.12 Storage

Over the winter months sugar beet is stored in managed heaps or 'clamps'. During the early part of the season, a beet clamp is simply an area where individual trailer-loads of beet have been tipped. These small piles of beet give the maximum surface area possible which assists with cooling the beet to help keep respiration losses low and the beet in a good state for rapid delivery to the processing factory.

Later in the season long-term clamps are built; many have concrete bases, or similar, with large straw bales on pallets for retaining walls, with the beet piled no more than 2.5m high. Ventilation is key here, and the pallets allow a good air-flow into the clamp which allows the beet to 'breath' and store well. It is important to note that a beet clamp is not a 'hospital' and if diseased, muddy and damaged roots go into a clamp, they will come out in a much poorer condition. Beet affected by root rots such as fusarium (*Fusarium oxysporum*) and violet root rot (*Rhizoctonia crocorum*) will deteriorate quickly within a clamp, and need to be removed prior to clamping. Similarly, frozen or frost damaged beet must not be used to build a clamp as they will quickly rot and risk the beet in the whole clamp being rejected at the factory.

If a self-propelled cleaner-loader, such as a Maus, is used, an 'A'-shaped clamp is required. These are commonly built on headlands where there is good access for cleaning and loading. The large surface area of these clamps is a concern during frosty weather, but these clamps and loading systems are, nonetheless, increasingly popular.

Once the longer term clamps have been built, a 'breathable' plastic sheet can be used to cover the clamp to provide protection against all but the most severe frosts.

If field and weather conditions permit, 'storing' beet in the field can be a very effective way of managing the crop during the winter months. Here, the beet simply remains unharvested awaiting a delivery slot from the processor. Sands and lighter soil types suit this system, as field conditions can be fit for harvesting even after very heavy rainfall. However, as mentioned above, a period of frost can lead to heavy crop losses, so this method is not risk-free.

15.3.13 Yield

Clean yield; 70 tonnes/ha.

Sugar; 16%–18%

15.4 Key points

Potatoes

- Potato markets have changed with a shift from fresh to processed consumption.
- New varieties with resistance to pests and disease are slow to become accepted by the processors.
- There is no longer a levy funded organisation that represents the interests of the potato industry or independent research.
- Chemical control options are under increased scrutiny.
- Soil protection and health is becoming an increasing problem with a drive toward regenerative agriculture.

Sugar beet

- Reform of the Sugar Regime has led to a significant decrease in the profitability of UK sugar beet.
- Protracted harvesting campaigns combined with difficult wet field conditions raises questions about the long-term sustainability of the crop.
- The public are less tolerant to rural infrastructure damage from large cultivator and harvesting equipment.
- As a levy-funded organisation, the BBRO continues to provide key information to all beet-industry stakeholders.
- Seed breeders are developing new genetic lines that will provide tolerance and resistance to a range of pests and diseases.

Sources of further information and advice

Further reading

Agriculture and Horticulture Development Board (AHDB). AHDB Variety Database. Available at: https://varieties.ahdb.org.uk/.

Agriculture and Horticulture Development Board (AHDB). AHDB BlightSpy. Available at: https://ahdb.org.uk/blightspy.

Agriculture and Horticulture Development Board (AHDB), 2021. Nutrient Management Guide (RB209) Section 1,2,4 & 5. Available at: https://ahdb.org.uk/rb209.

Agriculture and Horticulture Development Board (AHDB). AHDB Cultivating and planting potatoes. Available at: https://ahdb.org.uk/potato-planting.

Useful websites

AHDB Potatoes www.ahdb.org.uk/potatoes.

BBRO (www.bbro.co.uk) is a reliable and independent up-to-date source of a range of grower and industry specific information such as.

- Weed, pest disease and disease control tables.
- Recommended List of Sugar Beet Varieties, including ALS varieties.

British Sugar www.britishsugar.co.uk.

KWS www.kws.com.

Limagrain www.lgseeds.co.uk.

NFU www.nfuonline.com.

NIAB/CUPGRA www.niab.com/cupgra.

SES www.sesvanderhave.com.

Energy and industrial crops

16

16.1 Introduction

This chapter will introduce the global drivers for the decarbonisation of the energy system which, in turn, has resulted in greater production of energy crops. Domestic legislation in the United Kingdom will be introduced and analysis of data on crops grown for bioenergy will be summarised. Aspects of specific energy and industrial crops will be discussed.

16.1.1 Global drivers

The global system of energy production is in a state of transition from dependence on finite fossil fuels (e.g., gas, coal and oil) toward a greater reliance on renewable sources such as wind energy, solar power, hydroelectricity and, of most interest for our consideration in this chapter bioenergy. The decarbonisation of the global energy production system is driven by two primary concerns: climate change and energy security.

Climate change (see chapter 2) is caused by the emission of gases (such as carbon dioxide and others) which result in increased global temperatures through the greenhouse effect. Because the combustion of fossil fuels to produce energy is one of the major contributors to greenhouse gas emissions, the replacement of fossil fuels with renewable resources is a major global concern.

Energy security is a complex issue centred on the secure and affordable supply of energy, preferably through environmentally friendly means. As global energy prices can be affected by a range of international political and economic factors, many countries have sought to increase their own energy security by producing more of their energy domestically thus reducing their reliance on supplies from overseas. One way of doing so is through the production of energy from renewable sources.

As noted above, one of the principal renewable energy sources is bioenergy. This is a general term that refers to the production of electricity and/or heat from organic matter. Bioenergy can be categorised as biomass, biogas and biofuels. Biomass is used to produce heat through the combustion of crops (commonly Miscanthus and short rotation coppice—see below), wood fuel and waste materials; biogas is used to produce heat and/or electricity from crops, wastes and other materials through anaerobic digestion; and biofuels are derived from crops and used for transportation and other uses.

The use of bioenergy is controversial and is sometimes regarded as resulting in land use conflict between the production of energy and food; however, the usefulness of this characterisation is questionable. Notions of 'crop displacement' oversimplify a

Lockhart and Wiseman's Crop Husbandry Including Grassland. https://doi.org/10.1016/B978-0-323-85702-4.00020-0

complex issue which can, to some extent, be addressed by the use of 'flex crops', which can be used for either food or energy, multifunctional land use and the use of marginal land for energy crop production. Grown on a large scale, energy crops have the potential to affect the value of food and energy as commodities and to affect water availability in drier regions.

However, beyond energy, crops remain an important source for other industrial uses such as lubricant oils and chemicals used for plastic manufacture, for example. They have the potential to substitute for fossil hydrocarbons in these areas thus delaying the time when finite supplies run out. Natural fibres from tropical and temperate crops are replacing artificial fibres like polypropylene and nylon derivatives.

16.1.2 United Kingdom policy

As well as the global drivers discussed above, domestic legislation and policy in the United Kingdom also affects the energy system. One important example is the Climate Change Act (2008) which originally required an 80% reduction in carbon dioxide emissions (compared to 1990 levels) by the year 2050. This was amended in 2019 to require a 100% reduction known as 'net zero' by mid-century. Reaching this target will require all energy to be delivered in zero-carbon forms and come from low-carbon sources, including bioenergy.

Additionally, the Energy Act (2008) introduced economic incentives for renewable energy production, namely Feed-in Tariffs and the Renewable Heat Incentive. Although these measures have been important since their inception in 2010–11, they have been phased out between 2019 and 2022. Since 2020, a tariff has been offered through the Smart Export Guarantee to incentivise small-scale renewable electricity generation, including anaerobic digestion. It is important to note that installation costs for renewable energy developments have reduced over time and installations continue to rise in spite of policy changes.

However, it is important not to overstate the significance of energy crops in the UK. Studies have shown that the adoption of these crops so far has been limited; it is unlikely that extensive areas of agricultural land in the UK will be used for their production in the near future. Analysis of DEFRA data on crops grown for bioenergy can be summarised as follows. In 2020, less than 2.1% of arable land in the UK was used to grow bioenergy crops; these 121,000 ha comprised:

- 75,000 ha of maize used for anaerobic digestion;
- 29,000 ha of wheat used for biofuels;
- 8,000 ha of miscanthus used for biomass;
- 7,000 ha of sugar beet used for biofuels;
- 2,000 ha of short rotation coppice used for biomass.

16.1.2.1 Biofuels

In 2020, 36,000 ha of land were used for biofuel production; at 421 million litres, biodiesel production decreased by 27% from 2019 to 2020; at 326 million litres, bioethanol production increased by 24% in the same period.

16.1.2.2 Biomass

In 2019–20, 33,000 tonnes of Miscanthus were used in power stations, a decrease of 32% on the previous year; 7000 tonnes of short rotation coppice were used in power stations in 2019–20, a 74% decrease on the previous year (which had been higher than normal).

16.1.2.3 Anaerobic digestion

In 2020, anaerobic digestion produced energy totalling 1021 tonnes of oil equivalent, roughly equal to that produced in 2019; of this, electricity made up 953 and heat 68 tonnes of oil equivalent.

16.2 Energy crops

The main biomass crops are Short Rotation Coppice (mainly willow) and Miscanthus grown for pellets, billets or chips for use mainly in biomass power stations or in domestic or community heating systems. Crop wastes such as straw can be burnt in on-farm boilers to provide heating for grain drying.

16.2.1 Specific crops: miscanthus

Miscanthus species (sometimes erroneously called elephant grasses) are fast growing and high yielding tropical or sub-tropical perennial grasses which can survive and grow well in temperate climates. They are propagated commercially using rhizomes which are planted in rows and allowed to grow up each year before harvesting. They can produce commercial yields for 15–20 years before they need to be replanted. The species used in Europe is a hybrid *Miscanthus giganteus*.

16.2.1.1 Growth cycle

New shoots emerge in March. These grow to a maximum height (2.5–3.5 m in a mature crop) by late August and produce bamboo-like stems. The crop dies off during the autumn and loses most of its leaves, leaving only the cane. This is then harvested in February, usually with a modified forage harvester and, once dry enough, baled. It should yield between 10 and 15 oven-dried tonnes (ODTs) per hectare but has the potential to yield more if conditions are optimal.

Where can it be grown?

Miscanthus can be grown on a wide range of soil types, including reclaimed land where the annual leaf fall will add organic matter to the topsoil. It will grow readily in a wide range of temperatures but will yield best in areas of medium to high rainfall.

Planting

The rhizomes can be planted with a specialist rhizome planter or a modified potato planter after the site has been prepared. Herbicides are often used before planting to treat perennial weeds and the site may also need subsoiling in the autumn prior to planting to remove the effects of compaction. Rhizomes should be bought from a specialist producer in the UK to ensure they are disease free. Each piece should have at least three buds and should be planted at about 7 cm deep at a rate of 10,000 to 15,000 per hectare, depending on field conditions. The aim is to establish 10,000 plants per hectare but emergence can vary between 50% and 95%. The ideal time to plant is in March or April which gives a longer first year growing season and better development of the root systems which will have to support the plant for the next 15 or 20 years.

Nutrition

The offtake of a crop yielding 14 ODTs will be around 85 kg N/ha, 15 kg P/ha and 120 kg K/ha, so these should be replaced. It uses soil nutrients very efficiently and usually only requires maintenance dressings. Timing of fertiliser is not critical, although the logical time to apply the nitrogen would be in May, before rapid growth occurs.

16.2.1.2 Crop protection

Weed control in Miscanthus is relatively simple. A herbicide may be needed in year two, while the crop is establishing, for the control of grass and broad-leaved weeds and this must be done during the dormant period between harvest and the start of new growth in the spring. Miscanthus is very competitive against weeds during the growing season; the build-up of leaf litter also helps to supress weeds.

There are no serious diseases or pests that will affect the yield of the crop significantly. With an increasing area of the crop, there may be potential for aphid infestation carrying viruses such as barley yellow dwarf virus (BYDV).

16.2.1.3 Harvest and storage

The driest samples are achieved by late harvesting in early summer, but this compromises the following year's growth. The first year's growth is not usually harvested so, in the second year, the crop is cut and swathed with a forage harvester, allowed to dry and then baled. Bales can be stored short-term outside if stacked carefully. If the crop has been chipped it will need to be stored in a ventilated under-cover store; depending on length of storage, further drying may be required.

16.2.2 Short rotation coppice

Short Rotation Coppice (SRC) mostly uses willow (osier) or poplar, with willow being most popular. Unlike Miscanthus, it is harvested every 3 years instead of annually, to allow the crop to regrow into a small tree. However, it should last 10 years longer than Miscanthus if properly managed and newer varieties have the potential to give higher ODTs per hectare on an annual basis (up to 18 ODTs/ha/yr or 54 ODTs per ha per harvest). It needs specialist equipment for planting and harvesting and is usually used in the form of chips or billets (small rods 5–15 cm long). It is used in power stations or in on-farm or domestic boilers.

16.2.2.1 Growth cycle

The crop is harvested in the winter after leaf fall. New shoots emerge the following spring from the 'stool' or stump and are then allowed to grow for 3 years before harvesting again.

Where can it be grown?

It can be grown on a wide range of soils but needs sufficient moisture within a metre of the surface to grow to its fullest potential and will survive seasonal waterlogging. Its impact on the environment and landscape must be taken into account as it could be in the ground for 30 years and grow up to 8 m tall. There is also a lot of bare ground left after harvests and during the establishment phase so soil erosion risks must be considered.

Planting

Like Miscanthus, the land should be carefully prepared for good establishment with subsoiling if needed and perennial weed control with a herbicide. Bulky organic manure with low nitrogen is beneficial and rabbits should be kept out of the field by fencing for at least 2 years after planting. Unlike Miscanthus, there are several varieties available which have been tested by the Forestry Commission. Rods or cuttings are used which are then pushed into the ground by a step planter so they are left sticking vertically out of the soil in rows. A population of 15,000 cuttings per hectare is needed, although lower populations will mean thicker stems and larger billets. They are usually planted in twin rows with enough room between each set of twin rows to allow a tractor to pass through the crop. Planting can take place as early as the end of February but is more likely to be in March/April. The cuttings will produce three or four shoots which are allowed to grow through the summer and then cut back in the winter to encourage more shoots to form (similar to encouraging tillering in cereal crops). The field headlands and the 1.5 m tractor pathways can be grassed to prevent soil erosion.

Nutrition

No fertiliser is needed during the establishment year and it is almost impossible to spread fertiliser in years two and three of growth, so all must be applied during year one after a harvest. A crop yielding 30 ODTs will remove 90 kg/ha N, 55 kg/ha P and 72 kg/ha K.

16.2.2.2 Crop protection

A residual herbicide can be applied after planting and again at any time during dormancy if required. In years two and three the crop is very competitive with weeds and no herbicides should be required, indeed they would be impossible to apply.

Melampsora rust is the most important disease and willows are highly susceptible. Varietal resistance is being developed and it is recommended that four or five varieties should be planted randomly within a field to slow down the spread of the disease. Fungicides are not generally used.

Willow beetles are the main pests of willows and poplars and should be spot treated with a knapsack sprayer if in large colonies. An overall spray of insecticide is not practical and could be environmentally damaging.

16.2.2.3 Harvest and storage

As these are small trees that are being harvested, much more robust equipment is required than for Miscanthus. Forage harvesters are used with specialist headers, usually with two circular blades to slice through the bottom of the stems and a feeder mechanism to pull the cut shoots into the machine. The chips are blown into a silage trailer and carted off the field. Billets are produced using converted sugar cane harvesters which produce pieces of stem about 20 cm in length and 9 mm thick. Because they are thicker than chips, billets allow natural ventilation and will dry quite readily down to 30% MC in a heap outside. Chips may need to be stored on a drying floor to remove moisture by blowing cool air through them. The chips can heat up rapidly if left too wet.

16.2.3 Wetland crops

These are by-products of wetland management where areas are cleared of mainly excess reeds and rushes. This can be dried, compressed into briquettes and burnt as a raw material but other uses are being developed, including pyrolysation into a sustainable form of charcoal.

16.2.4 Crops for anaerobic digestion (AD)

Methane digester systems can use almost any wet organic substance to produce gas as long as the right conditions exist within the digester. Livestock farmers typically use a mixture of slurry with straw or maize. The agronomy of maize will be the same as for forage maize grown for silage (see section 18.1.1) but if it is being grown on an all-arable farm, artificial nitrogen rates will need to be higher than those used on dairy farms, because of the lack of manure or slurry applications. Maize grown for AD should also be grown to maximise dry matter per hectare, rather than starch, so narrower rows are often used to encourage tillering and increase height. Taller varieties should be selected. A range of other crops is also used in AD.

16.2.5 Woodland biomass

Two main products can be considered. The first is waste from forestry and sawmill activities and the second is biomass produced from woodland or forests grown specifically for this purpose.

Forestry waste (trimmings) can be used for wood chips while sawdust from mills can be compressed into pellets and burnt in boilers.

Stemwood produced from managed forestry or woodland is often better quality than that from short rotation coppice and no specialist equipment is needed for harvest. It is usually chipped or cut into manageably sized logs.

16.2.6 Liquid biofuel crops

The two main markets are for biodiesel and bioethanol as a petrol substitute. In the UK, the main crops grown for these are oilseed rape, cereals, sugar beet and fodder beet. They have the potential to cut carbon dioxide emissions substantially as well as reducing other pollutants such as exhaust particles.

There are no special agronomic requirements of these crops grown for biofuels. Any oilseed rape variety can be used for biodiesel and wheat is the favoured cereal for bioethanol as its starch yield is higher than that of barley.

16.3 Industrial crops

As well as for energy, crops can be grown for a range of other industrial purposes. These include:

- Lubricants and waxes
- Surfactants (wetting agents)
- Surface/paper coating
- Printing inks
- Pharmaceuticals and nutraceuticals
- Cosmetics
- Essential oils
- Fragrances
- Flavourings
- Dyes
- Packaging/compostable plastics
- Building materials
- Bio-composites
- Textiles
- Adhesives.

Starch and sugar crops like wheat and sugar beet, as well as bioethanol, can be used to produce carbohydrates for use in the manufacture of plastics, adhesives and surface coatings. Oilseed crops such as oilseed rape and sunflowers can, as well as biodiesel, produce lubricants, surfactants and slip agents. The main fibre crops such as flax, hemp and cereal straw can be used for making paper, textiles, insulation material and bio-composites for the automotive industry.

16.3.1 Lubricant oil crops

Oil crops should be able to produce erucamide so High Erucic Acid Rape (HEAR) varieties and Crambe are commonly grown. HEAR is grown in the same way as conventional double-low oilseed rape (see section 14.2) and Crambe is similar to spring oilseed rape in its requirements. Crambe (*Crambe abyssinica*) is a member of the mustard family. The oil contains about 58% erucic acid and the seed coat is high in glucosinolates so must be removed before the meal is suitable for animal feed; this can be used as a fibre source for low grade paper production.

The crop is planted at the same time as spring oilseed rape, flowers within 35 days and can be harvested 90 days after planting by the same techniques used for oilseed rape, i.e., combined direct, desiccated or swathed.

16.3.2 *Fibre crops*

There are two main crops of commercial interest to UK farmers: hemp and flax. Miscanthus, linseed straw and cereal straw can also be converted into useable, low grade fibre. Hemp core (shiv) gives a short length fibre for use in packaging, insulation and biocomposites whereas flax, if grown and harvested properly, can give high quality, long fibre for use in linen manufacture. Hemp must be licenced through the Home Office as it is classed as a narcotic.

16.3.2.1 *Hemp* (Cannabis sativa)

Hemp is a low input crop which can be grown on a wide range of soils but grows best in a pH of 6.5 or above. It can substitute for broad leaved break crops in a rotation and is very competitive against weeds. It is sown in the spring with a target population of 150 stems/m^2. Moderate amounts of nitrogen (80–120 kg/ha) can be applied onto the seedbed and, apart from a possible pre-emergence herbicide, no other crop protection is usually needed.

The crop should be cut in August and left in the field to ret (fibres are separated from the stem by the action of wetting and drying). Then it is rowed up and baled. Dual-hemp crops can be grown for fibre and oilseed, in which case they are combined and the straw left to ret and then baled normally.

16.3.2.2 *Flax* (Linum usitatissimum)

Flax is widely grown in Europe although commercial flax-processing plants, which produce long fibre suitable for linen, are limited in the UK. The agronomy is similar to linseed (see section 14.3). The crop is sown in the spring at a plant density of 2000 per m^2 (much higher than linseed) to encourage long straight stems; it receives maintenance dressings of P and K and often no nitrogen at all. The higher plant densities may encourage more disease, especially botrytis, so fungicides may have to be considered. Weed control is the same as for linseed.

Crops grown for long fibre are pulled out of the ground by special machines. This gives the maximum length of fibre which goes down into the main root. The crop is then retted like hemp, although there are some factories that can artificially ret in tanks. Short-fibre crops can be cut and baled after retting, similar to the hemp crop.

16.3.3 *Pharmaceuticals, neutraceuticals, essential oils and cosmetics*

These are small-scale crops usually grown on contract with a buyer and the acreages are often very limited. Commercial flower and fruit production is beyond the scope of this book, and of course many oils are imported from the tropics or sub-tropics, but the main potential crops in the UK are species such as lavender, mint, chamomile and a number of other herbs. Pharmaceuticals and neutraceuticals (body and skin care products) are mostly based on oilseed crops such as borage, evening primrose, Gold-of-Pleasure (Camelina), Echium, Calendula and Lunaria.

Many of these crops are very low yielding, e.g., borage averages around 0.4 t/ha and requires a high level of attention to detail for its agronomy, so farmers will expect good returns based on a realistic contract price. Unless margins match those of the mainstream arable crops, growers will usually not cultivate these lesser-known ones.

16.3.4 Carbohydrate crops

The main product from these crops is industrial starch used in many different ways, including for packaging, coatings for textiles and paper, adhesives, explosives, cosmetics, plastics and, of course, bioethanol (see above).

All crops have a carbohydrate content but the ones of most interest for commercially extractable starch or other products include maize, wheat, sugar beet, potatoes and barley. There may be the potential to develop more esoteric crops such as quinoa, sorghum and chicory for starch production.

The agronomy of these crops will be mostly the same as if they were being grown for food production.

16.4 Key points

- The transition in the global energy production system is driven by climate change and energy security.
- Bioenergy, as a renewable resource, is complex and sometimes controversial.
- In 2020, less than 2.1% of arable land in the UK was used to grow bioenergy crops.
- Miscanthus and short rotation coppice are grown for use as biomass crops for energy production.
- A wide variety of crops can be used as feedstocks for biogas production through anaerobic digestion.
- Woodland biomass can also be used for energy production.
- A variety of crops can be grown for the production of biofuels, such as biodiesel and bioethanol, and lubricant oils.
- Hemp and flax can be grown for the production of fibres.

Sources of further information and advice

Department for Environment, Food & Rural Affairs, 2021. Area of Crops Grown for Bioenergy in England and the UK: 2008–2020. Available at: https://www.gov.uk/government/statistics/area-of-crops-grown-for-bioenergy-in-england-and-the-uk-2008-2020.

El Bassam, N., 2010. Handbook of Bioenergy Crops. Earthscan, London.

Graham-Rowe, D., 2011. Beyond food versus fuel. Nature 474, 56–58.
Halford, N.G., Karp, A. (Eds.), 2011. Energy Crops. Royal Society of Chemistry, Cambridge.
Meakin, S., 2007. Crops for Industry: A Practical Guide to Non-food and Oilseed Agriculture. Crowood Press, Ramsbury.
Strezov, V., Anawar, H.M. (Eds.), 2019. Renewable Energy Systems from Biomass. CRC Press, Boca Raton.

Fresh produce crops

17

17.1 Growing fresh produce crops

Fresh produce is a term used to describe edible horticultural crops. The fresh produce sector is technically challenging; the range of crops is large and different production systems are employed for many of them. The crops are harvested fresh and, in contrast to arable crops, may be harvested regularly for a number of months. Some of the key differences between fresh produce and broad-acre arable crops are:

- Very diverse range of crops.
- High value crops.
- Intensive crop inputs.
- Labour use is high.
- Protected structures can be used for some high value crops.
- Mainly marketed fresh (some crops can be stored for months).
- Cold chain is important in distribution.

There have been major technological advances and changes in the fresh produce sector that have affected how fruit and vegetables are grown in the UK. Many large farms now specialise in a small range of fresh produce crops, growing produce both in the UK and abroad to supply multiple retailers for 12 months of the year. Codes of practice have been drawn up between growers and retailers to promote good practice, enable crop traceability and give assurance to the consumer. Over 80% of produce is now sold through multiple retailers and an increasing proportion of that is sold ready prepared. The demand for organic produce has increased but sales are sensitive to the general economy.

It is very important for the grower to know the requirements of the market, to produce only what is needed when it is needed, and of a suitable quality. It is usually advisable to arrange contracts with buyers, such as processors, freezers, pre-packers, multiple retailers or cooperatives in advance of growing the crop. Alternatively, all or part of the produce may be sold direct to the public at the farm gate, through pick-your-own or door-to-door delivery such as the 'box scheme' used by some organic growers.

There have been many developments in the farm-scale production of fresh produce that have resulted in a high level of mechanisation from seed to supermarket shelf. Seed choice and sowing and planting systems combine to result in a uniform crop that comes to maturity evenly and which often can be harvested at a single pass.

Lockhart and Wiseman's Crop Husbandry Including Grassland. https://doi.org/10.1016/B978-0-323-85702-4.00011-X

Husbandry techniques aim to optimise crop growth while reducing inputs to the benefit of the farmer, the environment and the consumer. Harvesting and postharvest methods aim to transfer the crop from field to customer as efficiently as possible and in the best condition. These developments may be summarised as follows:

- F1 hybrids have been developed to give uniformity and vigour.
- Seed is now available graded or pelleted to help with drilling, and treated or primed to aid and speed germination.
- In a number of crops young plants are now raised under protection in small modules or larger blocks allowing efficient transplanting and promoting quick establishment in the field.
- Use of bed systems in some crops can give good access to the crop for all tractor operations.
- Plastic mulches and fleece crop covers have helped to extend the growing season in the field.
- Protected growing structures have allowed intensive production of high value salad vegetable and fruit crops.
- Integrated crop and pest management are commonly used to reduce inputs by applying fertiliser and water only as needed for crop growth and targeting plant protection products at those weeds, pests and diseases actually present in the crop.
- Specialised harvesting rigs have enabled fast and efficient removal from the field; mobile packhouses and modern grading, washing and packaging facilities ensure the crop is then transferred from field on to distribution quickly while maintaining quality.

Limiting factors on the individual farm may be:

- Need for specialist machinery and structures.
- Unsuitable soils, poor drainage or lack of irrigation.
- Difficulty obtaining suitable casual staff and supervision.
- Lack of a suitable water supply for irrigation.

17.1.1 Value and volume

The UK is not self-sufficient in fresh produce. The proportion of crops marketed in the UK has generally declined in recent years due to increased imports and in 2020 UK-produced crops accounted for 56% of vegetables and only 16% of fruit marketed in the UK (Table 17.1). Fresh produce crops grown in the UK are of high value. For example, whilst the land in the UK used to grow fresh produce is only approximately 10% of that used for wheat, the value of fresh produce crops exceeds that of the wheat crop and in some years comes close to the value of the total cereal crop. Defra produces a summary of basic horticultural statistics each year where the area, volume and value of the main crops are summarised.

17.1.2 Assurance and food safety

A major challenge for growers of some fresh produce crops such as leafy salads or soft fruit is that they may be consumed fresh without being cooked. Any microbial contamination on the crop at harvest may be eaten by a consumer and there have been cases of food-borne illness linked to contaminated fresh produce crops around the world. The only way to prevent this happening is to prevent the contamination of the crop. The

Table 17.1 Area, value of UK production, tonnage and proportion of the UK market of fruit and vegetables supplied 2020 (Defra, 2021).

	Planted area (Ha)	Value of UK production (£ million)	Production (1000 t)	UK supply to the market (%)
Vegetables				
Field	118,131	1320.1	–	–
Protected[a]	874	349.8	–	–
Total	119,005	1669.8	2599	56
Fruit				
Field/ orchard[b]	29,586	993.2	–	–
Glasshouse[c]	217	51.4	–	–
Total	29,802	1044.6	657	16
Grand total	150,709	2714.4	3256	

[a]Including glasshouse crops and mushrooms.
[b]Including polytunnel grown soft fruit.
[c]Soft fruit.

main risks to the crop come from the use of manures as soil conditioners, contamination by livestock or wild animal manure, contaminated water sources, and equipment and worker hygiene. Growers are required by many customers to follow strict standards of production to reduce the risks of contamination. The most widely followed scheme in the UK is the Red Tractor Fresh Produce Scheme which has general standards for all crops.

17.1.3 Use of integrated pest management (IPM)

Many fresh produce crops are classified as minor or specialist crops. The costs involved for pesticide manufacturers to register new products is large and there has been a significant reduction in the number of pesticides approved for use by fresh produce growers. As a consequence, growers are increasingly relying on a combination of integrated pest management techniques (see Chapter 8). For example, the loss of suitable herbicides has led to mechanical weeding/hoeing being used in combination with herbicides in many crops. Where the number of available insecticides has been reduced the repeated use of the remaining products needs to be avoided to prevent development of resistance in the pest. The alternatives to pesticides are limited; one option is to use netting/mesh during risk periods to minimise damage, for example, to control cabbage root fly in brassica crops.

17.1.4 Protected production systems

Most fresh produce crops are field grown (Table 17.1) where growers have limited ability to manipulate the growing environment beyond irrigation, nutrition and the use of crop covers. The use of protected growing structures such as glasshouses and polytunnels allows a greater control over the growing environment. Both glasshouses and polytunnels are able to utilise solar gain, where sunlight heats up the growing area (i.e., the 'greenhouse effect') allowing crop growth when it is too cool in the field and extending the cropping season as well as enhancing growth rates and yield. Glasshouses can also be heated and illuminated to simulate summer conditions through winter and because glasshouses are closed structures carbon dioxide levels can be raised around the plants to maximise crop growth and production at high light levels.

Glasshouses are well suited to the use of biological control of pests as they are enclosed environments which retain insect predators. Consequently, glasshouse production generally utilises relatively few insecticides. Crops commonly grown in UK glasshouses are lettuces and long-season, vegetable-fruit crops such as tomatoes, peppers and cucumbers as well as soft fruit, especially strawberries. By using glasshouses, it is possible to achieve high yields compared to outdoor production through higher productivity and long harvesting seasons. For example, Defra data shows that the overall UK glasshouse production of cucumbers over the 2020 season was over 500 t/ha and for glasshouse tomatoes was approximately 350 t/ha.

Building and maintaining glasshouses is very expensive and growing in these structures requires specialist equipment and growing skills. Nevertheless, using heating and lighting means that some crops such as tomatoes can be produced all year round using these systems.

Polytunnels are a cheaper alternative to glasshouses and can be a temporary field structure over soil or a semi-permanent structure with soil-less substrate such as coir bags or hydroponics (Fig. 17.1). The environment under a polytunnel is less controllable than a greenhouse. The key differences are that the temperature will be warmer than the open field and rainfall will be kept off the crop. The structure is generally covered with a removable polythene film and it is common to remove the cover at the end of the season in autumn to prevent damage from high winds.

The polythene films can have additives that effect light transmission and can reduce temperature peaks in the growing environment preventing thermodormancy and improving working conditions. Some films can also inhibit fungal spore germination and reduce the occurrence of some insect pests. Polythene films need replacing after 4–5 years as their light transmission declines as they age and they also can become damaged. The use of polytunnels is widespread in soft fruit production and the majority of summer-grown soft fruit in the UK are grown under polytunnels.

The development of vertical farming systems where crops are grown in controlled environments using LED lighting, generally without any sunlight, is increasing and has potential to play a significant role in the production of some fresh produce crops. Commercial vertical farming systems generally utilise hydroponic or aeroponic systems where the roots of the plant are either suspended in a nutrient solution or misted with a nutrient solution using very little water. Plants are grown on shelves vertically stacked on top of each other with LED lights attached to the base of the shelf above

Figure 17.1 Strawberry polytunnels (Authors own, 2021).

giving a rack of 5–10 layers. The growing environment needs to be temperature controlled and ventilated using heating, ventilation and air conditioning (HVAC) systems. The energy costs of vertical farms can be high and alternative energy sources are often utilised to manage both the sustainability and running costs of the units.

Vertical farms can be located in redundant buildings or in newly built structures but are very expensive to develop. The major benefit in growing crops in a vertical farm are that the crops can be grown away from traditional field production areas, close to urban markets or processing facilities and can be produced all year round avoiding the need for importing crops. The total control of the growing environment may also enable crops to be grown with added-value traits such as enhanced flavour and nutrient content. A number of commercial units have been built next to distribution centres to maximise the efficiencies of the supply-chain. Crops that are well suited to vertical farming systems are lightweight, high value, fast growing crops. Leafy vegetables and herbs are well suited to commercial production using vertical farms although soft fruit crops may also have potential.

17.1.5 Automation

Labour requirements are high for many fresh produce crops and can account for a substantial proportion of the production costs. As labour availability and cost becomes more challenging growers try to automate or mechanise labour intensive stages where possible. The key stages traditionally for labour use in field crops was transplanting, weeding and harvesting.

17.1.5.1 Transplanting

Transplants are widely used in leafy vegetables and transplanting is the planting of young plants that have been propagated from seed in trays, often in a protected structure. Manually operated transplanting rigs relied on transplants being manually fed into the planting mechanism. These are being replaced with automated transplanters capable of planting at a much faster rate e.g., Planttape where transplants are mechanically fed into the mechanism (Fig. 17.2).

17.1.5.2 Weeding

The reduction in herbicides that are approved for use in fresh produce crops has led to increased use of mechanical weeding. The alternative to this is manual hoeing which is time consuming and hence expensive. Mechanical weeders can be manually steered or vision-guided hoes relying on camera based systems linked to the ability to move the tine or blade around individual plants. Vision-guided hoes are being used successfully with either fixed tines or responsive blades e.g., Garford robocrop (Fig. 17.3).

Figure 17.2 Planttape automated lettuce transplanter (Authors own, 2021).

Figure 17.3 Garford robo-crop weeder in a field of lettuce (Garford Farm Machinery Ltd., 2021).

17.1.5.3 Harvesting

Pea and bean crops grown for processing are harvested with specialist harvesters or viners that cut the whole plant and transport these to a drum that strips and opens the pods from the plant removing the peas or beans. Mechanical harvesters are widely used in root and bulb crops e.g., carrots and bulb onions, using share lifters similar to potato harvesting. Toplifters that pinch the foliage of plant and lift it out of the ground are used in early season carrots but have also been adapted to other crops where the foliage is strong including leeks, fennel and beetroot sold with green tops. Baby leaf crops are harvested using self-propelled harvesters that cut the leaf with either a 'bandsaw' or rotating discs before the leaf is passed up a belt to be manually packed into crates. Cabbage harvesters have been developed that grip the stem of the cabbage and the head is then cut off before being transported to either boxes or a manual deleafing platform.

These mechanical harvesting systems all rely on the crop being grown in a system that gives a uniform crop ready for a single cut/pass at harvest, whereas crops that are delicate e.g., wholehead lettuce and/or require grading/selecting at harvest e.g., broccoli and/or are picked multiple times e.g., strawberries, pose additional challenges to automation. The development of vision-guidance and robotic systems, combined with artificial intelligence-based control units, has allowed the development of precision harvesting equipment that can adapt to variable and delicate crops. These systems are in their infancy but will have a significant impact on labour requirements and production systems in the future.

17.2 Fresh peas

There are two main types of peas produced in the UK: *vining peas* are used for canning fresh (sometimes called 'garden' peas), freezing or drying and are normally grown under contract, or by cooperating groups. The contracting companies usually supply the variety they require and decide on times of sowing based on a 'heat unit' system. This is in order to spread the harvesting period to ensure a regular and constant supply to the processing factories. *Fresh peas (market pick or pulling peas)* make up ~5% of the UK crop and are grown to be sold fresh in the pod to be home-shelled or eaten whole (e.g., mange tout or sugar snap peas) or as shelled fresh peas.

Vining peas are best suited to the main arable areas and where there are local processing plants or markets (most pea growers are concentrated in East Anglia and Humberside), while most vining pea growers are members of cooperatives, often linked to a specific processing factory. The factory must ensure a continuous and reliable supply of produce so that it can operate efficiently. Growers must ensure, therefore, that the crop comes ready for harvesting over several weeks. This is achieved by: (a) growing different varieties with different maturity dates, or (b) staggering the drilling of the peas based on the calculation of accumulated heat units (AHUs). Thus, if the difference between 1 day and the next in East Anglia at harvest time is 11.5 AHUs, and the vining capacity of the cooperative is 40 ha per day, each block of 40 ha must be drilled 11.5 AHUs apart back in the spring. This could mean delays in drilling of several days if the weather is cold. This should result in the shared harvesters moving around the pea crops just at the right stage of ripeness. If there is a delay because of weather or breakdown, then the peas can often get too tough for freezing or canning. They can then be allowed to grow on for harvesting dry.

17.2.1 Soils and climate

Peas suit deep, free-working loams. Pea roots are very sensitive to the physical condition of the soil and compaction and waterlogging are very damaging so light sands and heavy clays are not suitable. Peas do not grow at temperatures below 4.4°C so overwintered fresh peas are only suited to frost-free coastal areas in the south of England.

17.2.2 Cultivations

Peas are sensitive to compaction and an adequate tilth is required for planting. Land can be autumn ploughed and weathered over winter or spring ploughed on lighter soils.

17.2.3 Drilling/planting and timings

Vining peas are drilled from February to May. Sequential plantings give a steady supply of crops over the harvest window. Fresh peas can be autumn planted and overwintered for an early crop or spring planted for later harvests.

17.2.4 Rotation

Peas can be a valuable break crop in cereal-based rotations. Peas are N-fixers and the crop debris contains high levels of N. Legumes (peas and beans) should be grown no more than 1 year in five to prevent the build-up of pests and diseases.

17.2.5 Pests and diseases

Peas can be attacked by a range of pests including pea and bean weevil, pea cyst nematode, aphids, moths, thrips, pea midge, silver Y moth and slugs. Damping off, foot and root rot, leaf and pod spot, downy mildew, powdery mildew, botrytis, virus and bacterial blight can affect the pea crop and may need controlling.

17.2.6 Weed control

Peas do not compete well with weeds and efficient weed control is essential to avoid yield loss. Perennial weeds should be dealt with before planting and thistles, oilseed rape and groundkeeper potatoes can be a particular problem. Nettles and thistles can interfere with hand picking and, in machine-harvested crops, weeds can lead to crop rejection from the processor. Inter-row cultivations may be effective when the peas are between the second and fifth node growth stage.

17.2.7 Nutrition

Nitrogen-fixing *Rhizobia* on the roots provide the crop with nitrogen. Fertiliser N or FYM has no benefit and can delay or suppress nodulation. For a soil with an index of 2, the fertiliser recommendations for fresh peas would be 85 kg/ha of P_2O_5 and 40–90 kg/ha of K_2O and for vining peas would be 40/ha of P_2O_5 and 20–40 kg/ha of K_2O.

17.2.8 Irrigation

Peas are sensitive to soil moisture deficits and irrigation may be beneficial from the beginning of flowering and during pod swelling.

17.2.9 Harvest and storage

Vining peas are harvested in June and July and are ready for harvesting when the crop is just starting to lose its green colour and the peas are still soft. The firmness of the peas is tested daily (near harvest time) with a tenderometer. The tenderometer reading (TR) for freezing peas is about 100 and for canning peas about 120. The crop must be cut as soon as possible. The pods are harvested using self-propelled vining machines that separate the peas from the pods. The shelled peas are rapidly transported to the processing plant for freezing or canning.

Fresh peas are harvested by removing the pods when the peas are at a required maturity. The peas can be either hand-harvested or mechanically harvested. The use of overwintered crops, crop covers and variety choice means that fresh peas can be harvested from May to September.

Fresh peas do not store well and deteriorate quickly in warm temperatures with much of their sugar turning to starch within a few hours. Peas should be cooled and distributed rapidly to retailers using the cold chain.

17.3 Broad beans

Broad beans are grown for freezing, canning and for the fresh market. There are two groups of beans based on flower colour, which is important for the processed crops, particularly canned broad beans. Coloured flower varieties give a pink-brown tinge to the beans and water during the canning process. White flowered varieties are used for both canning and freezing, varieties with coloured flowers are only used for freezing.

17.3.1 Soils and climate

Broad beans have a strong taproot and can cope with a wide range of soil types but require a pH of at least 6.0–6.5. If the pH is below 5.5 growth will suffer and an application of lime will be needed.

17.3.2 Drilling/planting and timings

The large seeds are likely to be damaged in some drills. Vacuum drills and some types of belt-fed drills are more suitable. For small areas the crops may be hand-sown. Retail crops can be sown November/December and overwintered to be harvested late May/June. The main crops are sown from February to May, for harvest from June to August.

Processing crops can be sown from early February to May as directed by the processors. Successional sowings give a regular supply of beans of the correct maturity over the harvest window from mid-July to the end of August. The normal row width is 45 cm with 13 cm spacing along the row.

17.3.3 Rotation

Broad beans can be a valuable break crop in cereal-based rotations. Beans are N-fixers and the crop debris contains high levels of N. Legumes (peas and beans) should be grown no more than 1 year in five to prevent the build-up of pests and diseases.

17.3.4 Pests and diseases

Pea and bean weevil, stem and bulb nematode, black bean aphids, bruchid beetle, downy mildew, chocolate spot, foot rot and rust are usually the main problems in broad beans.

17.3.5 Weed control

Broad beans do not compete well with weeds and efficient weed control is essential to avoid yield loss. Perennial weeds should be dealt with before planting and wild oats, thistles and groundkeeper potatoes can be a particular problem. Inter-row cultivations may be used if the crop is grown in wide rows.

17.3.6 Nutrition

Nitrogen-fixing *Rhizobia* on the roots provide the crop with nitrogen. Fertiliser N or FYM has no benefit and can delay or suppress nodulation. For a soil with an index of 2, the fertiliser recommendations for fresh broad beans would be 100 kg/ha of P_2O_5 and 50–100 kg/ha of K_2O.

17.3.7 Irrigation

The crop is sensitive to soil moisture deficit and higher yields are obtained from irrigating at early flowering and pod fill.

17.3.8 Harvest and storage

Hand-harvested crops are normally picked two or three times with sequential plantings picked over from May to August. Processing crops are machine-harvested by self-propelled viners in a similar way to peas but using direct cutting bean header. Processing crops are grown in the North of England and Scotland (to avoid bruchid beetle) and are harvested from mid-July to the end of August to fit in with pea harvesting. Processed crops are harvested according to tenderometer readings (TR) and crops for freezing have a TR of 110–140, for canning the TR should read 130 and over.

Broad beans deteriorate quickly at warm temperatures and pre-pack/retail crops should be cooled rapidly after harvest, while crops destined for canning or freezing should be transported quickly to the processor to prevent deterioration.

17.4 Green beans

Green beans (Dwarf French beans) are grown mainly for processing but there is a small retail market for fresh green beans. Green beans are usually classified according to pod width: Fine beans (8.0–9.0 mm pod width); Medium fine beans (9.0–10.5 mm pod width); and Medium fine/large beans (>10.5 mm pod width).

There are many varieties on the market and processors will give advice on the varieties they recommend. Consideration should be given to yield, quality and ease of harvesting, together with the growing environment on each individual farm.

17.4.1 Soils and climate

Green beans need to be grown in loams and the lighter types of soil. It is a late-sown, short-season crop and so requires a deep, free-draining, moisture retentive soil. Soils that cap, and those of a high organic matter content (peat soils), which will produce excessive vegetative growth should be avoided. A site offering shelter and a southerly aspect is ideal. Most of the crop is grown in the eastern counties south of the Wash.

17.4.2 Cultivations

The soil surface should be prepared so that following drilling it will be as level as possible to prevent problems when machine harvesting.

17.4.3 Drilling/planting and timings

Green beans are sown between mid-May and late-June, as early drillings can be damaged by late frosts. The seed testa is easily damaged and seed must be handled with care. Pneumatic precision or belt feed-drills are used with treated seed. The final plant population for processing is 30–40 plants/m^2 and for the fresh market 20–25 plants/m^2. The normal row width for mechanically harvested crops is 30–40 cm with 6–7 cm spacing along the row.

17.4.4 Rotation

Legumes (peas and beans) should be grown no more than 1 year in five to prevent the build-up of pests and diseases.

17.4.5 Pests and diseases

Green beans are subject to attack by a number of pests especially caterpillars of Silver Y moth, bean seed flies, aphids, slugs and cutworms. Halo blight, botrytis, rust and sclerotinia can be a problem in some seasons.

17.4.6 Weed control

Green beans do not compete well with weeds and efficient weed control is essential to avoid difficulties with machine harvesting. Perennial weeds should be dealt with before planting and thistles and groundkeeper potatoes can be a particular problem. If weeds are not controlled, the crop may be unsaleable. Inter-row cultivators/weeders can lead to irregular soil conditions and cause problems with machine harvesting. Weeds are controlled mainly through herbicides.

17.4.7 Nutrition

Green beans respond very well to nitrogen. In contrast to broad beans, the nitrogen fixing, *Rhizobium* bacteria that form nodules on green beans are not found in many UK soils so the seed/soil can be either inoculated when drilling or nitrogen fertiliser can be applied. For a soil with an index of 2, the fertiliser recommendations for green beans would be 120 kg/ha N, 100 kg/ha of P_2O_5 and 50–100 kg/ha of K_2O. To prevent fertiliser leaching and reduced germination a maximum of 100 kg N/ha should be applied at drilling. The remainder should be applied as required by the crop.

17.4.8 Irrigation

Green beans respond well to irrigation during early pod development but little benefit is observed when irrigating before the flowering stage. Overhead irrigation should be avoided when flowers are present as the wet petals stick to developing pods leading to fungal infections of the pods.

17.4.9 Harvest and storage

Green beans are harvested from July to October. Processing crops are harvested using one-pass mechanical harvesters. Retail crops can be machine or hand harvested. The harvested bean pods must be whole, undamaged, separated (not in clusters) and free of stems, leaves, soil and stones. As the crop matures, the pods lengthen rapidly and they then enlarge as the seeds develop. Maturity should therefore be judged according to pod width and not length. Flat-podded beans must be harvested before the development of seeds becomes obvious externally. The beans should not be allowed to over-mature or quality will be reduced. This is assessed by taking a random sample of 10 plants, taking the most mature seeds from the most mature pods on each plant and measuring the length of 10 seeds. As a guide, for all beans the length of one seed should be no more than the pod width.

Green beans deteriorate quickly at warm temperatures and pre-pack/retail crops should be cooled rapidly after harvest, crops destined for canning or freezing should be transported quickly to the processor to prevent deterioration.

17.5 Lettuce

Lettuces are grown for sale as a whole head or for further processing for foods such as sandwiches and salads. Most types are field grown with the main types being Iceberg/Crisphead, Cos/Romaine, Speciality/Continental and Little Gem. The traditional flat or butterhead lettuces are mainly grown under protection.

17.5.1 Soils and climate

Lettuce can be grown on a wide variety of soil types. An ideal soil has good water retention properties but is not too heavy as this may limit field access by machinery for planting, crop applications or harvest if wet. Very light soils may require frequent irrigation and additional fertiliser applications.

17.5.2 Cultivations

On heavy land it may be necessary to subsoil prior to ploughing and power harrowing before transplanting. Lettuce can be grown in beds or on the flat.

17.5.3 Drilling/planting and timings

All commercial lettuce in the UK is grown from transplants (Fig. 17.4) although it is possible to grow lettuce from drilled seed. New planting systems are utilising smaller plants such as the planttape system. Lettuce is susceptible to frost and can be grown without protection from the last frost in the spring to the first frost in the autumn. Early crops may be transplanted from February and will be covered in fleece or polythene for the first 2–4 weeks to reduce the risk of frost damage. Lettuce takes 6–10 weeks to mature depending on lettuce type and time of year with summer transplantings maturing most quickly. Plants are spaced at approximately 30–35 cm but closer plant spacing may be used for Little Gem.

Figure 17.4 Lettuce block for transplanting (Authors own, 2021).

17.5.4 Rotation

Due to the short production time, lettuce can have two or three crops grown in one field in a season. One year breaks are possible but longer breaks are beneficial particularly in preventing the build-up of pest populations.

17.5.5 Pests and diseases

Lettuce are attractive to a wide range of pests especially aphids, caterpillars and slugs. The diseases downy mildew, botrytis and sclerotinia are the main problems when growing the crop.

17.5.6 Weed control

The reduced number of active ingredients means that the control of weeds in lettuce has become difficult. Mechanical or hand weeding may be the only option where sites are badly affected with weeds. As a consequence, there is an increase in the use of vision-guided mechanical hoes which can remove weeds between plants within the row.

17.5.7 Nutrition

For a soil with an index of 2, the fertiliser recommendations based on Iceberg lettuce would be 160 kg/ha N, 150 kg/ha of P_2O_5 and 100−150 kg/ha of K_2O. Lettuce are not responsive to high levels of N and excess levels may reduce shelf life. Due to potential risk to health from having too much nitrate in a diet, there is a maximum level of nitrate allowed in lettuce leaves governed by legislation. The maximum value depends on season, method of production and lettuce type. Growers need to manage N to avoid excess applications. Trace element deficiencies should be identified by tissue analysis.

17.5.8 Irrigation

Lettuce has a very high water content (>95%) and irrigation has an important role in crop production. Sufficient water is needed after transplanting to help rapid crop establishment. Subsequent irrigation requirements will vary according to soil type, crop growth stage and rainfall. Growers can use irrigation-scheduling decision support systems to decide when and how much water to apply. Care is needed not to over irrigate crops, due to nutrient leaching and reduced postharvest quality. Under-irrigation will lead to reduced yield and may increase tip scorch in some crops. The water source used needs to be free of human pathogens that would contaminate the crop.

17.5.9 Harvest and storage

Lettuce are harvested by hand. Many are harvested using field rigs (Fig. 17.5) that pack and label the product in the field. The heads are selected for size and cut at the base of

Figure 17.5 Harvesting rig (Authors own, 2021).

the plant. The outer leaves may be left on (as protection in transport for processing heads) or removed and the head wrapped in a plastic bag before labelling in the field and dispatching to the multiple retailer. Lettuces are rapidly cooled then stored and transported at low temperatures in refrigerated vehicles.

17.6 Baby leaves and herbs

Baby leaf crops are leaf crops grown at high seed rates and harvested with three to five true leaves. They are grown mainly for further processing and sold washed or unwashed in bags although baby leaves can be sold loose or bunched. The greatest volume crop is spinach, grown both for salads and cooking, but a wide range of other crop species and types are also grown including brassicas, such as mizuna or mustard; red and green lettuce, including green oak leaf and lollo rosso; and herbs such as flat leaf parsley and coriander.

17.6.1 Soils and climate

Baby leaf crops will grow on most soils but frequent access to the crop is needed by machinery. Light free-draining soils are best suited to intensive production but attention to nutrition is needed as light soils may have low nutrient retention.

17.6.2 Cultivations

Fine seedbeds are required for uniform emergence of baby leaf crops. This can be achieved using power harrows following discing. Baby leaves are commonly grown in beds to minimise compaction and aid harvesting.

17.6.3 Drilling/planting and timings

Baby leaf crops are grown at high seed rates and seed cost is a significant production cost. For example, spinach will be drilled at 120–140 seeds/m^2. Crops develop quickly and can take 21–28 days from emergence to harvest. The size of the crop at harvest will dictate the market with crops grown for micro, baby and teen leaf specifications. Crops may be covered by fleece to hasten emergence at the start of the season.

17.6.4 Rotation

Ideally crops will be grown in a field 1 year in three. It is common for multiple crops to be grown in the same field within a year. Consideration should be taken of mixed fields, i.e., where leaves of different plant families such as brassicas and lettuces are being grown.

17.6.5 Pests and diseases

Baby leaves and herbs are attractive to a wide range of pests especially aphids, leaf miners, caterpillars and slugs. The contamination of the crop with insects is also a commercial issue. Seedling diseases can affect the crops such as *Pythium* and *Rhizoctonia* but the main disease is downy mildew which can lead to complete crop loss if untreated. Resistant lines are available for some of the larger volume crops like spinach but downy mildew evolves new strains rapidly and plant breeders are always playing 'catch up'.

17.6.6 Weed control

Due to the small plants and short production cycle soil steam sterilisation is sometimes carried out prior to the production of baby leaves. Some growers have also experimented with mulching to suppress weeds and this can work in some situations.

17.6.7 Nutrition

For a soil with an index of 2, the fertiliser recommendations based on baby leaf lettuce would be 40 kg/ha N, 150 kg/ha of P_2O_5 and 100–150 kg/ha of K_2O. Baby leaf crops will require less nitrogen than a mature full leaf crop but nitrogen needs to be adequate to avoid pale leaves. Due to potential risk to health from having too much nitrate in a diet, there is a maximum level of nitrate allowed in baby leaf leaves including spinach, lettuce and rocket, which is governed by legislation. The maximum value depends on season, method of production and crop. Growers need to manage N to avoid excess applications. Trace element deficiencies should be identified by tissue analysis.

17.6.8 Irrigation

Water is very important for germination and crop uniformity. Micro-sprinklers are commonly used as they can apply small amounts of water frequently. Large droplet sizes can damage the foliage so fine mists are often used. The short crop cycle means that baby leaves and herbs have shallow roots 20–50 cm deep. The water source used needs to be free of human pathogens that would contaminate the crop.

17.6.9 Harvest and storage

Baby leaves are rarely hand cut (except for bunched herbs) and machine harvesting is common. The crop is cut 1–3 cm above the soil surface using a trailed or self-propelled harvester. It is possible to reharvest a crop more than once following regrowth but this gives a poorer quality crop with variable leaf sizes. The leaves are then cooled rapidly and stored at high relative humidity often with misting or fogging.

17.7 Cabbages

Different types of cabbage are grown for year round availability. There are six types, defined by appearance and time of maturity.

1. *Spring greens* (Collards) are sold as loose leaf or open hearted heads. They are traditionally cut from overwintered crops from March to May but production is increasing through the year with winter crops grown on early sites in Cornwall and Kent.
2. *Early summer cabbage* (Primo) is sold as whole heads. They are produced from overwintered crops or spring planted crops and harvested from May.
3. *Late summer/autumn cabbage* is sold as whole heads. They are produced from late spring planted crops and harvested from July to October.
4. *Winter cabbage* (January King and green types) is sold as whole heads. They are produced from late spring planted crops and harvested from November to March.
5. *Winter white/red storage cabbage* is grown for retail as a whole head or for processing. Retail heads are approximately 0.5–1 kg whereas processing heads may be >3.5 kg. They are produced from late spring planted crops and harvested October onwards. The ability to store them for 6–9 months gives all year round availability.
6. *Savoy cabbage* is sold as whole heads. They are the most frost tolerant cabbage type with distinctive 'bubbled' leaves. They are produced from spring/summer planted crops and harvested from June to April.

17.7.1 Soils and climate

Cabbages can be grown throughout the UK but problems with ring spot are greater in areas with heavier rainfall. Cabbage can be grown on a wide range of soil types and grow well on moisture retentive soils. Lighter, sandier soil types may require irrigation. Heavy soils should be avoided with winter cabbage where machinery access will be needed in the winter months. Care is needed in light fenland soils where direct drilling can be affected by loss of soil through wind erosion. A soil pH of 7.0–7.5 is required, particularly where club root may be a problem. Lime should be applied well before planting/drilling. Capping soils should be avoided where crops are direct drilled.

17.7.2 Cultivations

Cabbages have moderate to deep roots. A firm soil with a good tilth is required. Compacted soils may require sub-soiling before ploughing and harrowing/discing. Cabbages can be grown in beds or on the flat with tramlines.

17.7.3 Drilling/planting and timings

Cabbages are predominantly produced from transplanted modules. Spring greens are the exception, being mainly direct drilled at high densities. Transplanting gives a better crop production programme and more even crop maturity. Where whole head crops are direct drilled, two seeds are sown together in each location at the final spacing for the crop, and the weaker plant removed by hoeing.

- *Spring greens* are mainly drilled direct in August, transplanted in September or direct drilled in April to June at its final crop spacing 38 × 13 cm.
- *Early summer cabbage* can be autumn drilled and overwintered, drilled in February or transplanted April/May at a final spacing of 46 × 23 cm for harvesting from May. Plastic covers can be used on early crops which advance harvests by up to 2 weeks.
- *Summer cabbage* is drilled in March/April or transplanted May/June, at a spacing of 46 × 23–46 cm, for harvesting July–October.
- *Winter cabbage* varieties are drilled/transplanted in May/early June for harvesting from mid-November onwards. The target spacing is 46 × 23–46 cm.
- *Winter white/red storage cabbage* is drilled April/May or transplanted May/June and harvested from October onwards. The target spacing is 46 × 36 cm.
- *Savoy cabbage* is drilled March to May, or transplanted April to July and harvested September onwards. The target spacing is 46 × 30–46 cm.

17.7.4 Rotation

Cabbage can fit well into a rotation leaving a high amount of residual nitrogen in the soil as the waste leaves are incorporated into the soil. Where club root is a problem, particularly on acid soils, brassicas should be grown no more than 1 year in four or five with a planned liming policy before planting. It is possible to grow brassicas continuously on a high pH soil.

17.7.5 *Pests and diseases*

Cabbage root fly can be a problem, particularly with summer and winter cabbage, and treatments will be required from April onwards. Aphids, caterpillars, cutworms, pollen beetle, cabbage stem weevil and slugs all attack cabbages. Wood pigeon is a particular problem on overwintered crops. Damping off and wirestem can be a problem at the seedling stage. Club root, downy mildew, dark leaf spot, ring spot, white blister and spear rot can all be a problem during the growing season.

17.7.6 *Weed control*

Cabbages can compete well with weeds due to their large leaves but established perennial weeds should be controlled prior to transplanting or drilling. Oilseed rape volunteers and potato groundkeepers can cause a problem in some crops. Mechanical weeding with manually steered or vision-guided hoes may be used where practical, which reduces the need for chemical control.

17.7.7 *Nutrition*

For a soil with an index of 2, the fertiliser recommendations for head cabbage would be 180–260 kg/ha N, 100 kg/ha of P_2O_5 and 150–200 kg/ha of K_2O. Storage cabbage would require 280 kg/ha N and spring greens between 180 and 270 kg/ha N but the same P and K recommendations would apply. Cabbage growing over the winter months may require split application of nitrogen fertiliser with no more than 100 kg/ha N at transplanting and the remainder to support later growth.

17.7.8 *Irrigation*

Irrigation can improve evenness of maturity but it is only economical to irrigate summer crops particularly on lighter soil types. Irrigation can be applied by overhead booms or trickle tapes.

17.7.9 *Harvest and storage*

Mechanical harvesting is becoming available for processing crops but currently most other crops are hand harvested. Cabbages can be packed on rigs or harvested in to crates. Loose outer leaves are stripped back and the head cut leaving approximately 10–15 mm of stem. Variable maturity of cabbage means that fields may be harvested a number of times with usually three or more passes. In the summer the field may be harvested every 3 days and in autumn/winter every 7 days. Cabbages can deteriorate in warm conditions and should be cooled rapidly after harvest. Long-term storage requires humidification equipment.

17.8 Broccoli

Broccoli (also known as Calabrese) is grown for freezing, processing and retail markets. The harvested head (spear) consists of closed flower buds which will open if the crop is left in the field too long after maturity. Green broccoli is the main type with a single terminal head. Secondary spears are usually more succulent and there are a number of sprouting broccolis where secondary spears are harvested including 'Tenderstem' types and Purple Sprouting Broccoli.

17.8.1 Soils and climate

Broccoli is grown around the UK from the East of Scotland to the SW of England and can be grown on a wide range of soil types. The crop grows well on moisture retentive soils and lighter, sandier soil types may require irrigation. Heavy soils should be avoided with later crops where machinery access will be needed in wet conditions. A soil pH level of 7.0–7.5 is required, particularly where club root may be a problem. Lime should be applied well before planting/drilling. Capping soils should be avoided where crops are direct drilled.

17.8.2 Cultivations

Broccoli does not root deeply. A firm soil with a good tilth is required. Compacted soils may require sub-soiling before ploughing and harrowing/discing. Broccoli can be grown in beds or on the flat with tramlines.

17.8.3 Drilling/planting and timings

Broccoli is planted from March to July and harvested from June to November/December and is produced predominantly from transplanted modules. Transplanting gives a better crop production Programme and more even crop maturity. Transplants are planted at spacings of 60 × 40 cm. Direct drilling is more common in Scotland. High-density crops can be sown direct to a stand. Modules are commonly raised under glass at commercial propagators from September to June and are ready for planting from March. Early crops are produced using plastic or fleece crop covers which can bring crops to maturity 2 weeks earlier. Use of covers needs managing as high temperatures under the covers can cause physiological damage and disease problems may develop in humid conditions. Crop covers need to be removed when heads are 10–15 mm in diameter, 4–6 weeks before harvest.

17.8.4 Rotation

Where club root is a problem, particularly on acid soils, brassicas should be grown no more than 1 year in four or five with a planned liming policy before planting. It is possible to grow brassicas continuously on a high pH soil.

17.8.5 Pests and diseases

Cabbage root fly can be a problem and treatments will be required from April onwards. Aphids, caterpillars, cutworms, pollen beetle, cabbage stem weevil and slugs all attack broccoli. Damping off and wirestem can be a problem at the seedling stage. Club root, downy mildew, dark leaf spot, ring spot, white blister and spear rot can all be a problem in the field.

17.8.6 Weed control

Established perennial weeds should be controlled prior to transplanting or drilling. Oilseed rape volunteers and potato groundkeepers can cause a problem in some crops. Mechanical weeding with manually steered or vision-guided hoes may be used where practical, which reduces the need for chemical control.

17.8.7 Nutrition

For a soil with an index of 2, the fertiliser recommendations for broccoli would be 165 kg/ha N, 100 kg/ha of P_2O_5 and 125−175 kg/ha of K_2O.

17.8.8 Irrigation

Irrigation is needed on light soils and often with later crops. Irrigation can damage the wax layers on the head increasing the risk of spear rot and should end 20 days before harvest.

17.8.9 Harvest and storage

Machine harvesters are being developed but currently crops are harvested with field rigs. The head is cut and trimmed by hand before packing into trays. Side shoots will be packed into field crates and taken to a packhouse for grading and final packing. Broccoli is cooled quickly after harvest to prevent deterioration e.g. using hydrocooling or icing. Care is needed as broccoli can be damaged at temperatures below 2°C. Broccoli does not store well and is dispatched into the supply chain within a few days of harvest (Fig. 17.6).

17.9 Cauliflowers

Cauliflower is grown for freezing, processing and retail markets. The harvested head is known as a curd. A creamy white curd is the most common cauliflower but green, purple and yellow colours are available along with baby and romanesco cauliflowers, the latter having a distinct geometric pattern to the curd.

Different types of cauliflower are grown for year round availability. There are four types, defined by time of maturity.

Figure 17.6 Iced broccoli (Authors own, 2021).

1. *Early summer cauliflower* is produced from overwintered or January sown modules for harvesting from late May to early July.
2. *Late summer and autumn cauliflower* is produced from modules transplanted from April to July and harvested from July to mid-November.
3. *Winter cauliflower* is transplanted in June and July and harvested from December through to early March. These cauliflowers are susceptible to frost and are grown on frost-free coastal land in Kent and the south-west of England and Wales.
4. *Winter hardy cauliflower* (spring heading) is transplanted in July, overwintered and harvested from March to June.

17.9.1 Soils and climate

Cauliflower can be grown on a wide range of soil types, growing well on moisture retentive soils. Lighter, sandier soil types may require irrigation. Heavy soils should be avoided with later crops where machinery access will be needed in wet conditions. A soil pH level of 7.0–7.5 is required, particularly where club root may be a problem and lime should be applied well before transplanting. Winter cauliflowers harvested from December to March are grown in frost-free coastal areas.

17.9.2 Cultivations

Cauliflower does not root deeply. A firm soil with a good tilth is required. Compacted soils may require sub-soiling before ploughing and harrowing/discing. Cauliflower can be grown in beds or on the flat with tramlines.

17.9.3 Drilling/planting and timings

The crop is mainly produced from 14 ml modules. Early summer production can be brought forward by using larger module/blocks giving a larger plant at transplanting or by using crop covers.

- *Early summer cauliflower* modules are either sown in October and overwintered in unheated glasshouses, or sown in January and grown in heated glasshouses. The modules are transplanted in March for harvesting from late May to early July. The target spacing is 50 × 50 cm.
- *Late summer* modules are sown February to April, transplanted April to May at a spacing of 60 × 40 cm and harvested from late July to early September.
- *Autumn cauliflower* modules are sown during late April to mid-June, transplanted June to July at target spacing of 60 × 40 cm and harvested from mid-September to mid-November (or the first frost).
- *Winter cauliflower* modules are sown May, transplanted June to July at a spacing of 60 × 50 cm and harvested from December to early March.
- *Winter hardy cauliflower* modules are sown in June, transplanted July at a spacing of 60 × 50 cm and harvested from early March until early June.

Use of covers needs managing as high temperatures under the covers can cause physiological damage and disease problems may develop in the humid conditions. Crop covers need to be removed when heads are 10–15 mm in diameter.

17.9.4 Rotation

Where club root is a problem, particularly on acid soils, brassicas should be grown no more than 1 year in four or five with a planned liming policy before planting. It is possible to grow brassicas continuously on a high pH soil.

17.9.5 Pests and diseases

Cabbage root fly can be a problem and treatments will be required from April onwards. Aphids, caterpillars, cutworms, pollen beetle, cabbage stem weevil and slugs all attack cauliflower. Damping off and wirestem can be a problem in propagation. Club root, downy mildew, dark leaf spot, ring spot, white blister, bacterial soft rot and black rot can all be a problem in the field.

17.9.6 Weed control

Established perennial weeds should be controlled prior to transplanting or drilling. Oilseed rape volunteers and potato groundkeepers can cause a problem in some crops. Mechanical weeding with manually steered or vision-guided hoes may be used where practical, which reduces the need for chemical control.

17.9.7 Nutrition

For a soil with an index of 2, the fertiliser recommendations for cauliflower would be 235 kg/ha N, 100 kg/ha of P_2O_5 and 125–175 kg/ha of K_2O. The winter hardy types growing over the winter months may require split application of nitrogen fertiliser with no more than 100 kg/ha N at transplanting and the remainder to support later growth.

17.9.8 Irrigation

Irrigation can improve evenness of maturity and it is important to irrigate autumn crops, transplanted in June and July, particularly on lighter soil types. Irrigation can be applied by overhead booms or trickle tapes.

17.9.9 Harvest and storage

The head is cut and trimmed by hand before packing into trays. The crop will be cut a number of times taking heads at the correct size for specifications but leaving the smaller heads to grow on. Cauliflower is cooled quickly after harvest to prevent deterioration. Care is needed as the heads can be damaged at temperatures below 2°C. Cauliflower does not store well and is dispatched into the supply chain within a few days of harvest.

17.10 Brussels sprouts

Brussels sprouts are the axillary buds growing in the leaf axils along the stem of the plant. They are grown for the fresh market and quick freezing, they can be sold loose, pre-packed or on the stem and the tops can also be sold rather like a small cabbage. Most varieties are now F1 hybrids offering a more uniform plant with an even distribution of buttons along the stem length and evenness of maturity (ideal for machine harvesting), compared with traditional open pollinated types. By planting six to seven different varieties which mature at different rates Brussels sprouts can be harvested from August to April.

17.10.1 Soils and climate

Brussels sprouts can be grown on a wide range of soil types and grow well on moisture retentive soils. Lighter, sandier soil types may require irrigation. Heavy soils should be avoided with later crops where machinery access will be needed in wet conditions. A soil pH of 7.0–7.5 is required, particularly where club root may be a problem. Lime should be applied well before planting/drilling.

17.10.2 Cultivations

A firm soil with a good tilth is required. Compacted soils may require sub-soiling before ploughing and harrowing/discing.

17.10.3 Drilling/planting and timings

Processed (freezing) crops may be direct drilled from March to April at close spacings. Transplanting of modules starts in April, until the first week in June at a target spacing of 60 × 38 cm. Later planting will lead to reduced yield even with late maturing varieties. Early varieties mature in 5–6 months, whereas late varieties mature in 10–11 months.

17.10.4 Rotation

The crop fits well into a cereal rotation on land that is clean and fertile. The earliest crops are cleared by the end of September and so it is possible to follow with winter wheat. Where club root is a problem, particularly on acid soils, brassicas should be grown no more than 1 year in four or five with a planned liming policy before planting. It is possible to grow brassicas continuously on a high pH soil.

17.10.5 Pests and diseases

Cabbage root fly can be a problem and treatments will be required from April onwards. Aphids, caterpillars, cutworms, pollen beetle, cabbage stem weevil and slugs all attack Brussels sprouts. Damping off and wirestem can be a problem at the seedling stage. Club root, downy mildew, powdery mildew, dark and light leaf spot, ring spot, white blister and canker can all be a problem in the field.

17.10.6 Weed control

Established perennial weeds should be controlled prior to transplanting or drilling. Oilseed rape volunteers and potato groundkeepers can cause a problem in some crops. Mechanical weeding with manually steered or vision-guided hoes may be used where practical, which reduces the need for chemical control.

17.10.7 Nutrition

For a soil with an index of 2, the fertiliser recommendations for Brussels sprouts would be 270 kg/ha N, 100 kg/ha of P_2O_5 and 150–200 kg/ha of K_2O. Nitrogen top-dressing may be required when recommended rates are high. The top-dressing is generally applied within 2 months of the base-dressing application.

17.10.8 Irrigation

Irrigation may be applied before drilling or after transplanting to aid establishment. Irrigating early and mid-season sprouts may increase yield and quality.

17.10.9 Harvest and storage

In earlier-maturing varieties due to be single-harvested in August and early September the crop is often 'stopped' by removing the growing point (or terminal bud) from the plant, or by destroying the bud with a sharp tap using a rubber hammer. This removes apical dominance and stimulates sprout growth producing a 5%–10% higher yield. Some varieties tend to be more cylindrical in set and stopping is not needed. Stopping may be done 4–10 weeks before harvest when half of the sprouts are 12 mm in diameter.

Figure 17.7 Brussel sprout mechanical harvester (Authors own, 2021).

The tops of the plant, which resembles a small open hearted cabbage can be removed and sold in October and November. The major part of the crop is machine harvested, usually beginning in August and extending until March, according to the variety and season (Fig. 17.7). Occasionally, the earliest maturing varieties are picked over by hand before the whole plant is cut for stripping. Where hand harvested, the crop is usually picked over three or four times during the season. In warm conditions sprouts can become discoloured and deteriorate and need to be cooled rapidly after harvest.

17.11 Bulb onions

Bulb onions are grown widely and fit well into arable rotations. The dominant type of onion grown is a Rijnsburger type. These are globe-shaped bulbs needing long day lengths to initiate bulbing. Brown onions are the most common type and are grown for cooking and smaller sizes are used for pickling. Red onions are increasing in popularity.

17.11.1 Soils and climate

Bulb onions can be grown on a wide range of mineral and peat soils provided that they are well drained, have a good available-water capacity, a pH of 6.5 or more and can provide a good seedbed. Bulb onions can be grown in many parts of the UK, but do best in the eastern and south-eastern counties where it is often drier at harvesting time, which helps promote good skin quality.

17.11.2 Cultivations

Onions develop best in loose, well-structured soil. Ideally the land should be deep ploughed and disced before planting. Capping soils should be avoided. In some fields destoning may be needed to remove large stones that could affect onion development and harvesting.

17.11.3 Drilling/planting and timings

Onions can be grown from seed or sets. Onion seed is prilled to aid precision drilling. Sets are produced from high-density crops producing small 'bulbs' 1–3 cm in size and stored over winter at moderate temperatures (10–15°C) to prevent bolting. Rijnsburger types initiate bulbing in mid-July regardless of planting date but larger plants in spring will be ready to harvest earlier in the summer.

Onions can be planted in autumn and over-wintered for an earlier harvest. Seed can be direct-drilled in August for harvest the following June/July or sets planted in October for harvest the following June/July. The majority of the UK crop is spring planted; some are direct drilled in February–March to harvest late August/September; and a significant proportion is grown from sets planted in January–March and harvested in late July/August. Onions can also be grown from multi-seeded modules propagated under glass in February for planting in April, but this is no longer common.

Onions are planted by precision drill in shallow double rows, 1–2 cm deep. A bed system is used with the seed spacing dictating bulb size. A target of 5–7 cm bulbs would require 50–100 plants/m^2. Sets are planted at similar spacings using specialist planters.

17.11.4 Rotation

Onions (and other alliums) should not be grown more often than 1 year in four, and ideally 1 year in six in the same field, because of the risk of nematode, onion fly and white rot disease.

17.11.5 Pests and diseases

The following can affect bulb onions: leaf blotch, downy mildew, onion white rot, neck rot, onion fly, cutworms, bean seed fly, thrips and bulb and stem nematode.

17.11.6 Weed control

Perennial weeds should, wherever possible, be controlled in the previous year. Bulb onions are not very competitive and grass weeds can be a problem.

17.11.7 Nutrition

Onions are responsive to P and K but response to N is poor. High K can increase keeping quality but excess N reduces keeping quality with soft tissues prone to rots. For a soil with an index of 2, the fertiliser recommendations for onions would be 110 kg/ha N, 100 kg/ha of P_2O_5 and 125–175 kg/ha of K_2O.

17.11.8 Irrigation

Onions have a shallow, poorly-developed root system with most root activity in the top 40 cm of soil. Adequate soil moisture is important for growth of new adventitious roots at the base of the bulb. An even supply of water to the bulbs is needed while they are developing. A period of drought followed by heavy applications of water can lead to the outer skins splitting, reducing quality. Late irrigation should also be avoided as it delays harvesting and gives rise to softer bulbs with more bacterial rots so irrigation is usually stopped when bulbs begin to mature approximately 3 weeks before harvest.

17.11.9 Harvest and storage

For long-term storage the crop is usually sprayed with maleic hydrazide, applied pre-harvest as a growth depressant at 10% leaf fall. Some stores use low levels of ethylene to suppress sprouting. Bulb onions are ready for harvesting when about 80% of the tops have fallen over. This is usually from June to September, depending on the time and method of planting. If left later, yields might increase slightly, but the outer skins are more likely to crack which can lead to disease loss in store.

Onion tops are removed in the field with a flail harvester at harvest time and may be left in the field for 1–2 days to dry in windrows or immediately lifted with adapted potato harvesters and either box stored or bulk-loaded into store.

Preparation for storage is completed in three separate stages: drying, curing and cooling. Throughout the process, temperature and humidity control are crucial for the control of quality. In the initial drying stage, warm air (25–30°C) is blown through ventilation ducts until the moisture is removed from the leaves and tops of the stems (approximately 3 days), to bring the onions to 'rustle dry' condition. In the curing stage, in the next 2–4 weeks, air at about 25°C is blown intermittently through the stack and the relative humidity is reduced to 65%. This removes the moisture from the bulb neck tissue and the skins should now be a golden colour. Once the bulbs are fully dried, in the cooling stage, the stack temperature is lowered slowly to prevent sprouting.

Stores can be cooled by refrigeration systems or by ambient cooling—which is where fans automatically draw in outside air when temperature is 3°C or more below crop temperature. Minimum temperature in ambient stores should be 5–8°C. Refrigeration at 0–1°C is essential for bulbs scheduled to be marketed from February to the end of May/early June. A controlled atmosphere is necessary for storage beyond this period. When the onions are removed from long-term storage, air of 2.5°C greater than the stack temperature is blown through the stack to prevent condensation forming on the onions. Onions are graded for quality and size before sale and marketable yield varies with conditions and timings from 30 to 50 t/ha.

17.12 Leeks

Leeks are alliums but do not bulb and are less pungent than onions. The crop can grow well around the UK and is available from July to May. Leeks are sold with the leaves trimmed into an inverted V, straight or stripped back. Leeks sold as pre-pack are usually smaller than those sold loose. Leeks can be harvested from the field over winter as they are tolerant of frosts.

The varieties change over the season with more rapid maturing varieties providing the early crops. Varieties that can hold well in the field are suited to overwintering.

17.12.1 Soils and climate

Leeks can be grown on a wide range of soil types, but the most suitable are sandy loam to sandy clay loam, silts and some peat based soils. It is best to avoid very light soils as these can spoil the crop if soil gets blown down into the leaves. Heavy soils may restrict access to the field in winter.

17.12.2 Cultivations

The land should be ploughed, bed formed, destoned or declodded before planting. Capping soils should be avoided with drilled crops.

17.12.3 Drilling/planting and timings

Leeks can be raised from direct drilled seed or blocks/modules. Leeks are planted from January until July and harvested from July to May. The new season starts with spring planted transplants harvested in July and August. Spring drilled crops are then harvested from August to December. Late spring drilled crops are overwintered and harvested from January until May. In general, crops drilled from January to April give crops before Christmas and May drillings give crops after Christmas. Early crops can be grown under plastic/fleece covers until mid-May to bring them forward 2 weeks.

Modules or blocks are machine planted and gapped up by hand. Blocks are bigger, giving an earlier crop but are more expensive. Seed is drilled using a precision drill in four rows across a bed 45 cm apart. Spacings between plants are usually 7.5–15 cm for modules and 4–7.5 cm for seed. Leeks are smaller at maturity at the higher density, being better suited to pre-pack crops.

17.12.4 Rotation

Leeks (and other alliums) should not be grown more often than 1 year in four, and ideally 1 year in six in the same field. Consideration should also be made of other non-allium crops in the rotation that are susceptible to bean seed fly, i.e., peas, beans, cucurbits.

17.12.5 Pests and diseases

Cutworms, bean seed fly, onion fly, leek moth and thrips can be a problem in the crop. Rust, foot rot, leaf blotch, white tip and white rot are the main disease problems in leeks.

17.12.6 Weed control

Perennial weeds should, wherever possible, be controlled in the previous year. Leeks compete weakly with weeds. Too many weeds can reduce yields and cause difficulty at harvest, particularly when harvesting mechanically. Mechanical or hand weeding may be required where sites are badly affected with weeds.

17.12.7 Nutrition

Leeks are responsive to nitrogen but applications should match crop growth as over-application of nitrogen can result in very soft crops that are very prone to disease. High levels of N at planting can reduce germination. Where crops are overwintered a split application of nitrogen can be applied as a top dressing in spring. Top dressing of a growing crop can lead to granules lodging in the leaves and damaging them, so care is needed to avoid this. For a soil with an index of 2, the fertiliser recommendations for leeks would be 170 kg/ha N, 100 kg/ha of P_2O_5 and 125−175 kg/ha of K_2O.

17.12.8 Irrigation

Leeks require adequate soil moisture for yield and a lack of soil moisture during crop development can lead to bolting in the crop. Care is needed with maintaining adequate soil moisture following drilling as the seeds and seedlings are sensitive to dry conditions. Irrigating newly planted modules can help wash soil around the modules improving contact with the soil.

17.12.9 Harvest and storage

Leeks can be harvested by hand or machine. Specialist leek harvesters are available that undercut the leeks and trim the leaves before placing them in bulk boxes. Leeks for pre-pack can be manually trimmed and the outer leaves stripped back. Roots need trimming back but care is needed not to damage the base plate of the leek. The leeks should be cooled and held between 2 degrees and 10°C prior to despatch. Leeks should not be harvested when frozen as they can be damaged by handling. The gap in harvest in May and June can be supplied by stored leeks. These require refrigerated, controlled atmosphere stores.

17.13 Carrots

Carrot production for human consumption in this country is now a large and specialised enterprise for most growers. Carrots are grown for many different markets—fresh, frozen and processing—and quality requirements of the buyers can be exacting. Uniformity and freedom from damage and disease are very important. Carrot out-grades and rejects may be fed to livestock.

There are six main types of carrot, based on shape. The cylindrical stump rooted *Nantes* types are popular for pre-packing and some processing; small finger carrots for freezing are usually *Amsterdam Forcing* types; small *Chantenay* types are used for pre-pack and canning; where large, uniformly-coloured carrots are required for slicing and dicing, *Autumn King* and *Berlicum* types are suitable. Long thin *Imperator* types are used for pre-pack or cut and polished to make a number of 'baby' carrots per root.

Varieties are also grouped by date of harvest: first earlies end of May, second earlies end of June and maincrop from July.

17.13.1 Soils and climate

The climate in most arable areas of the country is suitable for carrots; the main limiting factor is soil type, which should not restrict root growth, usually 50–75 mm deeper than the required length of carrot. Sandy loams, loamy sands and fen peats are ideal for carrots. Soils should be well drained but moisture retentive, stone free (to avoid 'fanging') and non-capping. Crops are grown around the UK from the north-east of Scotland to the south of England to provide continuity of supply.

17.13.2 Cultivation

To produce more than 70% of carrots of a desired size is very difficult, and every effort has to be made to prepare a really good, uniform seedbed. The ground will be destoned and beds raised by a bed former. Carrots can be grown on the flat or in beds. Beds are commonly used to prevent compaction and aid lifting.

17.13.3 Drilling/planting and timings

Coated seed is drilled with a precision drill, i.e., Stanhay or Mini Air (Fig. 17.6) into a moist seed bed. Seed rate will depend on the germination percentage and field conditions at sowing and the desired size of carrot at harvest. Seed rate ranges from approximately 60 seeds/m^2 for first earlies to 175 seeds/m^2 for maincrops. Chantenay types are drilled at very high rates of approximately 900 seeds/m^2. To prevent the carrots at the edge of a bed being oversized, seeds may be drilled at differential rates across the bed with more seed placed on the outside rows in a 60/40 split.

Figure 17.8 Crop covers over carrots (Authors own, 2021).

First earlies are sown under polythene in October. Second earlies are also sown under polythene from December to February (Fig. 17.8). Maincrop carrots are sown in the open from March to early July.

17.13.4 Rotation

A rotation of one crop in 5 years is preferable to avoid disease. Sugar beet and potatoes in the rotation can increase the incidence of violet root rot as they are also hosts of this disease.

17.13.5 Pests and diseases

Carrot fly, cutworms, willow-carrot aphids and nematodes, and the diseases violet root rot, black rot, leaf blight, cavity spot, scab and *Sclerotinia* rot are the main problems in the growing of the crop. Some varieties are more susceptible to cavity spot than others.

17.13.6 Weed control

The loss of active ingredients suited to carrot production means that increasingly machine and hand weeding may be needed to control weeds such as fool's parsley, hemlock and wild carrot, which are difficult to control with herbicides. Volunteer potatoes also pose a problem in carrots. Particular care is needed in managing weeds where crops are grown under plastic.

17.13.7 Nutrition

For a soil with an index of 2, the fertiliser recommendations would be 40 kg/ha N, 100 kg/ha of P_2O_5 and 125–175 kg/ha of K_2O. Leaf analysis should be taken for manganese and copper.

17.13.8 Irrigation

Irrigation is essential in most of the carrot growing regions, particularly on sandy soils and for early crops. Irrigation should be scheduled to avoid wet–dry cycles which can cause root splits and cracking.

17.13.9 Harvest and storage

The crop is harvested from June onwards. The roots are very easily damaged and careful handling is required at all stages. Early crops can be lifted using top-lifters, which pull the roots up by the leaves, and then top the roots. Top-lifters work well during the summer until the end of October, when the tops become too weak. From late October onwards, share-lifters (Fig. 17.9) are used which require the tops to be flailed-off first. To maintain freshness, carrots are often lifted at night, when it is cooler and the road network is clear allowing rapid transport to the packhouse.

Carrots can be lifted and held in long-term cold stores, but they lack the fresh appearance associated with newly lifted carrots. Most stored crop is left in the ground during the winter and lifted as required. Crops are either earthed-up or more commonly

Figure 17.9 Self-propelled carrot harvester (Authors own, 2021).

covered with black polythene sheets with a layer of straw 30 cm deep (approximately 100 t/ha). This protects the crop from frost and helps to stop regrowth in the spring. Lifting may continue until the crop becomes too woody or otherwise unsaleable, usually by early May. Yield can vary from 20 t/ha for early crops to well over 60 t/ha for maincrops, however rejects in grading can be 30% or more.

Maximum utilisation of the crop is important. Carrots are often washed and polished (using brushes to scrub the surface) to remove light blemishes of scab and cavity spot, increasing the proportion of class 1 crop.

17.14 Edible swede and turnips

Swedes and turnips are grown for sale as a whole root or for further processing as prepared vegetables, for example for soups.

17.14.1 Soils and climate

Swedes and turnips require uncompacted well-drained loams, silts and light clay loams. The crops can be grown around the UK but are mainly grown in areas with cool summers, such as the south-west, west and north of England, Wales and Scotland.

17.14.2 Cultivations

A fine, firm and moist seedbed is needed for uniform crop establishment.

17.14.3 Drilling/planting and timings

The crop is mainly direct drilled but early crops can be propagated in modules. Seeds are directly drilled in beds for mechanical lifting at a target spacing of 20–25 plants/m^2. Early crops are planted in April and harvested from July to September. The main crop is drilled from April to May in the north of the UK and May to July in the south for harvesting from September to March.

17.14.4 Rotation

Where club root is a problem, particularly on acid soils, swedes and turnips should be grown no more than 1 year in four or five after another brassica crop with a planned liming policy before planting.

17.14.5 Pests and diseases

Control measures may be necessary against cabbage root fly, flea beetle, aphids, caterpillars, cutworms, turnip sawfly and slugs. Hares, pigeons and rabbits can severely damage crops. Use of covers can be very effective at reducing pest damage, especially as there are very few approved pesticides for use in these crops. Club root, powdery mildew, downy mildew, *Phoma*, *Alternaria* and *Rhizoctonia* can all be a problem.

17.14.6 Weed control

Established perennial weeds should be controlled prior to drilling. Oilseed rape volunteers and potato groundkeepers can cause a problem in some crops. Mechanical weeding with manually steered or vision-guided hoes may be used where practical, which reduces the need for chemical control.

17.14.7 Nutrition

For a soil with an index of 2, the fertiliser recommendations for swede would be 70 kg/ha N, 100 kg/ha of P_2O_5 and 150–200 kg/ha of K_2O. Small edible turnips require lower rates of N.

17.14.8 Irrigation

In dry periods, irrigation will ensure even growth and quality of crop. Large variations in soil moisture can lead to growth cracks affecting quality.

17.14.9 Harvest and storage

Swedes and turnips can be hand pulled or lifted mechanically using a modified potato lifter and transported to the packhouse in crates or bulk boxes. Swedes are harvested from July to the following March. Late crops can be stored for several months in cold stores. Turnips store less well.

17.15 Strawberries

Strawberry production is widely distributed around the UK and has expanded partly due to the use of protection. Strawberries can be produced as field-grown crops covered with temporary structures such as Spanish or French tunnels, or under fixed structures such as glasshouses or fixed polythene clad tunnels. Open field-grown crops may also be covered with floating fleece to advance growth or insulated with straw, whilst they are still dormant in late winter, to delay harvest.

There are two main types of strawberry plant: June bearers and Everbearers. June bearers naturally initiate flower buds in autumn, become dormant overwinter and then produce flowers in late spring and produce fruit over 3–4 weeks in summer with the plants being used for 2–3 years. Everbearers naturally initiate flower buds in spring and summer and flower and produce fruit over 10–12 weeks during the summer. Everbearers may be cropped for 1 year and then pulled out.

By using cold stores, a range of planting dates, a mix of types/varieties and unheated polytunnels, strawberries can be produced from mid-May to late October/early November. Heated tunnels or glasshouses can extend production by a month at the start and end of the season but incur additional costs for heating. Early crops can be 'forced' with a double layer of floating fleece in the open or under polytunnels. At growing temperatures above 28°C strawberry growth may stop (thermodormancy) and fruit quality is reduced. The maximum growing temperature is 25°C. Crops can be cooled by removing covers or venting structures.

17.15.1 Soils and climate

Strawberries can be grown on a wide range of soils but very heavy clay soils can lead to poor aeration and waterlogging. Care is needed to avoid waterlogging as this can increase the spread of soil-borne diseases in winter and cause significant root death if soil becomes anaerobic. This can be avoided through field drainage and the use of raised beds.

Strawberries can also be produced using soil-less substrates, commonly coir mixes in bags or troughs. Bags can be placed on the ground or on raised 'table top' systems to ease picking (Fig. 17.10).

17.15.2 Drilling/planting and timings

Young plants are planted into the soil or substrate when dormant. A range of planting material can be used with the crown size influencing fruit yield. Larger crowns or

Figure 17.10 Table-top strawberry system in a polytunnel (Authors own, 2021).

waiting bed plants do not store well and are ideal for spring forcing. Misted tips or rooted runners cold-store well and are suitable for summer planting. Seed raised F1 hybrid Everbearers are becoming available and avoid the risk of disease carryover from the mother plant.

Dormant June bearers can be lifted when dormant in winter and held in a cold store at $-1°C$ before being sequentially planted in spring and summer. Fruit is ready to harvest approximately 2 months later and these are known as '60 day plants'. Everbearer varieties are usually planted in early spring.

Plant spacing depends on the size of starting material and time of planting but is generally 10 plants per metre in a double row.

17.15.3 Rotation

A wide rotation of five or more years will benefit yield. The minimum rotational break in soil-grown crops should be 3 years following a 3 year crop of strawberries, or 2 years following a 2 year crop of strawberries. Crops grown out of the soil, in bags or troughs, can be grown year on year.

Rotations including linseed, lucerne, flax, hops, peas, runner beans or potatoes should be avoided as these may increase levels of *Verticillium*. Care is also needed in the use of residual herbicides in preceding crops. Rotations are not needed where plants are grown in soil-less growing systems in permanent glasshouse or semi-permanent polytunnel structures.

17.15.4 Pests and diseases

The main pests are vine weevil, blossom weevil, spider mites, tarsonemid mite, caterpillars, capsids and aphids. Slugs can be a problem under plastic mulching. Powdery mildew, Botrytis, red core, crown rot and *Verticillium* wilt need to be controlled during the season.

Care is needed when applying pesticides under protection as some plastics will reduce the amount of UV light slowing the degradation of some active ingredients. Growing crops under protection helps to keep fruit dry and can reduce the occurrence of Botrytis and other fungal diseases. Protected structures are also suited to the use of biological control using insect predators. *Verticillium* wilt may be reduced through soil sterilisation prior to raising beds using steam or chemical fumigants.

17.15.5 Weed control

Perennial weeds should be controlled prior to mulching and planting. Polythene mulching of raised beds can help to control weeds. Hand weeding of weeds from planting holes may be needed during the growing season.

17.15.6 Nutrition

As strawberries are a long-season crop, nutrition can be fine-tuned using leaf analysis through the growing season. Nutrition in soil-less growing media can be applied through the trickle irrigation system, i.e., fertigation, but controlled release fertilisers can be used. The optimum levels of nutrients required, pH and conductivity are reported in RB209. In field production, with soil with an index of 2, the fertiliser recommendations would be 0–60 kg/ha nitrogen, depending on soil type and type/variety with Everbearers requiring more nitrogen, 40 kg/ha of P_2O_5 and 80 kg/ha of K_2O.

17.15.7 Irrigation

Fruit size and yield are affected by irrigation and this is important under protected structures and should be scheduled according to crop requirement. Trickle tape/drip irrigation is successful as it avoids wetting the fruit. Irrigation may be applied using overhead misting if crops are too hot during early growth. Overhead irrigation systems should be avoided during fruiting as this damages fruit and increases fungal disease, especially Botrytis.

17.15.8 Harvest and storage

Strawberries are hand harvested into punnets and removed quickly from the field and cooled to remove field heat. They are normally packed and labelled in a packhouse before distribution, using the cold chain, to retailers and markets.

17.16 Key points

- UK Fresh produce crops are high-value crops and have a farm gate value that exceeds the value of UK grown wheat.
- The UK produces just over 50% of the vegetables and 15%–20% of the fruit it consumes.
- The range of fresh produce crops is large and includes leaf, stem, root and fruit crops.
- Different production systems are employed for fresh produce crops including field production, soil-less systems and protected production.
- The supply of some field crops can be extended through early planting under covers, the use of transplants, and/or field storage.
- Some crops can be grown in protected structures to produce crops out of season.
- The use of automation is reducing the reliance on manual labour for some crops.

Sources of further information and advice

AHDB Horticulture have produced a series of crop walkers' guides that cover the crops in this chapter. These guides are available on-line at: https://ahdb.org.uk/knowledge-library.

CABI publish crop specific titles that cover crop genetics, physiology, pests and diseases and production. The following books give more detail about the fresh produce crops covered in this chapter.

Biddle, A.J., 2017. Peas and Beans, 27. CABI.

Brewster, J.L., 2008. Onions and Other Vegetable Alliums, 15. CABI.

Dixon, G.R., 2007. Vegetable Brassicas and Related Crucifers (No. 14). CABI.

Geoffriau, E., Simon, P.W. (Eds.), 2020. Carrots and Related Apiaceae Crops, 33. CABI.

Hancock, J.F., 2020. Strawberries, 34. CABI.

Ryder, E.J., 1999. Lettuce, Endive and Chicory. CABI.

Additional information on production systems and plant physiology can be found in the following books:

Agriculture and Horticulture Development Board, 2021. Nutrient Management Guide (RB209). AHDB.

Morgan, L., 2021. Hydroponics and Protected Cultivation: A Practical Guide. CABI.

Welbaum, G.E., 2015. Vegetable Production and Practices. CABI.

Wien, H.C., Stützel, H. (Eds.), 2020. The Physiology of Vegetable Crops. CABI.

Websites

The British Growers Association is the umbrella group of a range of crop-based associations across the UK fresh produce sector. www.britishgrowers.org.

Defra Basic Horticultural statistics are updated every year and give a breakdown by crop for area, tonnage and value. https://www.gov.uk/government/collections/horticultural-statistics.

Arable forage crops

18

18.1 Crops grown for ensiling

18.1.1 Forage maize

Aside from grass, forage maize has become the most important arable forage crop grown in the UK. In the year 2000 the area sown was just over 100,000 ha, expanding to almost 250,000 ha by 2020, approximately 70% of which was for ensiling for livestock feed and 30% of which was for biogas production. A very small area is grown for grain maize which is dried for use in the bird and pet food market or crimped for use in livestock rations.

The first introduction of maize to the UK at the beginning of the twentieth century was largely unsuccessful due to the fact that the varieties introduced from America did not mature early enough for UK conditions, compromising starch yields. Effective harvesting and ensiling methods were also not well developed at this time.

The introduction of high yielding, early maturing hybrid varieties, with high starch percentages, together with the advent of high capacity, contractor-based growing and harvesting machinery systems has facilitated the rapid expansion of the area of the crop grown in the UK since the 1970's.

With its high starch content and good energy and intake characteristics maize silage has become the major forage component in many dairy cow rations over the last few decades but is also commonly used as a high energy feed for beef cattle and sheep.

Despite its popularity, maize is an expensive crop to grow and requires attention to detail right through from ground preparation to harvesting to achieve optimum dry matter yields and nutritional value.

18.1.1.1 Site selection

Maize has subtropical origins and photosynthesizes using a C4 process, thereby differing from the majority of UK crops. This means it is very efficient in producing a lot of energy in hot, dry conditions which are by no means guaranteed in the UK.

The UK is categorised into both suitable and marginal areas for growing maize based on available heat units, calculated from minimum and maximum air temperatures. The South and East of England are considered most suitable, having higher summer temperatures. On more marginal sites, selection of earlier maturing varieties (requiring less heat units to reach maturity) and techniques such as drilling under biodegradable film will help. Maize is best grown on fields with a southerly aspect

Lockhart and Wiseman's Crop Husbandry Including Grassland. https://doi.org/10.1016/B978-0-323-85702-4.00004-2

at lower altitudes as they receive more heat. Fields with steep slopes should be avoided because of the risk of soil erosion especially on very light soils.

Maize is well suited to light to medium textured soils with good drainage. Growing maize on poorly drained soils is not advisable; due mainly to the fact that in a wet autumn harvesting difficulties may be encountered with the risk of damaging soils and impacting on following crops.

Ideally, forage maize should form part of a rotation of crops, preferably with crops from different plant families to disrupt pest and disease life cycles. Repeated cropping can lead to problems with soil structure, build-up of some harmful weed species such as black nightshade (*Solanum nigrum)* and with residue-borne diseases such as maize eyespot (*Kabatiella zeae*) and maize smut (*Ustilago maydis*).

18.1.1.2 Seedbed preparation

Despite its large seed size, maize requires a fine, consolidated seedbed to aid establishment and rapid development of a large canopy to coincide with late spring/early summer when most sunlight is available.

With the high demands placed on the root system over a relatively short growing season, maize is very sensitive to compaction and this should always be alleviated by subsoiling providing soil conditions allow. Traditionally most fields have been ploughed followed by a secondary cultivation for maize establishment, ideally over winter prior to precision drilling in the spring on the heavier soil types. Rolling after drilling is also advised. Where maize fields are to receive large quantities of organic manures, ploughing may be delayed until the spring, to facilitate application and immediate incorporation. More recently, both minimum tillage and strip tillage methods of establishment have been used successfully in maize given appropriate soil conditions, both of which will help to reduce costs and the risks of soil erosion.

18.1.1.3 Choice of variety

The selection of a suitable variety is a critical factor in determining the success of a maize crop. Each variety requires a specific number of heat units to reach maturity. Early maturing varieties require less heat units and therefore have a shorter growing season than later ones, so their total yield is lower, but the risk of poor harvest conditions is reduced.

The BSPBs (The British Society of Plant Breeders) organises independent Descriptive List testing for maize varieties. Varieties are allocated to one or more of three Lists—Favourable, Less Favourable and Very favourable. The Favourable List is for growers who are in suitable growing locations but for certain reasons such as drilling follow on crops need to find a balance between maturity and yield. The Less Favourable List is for growers in marginal areas whose major concern is growing a crop to reach acceptable maturity by a given date. The Very Favourable List includes some later maturing varieties which may be better suited to biogas production. It is important that a grower identifies which list is most suitable for their situation before comparing individual varieties.

Within each List, produced annually but based on 5 years of trials from a range of locations, maize varieties are compared in terms of crop dry matter and quality characteristics such as ME (metabolisable energy), starch and cell wall digestibility, and agronomic characteristics such as lodging and disease resistance.

18.1.1.4 Seed treatment

Fungicide seed treatments are used to protect against damping off and seedling blight diseases during establishment. Seed treatments are also available that contain a bird repellent to reduce seed losses and trace elements to boost early growth.

18.1.1.5 Seed rate and sowing date

Optimum seed rates have historically been between 100,000 and 120,000 seeds/ha but more recently 5%–10% reductions in seed rates have been shown to increase starch yields by increasing light interception and grain content. Row width is commonly 75 cm but reduced row widths of 50 or even 37.5 cm can also be used. The closer row spacing produces a denser crop with higher fresh weight yields suited to favourable sites. Thicker crops also show a faster dry-down over standard row widths, but care should be taken to avoid excess plant numbers, which may cause lodging.

Maize seed is best sown at between 2.5 and 7.5 cm deep, depending on soil moisture content. Deeper sowing is preferred where damage from rooks and crows is likely to be an issue.

Sowing date is important and drilling should start when the soil at 10 cm depth has reached a temperature of 8–10°C for five consecutive days (decrease to 6°C for planting under degradable film). This results in the majority of maize crops in England being drilled in the latter half of April, therefore, emerging after the risk of frost damage has passed.

18.1.1.6 Sowing maize under plastic film

Photodegradable plastic film can be used, particularly in marginal areas to accelerate maturity and increase dry matter yields and nutritional value but can also be of benefit to increase yields further at favourable sites. Specialised drills sow the seed, apply a residual herbicide and lay a thin layer of degradable plastic. This system, does however, add significantly to the growing costs so must be considered carefully on a site-specific basis.

18.1.1.7 Undersowing forage maize

Soil erosion and nutrient leaching over winter are a high risk from maize stubbles left bare over winter on light soils. Indeed, leaving maize stubbles bare over winter may not be an option in the future. However, it is very often too late to establish a catch crop successfully after the maize is harvested. Therefore, the practice of interrow drilling with grass mixtures is becoming more popular as a means of retaining soil and

Figure 18.1 Maize stubble with interrow grass mixture.
Source: Field Options, 2021.

nutrients and providing some additional forage for grazing or silage as shown in Fig. 18.1. Grass mixtures at reduced seed rates containing species such as tall fescues alongside ryegrasses have proven most successful as they are less competitive. Sowing the grass mixture with a specilaised drill that sows three bands of grass between the maize rows when the maize crop is at the 6 to 8 leaf stage is the most popular technique to reduce any competitive effect of the grass on maize yields. The soil is also in a better state of repair in spring to accommodate organic manure/slurry applications and establish a following crop.

18.1.1.8 Fertilisers

Soil pH is an important consideration for maize growers since the crop is often grown on light or sandy soils which have a predisposition to acidity which the crop does not tolerate well. If lime is required it is best applied in the autumn or on the ploughing prior to final seedbed preparation in the spring. The field should be limed to a target pH of 6.5.

Maize ground often receives heavy applications of farmyard manures and slurries and these must be taken into account by measuring or estimating their nutrient content and recording the amounts applied in order to accurately determine additional inorganic fertiliser requirements. This also ensures that the Nmax limit for the crop of 150 kg ha N is not exceeded in designated Nitrate Vulnerable Zones (NVZ's). Closed periods for spreading in NVZ's must also be adhered to. Detailed guidance can be found at www.gov.uk/nitrate-vulnerable-zones.

To encourage rapid early growth, it is common for all of the phosphate and a proportion of the nitrogen required (12–15 kg/ha) to be placed below the seed at drilling in the form of mono-ammonium phosphate (MAP) or diammonium phosphate (DAP). Phosphate is not highly mobile in the soil and this early placement application reduces the risk of deficiency occurring during rapid early growth even where soil indices are adequate. The remainder of the N can be top-dressed when the crop emerges and is at the two to four leaf stage. Later dressings risk prills being caught in leaf axils causing scorch. Potash should be applied before seedbed preparation and thoroughly incorporated.

The current nutrient recommendations for forage maize are summarised in Table 18.1.

18.1.1.9 Micronutrients

Maize is most likely to show deficiencies in Iron, Copper, Manganese, Zinc and Boron and co-formulated foliar feed products are often applied at the 8–10 leaf stage to address these. Deficiencies may show even when levels in the soil are adequate as the plant struggles to acquire them in sufficient quantity to support the rapid canopy growth.

18.1.1.10 Crop protection

Effective early weed control, ideally before the 4-leaf stage of the crop, is important to avoid yield loss in maize. Therefore, the complete removal of weeds using a total herbicide prior to establishment is advisable either pre-cultivation or ideally after the production of a stale seedbed prior to drilling. This should be followed with a pre-emergence residual herbicide. For subsequent flushes of weeds a post-emergence contact herbicide may also be required. Care must be taken when spraying the crop in warm conditions, during which it is easily scorched.

As the area of maize grown has increased, the prevalence of disease in the crop has inevitably increased. Maize eyespot (*Kabatiella zeae*) inoculum overwinters on plant

Table 18.1 Nutrient recommendations (kg/ha) for forage maize.

	SNS[a], P or K index						
	0	1	2	3	4	5	6
Nitrogen (N)	150	100	50	20	0	0	0
Phosphate (P_2O_5)	115	85	55	20	0	0	0
Potash (K_2O)	235	205	175 (2−) 145 (2+)	110	0	0	0

These recommendations have been based on Nutrient Management Guide (RB209) Section 18.3 updated in January 2021 by the Agriculture and Horticulture Development Board.
[a]Soil Nitrogen Supply Index.

residues and is particularly problematic in continuous maize situations when cool, wet conditions occur. Maize varieties are rated for their susceptibility to eyespot on the descriptive list and in high risk situations a protectant fungicide can be applied at the 4 to 6 leaf stage of the crop. Northern corn leaf blight (*Helminthosporium turcicum)* inoculum also overwinters on residues from infected crops and is most likely to occur in warm, humid seasons. In high risk situations a protectant fungicide can be used. Smut (*Ustilago maydis*) spores overwinter in the soil or on residues from previously infected crops for up to 10 years, causing sporadic infections particularly where there is mechanical damage to the crop or it is suffering from stress. However, it rarely has economic consequences and no fungicide treatments are available. Fusarium build up in maize crops can pose a threat to subsequent wheat crops especially if minimum tillage techniques are practised.

Pests are most likely to be a problem when maize is sown after grass in the rotation. Frit fly (*Oscinella frit)* and leatherjackets (*Tipula* spp) can be a problem after short term grass leys and the risk can be reduced by ploughing out the grass in the autumn to create a break before maize is planted in the spring. Wireworm (*Agrostis* spp) may be a problem after long term grass after which a much longer 1–2 year break is required to reduce the risk. In a wet spring slug damage may also pose a threat during establishment. For some growers damage from rooks and crows at the seedling stage is a significant problem. Seed should be buried well, bird scaring techniques employed and seed treatments with a bird repellent effect should be chosen.

18.1.1.11 Harvesting

The optimum dry matter content at harvest is 28%–35% which usually occurs in late September/early October. The crop should be assessed regularly during September as harvesting above or below the optimum dry matter will potentially compromise dry matter yield and reduce feed value and quality. It can be most accurately assessed by drying a representative sample in an oven or microwave but can also be estimated in the field. When it is at the correct stage, the leaves at the top of the plant will be becoming pale and papery and the leaves level with the cob will be beginning to turn brown. When a plant is picked and the stem twisted, no juice will emerge. A milk line test can also be carried out by breaking the cob open half way down and removing one of the grains. The milk line is where the solid starch (yellow) ends and the liquid milk (white) begins. The ideal is to have 75%–100% of the kernel yellow in colour and for no liquid to be excreted when pressed.

Harvesting is frequently carried out by contractors using self-propelled harvesters. It is important that the crop is well chopped to 12–18 mm in length and the grains are well cracked to aid digestion by livestock. Because of the high level of soluble carbohydrates located in the stems at harvest additives are less likely to be required than when ensiling grass. For further detail on the conservation of forage crops see Chapter 22.

18.1.1.12 Yields

The yield and quality characteristics of forage maize are summarised in Table 18.2 in comparison to grazed grass and grass silage.

Table 18.2 Maize yield and quality characteristics in comparison to grazed grass and grass silage.

	Fresh weight yield t/ha	Dry matter %	Utilised dry matter yield t/ha	ME[a] MJ/kgDM	CP[b]%
Grazed forages					
Grazed grass	70.6	17	9.36	12.0	20.0
Grazed grass (20%White Clover)	70.6	17	9.36	12.0	22.0
Grazed grass (old Pasture)	55.9	17	7.13	11.0	16.0
Grass silages					
Grass silage (1st cut)	21.5	27	4.93	11.3	15.0
Grass silage (2nd cut)	15.2	29	3.74	11.0	14.0
Grass silage (3rd cut)	7.9	29	1.96	10.8	14.0
Maize options					
Maize	43.3	30	11.05	11.0	9.0
Maize under plastic	43.8	32	11.90	11.0	9.0

These figures are based on the Kingshay Forage Costings Report 2020. Grass yields in particular may vary substantially according to growing conditions. Published figures regarding the yields and composition of all these crops vary substantially and farm experience too can be extremely variable. The figures presented represent the potentials in average conditions with near optimum inputs.
[a]Metabolisable energy.
[b]Crude Protein.

18.1.1.13 Cob-only maize harvesting options

A small proportion of maize is harvested using harvesters equipped with a picker header which harvests the cob, husk and some leaf material which is chopped and ensiled resulting in a very high energy ensiled product known as corn cob mix (CCM). The remaining stem material is spread on the field.

18.1.1.14 Environmental concerns

The practice of interrow sowing with grass has already been described as a means of retaining soil and nutrients following a forage maize crop.

Practices such as maintaining land drainage systems, using early maturing varieties to facilitate an early harvest and cultivating and ploughing, if possible, soon after harvest to remove wheelings and compaction and encourage infiltration, are also important components of good soil management which are particularly relevant to maize growers and should help to mitigate some of the problems associated with the crop.

18.1.2 Other cereals as silage crops

Fermented wholecrop silage provides a good alternative source of starch where maize growing is marginal and also offers some flexibility as two grass cuts will have already taken place so a decision can be taken on how much wholecrop is required with any excess left for combining as a grain crop at maturity. The benefits of an early harvest also leave the grower with a wider range of following crop opportunities including grass reseeding, than maize allows with its late harvest date.

Any cereal can be used for fermented wholecrop, which is normally harvested at 35–45% dry matter when the cereal grain is 'cheesy' ripe. Alternatively, the crop can be cut later when it is more mature and higher in starch at around 75–85% dry matter and treated with urea or enzyme products which produce ammonia gas to preserve the crop (urea or Alka wheat).

The management of the crop depends on the cereal used but is similar to a crop taken to full maturity (see Chapter 13) although consideration should be given to the timing of later fungicide applications due to harvest intervals.

18.1.3 Legumes grown for ensiling

Peas are a good source of home-grown protein. Limited specialist varieties of peas grown for forage are available but good yields may also be obtained from conventional combining pea varieties.

The management of peas grown for forage is very similar to those grown for combining dry (see Chapter 14), although less crop protection inputs are likely to be used as for grazing or ensiling there is less need to control weeds, pests and diseases unless they are seriously compromising yield. As no nitrogen is required either inputs are relatively low.

Peas for ensiling are normally harvested when the first pods are formed but the crop is still flowering. The yields are not particularly high (6–8 t DM/ha) but a benefit of peas is their speed of growth. A crop sown in early May with average rainfall could be harvested in mid July at about 8 tDM/ha and 20% crude protein. Earlier sown crops could be harvested in June.

Commercially available arable silage mixtures of spring cereals with legume crops to boost protein content are increasingly popular e.g., spring barley or spring oats with peas. The peas benefit from the structural support offered by the cereal and the starch content of the cereal also aids a more stable fermentation.

Vetches are a traditional legume crop. Winter and spring varieties are available but spring varieties are more widely used for forage, most commonly, mixed with spring cereals such as oats or barley.

White or blue lupins can also be used in mixtures for wholecrop, producing high quality protein. They can also be harvested dry for animal feed.

The yield and quality of wholecrop silage and legumes are summarised in Table 18.3 in comparison to grazed grass and grass silage.

Table 18.3 Wholecrop and legume yield and quality characteristics in comparison to grazed grass and grass silage.

	Fresh weight yield t/ha	Dry matter %	Utilised dry matter yield t/ha	ME[a] MJ/ kgDM	CP[b] %
Grazed forages					
Grazed grass	70.6	17	9.36	12.0	20.0
Grazed grass (20%White Clover)	70.6	17	9.36	12.0	22.0
Grazed grass (old Pasture)	55.9	17	7.13	11.0	16.0
Grass silages					
Grass silage (1st cut)	21.5	27	4.93	11.3	15.0
Grass silage (2nd cut)	15.2	29	3.74	11.0	14.0
Grass silage (3rd cut)	7.9	29	1.96	10.8	14.0
Wheat options					
Wholecrop	31.9	40	10.84	10.3	10.0
Wholecrop urea treated	20.0	65	11.05	10.4	14.0
Alka wheat	11.3	75	7.65	13.0	15.0
Protein crops					
Forage peas	22.3	30	5.70	9.9	18.0

These figures are based on the Kingshay Forage Costings Report 2020. Grass yields in particular may vary substantially according to growing conditions. Published figures regarding the yields and composition of all these crops vary substantially and farm experience too can be extremely variable. The figures presented represent the potentials in average conditions with near optimum inputs.

[a]Metabolisable energy.

[b]Crude Protein.

18.2 Root crops grown for lifting and storage or grazing in-situ

18.2.1 Fodder beet

Fodder beet is a biennial crop which is utilised in its first year for the storage root it produces prior to flowering in its second year. The root contains high yields of high energy (low protein), palatable and digestible forage and can be used for lifting and feeding or for grazing in-situ when outwintering cattle or sheep and can also be grown as a feedstock for anaerobic digestors (often termed energy beet) or for bioethanol production. Occasionally fodder beet is also grown as a cash crop, for sale, and there has been some increase in interest in it in areas where sugar beet growing is no longer possible due to lack of processing facilities but lifting equipment is still available. Many aspects of growing the crop are the same as for sugar beet and similar attention to detail is required to achieve high dry matter yields.

18.2.1.1 Site selection

The crop is best suited to light and medium bodied, free draining soils; it doesn't perform well in wet and heavy soils and harvesting or grazing on wet soils is likely to cause damage. Soil should be limed to pH 7.0 and due to the risks of beet cyst nematode, rhizomania and the build-up of weed beet it should not be grown more than 1 year in three on the same field.

18.2.1.2 Seedbed preparation

Following ploughing, preferably over winter to keep moisture loss to a minimum in spring, a fine firm seedbed should be prepared for a precision drill.

18.2.1.3 Choice of variety

Varieties should be selected on the basis of whether the crop is to be grazed in-situ or lifted and stored, with higher dry matter varieties being more suitable for lifting with sugar beet harvesting equipment for storage. Lower dry matter varieties tend to grow further out of the ground and are better suited to grazing. In reality the market is dominated by intermediate varieties with around 50%–60% of the root above ground which can be grazed or lifted with leaf-lifting equipment resulting in cleaner roots for storage.

Rhizomania tolerance is also now available in a number of varieties.

18.2.1.4 Seed treatment

The seed is pelleted and fungicide seed treatments are applied for the control of seedling diseases. Insecticide seed treatments for the soil pest complex are also currently available.

18.2.1.5 Seed rate and sowing date

Seed rates of 120–130,000 seeds per ha should be used to achieve target plant populations of 90–100,000 plants per ha.

Drilling can take place when soil temperatures stabilise at 7°C at drilling depth (typically early to mid April). Sowing too early can result in bolting (producing a flowering head in its first year) and sowing late will compromise yield. The crop should be precision drilled on 45–55 cm row widths with 15–20 cm spacings to a depth of 2–3 cm.

18.2.1.6 Fertilisers

Fodder beet is a demanding crop in terms of nutrients. Having been descended from sea beet, fodder beet responds to sodium, and agricultural salt is often used as a basal fertiliser. About 400 kg/ha (supplying 200 kg/ha Na_2O) should be applied and worked into the soil at least 2 weeks before the crop is drilled on all except fen silt and peat soils. The phosphate and potash can also be applied at this time, if not before. If salt is not applied potash fertiliser should be increased by 100 kg/ha K_2O.

The use of slurry/farmyard manure will be beneficial but must be taken into account by measuring or estimating the nutrient content and recording the amounts applied in order to accurately determine additional inorganic fertiliser requirements. This also ensures that the Nmax limit for the crop of 120 kg ha N is not exceeded in designated Nitrate Vulnerable Zones (NVZ's). Closed periods for spreading in NVZ's must also be adhered to. Detailed guidance can be found at www.gov.uk/nitrate-vulnerable-zones. 30–40 kg/ha of the inorganic nitrogen can be applied immediately after drilling with the remainder applied at the three to seven leaf stage. Soils at risk of sulphur deficiency also require 25 kg SO_3/ha.

Up to 150 kg/ha of Magnesium is also needed on soils with Mg indices of 0 or 1. General fertiliser recommendations are shown in Table 18.4.

18.2.1.7 Micronutrients

Boron and manganese deficiencies are common in fodder beet on light soil types and can be addressed using foliar feeds at the six to eight leaf stage.

18.2.1.8 Crop protection

Weeds can have a big impact on yield and it is important that they are controlled at an early stage using a programme of pre- and post-emergence residual and contact herbicides. Interrow cultivators can also be used for weed removal. Weed beet can also be a problem competing with the crop and reducing yields. Depending on their prevalence within the crop, tractor hoeing, hand pulling, weed wiping and cutting are recognised methods of removal.

A number of pests such as millipedes, symphylids, spring-tails, pygmy beetles, flea beetles, thrips, capsid bugs and leaf miners can affect establishment, some of which

Table 18.4 Nutrient recommendations (kg/ha) for fodder beet.

	SNS[a], P or K index						
	0	1	2	3	4	5	6
Nitrogen (N)	130	120	110	90	60	0–40	0
Phosphate (P_2O_5)	120	90	60	0	0	0	0
Potash (K_2O)	400	370	340 (2−) 310 (2+)	190	0	0	0

These recommendations have been based on Nutrient Management Guide (RB209) Section 18.3 updated in January 2021 by the Agriculture and Horticulture Development Board.

[a]Soil Nitrogen Supply Index.

can be controlled by the use of insecticide seed treatments. Fodder beet is also susceptible to beet cyst nematode which can cause severe yield losses. Extended rotations are the only means of control.

Rhizomania is a serious root-infesting viral disease (Beet Necrotic Yellow Vein Virus) which is now endemic in the UK and can reduce yields by up to 50%. The risk is reduced by growing tolerant varieties. Seedling diseases are largely controlled by the use of fungicide seed treatments. The aphid transmitted disease, virus yellows can be very serious in some years, following mild winters when an early infestation by Peach Potato aphids (*Myzus persicae*), the main vector, occurs. In the absence of approved systemic seed treatments for aphid control, foliar insecticides may be needed, although resistance to these in the Peach Potato Aphid is prevalent. The foliar diseases Powdery mildew, Beet rust, Ramularia and Cercospora may require fungicide treatment in late summer/early autumn but growers should be mindful of harvest intervals.

18.2.1.9 Harvesting

Harvesting usually takes place in Nov/Dec. Low dry matter varieties with a high percentage of the root above ground can be harvested with top or leaf lifters resulting in relatively clean roots for feeding whole. Higher dry matter varieties for longer term storage with less root above ground will need to be lifted using a sugar beet harvester and may require cleaning before storage. Fodder beet can be stored uncovered on a concrete base for a short period but longer-term storage requires more careful management. The clamp should be maintained at a temperature of 3–6°C, heating should be avoided, although the beet should be adequately protected against frost. A straw covering or plastic sheet can be used, but a strip along the top of the clamp should be left uncovered to aid ventilation, except in freezing conditions.

Cleaned, chopped beet can also be successfully ensiled with other forage crops such as maize.

18.2.1.10 Grazing in-situ

Grazing in-situ has become more popular in recent years as the most profitable means of utilisation as it reduces harvesting and storage costs and simplifies feeding. A strip grazing technique will optimise utilisation. The livestock should be transitioned on to the crop over a period of 2–3 weeks and dry matter yield should be measured in order to accurately allocate the correct amount of crop on a daily basis. Water and a supplementary source of fibre must also be provided.

18.2.1.11 Yield

Fodder beet yield and quality is summarised in Table 18.5. Dry matter yields of tops may provide an additional 2–3 t/ha.

The removed tops can be fed in the field after harvesting is complete and, normally, after wilting to reduce the soluble oxalate content which may cause scouring in livestock.

Table 18.5 Fodder beet yield and quality characteristics.

	Fresh weight yield t/ha	Dry matter %	Utilised dry matter yield t/ha	ME[a] MJ/ kgDM	CP[b] %
Fodder beet	87.5	16	11.90	13.0	12.5

These figures are based on the Kingshay Forage Costings Report 2020. The figures presented represent the potentials in average conditions with near optimum inputs.
[a]Metabolisable energy.
[b]Crude Protein.

18.2.2 Swedes and turnips

Swedes and maincrop turnips can be grown as a high energy forage for both cattle and sheep and are most popular in the northern and western parts of the United Kingdom. Swedes are usually the higher yielding crop. Turnips are usually grown in less fertile soils at higher elevations and are lower yielding with slightly lower dry matter contents. Swedes and turnips are only recommended for cattle with fully mature teeth and are not suitable for broken mouthed sheep. The growing of swedes and main crop turnips is similar, hence they are described together here.

18.2.2.1 Site selection

The crops like a cool, moist climate without too much sunshine. Well-structured soils with a pH of 6.5 are preferred as the soil borne disease, Clubroot (*Plasmodiophora brassicae*) is more prevalent in soils of low pH. A minimum interval of 3 years between brassica crops is also advised due to Clubroot risk.

18.2.2.2 Seedbed preparation

A fine, firm and level seedbed should be prepared using the minimum number of passes possible to avoid compaction and moisture loss.

18.2.2.3 Choice of varieties

Low dry matter varieties will bulk earlier but high dry matter varieties will be more winter hardy for utilisation after Christmas or for storage. Clubroot and Powdery mildew (*Erysiphe cruciferarum*) tolerance are also desirable characteristics to look for in a variety.

18.2.2.4 Seed rate and sowing date

Ideally, swedes should be established using a precision drill to a depth of 1–2 cm with a row width of 38–42 cm and spacing within the row of 15 cm at a seed rate of 350–850 g/ha of graded and pelleted seed. Direct drilling methods or broadcasting of natural seed can also used (more commonly for turnips) but a significant increase in seed rates is required to compensate for poorer establishment. The crop can be sown from early April when soil temperatures are above 10°C through to late June depending on the required utilisation period.

18.2.2.5 Manures and fertilisers

Fertiliser should be applied to the seedbed and worked in thoroughly. The use of slurry/farmyard manure will be beneficial but must be taken into account by measuring or estimating the nutrient content and recording the amounts applied in order to accurately determine additional inorganic fertiliser requirements. Closed periods for spreading in NVZ's must also be adhered to. Detailed guidance can be found at www.gov.uk/nitrate-vulnerable-zones. Soils at risk of sulphur deficiency also require 25 kg SO_3/ha. Magnesium fertiliser is only justified at soil index 0, where 50−100 kg MgO/ha should be applied every 3 or 4 years.

The average fertiliser requirement based on typical yield is summarised in Table 18.6.

18.2.2.6 Micronutrients

Boron deficiency may occur in some soils and a soil sample to determine boron availability is advisable. Foliar micronutrient products can be used to address any deficiency at the four to six leaf stage when sufficient leaf is present for uptake.

Table 18.6 Nutrient recommendations (kg/ha) for swedes, forage rape and turnips.

	SNS[a], P or K index						
	0	**1**	**2**	**3**	**4**	**5**	**6**
Swedes and turnips (65 t/ha lifted)							
Nitrogen (N)	100	80	60	40	0−40	0	0
Phosphate (P_2O_5)	105	75	45	0	0	0	0
Potash (K_2O)	215	185	155 (2+) 125 (2−)	80	0	0	0
Swedes, forage rape and turnips (grazed)							
Nitrogen (N)	100	80	60	40	0−40	0	0
Phosphate (P_2O_5)	85	55	25	0	0	0	0
Potash (K_2O)	110	80	50 (2+) 20 (2−)	80	0	0	0

Adjustments can be made where higher or lower yields are expected. The recommendations for grazed crops assume that all manures are returned. These have been based on Nutrient Management Guide (RB209) Section 18.3 updated in January 2021 by the Agriculture and Horticulture Development Board.
[a]Soil Nitrogen Supply Index.

18.2.2.7 Crop protection

The stale seedbed technique, using cultivation or a non-selective herbicide to kill weeds prior to drilling is advised and can be used in combination with a pre-emergence application of a residual herbicide if high weed pressure is expected but options are very limited. Interrow cultivations are also carried out by some growers.

Flea beetle can be an issue during establishment for which no seed treatments are currently available so the crop must be closely monitored and a foliar insecticide applied if damage is significant, although resistance to these products is an issue. Slugs may also require treatment with pellets in a wet spring. Saw fly, Cabbage white butterfly and Diamond back moth may also require treatment with insecticides at a later stage.

The main disease threats are Clubroot, the risk of which can be reduced by maintaining high soil pH, rotation and varietal tolerance and Powdery mildew for which varietal tolerance is also available. Fungicide treatments for Powdery mildew are available if required but are not widely used.

18.2.2.8 Harvesting swedes

Swedes and turnips can be grazed in situ in the field, but both crops can deteriorate under wet and frosty conditions. A strip grazing technique will optimise utilisation. The livestock should be transitioned on to the crop over a period of 2–3 weeks and dry matter yield should be measured in order to accurately allocate the correct amount of crop on a daily basis. Water, a supplementary source of fibre and minerals must also be provided.

Crops to be lifted in autumn should be handled carefully and not clamped in wet conditions. Clamps should be no higher than 2 m and adequately protected against frost. Because of the low dry matter content and probable soil contamination, the tops are usually wasted. However, they are quite high in protein (Table 18.7).

18.2.2.9 Yield

Table 18.7 Swede and maincrop turnip yield and quality characteristics.

	Fresh weight yield t/ha	Dry matter t/ha	ME[a] MJ/kgDM	CP[b] %
Maincrop turnips	50–60	5.5–6	10–11	12.8–13.1
Swedes	70–90	7–10	12.8–31.1	10–11

These figures are based on the Kingshay Forage Costings Report 2020. The figures presented represent the potentials in average conditions with near optimum inputs.
[a]Metabolisable energy.
[b]Crude Protein.

18.3 Leafy crops grown for grazing

18.3.1 Kale, forage rape and hybrids (leafy brassicas)

Kale is a high yielding crop which is also high in crude protein with a flexible utilisation period such that it can be used for out-wintering, to extend the grazing season or for buffer feeding in dry summers. Kale can also be zero grazed successfully. Forage rape is a faster growing but lower yielding leafy high crude protein brassica which is less winter hardy and is popular for finishing lambs or flushing ewes but can also be grazed by cattle. Kale/forage rape hybrids are also increasingly popular combining some of the beneficial characteristics of both crops. Due to the similarities in their management, all three crop types are considered together in this section.

18.3.1.1 Site selection

For grazing it is essential that it is grown on well drained fields. A pH of 6.5 or above is preferred as the soil borne disease, Clubroot is more prevalent in soils of low pH. A minimum interval of 3 years between brassica crops is also advised due to Clubroot risk.

18.3.1.2 Seedbed preparation

Crops can be direct drilled into stubble or into a grass ley which has been sprayed off with a non-selective herbicide. Indeed, it can provide a useful 'break' from some of the pests associated with grassland prior to a grass reseed in spring. The advantages of this method of establishment are that it conserves moisture at sowing time, leaving a much firmer surface for grazing and also disturbs less weeds seeds.

More conventional cultivations can also be used to prepare a seedbed and the seed then drilled or broadcast. Good consolidation is important, for moisture conservation and good seed—soil contact.

18.3.1.3 Choice of varieties

Varieties vary in height with taller types better suited to grazing by cattle and shorter types better suited to sheep. Leaf to stem ratio and winter hardiness may also influence variety choice. Clubroot tolerance is also available and should be considered where brassicas are frequently grown or land is known to be infected.

18.3.1.4 Seed rate and sowing date

Kale can be sown from early April when soil temperatures reach 10°C until mid July depending on the required utilisation period. Forage rape can be sown from the end of May through to August.

Conventionally drilled crops should be sown at a seed rate of 4—5 kg/ha but this should be reduced for precision drilled crops or increased for broadcast crops.

18.3.1.5 Fertilisers

Leafy brassicas are fast-growing crops which respond well to adequate nutrition. Fertiliser should be applied to the seedbed and worked in thoroughly. The nitrogen application can be split with 60% applied to the seed bed and the remainder top dressed when the young plants have up to 5 leaves. The use of slurry/farmyard manure will be beneficial but must be taken into account by measuring or estimating the nutrient content and recording the amounts applied in order to accurately determine additional inorganic fertiliser requirements. Closed periods for spreading in NVZ's must also be adhered to. Detailed guidance can be found at www.gov.uk/nitrate-vulnerable-zones. Soils at risk of sulphur deficiency also require 25 kg SO_3/ha. Magnesium fertiliser is only justified at soil index 0, where 50–100 kg MgO/ha should be applied every 3 or 4 years.

The average fertiliser requirements based on typical yields for grazed crops are summarised in Table 18.8. Adjustments can be made where higher or lower yields are expected by estimating crop offtake. The recommendations for grazed crops assume that all manures are returned.

18.3.1.6 Crop protection

The stale seedbed technique, using cultivation or a non-selective herbicide to kill weeds prior to drilling is advised and can be used in combination with a pre-emergence application of a residual herbicide if high weed pressure is expected but options are very limited. Interrow cultivations are also carried out by some growers, particularly on organic farms.

Flea beetle can be an issue during establishment for which no seed treatments are currently available so the crop must be closely monitored and a foliar insecticide applied if damage is significant, although resistance to these products is an issue. Slugs may also require treatment with pellets in a wet spring. Saw fly, Cabbage white butterfly and Diamond back moth may also require treatment with insecticides at a later stage.

Table 18.8 Nutrient recommendations (kg/ha) for grazed kale and forage rape.

	SNS[a], P or K index						
	0	**1**	**2**	**3**	**4**	**5**	**6**
Kale							
Nitrogen (N)	130	120	110	90	60	0–40	0
Phosphate (P_2O_5)	80	50	20	0	0	0	0
Potash (K_2O)	200	170	140 (2+) 70 (2–)	70	0	0	0
Forage rape							
Nitrogen (N)	100	80	60	40	0–40	0	0
Phosphate (P_2O_5)	85	55	25	0	0	0	0
Potash (K_2O)	110	80	50 (2+) 20 (2–)	80	0	0	0

These recommendations have been based on Nutrient Management Guide (RB209) Section 18.3 updated in January 2021 by the Agriculture and Horticulture Development Board.
[a]Soil Nitrogen Supply Index.

The main disease threat is Clubroot, the risk of which can be reduced by maintaining high soil pH, using a minimum one in three rotation and choosing tolerant varieties. Alternaria and Powdery mildew can also affect crops but do not require treatment in grazing crops.

18.3.1.7 Grazing in-situ

A strip grazing technique will optimise utilisation. The livestock should be transitioned on to the crop over a period of 2–3 weeks and dry matter yield should be measured in order to accurately allocate the correct amount of crop on a daily basis. Water, a supplementary source of fibre and minerals must also be provided.

Crops can also be zero grazed by cutting with a forage harvester and then fed from a forage box or from behind a barrier. This will help minimise the wastage which occurs during grazing.

18.3.1.8 Yield

Kale and forage rape yield and nutritional qualities are summarised in Table 18.9.

18.3.2 Stubble turnips

Stubble turnips are fast growing (12–14 weeks), palatable and cheaper to grow than the other forage brassicas for grazing but dry matter yield is lower. The name stubble turnips has arisen from the practice of growing them as a catch crop for winter feed on the stubble following an early harvested cereal crop. However, they can also be sown earlier to provide a useful late summer/autumn feed.

18.3.2.1 Site selection

As the crop is to be grazed a free draining soil type with a pH of 6.5 is preferred. A minimum interval of 3 years between brassica crops is also advised due to the risk of the soil borne disease, Clubroot.

Table 18.9 Kale and forage rape yield and nutritional qualities.

	Fresh weight yield t/ha	**Dry matter %**	**Utilised dry matter yield t/ha**	**ME[a] MJ/kgDM**	**CP[b] %**
Grazed kale	69.2	13	6.75	11.0	16.8
Forage rape	38.5	13	3.75	10.5	19.0
Stubble turnips	50.0	9	3.15	11.0	18.0

These figures are based on the Kingshay Forage Costings Report 2020. The figures presented represent the potentials in average conditions with near optimum inputs.
[a]Metabolisable energy.
[b]Crude Protein.

18.3.2.2 Seedbed preparation

If stubble turnips are to be sown as a catch crop after a grass ley, plough-based cultivations are appropriate but if they are drilled following an arable crop, a cereal for example, then minimal cultivations may be sufficient. Direct drilling can also be used in either situation and broadcasting into stubbles is another low-cost option but its success depends on the evenness of straw distribution and moisture levels. In all cases, it is vital that soil moisture is not lost and rolling is advised where possible.

18.3.2.3 Choice of varieties

Bulbing types produce a tankard shape root which protrudes from the ground and is more easily eaten. Tetraploid varieties produce a sweeter and more palatable root. Leafy types with round bulbs are better suited to late summer and early autumn grazing. There are some varietal differences in susceptibilities to Clubroot, Powdery mildew and *Alternaria* diseases.

18.3.2.4 Seed rate and sowing date

Stubble turnips can be sown in April when soil temperatures are above 10°C for late summer use (e.g., finishing lambs) through to the end of August (end of July in the North). Seed rates of 4–5 kg/ha of natural seed are appropriate for drilled crops but higher rates should be used to account for increased losses if broadcasting.

18.3.2.5 Fertilisers

The crop will respond to a nitrogen application to encourage rapid early growth. The fertiliser should be worked into the seedbed or top dressed shortly after emergence.

The fertiliser requirements based on typical yields for grazed stubble turnips are summarised in Table 18.6. At index 0 and 1 no more than 75 kg/ha N should be applied to stubble turnips.

18.3.2.6 Crop protection

Spraying with a non-selective herbicide prior to establishment is essential if direct drilling into a grass ley and desirable in all other situations where weeds are present. When established after a cereal crop particularly after minimal cultivations, volunteers can be prolific and compete with the young plants in which case an approved graminicide may be required at an early stage.

Flea beetle can be an issue during establishment for which no seed treatments are currently available so the crop must be closely monitored and a foliar insecticide applied if damage is significant, although resistance to these products is an issue.

The main disease threat is Clubroot, the risk of which can be reduced by maintaining high soil pH, using a minimum 1 in 3 rotation and choosing tolerant varieties. Alternaria and Powdery mildew can also affect crops but do not require treatment in grazing crops.

18.3.2.7 Grazing in-situ

Crops are usually ready for grazing 12–14 weeks after drilling and can hollow out if left too long. In common with other grazed brassicas a strip or paddock grazing technique will optimise utilisation. The livestock should be transitioned on to the crop gradually and dry matter yield should be measured in order to accurately allocate the correct amount of crop on a daily basis. Water, a supplementary source of fibre and minerals must also be provided.

18.3.2.8 Yield

The figures for yield and nutritional qualities of stubble turnips are included in Table 18.9.

18.3.3 Winter cereals as grazing crops

Rye can be sown in September/October after cereals or maize as a catch crop to deliver winter grazing for sheep or early spring grazing for cattle prior to turning out on to grass. Strip grazing is preferred to optimise the utilisation of the crop. Zero grazing is also an option to reduce soil damage. A spring sown forage crop can then follow once the rye has been grazed.

Average dry matter yields of around 3.5 t/ha are possible from rye.

18.4 Key points

- Forage crops can be beneficial, not only in reducing bought-in feed costs but also in diversifying both arable and livestock crop rotations.
- Outwintering livestock on forage crops can reduce capital investment and feed and labour costs.
- The selection of a suitable site and an appropriate choice of crop type for the livestock intended to utilise it is crucial.
- Forage maize, aside from grass, is the most widely grown forage crop, popular for its high starch content, the majority being used for livestock but an increasing proportion being used for biogas production.
- Forage maize is a demanding crop which is marginal in terms of its suitability for the UK climate and therefore requires attention to detail in its management.
- Other cereals, often in mixture with legumes can also be used to produce silage as a flexible and less demanding alternative to maize.
- Some crops such as fodder beet offer flexibility to graze in-situ or to lift and store.
- Leafy brassicas are a relatively cheap, quick growing option that can be used to extend the grazing season or for outwintering.

Sources of further information and advice

Further reading

Agriculture and Horticulture Development Board (AHDB), 2018. Using brassicas for better returns. Available at: https://ahdb.org.uk/knowledge-library/grazed-brassica-forage-crops-cattle-sheep.

Agriculture and Horticulture Development Board (AHDB), 2018. Growing and feeding maize silage for better returns. Available at: https://ahdb.org.uk/knowledge-library/growing-maize-silage.

Agriculture and Horticulture Development Board (AHDB), 2021. Nutrient management Guide (RB209) section 3. Available at: https://ahdb.org.uk/rb209.

British Society of Plant Breeders (BSPB), 2021. Descriptive lists for maize. Available at: https://www.niab.com/research/agronomy-and-farming-systems/variety-evaluation-and-management/bspb-forage-maize. (Accessed 20 August 2021).

DEFRA, 2020. Farming Statistics – provisional arable crop areas, yields and livestock populations at 1 June 2020 United Kingdom. Available at: https://www.gov.uk/government/statistics/farming-statistics-crop-areas-and-cattle-sheep-and-pig-populations-as-at-1-june-2020-england. (Accessed 20 August 2021).

Kingshay, 2020. Kingshay Forage Costings Report 2020.

Websites

Seed Suppliers Often Provide Useful Information and Comparisons of Crop Types and Their Management as Well as More Specific Information on Varieties.

Barenburg. www.barenburg.co.uk.

Cotswolds seeds. www.cotswoldsseeds.com.

DLF. www.dlf.co.uk.

Field Options, 2021. www.field-options.co.uk.

Germinal. www.germinal.co.uk.

LG Seeds. www.lgseeds.co.uk.

Maize Growers Association. www.maizegrowersassociation.co.uk.

Introduction to grass production/characteristics of grassland and the important species

19

19.1 Types of grassland

Grass is the United Kingdom's most important crop and grassland (including rough grazing, heath and moorland) is a valuable natural resource, covering more than 67% of the area utilised by agriculture. Recent Defra statistics show the approximate areas of the main types of grassland (Table 19.1).

Grassland can be broadly classified as described below.

19.1.1 Uncultivated grasslands

These represent about 41% of the total grassland area. They consist of the following types.

19.1.1.1 Rough mountain and hill grazing

The plants making up this type of grassland are not of great agricultural value. They consist mainly of fescues, bents, *Nardus* (mat grass) and *Molinia* (purple moor grass) as well as cotton grass, heather, bracken and gorse. This grassland is usually in what are called Less Favoured Areas (LFA) including disadvantaged areas (DA) and severely disadvantaged areas (SDA). These are regions where conditions are difficult for production. Many of these areas are nationally important for biodiversity as well as for archaeology, recreation and for management of resources such as water and carbon storage. Traditionally, sheep, beef cattle and ponies were farmed with grouse shooting and deer stalking increasing in importance from the mid-nineteenth century. Burning on a regular planned basis at between 10 and 15 year intervals has been common practice and can have beneficial results, both for grazing stock and wildlife though this practice is currently being reviewed. In the past these upland farms have been able to receive a range of farming support payments but some of these payments did increase stocking rates which subsequently had a deleterious effect on sward composition. Subsequently a range of agri-environment schemes have been introduced to help improve biodiversity.

Lockhart and Wiseman's Crop Husbandry Including Grassland. https://doi.org/10.1016/B978-0-323-85702-4.00013-3

Table 19.1 Areas of grassland in the UK.

	'000 ha
Uncultivated grassland	
• 'Sole right' rough grazing	3986
• Commons	1197
Cultivated grassland	
• Permanent pasture	6207
• Temporary grassland or 'leys'	1193

These areas need to be sustainably managed to maintain their environmental, resource and landscape value. These are aspects that will be covered in England by the new Environmental Land Management Schemes to be introduced in 2024.

19.1.1.2 Lowland heaths

Heather, bracken and grasses are the main species to be found here. Sheep's fescue is often the dominant grass on the predominantly acid soils. These heaths are to be found in south and east England and some of them have been reseeded. Traditional management involved regular burning and grazing by cattle and ponies, but nowadays many heaths are not grazed at all.

19.1.1.3 Calcareous downland

This occurs predominantly in southern England. Apart from herbs and broad-leaved flowering plants, grasses such as sheep's fescue and erect brome are found on these chalk and limestone soils. The traditional management of these grasslands involved predominantly grazing by both cattle and sheep. Unimproved areas of calcareous downland are now restricted to the steeper slopes and can be very species rich.

19.1.1.4 Wetland and fen areas in east and south-west England

Any un-reclaimed areas are mostly poorly drained and are dominated by water-loving plants such as cotton grass, rushes and sedges. Fens in eastern England, when drained, are associated with intensive arable cropping. In south-west England areas such as the Somerset Levels have been partly improved by reseeding but are frequently inundated by floodwater.

19.1.1.5 Maritime swards

These consist of salt marshes and coastal dune areas with marram and cord grasses. The 'machair' areas of coastal Scotland and the Western Isles contain a wide variety of rare plants and offer a unique and valuable habitat for wildlife.

19.1.2 Cultivated grasslands

These represent about 59% of the total grassland area. They consist of the following.

19.1.2.1 Permanent pasture

The statutory definition of permanent pasture is grassland which is more than 5 years old. An alternative one would be 'grassland not normally included in an arable rotation'. Since 1984 the area of permanent pasture has increased by 21%. The agricultural value of a permanent pasture is dependent mainly on its content of perennial ryegrass and white clover. A sward containing more than 30% of its annual dry matter production as perennial ryegrass would be considered very productive and would not normally be thought to require much improvement. Other, less productive grasses, such as Yorkshire fog, bents (*Agrostis* spp.) and meadow grasses (*Poa* spp.) make up the balance. Although much less productive than the ryegrasses, these constituents perform a useful function in that they create a 'thatch' over the surface of the soil which, if well developed, has the capacity to carry stock physically and reduces the severity of treading or 'poaching' which can occur in wet conditions. In addition to clover and other legumes, a large number of broad-leaved plants can occur. Many of these, such as buttercup, docks, nettles, and thistles are weeds and have very little forage value, other weeds are also poisonous such as ragwort. Some other species, however, such as yarrow, burnet and ribwort (plantain) are termed forage herbs and have been shown to have beneficial effects on grazing livestock as well as helping to increase soil carbon.

The Countryside survey found that between 1998 and 2007 there was a decline in number of species present on average in improved permanent grassland from 15 down to 12. It is hoped that the introduction of all the agri-environment schemes over the years will have improved species numbers.

There are now approx. 3% of species-rich permanent pastures present in the UK, some of these are subject to statutory protection such as Sites of Special Scientific Interest (SSSIs). Others have been targeted by one of a number of agri-environment schemes. Management strategies for such pastures are normally designed to maintain soil fertility at low or decreasing levels, in order to maintain or increase the species diversity and structure.

19.1.2.2 Temporary grassland or 'leys'

These are temporary swards which have been sown to grass or grass/clover mixtures for a period of up to 5 years. A less precise definition is 'grass in an arable rotation'. Recently there has been quite a reduction in acreage of short-term leys partly due to costs involved in their establishment. With the current concerns about low soil carbon content in arable soils there is growing interest again in including leys in the rotation. Leys represent an important opportunity for the restoration of soil fertility to arable land. Fields which have been grazed, or which have included significant quantities of forage legumes, will contribute substantial amounts of available nitrogen to the following crops. Opportunities exist also for the reduction of some of the weed, pest and disease problems associated with arable farming. In organic systems clover leys are one of the main ways in which soil fertility can be enhanced for subsequent arable cropping.

19.2 The nutritive value of grassland herbage

The great value of grass and indeed of other forages is their potential to provide cheap (often the cheapest) sources of energy and protein for farm livestock. Many calculations have been carried out to demonstrate their value and, in particular, the way in which good grassland management, and the achievement of high yields of digestible nutrients, can bring about worthwhile improvements in livestock performance, and a reduction (or complete elimination) of the need to buy in other feeds.

19.2.1 The effects of grass maturity

Short, leafy grass is rich in protein and highly digestible. As it matures the proportion of yield represented by cell walls increases whereas the proportion of cell contents decreases. This results in an increase in the percentage of fibre and a decrease in the percentage of crude protein with maturity. As a result of the much greater leaf and stem area of the mature plant, and the greater opportunity for photosynthesis, the sugar (water soluble carbohydrate or WSC) percentage also increases with maturity. This explains why it is usually easier to make a well-fermented silage from mature grass and why it is more difficult to do so from young, leafy grass.

19.2.2 Digestibility ('D value')

Digestibility is one of the main characteristics which determines the feeding value of herbage plants, i.e., the amount of the plant which is actually digested by the animal. In the past, the digestibility of forages was only determined by in vivo animal feeding trials. With the development of the laboratory in vitro technique which simulates rumen digestion, it is possible for the digestibility of any species of herbage plant to be assessed very much more easily and cheaply without using animals directly.

The digestibility of herbage plants is expressed as the 'D value' or DOMD. It is the digestible organic matter expressed as a percentage of the dry matter. High digestibility values are normally desirable as it means that animals are able to obtain large amounts of nutrients from the herbage being fed. It is now accepted that digestibility is a major factor affecting intake. Animals usually eat more of a highly digestible feed. However, other factors, such as sward height and density in a grazing situation and the pH, chop length and dry matter percentage of silages are also important. Dry matter yield increases as the plant grows; the proportion of leaf (the most nutritious and digestible part of the plant) decreases and the proportion of stem (more fibrous, lignified and less digestible) increases. This causes a gradual decline in the D value of the whole plant with maturity.

The rate of decline depends on the grass species and the variety. Cocksfoot, for example, has a lower D value than the ryegrasses. The D value of particular varieties on any date is usually determined by the earliness or lateness of flowering. The D value of first cuts normally declines at a rate of between 3 and 5 units per week. The rates of decline in timothy and Italian ryegrasses are slower than in the perennial ryegrasses. D values also differ between grass and legume species. For example, white clover, when young, has a D value of about 80 and the rate of decline is only about 0.8 units per week. Red clover and lucerne are substantially less digestible and their rates of decline are between 2.5 and 2.8 units per week.

When cutting for silage, higher yields will result when the crop is cut at a more mature stage of growth. Many farmers aim to cut grass at or before it achieves a D value of 67. When making field-cured hay, a D value of 60 or even less is a more realistic expectation. This reflects the more mature nature of the crop which can be expected by the time the weather becomes suitable for hay making (if it ever does!). Information about the relative D values of different species and varieties of grasses can be found in the *Recommended Grass and Clover Lists* leaflet for England and Wales and similar information published by the devolved administrations in Scotland and Northern Ireland.

The approximate conversion of D values to ME (metabolisable energy) data for rationing purposes can be accomplished by the application of the following simple equations:

For fresh and dried grass and hay, ME (MJ/kgDM) $= 0.15 \times$ DOMD.

For grass silage, ME (MJ/kgDM) $= 0.16 \times$ DOMD.

19.3 Identification of grasses

19.3.1 Grass species of economic value

Over 150 different species of grasses can be found growing in this country, but only a few are of importance to the farmer. They are: short duration ryegrasses such as Italian ryegrass and longer term grasses such as perennial ryegrass, timothy, cocksfoot and meadow fescue. With the exception of the short duration ryegrasses, most of the grasses described in Table 19.2 may be found in permanent pastures, either as indigenous types or bred varieties which were sown when the pasture was established. Before it is possible to recognise plants in a grass field, it is necessary to know something about the parts which make up the plant. Most grasses have one or two distinctive vegetative characteristics that can aid identification. A hand lens is often required to see some of the leaf characteristics. Once the grass is flowering correct identification is much easier.

Table 19.2 Recognition of main sown grass species in the vegetative stage.

	Italian ryegrass	**Perennial ryegrasses**	**Meadow fescue**	**Cocksfoot**	**Timothy**
Leaf sheath	Definitely split Red/pink at base	Split or entire Red/pink at base	Split	Entire at first, later split	Split Pale at base
Blade	Rolled in shoot[a] Broad Margin Smooth Dark green	Folded in shoot Narrow Margin Smooth Dark green	Rolled in shoot Narrow Margin rough Lighter green	Folded in shoot Broad Margin rough Light green	Rolled in shoot Broad Margin Smooth Light green
Lower side of blade	Shiny	Shiny	Shiny	Dull	Dull
Ligule	Blunt	Short and blunt	Small, blunt, greenish-white	Long and transparent	Prominent and membranous
Auricles	Medium size and spreading	Small, clasping the stem	Small, narrow and spreading	Absent	Absent
General	Veins indistinct Not hairy	Not hairy	Veins appear as white lines when held to light Not hairy	Not hairy	Base of shoots may be swollen Not hairy

[a]Identification of hybrid ryegrass is more difficult; it can show a mixture of vegetative features similar to both Italian and perennial ryegrasses. Round tillers predominate.

19.3.2 Identification of vegetative parts of main sown grass species

19.3.2.1 Stems

There are two types of stems found on grass plants. They are:

- The *flowering stem or culm.* This grows erect and produces the flower. Most stems of annual grasses are culms.
- The *vegetative stem.* This does not produce a flower and has not such an erect habit of growth as the culm. Perennial grasses have both flowering and vegetative stems.

19.3.2.2 Leaves

These are arranged in two alternate rows on the stem and are attached to the stem at a node. Each leaf consists of two parts (Fig. 19.1):

- The *sheath* which is attached to the stem.
- The *blade* which diverges from the stem.

The leaf sheath encloses the buds and younger leaves. Its edges may be joined (entire) or they may overlap each other (split) (Fig. 19.2). If the leaves are rolled in the leaf sheath, the shoots will be round (Fig. 19.3), but when folded the shoots will be flattened (Fig. 19.4). This characteristic is important when identifying different grasses. Perennial ryegrass shoots are flattened unlike Italian ryegrass where leaves are rolled in the stem.

19.3.2.3 Other structures

At the junction between the leaf blade and leaf sheath is the *ligule*. This is an outgrowth from the inner lining of the sheath (Fig. 19.5). *Auricles* may also be seen on some

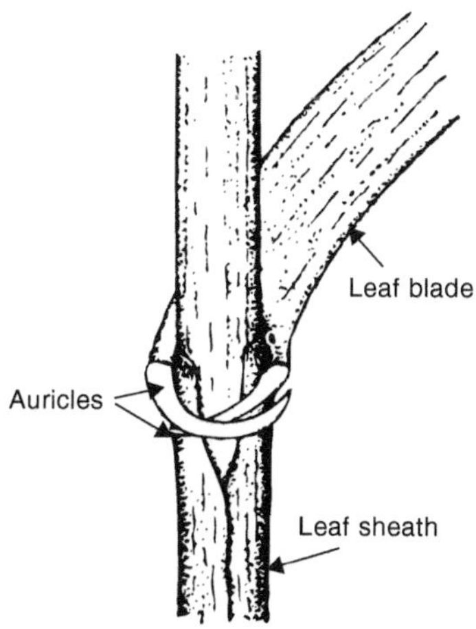

Figure 19.1 Parts of the grass leaf.

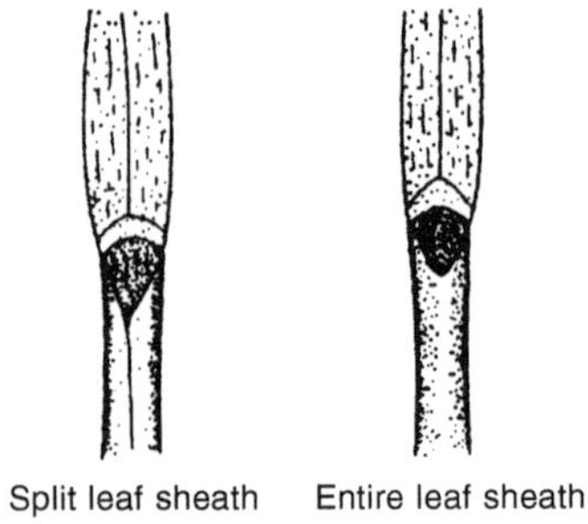

Figure 19.2 Parts of the grass leaf: sheath.

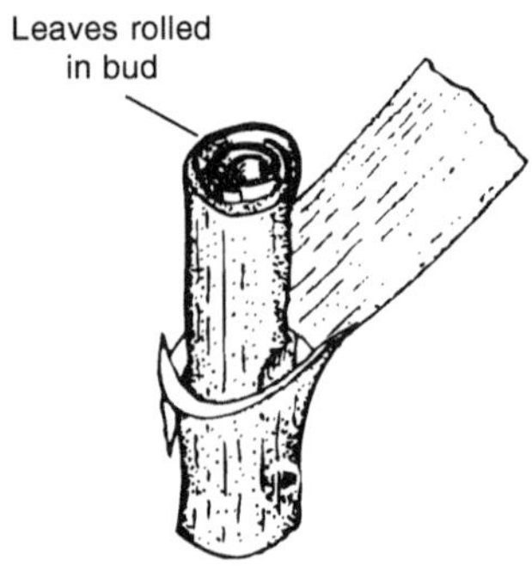

Figure 19.3 Parts of the grass leaf: rolled leaves leading to round shoots.

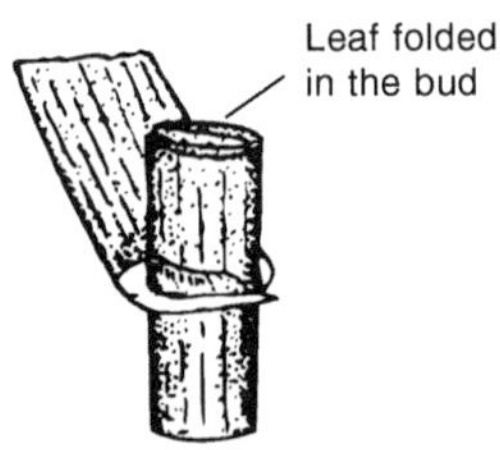

Figure 19.4 Parts of the grass leaf: folded leaves leading to flattened shoots.

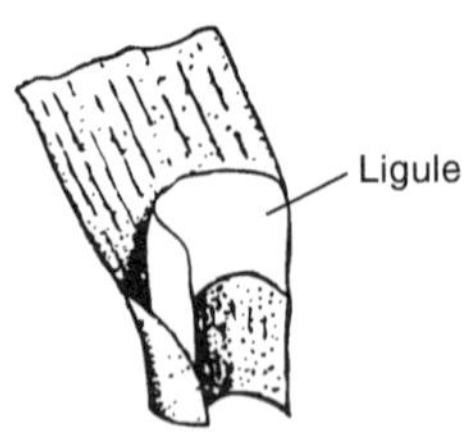

Figure 19.5 Parts of the grass leaf: ligule.

grasses where the blade joins the sheath. These are a pair of claw-like outgrowths (see Fig. 19.1). Ryegrasses have small auricles unlike timothy which has none.

In some species, the leaf blade will show distinct veins when held against the light as seen in meadow fescue. Other important characteristics include stem base colouration which are red/pink in the ryegrasses.

Stem/root features can be important when identifying non-sown grass species. Some of these grasses have above ground creeping stems or *stolons* and others have below ground stems or *rhizomes* (see Chapter 5 for further details).

19.3.3 The inflorescence or flower head

The inflorescence consists of a number of branches called spikelets which carry the flowers. There are two types of grass inflorescence:

- *Spikes*, where the spikelets are attached to the main stem without a stalk as in the ryegrasses (Fig. 19.6).
- *Open panicles*, where the spikelets are attached to the main stem by a stalk (Fig. 19.7).

In some grasses the spikelets are attached to the main stem with very short stalks to form a dense type of inflorescence termed 'spike-like' (Fig. 19.8) such as in timothy. The spikelet is normally made up of an *axis*, bearing at its base the *upper* and *lower glumes* (Fig. 19.9). Most grasses have two glumes. Above the glumes, and arranged in the same way, are the *outer* and *inner palea*. In some species these palea may carry *awns* which are usually extensions from the palea (Fig. 19.10). The presence of awns on the ear is the main feature distinguishing between Italian and perennial ryegrass. Italian ryegrass ears are awned whereas perennial ryegrass is awn less (Fig. 19.15).

Figure 19.6 A spike inflorescence (e.g., ryegrass).

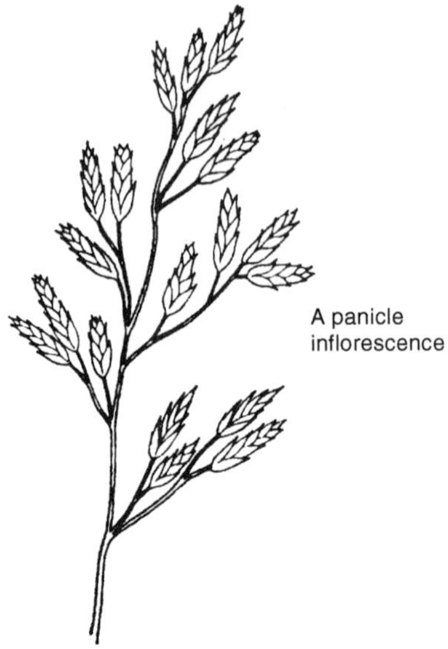

Figure 19.7 A open panicle inflorescence (e.g., fescue).

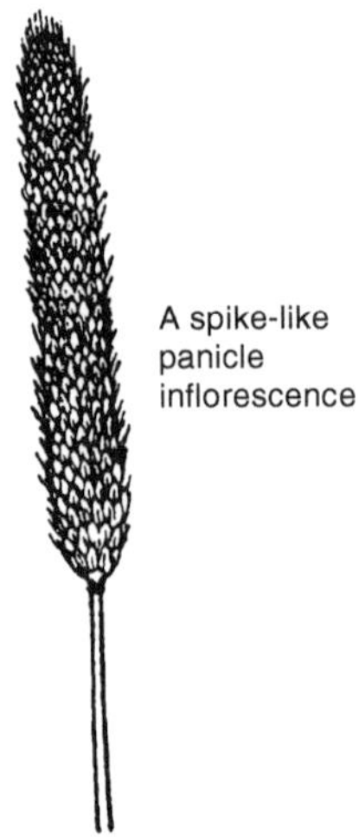

Figure 19.8 A spike—like panicle inflorescence (e.g., Timothy).

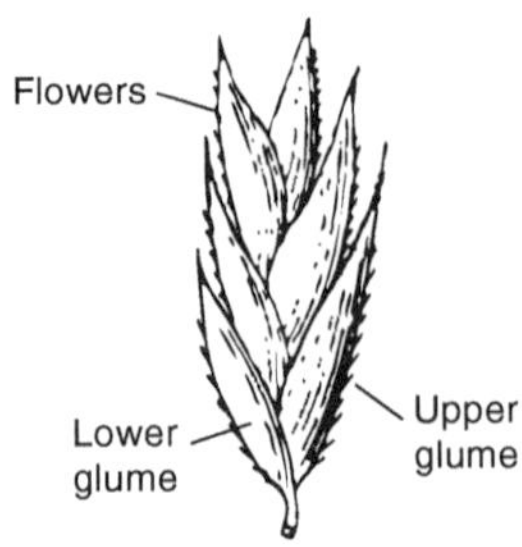

Figure 19.9 A grass spikelet.

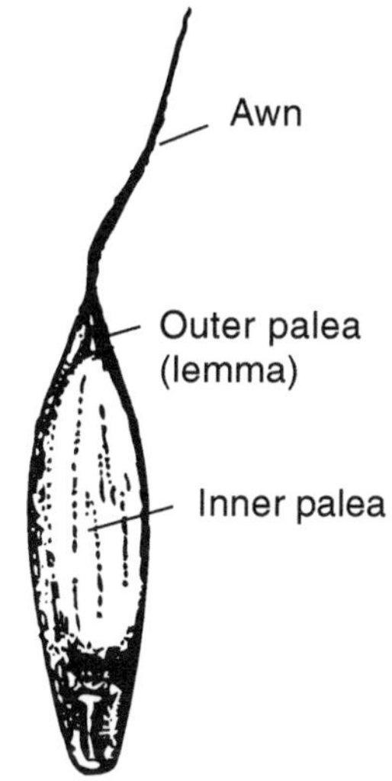

Figure 19.10 The palea.

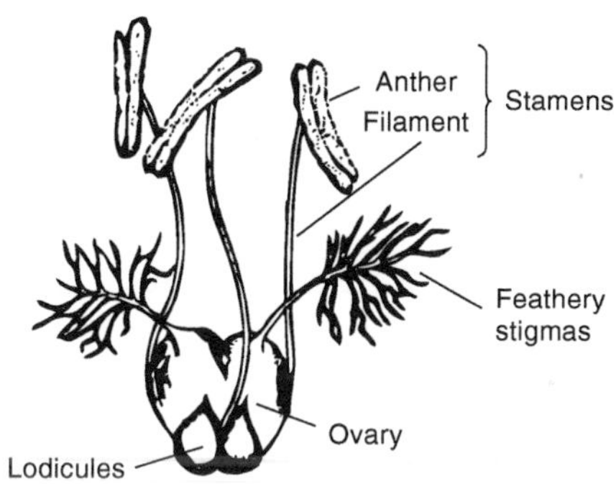

Figure 19.11 The flower of a grass.

Between the 2 palea is the flower. The flower consists of three parts (Fig. 19.11):

- The male organs—the *stamens.*
- The female organ—the rounded *ovary* from which arise the feathery *stigmas.*
- A pair of *lodicules*—at the base of the ovary. These are indirectly concerned with the fertilisation process, which is basically the same in all species of plants (Fig. 1.22).

19.4 Identification of legumes

The important perennial forage legumes used by farmers in the UK are white and red clover and lucerne. Sainfoin, bird's-foot trefoil and alsike are less important though are increasingly being included in herbal ley mixtures and in some mixtures recommended for use in agri-environment schemes. A detailed description of each is given in Table 19.3.

Table 19.3 How to recognize the important perennial forage legumes.

	Leaves, etc.	Stipule	General	Species
Mucronate tip	Centre leaflet with prominent stalk	Broad, serrated and sharply pointed	May be hairy	Lucerne
	Leaflets serrated at tip 6–12 pairs of leaflets, plus a terminal one	Thin, finely pointed	Stems 30–60 cm high Slightly hairy	Sainfoin
	Trifoliate (or pentafoliate if including the stipules)	Leaflet-like	Not hairy	Bird's-foot trefoil (yellow and orange/red flowers)
No mucronate tip	Trifoliate, dark green with white half-moon markings on upper surface	Membranous with greenish-purple veins pointed	Hairy	Red clover
	Trifoliate, serrated edge with or without markings on upper surface	Small and pointed	Not hairy	White clover
	Trifoliate, serrated edge, no leaf markings	Long tapering point, never red-veined	Not hairy	Alsike clover (pink or white flowers)

19.4.1 Leaves

With the exception of the first leaf (which may be simple), all leaves are compound. In some species the mid-rib is extended slightly to form a mucronate tip. Other features on the leaf may be serrated margins, the presence or absence of marks, colour, and hairiness (Figs. 19.12 and 19.13). The leaves are arranged alternately on the stem, and they consist of a stalk which bears two or more leaflets according to the species (Fig. 19.14).

19.4.1.1 Stipules

These sheath-like structures are attached to the base of the leaf stalk. They vary in shape and colour (Fig. 19.12).

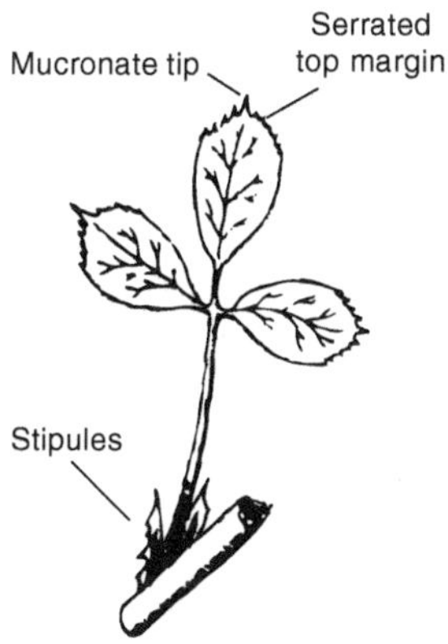

Figure 19.12 Legume leaf features.

Figure 19.13 Legume leaf feature variations.

19.4.1.2 Flowers

The flowers are brightly coloured and, being arranged on a central axis, form an indefinite type of inflorescence (see Fig. 19.14).

19.5 Grasses of economic importance

19.5.1 Types of grass varieties

Details of good quality, reliable varieties for England and Wales, from UK and European plant breeders, can be found in the *Recommended Grass and Clover Lists.* The lists are published annually using data from trials carried out by various organisations including NIAB-TAG, IBERS and SRUC. The trials data is evaluated by a panel of experts. Farmers in Scotland and Northern Ireland have access to information of a similar quality for their own region. Seed catalogues and the online information provided by reputable seed merchants can also be a valuable source of information.

Figure 19.14 Parts of the legume.

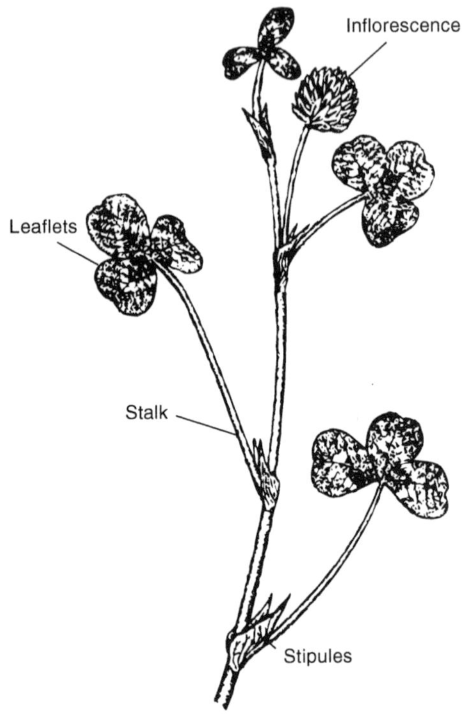

19.5.2 Details of individual species

19.5.2.1 Short duration ryegrasses

This group consists of varieties of three species: Westerwolds, Italian and hybrid ryegrasses.

19.5.2.2 Westerwolds ryegrass

This is an annual and the quickest growing of all grasses. A good crop can often be obtained within 12–14 weeks of sowing. It should not be undersown. It performs best when direct sown in the spring and summer, as it is not winter–hardy, except in very mild districts. It can be sown as a green manure crop.

19.5.2.3 Italian ryegrass (Fig. 19.15)

This is short-lived (most varieties persist for 18–24 months) and very quick to establish. Sown in the spring, Italian ryegrass can produce good growth in its seeding year and an early grazing the following year. For optimum production it is best sown in summer or early autumn and will then produce a full crop in the following year. It does well under most conditions but responds best to fertile soils and plenty of nitrogen. Many currently available varieties have good resistance to mildew with mostly moderate resistance to ryegrass mosaic virus. Like many ryegrasses, its winter-hardiness can be improved when surplus growth is removed in the autumn.

Figure 19.15 Flowering inflorescence of perennial ryegrass (*left*) and Italian ryegrass (*right*).

19.5.2.4 Hybrid ryegrass

Some varieties in this group are similar to the less persistent Italian ryegrass varieties and others are more similar to the perennial ryegrasses. A feature of hybrid ryegrasses is their increased longevity (3–4 years) compared with Italians, and good resistance to diseases including mildew and ryegrass mosaic virus. Most hybrid varieties are tetraploids, however, annual yields are usually lower than for the Italians.

19.5.2.5 Perennial ryegrass—early, intermediate and late varieties (Fig. 19.15)

The *Recommended Grass and Clover Lists* leaflet classifies perennial ryegrass varieties according to date of heading (the date when 50% of the ears in fertile tillers have emerged) as 'early', 'intermediate' and 'late'. Early heading varieties are usually associated with early spring growth and a less densely tillered plant with an erect habit of growth. These varieties are not very persistent. Late heading is associated with densely tillered plants which generally means good persistency. They have a prostrate growth habit and are more suitable for grazing. Intermediate varieties combine the advantages of both the previous groups of varieties, namely relatively early spring growth but later flowering and good persistency.

Further information given by the recommended lists concerns the comparative dry matter yields under cutting or simulated grazing regimes, the approximate heading dates for the first cut, D values, and a 'ground cover index' which gives an indication of persistency. Useful details are also given about the resistance of individual varieties to disease. It should be remembered that these recommended lists are amended and updated with new information each year as plant breeders introduce improved varieties.

Perennial ryegrass forms the basis of the majority of long leys and is the most important grass found in good permanent pasture, perennial ryegrass is, depending on the variety, also used in medium term leys. It is quick to establish and yields well in the spring, early summer and autumn. It does best under fertile conditions and responds well to nitrogen. Crown rust and *Drechslera* diseases are a problem—particularly in the south-west of England. The choice of a resistant variety is the best method of control.

Slow growing varieties have also been developed suitable for amenity and for sports turf use.

19.5.2.6 Diploids and tetraploids

There are now a similar number of recommended tetraploid as diploid varieties of perennial ryegrass. Tetraploid varieties have bigger cells and twice the number of chromosomes compared with diploids. They also have larger seeds and bigger leaves and establish quicker than diploids. The *Recommended Grass and Clover Lists* leaflet gives details of whether varieties are diploid or tetraploid. Seed merchants' catalogues usually provide similar information. Tetraploid varieties are often included in mixtures to improve palatability because of their high sugar content. The dry matter content of tetraploids is usually lower than those of diploids and their persistency and winter hardiness are also often inferior. As a result of being less 'aggressive' with fewer tillers, tetraploids make excellent companion grasses for legumes, especially white clover. Their high sugar content also makes them ideal for conservation as silage, particularly when the grass is wilted. They are often the first choice, too, for the very high dry matter (<65% DM) 'haylage' which has become so popular as a roughage feed for horses. However, when a ley is being established specifically for making hay, some farmers prefer to stipulate diploid varieties, since tetraploids may prove more difficult to dry. However, tetraploid varieties are sometimes preferred by the buyers of hay for the equine trade.

19.5.2.7 Festulolium

This is a cross between Italian or perennial ryegrass and meadow or tall fescue. The characteristics of the variety will depend on the cross taking attributes from both species.

These grasses appear to be high yielding with good disease resistance and feed quality. They have also been found to be quite persistent and stress tolerant.

19.5.2.8 Timothy

Timothy is fairly slow to establish and is not particularly early in the spring (Fig. 19.16 and Table 19.2). It is less productive than the other commonly used grasses, but is very palatable, although its digestibility is not as good as that of the ryegrasses. It is often included in grazing leys with perennial ryegrass. Timothy is winter–hardy and does well under a wide range of conditions, except on very light dry soils.

19.5.2.9 Cocksfoot

This is quick to establish and fairly early in the spring (Fig. 19.17 and Table 19.2). It is a moderately high-yielding grass, but unless heavily stocked, most of the existing varieties soon become coarse and unpalatable. Cocksfoot is a deep-rooting grass and therefore often used in dry or drought-prone situations.

Cocksfoot does not need fertile soil conditions and produces well into late summer and autumn. Most varieties show good winter-hardiness. Cocksfoot should be considered as a special-purpose grass on drier, lighter soils in areas of low rainfall. Cocksfoot varieties are not as digestible as the other important grasses.

Figure 19.16 Timothy inflorescence.

Figure 19.17 Cocksfoot inflorescence.

19.5.2.10 Meadow fescue

Meadow fescue is only included in a few seed mixtures (Fig. 19.18 and Table 19.2). This is because the once famous timothy/meadow fescue mixture is no longer popular; it is not so productive as the perennial ryegrass-based swards. Meadow fescue is slow to establish, but fairly early to start growth in the spring and has a high digestibility.

19.5.2.11 Other grasses including non-sown grasses

Tall fescue (Fig. 19.18) can be useful for early grass in the spring, it is quite stress tolerant but often becomes unpalatable to stock. It is very hardy and can be grazed in the winter. When sown it can be very slow to establish. Because of its long growing season it is sometimes used for green crop drying enterprises. The small fescue grasses, such as creeping red fescue and sheep's fescue, are useful under hill and marginal land conditions, and in some situations, they will produce more than perennial ryegrass. They have little practical value under lowland farm conditions. They are typical

Figure 19.18 Fescue inflorescence.

'bottom' grasses, and they produce a close and well-knit sward. Chewing fescue is often included in seed mixtures for lawns and playing fields. Strong creeping red fescue, as its name implies, is a very resilient grass and a frequent constituent of grass horse gallops. It is also sometimes included in grass mixtures for horse paddocks because of its hard wearing capability.

The brome grasses occur in both permanent pasture (e.g., meadow brome) or as weeds of arable land. Improved varieties of bromes have been introduced in the past from New Zealand but have proved unsuccessful due to lack of persistency under UK conditions.

Rough meadow-grass (Fig. 19.19) is indigenous to the majority of soils, but it prefers more moist growing conditions. In the later years of a long ley, it can make a useful contribution to the total production of a sward if white clover has been suppressed by heavy nitrogen fertilising. Smooth meadow-grass (Kentucky blue grass) spreads by rhizomes unlike rough meadow-grass that spreads by stolons, it can withstand quite dry growing conditions. It is very persistent and hard wearing and is sometimes included in mixtures for horse paddocks or gallops. Annual meadow-grass is a common weed of grassland particularly if the ley has been badly poached.

Figure 19.19 Rough meadow-grass.

Crested dog's-tail (Fig. 19.20) is not very palatable because of its wiry inflorescence. It can be found in unimproved and semi-improved grassland particularly neutral soils. It does not grow well in waterlogged soils. These days it is only used in seed mixtures for lawns, playing fields and sometimes for horse paddocks.

Figure 19.20 Crested dog's-tail.

Figure 19.21 Sweet vernal grass.

Sweet vernal grass (Fig. 19.21) is the grass that can make hay smell sweet. It is commonly found in acidic permanent pasture that has been cut regularly for hay as it is one of the earliest grasses to flower and seed. It is not very responsive to the application of nitrogen and tends to have a low mineral content.

The bents (*Agrostis* spp.) (Fig. 19.22) are very unproductive and unpalatable grasses and, under most conditions, they can be considered as weeds. Creeping and common bent are very commonly found in permanent pastures where they perform a useful function in creating a 'thatch' which can help to reduce poaching. Browntop bent is often included in mixtures for lawn and amenity turf and Highland bent is a useful constituent of grass gallops, racecourses and amenity areas.

Yorkshire fog (Fig. 19.23) is one of the most common un-sown grasses found in permanent grassland. It can be quite high yielding though extremely unpalatable except when very young. It tolerates a wide range of growing conditions and is especially prevalent where the soil is acidic.

Tall oatgrass (Fig. 19.24) is another common grass of permanent pastures especially in neutral and fertile soil conditions. It is not very tolerant of grazing and is mostly found where grazing is lax and in field margins.

A guide to the recognition of some of these grasses are given in Table 19.4. More detailed descriptions and illustrations can be found online and in standard reference books.

(For identification of other grass species see Chapter 5).

Figure 19.22 *Agrostis* spp.

Figure 19.23 Yorkshire fog.

Figure 19.24 Tall oatgrass.

Table 19.4 Vegetative characteristics of other grass species.

Grass species	Distinctive plant characteristics (note none of these grasses have auricles.)
Yorkshire fog	Leaves rolled in stem, plant softly hairy, pink stripes at base of stem.
Tall oatgrass	Leaves rolled in stem, long pointed leaves which are prominently keeled, plant slightly hairy. Brown slightly swollen stem base
Creeping bent	Leaves rolled in stem and leaves have an angular shape. Stolons present and often have some red colouration at nodes.
Smooth meadow-grass	Leaves folded in stem. Leaves have a double main vein (tramline) with a boat shaped leaf tip. Plant is rhizomatous.
Red fescue	Shiny bristle like leaves. Red stem base and rhizomes present
Sweet vernal grass	Leaves folded, plant aromatic and slightly hairy. Hairy ligules.
Crested dog's-tail	Leaves rolled in stem. Leaf blade pointed. Leaf sheath yellowish at base.

19.6 Forage legumes of economic importance

The clovers of greatest agricultural importance are red clover and especially white clover. Red clovers are used in short leys for conservation. White clovers are used for longer duration grazing leys where they act as 'bottom plants'; with their creeping habit of growth they knit the sward together and help to keep out weeds. Although the majority of clovers are palatable, with a high feeding value, they are not so productive as the grasses and they should not be allowed to dominate the sward (aim for 30% clover).

The digestibility of red clover and most other legumes, except white clover, is not as high as that of the grasses, but their voluntary intake at equal digestibility is higher. Because of the ability of legumes to fix atmospheric nitrogen through the action of the *Rhizobium* bacteria in their root nodules, a balanced mixture of grasses and clover enables farmers to economise on fertiliser nitrogen inputs and will leave a useful legacy of nitrogen for the following crop. This is particularly important in the case of organic and other low input farms, where clover-based leys are one of the most important ways of naturally enhancing the fertility of the soil prior to arable cropping. Estimates of the quantity of nitrogen supplied to a mixed sward by a strong stand of clover, under UK conditions, vary from 150 to 200 kg N/ha/year.

19.6.1 Red clovers

Red clovers are usually a constituent of short term leys for cutting for hay or silage (Fig. 19.25 and Table 19.3). Normally the mixture would also contain Italian or hybrid ryegrasses or timothy although red clover can be sown alone. Although used more for conservation, some of the more persistent improved varieties are useful for aftermath grazing. In recent years red clover leys have become very popular for fattening store lambs. The Institute of Biological, Environmental and Rural Sciences (IBERS) at Aberystwyth though mostly recognised for all their breeding work on perennial ryegrass have also been involved with breeding the Aber red clover varieties. Some tetraploid varieties are also available. There are two main types:

- *Early red clover. Merviot* is the most commonly used variety and the industry standard. It has good resistance to *Sclerotinia* (clover rot) but is susceptible to stem nematode.
- *Late red clover.* These varieties are later to start growth in the spring than the early red clovers. Yields can be about 10% less than the early varieties but are more persistent.

Figure 19.25 Red clover.

19.6.2 White clovers

These should be regarded as the foundation of the grazing ley (Fig. 19.26 and Table 19.3). They are not so productive as the red clovers but are much more persistent. There are three types, classified according to leaf size.

- *Large-leaved white clovers.* These white clover varieties have become popular in recent years due to their ability to survive in conditions of high fertiliser nitrogen application and are suitable for inclusion in mixtures for dairy cow grazing and cutting.
- *Medium-leaved white clovers.* These are normally included in leys where a variety of different managements are likely, both cutting and grazing.
- *Small-leaved white clovers.* These are used in the longer term ley and are particularly favoured for sheep grazing.

19.6.2.1 Clover blends

Since the management applied to leys containing white clovers can vary so much (sheep/cattle grazing, silage/hay cutting) it has become common practice to include blends of varieties of white clover from several of the categories described above, in ley mixtures. This practice is designed to ensure that one or more varieties can contribute well under each of the different management systems imposed.

Figure 19.26 White clover.

Lucerne (or alfalfa)

This is a very deep-rooted legume and it is therefore useful on dry soils, although it can be grown successfully under a wide range of alkaline soil conditions provided drainage is good (Fig. 19.27 and Table 19.3). For best establishment lucerne should be sown in the spring (May is the ideal month) using *Rhizobium* inoculated seed. Under sowing or direct sowing after winter barley harvest are less satisfactory. Lucerne can be sown alone or with a companion grass such as cocksfoot or meadow fescue. It yields best, however, as a single species stand. In the UK, lucerne is mainly used for silage or haymaking but it is also very suitable for zero grazing in dry summers. Lucerne also forms the basis of several large-scale green crop drying ventures. The main problems associated with lucerne growing in the UK concerns Verticillium wilt, a soil-borne fungal disease and stem nematode. There is currently no recommended list for Lucerne. Once established lucerne can be productive for 3 or 4 years.

19.6.3 Sainfoin

The growth characteristics of this most traditional of forage legumes are similar to those of Lucerne (Fig. 19.28 and Table 19.3). Generally, it yields less well than lucerne or red clover but can do reasonably well on dry calcareous soils. Traditionally it was used for hay (especially valued for horses) and the aftermath growth was used for fattening lambs. Sainfoin is currently the subject of much scientific research as it has two unusual qualities; it is non-bloating which is unusual for a legume and it is also a natural anthelmintic. The most commonly used companion grass is meadow

Figure 19.27 Lucerne.

Figure 19.28 Sainfoin.

fescue but a late heading tetraploid perennial ryegrass can also give good results. Spring (April/May) sowing is preferred and sainfoin will establish well both direct sown and undersown in spring cereals.

19.6.4 Bird's-foot trefoil or lotus

Bird's-foot trefoil is indigenous in many lowland pastures and is often an indication of an old permanent pasture. Though rarely sown in the past it is becoming increasingly popular as a constituent of grazing mixtures (Table 19.3 and Fig. 19.29). In common with sainfoin it is a non-bloating legume and also has some anthelmintic properties.

19.6.5 Alsike

Alsike is sometimes included as a substitute for part of the red clover content of a ley (Table 19.3). It is more tolerant of adverse growing conditions, such as acidic or wet soils, than red clover. It should never be used for horse hay as it can cause photosensitivity and even liver damage.

19.6.6 Other forage legumes

Wild red clover and yellow suckling clover are indigenous species in many pastures but are of low productivity. Crimson clover and black medick (also known as trefoil or yellow trefoil but not to be confused with the perennial bird's-foot trefoil) are

Figure 19.29 Bird's-foot trefoil.

sometimes sown as 'winter annuals' for subsequent hay or silage crops. Sometimes these species are mixed with Italian ryegrass, or even sown as a green manure crop to increase the soil nitrogen supply for a spring-sown crop.

19.7 Herbs

These are deep-rooting plants which can help to improve soil structure and are generally beneficial to pastures. However, to be of any value they should be palatable, in no way harmful to stock, and they should not compete with other species in the sward. They have a high mineral and trace element content which may benefit the grazing animal.

Yarrow, chicory, ribwort plantain and burnet (Figs. 19.30, 19.31, 19.32 and 19.33) are the most useful of the many herbs which exist. They can be included in a seed mixture for a grazing type of long ley. They are not cheap, however, and one or more of them may well establish in a sward of its own accord. Chicory is occasionally sown as a crop specifically for fattening lambs.

19.8 Grass and legume seed mixtures

19.8.1 Traditional mixtures

Grasses and legumes may be sown as single species or even single varieties but are usually sown in a mixtures. Mixtures may produce slightly higher yields than those

Figure 19.30 Yarrow.

Figure 19.31 Chicory.

achieved from individual varieties partly due to differing growth habits and differences in disease susceptibility. Unfortunately, if a seed mix is large what can happen is that the dominant species can take over to the detriment of other species, this will of course depend on the management of the sward. However, these more complex mixtures were

Figure 19.32 Ribwort (Plantain).

Figure 19.33 Burnet.

Table 19.5 Example of a currently available 'Cockle park' type mixture.

% seed content and plant species	Characteristics of mixture
8.5% perennial ryegrass early diploid 11% perennial ryegrass intermediate diploid 18% perennial ryegrass intermediate tetraploid 11% perennial ryegrass late diploid 11% perennial ryegrass late tetraploid 9% cocksfoot 7% meadow fescue 3.5% Timothy 10.5% white clover varieties	The range of earliness/lateness of species and varieties should give a slightly flatter seasonal growth pattern. Though only a small amount of Timothy, note the seed, is much smaller than perennial ryegrass seed. Mixtures like this can be less affected by drought conditions.

quite common practice for many years and so, for example, the traditions of the 'Cockle Park' and 'Clifton Park' mixtures and those devised by Frank Newman Turner survive even to this day. Indeed, the need to re-establish some of the species rich grasslands has created a resurgence of interest in them and mixtures are available from many seed merchants (Table 19.5).

19.8.2 Modern seed mixtures

The following section contains a few examples of mixtures which would be suitable for mainstream agricultural and organic use. It is strongly recommended that the current issue of the *Recommended Grass and Clover Lists* leaflet, together with information from reputable seed houses, be consulted for the most up-to-date information. When choosing seeds mixtures look at the species/varieties in the mixture, their characteristics whether the mixtures include varieties on the recommended lists as well as considering what the new ley is going to be used for, whether just for grazing or for cutting or a combination of the 2 and how long it is required to last. A recent survey by the Agriculture and Horticulture Development Board found that most seed mixtures used were chosen for longer term grazing and cutting. Only a limited number were chosen only for cutting or only for grazing. Organic producers should also check for the availability of organically produced seed and use it if available at the appropriate inclusion rate.

19.8.2.1 Short-term mixtures

Westerwolds ryegrass is only suitable for a 1-year crop and is normally sown in the spring for grazing or cutting in the same year. The majority of short term (1 or 2 year) mixtures incorporate Italian ryegrass varieties because of their very high yield potential. They are suitable for early grazing and for cutting for hay or silage. The inclusion of early perennials or hybrids is often practised in order to extend the potential

Table 19.6 Examples of short-term leys (1–2) years.

% seed content Plant species	Ley characteristics
100% westerwolds ryegrass or 50% westerwolds ryegrass 50% Italian ryegrass	A 1-year ley ideal for sowing in spring or autumn; not suitable for undersowing; it should produce two good silage cuts within the year. Expensive option but has been excellent where waterlogging has damaged previous grass crop.
50% Italian ryegrass diploid 50% Italian ryegrass tetraploid or 50% Italian ryegrass 50% hybrid tetraploid ryegrass	These leys can be productive for up to 2 years (some extended to three where conditions suitable); usually autumn-sown after cereals or maize; suitable for early bite grazing and subsequent silage cuts.
50% Italian ryegrass tetraploid 25% Italian ryegrass 25% red clover	Suitable for conventional or organic use over 2 years for silage, hay and aftermath grazing; select diploid ryegrass varieties for haymaking.

of the ley into a third year. However, yields will almost certainly suffer in year three as the Italians will have died out. Tetraploid varieties with high sugar contents are favoured for silage and haylage production. Where it is intended to make traditional hay regularly, diploid varieties are sometimes preferred. The most suitable types of clover are the large-leaved white clovers and/or red clover. Example mixtures are shown in Table 19.6.

19.8.2.2 Medium-term cutting/grazing mixtures

This type of ley typically will be taken for one or more silage cuts and followed by grazing. Early, intermediate, and late perennial ryegrasses will all be suitable but obviously early heading perennials would be favoured for early cutting. These mixtures may also contain late perennial ryegrasses which are most suited to grazing. It is normal to include both tetraploid (more palatable and better for making silage) together with diploid (more persistent) varieties. A higher seed rate may be required when sowing mixtures which contain tetraploids. The white clover is a blend including both large- and medium-leaved varieties. An example mixture is given in Table 19.7.

19.8.2.3 Long-term grazing mixtures

Intermediate and late perennial ryegrasses predominate in such mixtures and some tetraploids and timothy can be included for improved palatability. A white clover blend with small- medium- and large leaved varieties should be used, especially where grazing may be carried out by sheep as well as cattle. Example mixtures are given in Table 19.8.

Table 19.7 Example of a medium-term ley (up to 5 years).

% seed content Plant species	Ley characteristics
30% hybrid ryegrass 28% intermediate perennial ryegrass 22% intermediate perennial ryegrass tetraploid 13% festulolium 7% white clover	A 'dual-purpose' type mixture suitable for one or two cuts of silage followed by aftermath grazing. For later cutting or more grazing, the proportion of hybrid ryegrass varieties could be reduced and replaced with late perennial ryegrass varieties.

Table 19.8 Examples of long-term leys (over 5 years).

% seed content Plant species	Ley characteristics
20% intermediate perennial ryegrass 20% intermediate perennial ryegrass tetraploid 20% late perennial ryegrass 20% late perennial ryegrass tetraploid 15% Timothy 5% white clover	A long-term mixture suitable for intensive grazing by dairy cows.
93% late perennial ryegrass 7% small leaved white clover	An extreme long-term pasture type ley for grazing by sheep or horses
16% intermediate perennial ryegrass 32% late perennial ryegrass 40% late perennial ryegrass tetraploid 5% Timothy 5% white clover blend 2% bird's-foot trefoil	A long-term mixture. Bird's-foot trefoil is included to reduce the problem of bloat. It may also have some anthelmintic benefit. May also be suitable as a productive sward for equine use, needing little artificial nitrogen.

19.8.2.4 *Herbal leys*

These leys are usually longer ley types. They include a greater variety of species than the examples of long-leys given here. The species included will depend on use of the ley and soil conditions where growing. An example is given in Table 19.9.

19.8.2.5 *Mixtures including lucerne and sainfoin*

Lucerne and sainfoin may be sown with a grass companion but there is some evidence that lucerne, in particular, yields better as a single species sward. Timothy, cocksfoot and meadow fescue are suggested as suitable species for inclusion. Examples of mixtures are shown in Table 19.10.

Table 19.9 Herbal ley mixture example.

% seed content by weight Plant species	Ley characteristics
17% festulolium 20% perennial intermediate ryegrass 20% perennial late ryegrass tetraploid 13% cocksfoot 13% Timothy 7% meadow fescue 4% white clover 2% red clover 1% alsike clover 2% chicory 1% plantain	A simpler mixture than many herbal ley seed mixtures available. Some deep rooting species will ensure ley more drought tolerant than a straight ryegrass/clover ley

Table 19.10 Examples of mixtures incorporating lucerne or sainfoin.

% seed content Plant species	Characteristics of mixture
92% sainfoin 5% meadow fescue 3% Timothy	This mixture could last for about 3 years.
100% lucerne Or 73% lucerne 27% grasses	Timothy, cocksfoot or meadow fescue are suitable companion grasses for lucerne. Lucerne will persist for up to 5 years in UK conditions.

19.8.2.6 Mixtures suitable for equine use

Table 19.11 gives some examples of mixtures suitable for equine use.

For native ponies, and other situations where productive ryegrass pastures may risk weight issues and laminitis, the perennial ryegrass can be replaced by increased use of fescues, smooth meadow-grass and the addition of cocksfoot, rough meadow-grass and crested dog's-tail in descending order of inclusion.

Table 19.11 Examples of mixtures suitable for equine use.

% seed content by weight Plant species	Ley characteristics
30% intermediate perennial ryegrass 30% late perennial ryegrass 20% creeping red fescue 12% Timothy 8% smooth meadow-grass	A permanent pasture for use where productivity is required, such as for studs. Provides some 'bottom' to the ley to resist damage from hooves. Timothy can provide some drought resistance, while smooth meadow-grass is another palatable species popular for equine use.

% content by weight Plant species	Ley characteristics
50% creeping red fescue 25% dwarf perennial ryegrass 25% smooth meadow-grass 50	An example for use on gallops to provide a dense turf capable of resisting damage from flying hooves, providing some 'spring' reducing jarring to limbs, and having some ability to recolonise divots from hoof fall. Not suited for polo pitches due to rhizomes (use newer dwarf perennial ryegrass varieties tolerant to very close mowing).

There are many new varieties of ryegrass with rapid establishment in cool climates, developed for sports turf; they are in use for racecourses and polo grounds to repair, renew and renovate swards that have heavy use. For winter National Hunt racecourses, pre-germination of seed enables divots to be renovated within 2 weeks in temperatures as low as 3°C. Some of these sports turf varieties will even germinate in soil temperatures as low as 3.5°C.

19.8.2.7 Mixtures suitable for agri-environment schemes

Many of the current agri-environment schemes stipulate number of grass, legumes and herb species to be included in seeds mixture. There are various recommendations depending on the target aim. For example, for a multi species ley the recommendations are for there to be at least 5 grass species (making up no more than 75% of seed weight) plus 3 legumes including bird's-foot trefoil and 5 herbs or wildflowers. The aim of this mix is to produce a sward that will provide habitat and food for invertebrates and help improve soil structure and water infiltration. Table 19.12 gives one example of a multi species mixture.

Table 19.12 An example of a multi species ley mix.

% seed content by weight Plant species	Ley characteristics
30% perennial ryegrass 9% meadow fescue 9% cocksfoot 13% Timothy 7% festulolium 5% red clover 6% white clover 4% alsike clover 2% bird's-foot trefoil 5% sweet clover 10% herbs and flowers including chicory, plantain, burnet, sheep's parsley, yarrow and knapweed	In alkaline soils lucerne and sainfoin are often included in the mix. Many herbs can struggle to establish in wetter growing conditions where herbs can be outcompeted by sown grass species. These leys can be very palatable, and the inclusion of herbs can improve the trace elements availability of the sward.

19.9 Key points

- Upland grass areas though relatively unproductive agriculturally are becoming more important for water management and carbon storage as well as for recreation and biodiversity.
- Most grasses can be identified from their different leaf, ear and stem characteristics.
- When reseeding choose appropriate seed mixtures for the required use whether for cutting, grazing or both.
- There is increasing interest in herbal and species rich leys.
- Environmental Land Management Schemes are likely to encourage the adoption of more mixed species grass swards.

Sources of further information and advice

Anon., Annual, Recommended Grass and Clover Lists for England and Wales. Agriculture and Horticulture Development Board.

Anon., 2019. Grassland Reseeding Guide. Agriculture and Horticulture Development Board.

Blakesley, D., Buckley, P., 2016. Grassland Restoration and Management. Pelargic Publishing.

Frame, J., Laidlaw, A.S., 2014. Improved Grassland Management, second ed. Crowood Press.

Frame, J., 2005. Forage Legumes for Temperate Grassland. CRC Press.

Gilliland, T.J., Hennessy, D., Griffith, V., 2011. Studies into the dynamics of perennial ryegrass (*Lolium perenne* L.) seed mixtures. Irish Journal of Agricultural and Food Research 50, 99–112.

Growing interest in herbal leys, grass and forage manager. In: Gregson, S. (Ed.), British Grassland Society 142, 5.

Locking in carbon with grassland, grass and forage manager. In: Gregson, S. (Ed.), British Grassland Society 142, 8–9.

Hopkins, A. (Ed.), 2000. Grass: Its Production and Utilization, third ed. Blackwell Science.

Hubbard, C.E., Hubbard, J.C.E., Sampson, J., 1992. Grasses: A Guide to Their Structure, Identification, Uses and Distribution in the British Isles, third ed. Penguin Science.

Price, D., 2016. A Field Guide to Grasses, Sedges and Rushes, the Species Recovery Trust.

Sheldrick, R.D., Newman, G., Roberts, D.J., 1995. Legumes for Meat and Milk. Chalcombe Publishing.

Stiles, W., 2017. Multi-species Grassland is it Time to Consider Your Roots, IBERS. Aberystwyth University.

Weddell, J.R. (Ed.), 1997. Seeds of Progress. British Grassland Society Occasional Symposium No 31.

Websites

www.ahdb.org.uk.
www.britishgrassland.com.
www.bsbi.org.
www.cotswoldseeds.com.
www.countrysidesurvey.org.uk.
www.gov.uk/government/statistical-data-sets/agriculture-in-the-united-kingdom.

Establishing and improving grassland

20

20.1 Establishing leys

Newly established leys, containing almost 100% improved grass, legume, and forage herb species and varieties, constitute the most productive and nutritious form of grassland and so it is appropriate to commence this chapter with a section on their establishment.

20.1.1 *Terminology*

The following refer to the various practices associated with sowing grass seeds:

- *Direct reseeding.* Sowing grass seeds again in a field which has previously contained grass without an arable or alternative forage crop break.
- *Direct sowing.* Sowing grass seeds in spring or autumn without any form of cover crop.
- *Undersowing.* Sowing grass seeds with a cover crop, e.g., spring cereals, with or without legumes such as peas (usually at a reduced seed rate) which may be harvested for grain or as a wholecrop silage, leaving the undersown ley established in the stubble. Alternatively, a light cover crop of cereals or forage brassicas such as rape may be grazed rather than harvested.
- *Oversowing.* The introduction of seeds by surface broadcasting without completely destroying the old sward, in some cases associated with raking or harrowing (for example using the "Opico" or "Einbock" grass harrows).
- *Sod or slot seeding.* A form of direct drilling, often referred to as "stitching in," associated with the improvement of a ley or permanent pasture by the introduction of new species without cultivation and the use of specialist drills such as a Moore Unidrill, an Aitchison T-Sem Drill or a Cross Slot Drill.

20.1.1.1 *When is a reseed necessary?*

Many farms rely heavily, if not entirely, on long term or permanent pastures which, when well managed, can be both productive and resilient and the decision to reseed is not one that is made lightly. The process can be costly, takes land temporarily out of production and there is also a risk of partial or complete failure. On the other hand, reseeding is often a means of greatly increasing the potential production of grassland and can represent a very sound investment.

Some farmers reseed as much as 10%−15% of the farm each year as part of a planned reseeding programme to capitalise fully on the improved species and varieties

Lockhart and Wiseman's Crop Husbandry Including Grassland. https://doi.org/10.1016/B978-0-323-85702-4.00024-8

available. They also recognise that previously reseeded pastures often gradually become less productive over time following repeated grazing and cutting, however well managed. Others grow grass in a rotation with other forages or arable crops resulting in a given proportion of new reseeds being sown annually.

Otherwise, the decision to reseed is normally based on the proportion of the most desirable plant species recorded falling below a critical threshold, usually 50% of ground cover, or the perceived poor performance of individual fields or groups of fields. Until recently, this was a decision that was often made somewhat intuitively but is now one that can be based on more accurately identifying underperforming leys on farms by measuring grass growth regularly.

The cause of poor current performance should be determined as far as possible at the outset and frequent causes of poor grass productivity such as soil compaction, poor soil fertility and weed infestation either diagnosed and corrected or dismissed by following the flow chart (Fig. 20.1). Once the decision has been made to reseed, the timing, technique and management need to be considered carefully.

20.1.2 Spring sowing

If directly sown without a cover crop, the maiden seeds can give valuable production in the summer months. Establishment can be enhanced by stock being able to graze the developing sward within a few weeks of sowing and encouraging the development of a thick, well tillered ley. This is not possible to the same extent with autumn sowing because of the subsequent deterioration of grazing conditions and slowing down of plant growth. A limiting factor with spring sowing may be moisture, and in the drier districts the seeds should be sown in March if possible. The plant should thus establish itself sufficiently well to withstand a possible dry period in early summer. Due to the time taken for reseeding to be carried out and the establishment phase, there can be a loss of total production in the sowing year, although the reseed should comfortably compensate for this in subsequent years.

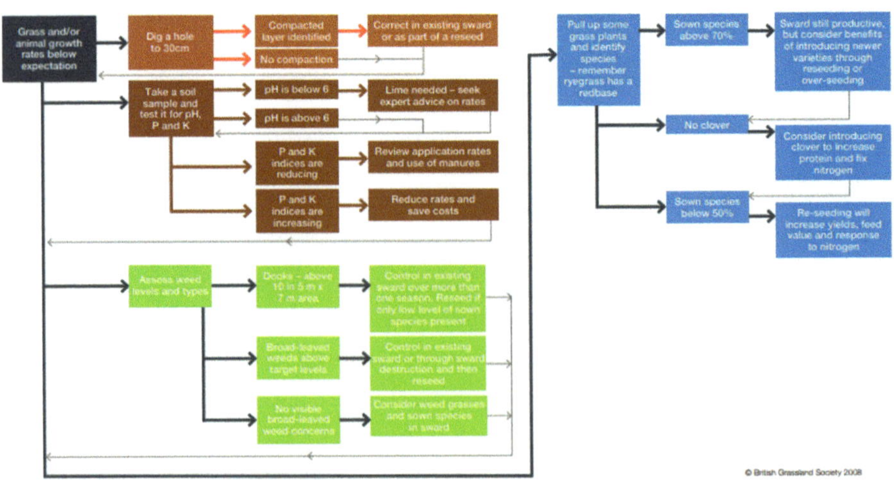

Figure 20.1 Pasture improvement flow chart (British grassland Society).

Undersowing in a cereal crop in the spring means that the greatest possible use is being made of the field in terms of total production, although grass establishment may not always be so good, especially if the cereal crop lodges prior to harvest. Weed problems in a ley are often reduced by undersowing to cereals.

20.1.3 Late summer/early autumn sowing

Leys are often sown at this time of year as the reseed can follow the harvesting of the preceding crop and any loss of production associated with the establishment phase tends to be less marked during the late summer and autumn months. Provided satisfactory establishment is achieved, early growth is ensured in the spring as well as excellent yields in the following full harvest year. There is usually some rain at this time, heavy dews have started again, and the soil is warm. When clovers are included in the seed mixture, sowing must be undertaken before the onset of frosts, probably no later than August. Whether earlier sowing is possible depends on when the preceding crop in the field is harvested as well as soil moisture levels following a dry summer. Winter barley is the ideal crop to follow because of its early harvest. Sowing later than September is not advised and can lead to poor establishment.

20.1.4 Direct sowing

Reference has already been made (in Chapter 8) to the cultivation necessary in order to prepare the right type of seedbed for grass seeds. It must be emphasised that grass and clover seeds are small, should be sown at shallow depths, and that a fine, firm seedbed is necessary.

With ample moisture the seed can be successfully broadcast onto the soil surface. This should be on to a firm ribbed-rolled surface, so that the seeds fall into the small furrows made by the roller. Most fertiliser distributors can be used for broadcasting seeds, and seeds and fertiliser are sometimes broadcast together. Seeds may then be covered by taking a single pass with a light harrow or grass harrow or they can simply be rolled in.

In drier areas, and on light soils, drilling to a depth of no more than 2 cm is preferred. The seed is then in much closer contact with the soil. The 10 cm coulter spacing of a cereal seed drill should give a satisfactory cover of seeds but, for improved results, cross-drilling i.e., two drill runs perpendicular to each other could be considered. After broadcasting or drilling, rolling is essential to complete the whole operation. Where severe weed problems are anticipated e.g., unavoidable carry-over of docks, establishing the grass seeds alone prior to applying a selective herbicide for weed control and subsequently broadcasting or oversowing clover seeds have frequently proved successful.

At soil index 2, 50 kg/ha each of phosphate and potash can be broadcast (up to 120 kg/ha at index 0) and worked in during the final seedbed preparation. For spring sowing, 60 kg/ha of nitrogen is also generally advised unless legumes are being sown. The application of nitrogen in the seedbed may adversely affect legume establishment and is usually best avoided when legumes are sown. Some farmers also use up to

30–50 kg/ha N on seeds sown in August, but this is only necessary if there is a Low or Moderate Soil Nitrogen Supply.

20.1.5 Direct drilling

Grassland can be direct reseeded by direct drilling. Although it is not necessarily cheaper than conventional reseeding, the practice does allow a much quicker turnaround from the old grass to the new sward. It also reduces the risk of subsequent poaching since the soil should be firmer following a lack of disturbance by cultivation. Glyphosate is a desiccant herbicide which can be used to kill off or suppress the existing vegetation prior to the introduction of the new seeds. Cropping with spring-sown direct drilled kale, stubble turnips or forage rape prior to direct drilling with grass seeds in late summer or autumn ensures the best possible establishment from this technique. Another option, particularly if the old ley is thick and matted, is to spray in the autumn and allow the old sward to die off during the winter months, prior to direct drilling in the spring. Rolling immediately following drilling is particularly beneficial as it ensures good soil contact with the seeds. Slugs, leatherjackets, and frit fly pose a potentially serious threat, especially when no break crop is grown and when sowing takes place in the autumn. Growing a break crop or sowing in spring reduces the threat of pest attack but a substantial area of leys is direct drilled into cereal stubbles in most autumns with very satisfactory results.

"Sod seeding," sometimes also referred to as "strip seeding" or "stitching in," usually refers to the direct drilling of new seeds without the complete kill of the old sward. Specialist drills were devised for this purpose, which in some instances incorporate a "band" spraying capability sufficient to kill off or suppress the old sward adjacent to the strip where the new seeds have been sown.

Sod or strip seeding is also a potentially useful way of introducing white clover into an established perennial ryegrass pasture.

20.1.6 Undersowing

Good establishment can be achieved when undersowing spring cereals. The cereal is generally sown first and followed immediately by the grass seed mixture, which is drilled or broadcast. The cereal is usually sown at no more than two-thirds of the normal seed rate. With more vigorous species such as Italian ryegrass and red clover, it is possible to broadcast the seed after the cereal is established and is at approximately the three leaf stage. This may help to prevent excessive growth of the undersown species which, particularly in a wet season, may interfere with the harvesting of the cereal. Another useful technique is to harvest the cover crop of cereals for silage several weeks before the crop is ripe for combining when the presence of the undersown grasses and clovers does not create harvesting problems. The undersown plants may even enhance the nutritive value of the silage as well as the fact that the earlier harvest allows more time for the ley to establish fully before winter.

An alternative is undersowing to a mixture of cereals and peas (grain or forage varieties are both suitable) which may be harvested for silage with a superior protein content.

Weed control in undersown crops is sometimes difficult, especially if clovers have been included in the mixture. If peas are being used as the cover crop then a stale seedbed prior to sowing is the most logical option. Undersowing in autumn sown cereals is not normally advised due to the competitive nature of the cover crop. However, in an organic cropping situation with a less densely tillered cereal crop, undersowing an autumn sown cereal in spring has been shown to be successful in conjunction with operations such as the use of light harrows for weed control which can also rake in and cover the grass seeds.

20.1.7 Weed, pest and disease control in establishing leys

Seedling diseases such as "damping off" (*Pythium*) may adversely affect the establishment of some grasses. Good seedbed preparation and sowing in favourable conditions are strongly advised.

Particularly important pests are slugs which can cause damage at any time of year, but particularly in wet conditions and almost always when sowing in the autumn or sod/slot seeding. Ploughing reduces the risk of damage as does the creation of a fine, firm seedbed. If necessary, slug pellets can be applied to the surface of affected fields after test baiting.

Leatherjackets (larvae of the crane fly or "daddy longlegs") can cause significant damage or even complete crop failure, especially when direct reseeding (grass after grass) is being practiced in the spring. Using a break crop or summer ploughing can help control leatherjackets, as can rolling to consolidate soil. Frit fly larvae can also be present in old turf in substantial numbers and can seriously damage grass leys sown at any time of year (see Chapter 7). Growing a summer break crop such as brassicas or reseeding before July, when the main egg-laying period starts, offer potential solutions. Wireworm (the larvae of the click beetle) can pose an additional challenge, especially following long-term grassland. Growing one or more break crop, double rolling to consolidate the seed bed or even adopting the practice of cultivating and then leaving the soil fallow in July and August (occasionally referred to as a "bastard fallow") offer possible means of control. These pests are not confined to direct reseeds and can also be present in significant numbers in arable situations, particularly in grassy stubbles, so vigilance is key, especially following the banning of chlorpyrifos as a pesticide.

Control of annual broad-leaved weeds in new grass leys is relatively straightforward. Careful cultivation, the choice of appropriate species and varieties and appropriate use of lime, fertilisers and manures should give the reseed the best possible start. As long as the population of annual weeds does not threaten the establishment of the reseed from the outset, the pasture can simply be topped following the first grazing and that should suffice. The most important weed to control in a new reseed is chickweed, particularly in autumn sown leys, as this can grow extremely vigorously if left unchecked and cause significant loss of plants through competition and smothering. Unlike most other annual broad-leaved weeds, topping is ineffective to control chickweed but harrowing can be carried out as a control measure in well-established reseeds.

Perennial weed problems should be anticipated in the previous crop or sward and the aim has to be to avoid carry-over into the new reseed. Problem weeds can either be controlled by means of a selective herbicide in the previous crop or by the use of a nonselective herbicide such as glyphosate as part of the reseeding operation.

Herbicides can be used, if required, in the establishing ley (see Table 20.1). The most important point is to ensure that grasses are at least at the two to three leaf stage of growth and that legumes have at least one true leaf and to remember to use a legume-safe herbicide when spraying grass/clover mixtures.

20.1.8 Early management

Leys can usually be grazed within 6–8 weeks of sowing, when the grass is ~10–15 cm tall, if ground conditions are suitable. Adopting a rotational or "on-off" grazing system and grazing to 4–6 cm before allowing regrowth to 10–15 cm and repeating the grazing will result in better and quicker tillering of new shoots and minimise the risk of damage due to trampling and up-rooting as well as selective grazing. Sheep are generally preferred to cattle because of their lighter weights, especially following late summer sowing. Grazing an August sown ley with sheep in October can be very beneficial as it encourages the grasses to tiller. Grazing a spring sown ley in June or July with cattle should be possible if conditions are dry enough. Grazing in wet conditions should be avoided, however, as it will inevitably lead to poaching, a loss of sown species and the early ingress of weeds.

20.2 Grassland improvement and renovation

20.2.1 Problem identification in existing grassland

This section will deal mainly with the improvement and renovation of long-term pastures which have deteriorated, although this obviously does not preclude reliance on similar operations to those already described. Full reseeding, as described above, is an expensive operation and should only be carried out when absolutely necessary or when the benefits in terms of improved yield and nutritive value are clear cut. Evidence of deterioration should be sought from the species composition as well as from the annual and seasonal yields if the pasture is measured regularly. The problems may be obvious (e.g., infestation with such weeds as docks, thistles, nettles, and ragwort) or slightly more difficult to discern (e.g., a general lack of productivity in comparison with other pastures).

The first step should be to undertake a detailed botanical analysis of the species present in the field so far as is possible. The obvious points to note are the presence or absence of significant quantities of perennial ryegrass. If it is possible to identify ~25%+ of the grass tillers in the sward as perennial ryegrass then it should be possible to improve the sward by changing the management. If, on the other hand, there is hardly any perennial ryegrass present, then some form of sward renewal by reseeding or, at least, a renovation operation, may be necessary. However, before

Table 20.1 Some common weeds and their control in grassland (herbicides approved at the time of publication—always check current recommendations).

weed	Control
Bracken	An increasing problem, especially in hill areas, resulting in loss of agricultural production and environmental value of grasslands. The fronds (unfurling leaves) should be cut or crushed twice a year when they are almost fully open. The soil could be ploughed deep and then cropped with stubble turnips, rape or kale before reseeding. Asulam has historically given effective control, applied from the air or from the ground, but, since the 2013 bracken control season, it has only been available under the terms of a succession of annual Emergency Authorisations (EAs). Glyphosate can be used as a nonselective spray or by contact with a weed wiper.
Buttercups	Indicative of poorly drained and low fertility soils. Improved soil aeration and drainage as well as soil fertility will lessen this problem. Spray with MCPA. The bulbous buttercup is the most resistant type.
Chickweed	A problem annual weed in many grasslands, especially new leys. Its ability to germinate in both spring and autumn mean it capitalises on any gaps appearing in grassland. Several chemicals will control chickweed including mecoprop-P and fluroxypyr (note that they kill clover).
Docks	The most problematic weed in many intensively managed grasslands, particularly silage fields receiving heavy slurry and manure applications. Grazing can be helpful. Triclopyr, fluroxypyr and mixtures, as well as the clover safe amidosulfuron, will give good control if applied at the rosette stage, although treatment may need repeating. If reseeding a heavily infested field, the old sward could be treated with glyphosate before ploughing.
Nettles	Another common weed that can quickly develop into large tufted patches. Spray with triclopyr or fluroxypyr alone or in mixtures; glyphosate can also be used through a weed wiper.
Ragwort	Ragwort often grows in poached or overgrazed grassland and can be poisonous to cattle and horses, particularly in hay or silage. Grazing with sheep in winter can be helpful. Spray with 2,4-D or MCPA in early May or in the autumn when in the rosette stage. Both cut and sprayed plants are palatable to livestock and can increase the risk of consumption and poisoning.
Rushes	An increasing problem in wet or poorly drained grassland. If possible, drainage should be improved. Both hard and jointed rush should be cut more than once a year if ground conditions allow. Apply glyphosate with a weed wiper when the rushes are growing well. Otherwise, spray with MCPA, 2,4-D or triclopyr.
Thistles	Both creeping and spear thistle appear in grassland following overgrazing in winter in particular. Overgrazing should be avoided. Cutting in July is very effective for control of the biennial spear thistle. Spray with clopyralid or clopyralid + triclopyr at the early bud stage.

any operations are undertaken it is necessary to check on a few of the important "basics."

The Environmental Impact Assessment regulations protect rural land that is uncultivated or seminatural from changes in agricultural activities that might cause damage by increasing productivity. Uncultivated land is land that has not been cultivated in the last 15 years by physical means, such as ploughing or an activity that breaks the soil surface, or chemical means, such as adding fertiliser or soil improvers. Seminatural land includes priority habitats, heritage or archaeological features, or protected landscapes. It is usually land that has not been intensively farmed, such as unimproved grassland or lowland heath.

The appropriate Government Agency must decide if the proposal to change the use of land is likely to have a significant effect on the environment. This decision, known as an "EIA screening decision," must be applied for before any changes to rural land are carried out. Improving or renovating grassland would be viewed as "increasing productivity of land for agriculture" and a screening decision is required if it is proposed to affect uncultivated or seminatural land by:

- disrupting the soil surface by ploughing, tine harrowing or rotovating
- increasing the use of fertiliser or soil improvers including lime
- sowing seed that will increase grassland productivity
- draining land
- clearing existing vegetation or scrub equal to or above an area of 2 ha, either physically or using herbicides
- increasing stock density that will result in improved vegetation from grazing.

20.2.2 Drainage

Evidence of poor drainage should be fairly easy to find. Obviously, standing water and rush infested patches are important indicators; also blocked ditches and drainage outfalls which are failing to run. Digging a shallow pit for soil examination may be worthwhile. The grey and rusty red mottling in the topsoil (the top 30 cm or so), known as "gleying," will indicate seasonal waterlogging and a probable need for improved drainage although it may not be necessary to install a complete new pipe system. Frequently there are drains in a field already which just need unblocking or renewing. In other cases, where the soil contains 30% or more clay, it may be possible to set out a "mole" system (Chapter 8), directing the mole channels into main drains covered with permeable backfill. Improving the drainage of a wet grass field has been shown to be capable of improving dry matter yields by up to 30% and can increase the length of the grazing season by up to a month.

20.2.3 Soil structure and health

A healthy, well drained and aerated soil is a fundamental key to productive grassland. A Defra funded survey of 300 grassland fields undertaken recently indicated that 60% had imperfect soil structure with evidence of weed invasion, poor sward quality, and reduced grass yields.

Regardless of soil type, stable structural units or aggregates with good drainage and aeration are essential for the growth and survival of productive species. It has been estimated that soil compaction can cause a 10%–20% reduction in grass yield, restricting drainage and plant rooting as well as reducing nutrient uptake. Grazing livestock as well as farm machinery can place soil structure under pressure, especially in wet conditions, resulting in soil compaction and impeded drainage and aeration. This in turn increases the presence of weed species, increases the risk of run-off and erosion as well as nutrient loss and reduces the number of grazing or working days available.

Soil aerators such as sward slitters (0–10 cm soil depth) and sward lifters ("topsoilers" rather than the more conventional "subsoilers"—10–35 cm soil depth) can be used when soil moisture conditions are appropriate to correct such problems. Diagnosis of soil problems can be carried out simply by digging a series of shallow pits and examining the soil for obvious signs of physical damage both in the top 10 cm, largely due to grazing livestock, and deeper in the soil profile, usually due to machinery traffic and cultivation.

More importantly, soil compaction can be avoided by minimising grazing in wet conditions, ensuring a good network of tracks and gateways to limit traffic in fields, siting drinking water appropriately and adjusting grazing rotations as well as using back fences to limit trampling of grazed areas.

20.2.4 Soil pH

Acidity can also have a major effect on the productivity of grassland. Defra recommend a target pH 6 for long-term grassland. However, in the "Park Grass" classical experiment at Rothamsted Research it has been shown that maintaining a soil pH between 6.5 and 7.5 increases grass dry matter yields by up to 28% when compared to allowing it to fall to or remain at a pH 5. Furthermore, in the most acid areas the species composition has deteriorated to almost 100% Yorkshire fog.

Correcting the soil pH by the application of agricultural lime is very straightforward and cost-effective. Current recommendations are to maintain a minimum soil pH 6.0 for grassland on mineral soils (5.3 on peaty soils) according to AHDB (2021) in their *Nutrient Management Guide* RB209 Section 1: *Principles of nutrient management and fertiliser use.*

20.2.5 Phosphorus and potassium

Routine soil samples should be taken to check on soil phosphorus and potassium as well as pH. Visual evidence may include grass with very poor stunted growth and discolouration in the leaves. Correction of low (index 0 or 1) phosphorus and potassium levels should be carried out as recommended by AHDB (2021) in their *Nutrient Management Guide*—Section 3: *Grass and Forage Crops* RB209.

20.2.6 Control of perennial weeds

The control of perennial weeds such as docks, thistles, nettles, and ragwort should fundamentally be achieved by reliance on good grassland management and the maintenance of dense, productive grassland. Cultural control such as cultivation and pasture topping can also incorporate the use of an effective herbicide and result in a markedly beneficial effect on the productivity of the sward (see Chapter 5). Control can also (in theory) be enforced under the provisions of the MAFF Weeds Act (1959) which covers five specific injurious perennial weed species, spear thistle (*Cirsium vulgare*), creeping or field thistle (*Cirsium arvense*), curled dock (*Rumex crispus* L.), broad-leaved dock (*Rumex obtusifolius* L.), and ragwort (*Senecio jacobaea* L.), all of which are commonly found in grassland.

Regular topping to prevent weeds setting seed is also important (a mature dock plant can set up to 60,000 viable seeds in a year). In the case of ragwort there is often no alternative to pulling by hand, removing and burning, since it is so poisonous and becomes very attractive to animals when it has been sprayed or mown. Sheep grazing during the autumn and winter periods can also help to control ragwort plants in their first winter "rosette" stage.

The information contained in Table 20.1 indicates the appropriate herbicides for use on established grassland. Particular care should be taken where clovers or other forage legumes are a valued sward component, to select a suitable legume-safe herbicide.

20.3 Improving a sward by changing the management

In the case of a sward which contains significant quantities (25%+) of perennial ryegrass and other desirable species, which are uniformly distributed, it should be possible to improve the species composition by management without the need for the introduction of seeds. The first step is to ensure that the basic good management practices, as set out above, are in train. Then increase the annual application of nitrogen fertiliser to near optimum with a corresponding increase in the intensity of utilisation, either by intensive grazing or by cutting early for silage or, preferably, both. Avoid cutting late for hay as this enables the poorer grasses to come to maturity and to set seed. Autumn grazing by sheep, although it may to some extent reduce the earliness of the spring flush in the following year, will certainly reduce the incidence of the poorer bents and meadow grasses. It is important to avoid both undergrazing in late spring and summer and overgrazing in winter as well as excessive poaching. Regular defoliation on a rotational basis to a sward height of 4–8 cm (depending on animal species), plus judicious inputs of nitrogen fertiliser, should substantially favour the perennial ryegrass component of the sward over time and lead to a gradual improvement.

20.4 Improving a sward by renovation

In the case of a sward which contains few species of value it will be necessary to introduce new seeds, specifically of perennial ryegrass and white clover, and to encourage them to establish and thrive. The advantage of this technique is that it is cheaper than complete sward renewal through a direct reseeding operation. Also, it is possible to retain some of the "thatch" supplied by the old turf which is so valuable in physically carrying stock and reducing poaching. Such operations are usually most successful in August. September sowing can lead to disappointing establishments of both grass and clover.

There are two basic approaches. One is to introduce the new seeds by way of one of the sod-seeding machines described above under the section on direct drilling. Examples of these machines currently in use are the Moore Unidrill, Aitchison T-Sem Drill or the Cross Slot Drill as well as the slightly less sophisticated machines such as the Opico or Einbock grass harrows with seed boxes. A degree of sward suppression may be achieved by an application of glyphosate, at a reduced rate, or the sward may be severely grazed prior to the operation. The application of a suitable compound fertiliser, rolling, and the application of pesticides to control slugs as previously described should complete the operation.

The other approach involves a very severe grazing and/or vigorous surface cultivation to expose soil and obtain a tilth. This can most conveniently be undertaken with disc harrows. Broadcasting seeds and fertilisers then takes place and the operation is concluded by rolling as previously described.

20.5 Fertilisers and manures for grassland

Grass, like all other crops, needs mineral nutrients for its establishment, maintenance, and production. Nitrogen, phosphorus, and potassium are the most important but lime has an overwhelming influence on nutrient availability and soil pH must be addressed before paying attention to individual nutrients.

20.5.1 Liming grassland

Soil acidity is probably the biggest single factor adversely affecting the productivity of grassland and it is one of the most simple to correct. The ideal pH for grass leys grown in a grass-arable rotation is 6.2 on mineral soil (5.8 on peaty soil) and for long term or permanent pastures is 6.0 on mineral soils (5.3 on peaty soils). On the other hand, the target pH for brassica forage crops is higher at 6.5 and for a legume such as lucerne is 6.5–7. Agricultural limes are relatively cheap and extremely cost-effective. The usual dressing is about 5 t/ha (depending on the original pH, target pH, and soil type), and the liming materials that are used on arable land are equally suited to grassland. Magnesian or dolomitic lime can, in addition, have a useful effect on the soil magnesium level if it is deficient.

Other materials sometimes favoured for use on grassland include calcified seaweed and granular lime. Like any other liming materials, they should be evaluated solely on the basis of their neutralising value (NV). Such an evaluation may make them appear expensive; however, availability in bagged form or the ability to apply with a farm fertiliser spreader may make them attractive in some circumstances. When seeding a new ley it is essential to have the soil tested for pH and, should lime be necessary, the best advice is to apply it on the ploughed furrow and work it into the seedbed before the seeds are sown. When direct drilling into an old sward it is advisable to check the pH of the top 2–3 cm of soil as this is often more acid than the rest of the soil profile and to lime at least several months in advance of drilling if possible.

20.5.2 Nitrogen

20.5.2.1 Optimum nitrogen

Nitrogen is the most important nutrient required by grasses. When applied in the form of inorganic fertilisers, nitrogen has an immediate and very positive effect on grass growth provided there are no other limiting factors, as seen in Fig. 20.2.

The yield response declines as nitrogen input increases and there is an optimum or target level (Target *N*) that could be applied for any given site after the grass growing potential (Table 20.2), the nitrogen supplied by the soil (Table 20.3) and all other sources of nitrogen have been allowed for and a target yield established.

It is often possible to observe very strong correlations between the application of nitrogen fertilisers, farm stocking rates and ultimate profitability. Nitrogen fertilisers have been the central pivot to the development of intensive grass-based livestock systems but have also been responsible for the creation of some environmental problems.

Many field trials have been carried out to observe the extent to which grasses respond to nitrogen. In the absence of any other limiting factors such as poor drainage, soil acidity or lack of one of the other important nutrients, it is probably true to say that summer rainfall and soil type have the major impact on this response. Based on the "GM 20" series of trials, one of the main suggestions was the subdivision of the UK into five "grass growth classes" according to their theoretical ability to grow grass.

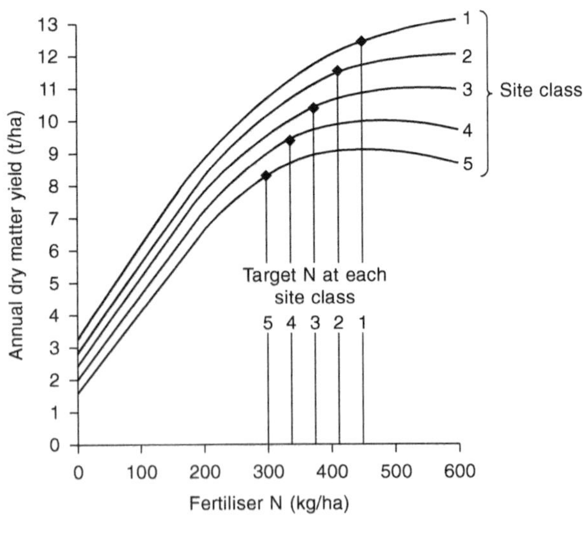

Figure 20.2 Yield and response of grass to fertiliser nitrogen according to Grass Growth Class (1 - Very good; 5 - Very poor) showing target yield and target rate of fertiliser nitrogen.

Table 20.2 Determining grass growth classes.

Soil available water	Soil types	Rainfall (April–September inclusive)		
		Up to 300 mm	**300–400 mm**	**Over 400 mm**
Low	Light sand soils and shallow soils (not over chalk)	Very poor	Poor	Average
Medium	Medium soils, deep clay soils and shallow soils over chalk	Poor	Average	Good
High	Deep silty soils, peaty soils and soils with groundwater (e.g., river meadows)	Average	Good	Very good

Notes:
1. Mean summer rainfall (April–September) is usually about half of annual rainfall.
2. For sites above 300 m altitude where lower temperatures restrict growth, move down one GGC, e.g., Good becomes Average.
Source AHDB, 2021.

Table 20.3 Determining Soil Nitrogen Supply (SNS) status of grassland.

Previous management	Previous nitrogen use (kg N/ha)	SNS status
Long-term grass, high input. Includes: • grass reseeded after grass or after 1 year arable • grass ley in second or later year	Over 250 kg 100–250 kg or high clover content Up to 100 kg	High Moderate Low
First year ley after two or more years arable (last crop potatoes, oilseed rape, peas or beans, NOT on light sand soil)	NA	Moderate
First year ley after two or more years arable (last crop cereal, sugar beet, linseed or any crop on a light sand soil)	NA	Low

Notes:
1. Previous nitrogen use refers to typical fertiliser and available manure nitrogen used per year in the last 2–3 years.
2. The nitrogen values in recommendation tables assume a Moderate Soil Nitrogen Supply (SNS) status and so adjustments need to be made for high or low SNS: increase total fertiliser nitrogen input by 30 kg/ha in a low SNS situation; decrease total fertiliser nitrogen input by 30 kg/ha in a high SNS situation.
Source AHDB, 2021.

In the most recent (2021) edition of the AHDB Nutrient Management Guide (RB209) this concept is still applied, but in simplified form, setting out "grass growth classes" dependent on soil type and average (April–September) rainfall (Table 20.2). These grass growth classes give an indication of potential annual yield as well as predicted response to applied nitrogen and nitrogen recommendations have been formulated for each site class.

Nitrogen requirement and uptake can be heavily influenced by weather conditions, so it is important to reevaluate nitrogen fertiliser requirements throughout the season. As nitrogen is potentially liable to be lost from the soil, the aim must be to target fertiliser timing to meet crop nitrogen demand. Most nitrogen should be applied in spring or early summer, when sward demand is greatest. If grass growth is restricted because of drought, the use of N should not continue and should be reduced once growth restarts following rain.

Current nitrogen recommendations are based on the requirement of the crop, which is then adjusted for the amount of nitrogen available from the soil, known as the Soil Nitrogen Supply (SNS). The method used to assess SNS in grassland systems is different from that used in arable or vegetable cropping systems. Three levels of SNS status are recognised: low, moderate, and high (Table 20.3). Fields with a low SNS need more additional nitrogen than fields with a moderate or high status.

Nitrogen fertiliser, organic manure use, and management history in the last 1–3 years are of most importance for determining the SNS status. So, for example, fields which have been intensively grazed for several years will be likely to have high levels of soil nitrogen because of accumulations due to the excreta of grazing animals and the use of organic manures; furthermore, nitrogen losses from leaching will be low due to the perennial plant cover. Conversely, fields which have recently been sown to a ley after several years in arable cropping will be likely to have low levels of soil nitrogen.

There are many more factors impinging on the farmer's decisions about the optimum rate of nitrogen fertiliser in different circumstances. This includes the enterprise carried (dairy, beef, or sheep), the levels of purchased feeding stuffs employed and the stocking rate in terms of livestock units (LUs or GLUs) per hectare. By today, a key factor is a knowledge of the total annual yield anticipated in tonnes of DM/ha/annum.

The following steps are advised:

1. Select the correct Soil Nitrogen Supply
2. Select the correct Grass Growth Class
3. Select the target annual DM yield
4. Determine the precise quantity of *N* to be applied based on whether the field is to be grazed, or harvested for hay or silage.

The specific requirements of a range of agri-environment schemes as well as the restrictions imposed in nitrate vulnerable zones (NVZs) also need to be borne in mind as does the Defra "Code of Good Agricultural Practice," *Protecting our Water, Soil, and Air.*

20.5.2.2 Grass and legume swards

In the absence of fertiliser nitrogen, legumes can make a significant contribution to the nitrogen requirements of the grass crop. The *Rhizobium* bacteria in the root nodules fix atmospheric nitrogen and, particularly as they die and break down, some of this nitrogen is made available for the companion grass plants. The amount of nitrogen fixed will be proportional to the amount of clover in the sward, as well as to pH, soil temperature and rainfall. With ~30% of a mixed grass based sward present as white clover, a reasonable expectation is for a contribution of up to 180 kg N/ha/annum. Higher levels (up to 300 kg N/ha) are known to be possible with a greater clover percentage and under ideal growing conditions, but total yield is likely to be reduced because of the high proportion of lower yielding clover.

Lucerne and red clover crops have no requirement for N fertiliser when grown as a pure stand, apart from that needed for establishment in low SNS soils (up to 50 kg N/ha). Generally, no nitrogen fertiliser should be required on mixed grass and clover or grass and lucerne swards with an appreciable legume content either. If nitrogen fertiliser is applied, then the clover contribution may well decrease although it is common for farmers with white clover rich swards to apply some nitrogen (up to 50 kg/ha) early in the season, to encourage grass growth before the growth of clover commences in order to provide early spring grazing, and a similar quantity in August to generate more autumn grass. Modern white clover varieties, particularly the large- and very large-leaved varieties, will survive well in a nitrogen fertilised sward and continue to supply the nutritional benefits for which they were also sown.

The key in mixed grass and legume swards is to achieve and maintain the correct balance between the grass and legume components. The decision whether to cut or graze such pastures, the seasonality and severity of both grazing and cutting as well as the grazing system adopted (rotational grazing being essential for sheep if the legume is not to be grazed out), in addition to the strategic application or withholding of nitrogen fertiliser applications, are all management tools that can be used to manipulate the relative proportions.

With the increased current interest in mixed species swards (frequently referred to as multi-species swards or herbal leys), managing a greater diversity of grass and legumes species, as well as herbs, becomes even more challenging.

20.5.2.3 Organic manures and slurry

On most farms, especially those with livestock enterprises, there is an opportunity to apply organic manures or slurry which can complement, or even replace in part or as a whole, the use of inorganic fertilisers to the grass and forage area. Organic manures contain less readily available nitrogen but still have a substantial effect on yield. These are best applied to the conservation (hay or silage) areas where their contents of phosphate and potash will be most valuable in replenishing that removed by cutting. Application to grazed areas can be more tricky because of rejection of grass by grazing livestock but the application of slurry by direct injection into the soil or by the use of trailing shoe applicators can avoid this.

Organic manures may also be applied to fields intended for forage maize, kale, reseeding, etc.

It is essential, from an environmental as well as an economic point of view, to take account of the nutrient content of these materials when calculating the optimum inputs of inorganic fertilisers. Organic manures and slurry are some of the most important sources of nitrate contamination of surface and ground water. Within Nitrate Vulnerable Zones (NVZs) there are specific regulations governing the amounts of nitrogen in organic form which can be applied to agricultural land and the times of year when they may be applied.

Important references are contained in the Defra publication, *Guidance for Farmers in Nitrate-Vulnerable Zones.*

Nitrogen losses from organic sources can be minimised by application in the spring and summer rather than the autumn and winter, and such practices as trailing shoe applicators and soil injection will reduce volatilisation losses. Nitrification inhibitors have been investigated but their effects are often variable, and their widespread use is precluded by cost. Dilution, acidification, aerobic, or anaerobic digestion and separation using membrane filtration are also techniques which have been investigated and which show promise.

20.5.2.4 Timing of nitrogen applications

Grass growth is very seasonal with the main period of rapid growth in the UK falling between April and June. It appears sensible therefore for the main applications of nitrogen fertiliser to be made during this period. Two-thirds of the target nitrogen application before the end of May is a good "rule of thumb" under lowland conditions, particularly in drier regions of the country. The usual principle is for moderately large or frequent applications early in the season followed by smaller applications after each cut or grazing. Further suggestions concerning the timing of applications for grazing and cutting situations is provided in Table 20.4 as a supplement to the AHDB Nutrient Management Guide RB209 annual recommendations.

The timing of the first nitrogen application is always an area for discussion since early spring grass is invaluable on all farms. Monitoring soil temperatures is quite straightforward with a simple soil thermometer, and to introduce the first nitrogen application when the soil temperature at 10 cm reaches 5°C is a logical approach, provided ground conditions are suitable and the weather forecast is reasonably good for the next few days.

The latest date recommended for the application of nitrogen fertiliser is mid-August. In most seasons further applications are unnecessary since warm soils and autumn rainfall frequently result in nitrogen being mineralised from the large organic reservoirs within many soils. A slightly later application by farmers with mainly clover swards is sometimes advocated, to improve the supply of autumn grazing.

20.5.3 Phosphorus and potassium

In a grazing situation there is frequently no need to apply any because of the frequent returns of dung and urine. Only in deficient soils (soil indices 0 and 1) will any response be observed. In the case of potassium there is an important connection with the incidence of hypomagnesaemia or grass staggers. High levels of potassium in the herbage impedes the uptake of magnesium. This is a very serious condition affecting all classes of stock which can cause substantial loss of production or, often, sudden death. The main periods of risk are in the spring, especially during cold, wet

Table 20.4 Fertiliser nitrogen recommendations for grazing swards and for swards cut for silage.

Date	Grazing swards	Swards cut for silage
February/March or when soil temperature at 10 cm reaches 5°C and when soil and weather conditions are suitable	Apply 30−40 kg N/ha; only apply in February where March grazing is possible; if grazing is carried out in March apply a further 30−40 kg N/ha	Apply 30−40 kg N/ha; if an "early bite" grazing is taken reduce the precutting application by 25 kg N/ha
April/May	Continue to apply nitrogen monthly at 30−50 kg N/ha; apply after each grazing on a rotational system; where continuous grazing is practiced either apply monthly to the whole area or weekly to 25% of the area	Apply the balance of N to bring the total up to 120 kg N/ha (110 kg N/ha where SNS levels are high and 130 kg where low); this application should be carried out more than 6 weeks before the planned date of cutting; as soon as the first cut is cleared apply up to a further 90 kg N/ha for the second cut (70 kg N/ha where SNS levels are high and 110 kg where low); if the field is to be grazed continue with grazing recommendations
June/July/August	Extra grazing is normally introduced at this time after first cut silage; continue to apply 30 kg N/ha monthly until the target N level is reached; in very dry conditions nitrogen application should be discontinued; it is not recommended to apply nitrogen fertiliser after mid-August	If a third or even fourth cut of silage is to be taken apply up to 70 kg N/ha prior to the third and 30 kg N/ha for the fourth; otherwise continue as for the grazing recommendations

Notes:

1. The recommendations above are most appropriately applied to intensively managed grassland where the aim is to produce up to 12−15tonnes DM/ha/annum. In these situations the AHDB (2021) *Nutrient Management Guide* recommends annual applications of between 200 and 340 kg N/ha for cutting and between 150 and 340 kg N/ha for grazing, in "average" SNS and grass growth class conditions, depending also on stocking rates and the levels of purchased feeds used. For more extensive beef and sheep systems the use of N on grazing land, especially, is much less and in some cases none is recommended to be used at all.
2. Urea is an acceptable fertiliser material to use in the period February−May or before soils become too dry. After this, ammonium nitrate and its derivatives usually give better results.
3. For hay crops nitrogen applications of 40−100 kg N/ha are usually sufficient, depending on the SNS. Higher levels will give very heavy cuts which will be difficult to dry prior to baling.
4. Slurry is most appropriately applied early to the cutting fields, but care is needed to avoid contamination which may adversely affect silage fermentation. Always take account of the contribution of nutrients from this source when deciding on individual field applications.
5. For ley establishment between April and mid-August apply up to 60 kg N/ha. For autumn establishment nitrogen is not usually necessary except when direct drilling.
6. Where the clover content of a sward is high then applications of nitrogen should be restricted in order to maintain it to no more than 50 kg N/ha in March followed by a further 50 kg N/ha in August if extra autumn grazing is desired.

spells after turnout. Suckler cows are particularly at risk in the autumn. The best advice is to avoid any applications of potash fertilisers in the spring months except where the soil is seriously deficient (index 0). Where needed, potash can usually be applied safely from June onwards. The other remedy is to feed a suitable magnesium supplement throughout the risk period. This can be variously given as a feed supplement, lick, added to the water supply or as an ingested bolus. Another option is to dust the pasture regularly with feed-grade calcined magnesite, but this can be readily washed away by rainfall. A magnesium based pasture spray is another option.

On cut swards phosphorus and potassium are both required in large amounts, especially potassium, where optimum annual dressings for frequently cut swards may total 200–300 kg/ha K_2O. Table 20.5 sets out the phosphorus and potassium, recommendations.

20.5.4 Other nutrients

20.5.4.1 Magnesium (Mg)

It is important that herbage contains sufficiently high level of magnesium if the risk of hypomagnesaemia (grass staggers) is to be reduced. At soil magnesium index 0, the recommendation is to apply 50–100 kg MgO per ha every 3 or 4 years. The uptake of magnesium by herbage decreases as nitrogen and potash increase, however, and hypomagnesaemia can occur when soil magnesium appears to be adequate. If there is a risk of hypomagnesaemia, 100 kg/ha MgO may be justified or, alternatively, if soil pH requires correction, application of magnesian limestone would be most convenient and cost-effective. It should be borne in mind that direct treatment of livestock with boluses or supplementation may also be needed to avoid hypomagnesaemia.

20.5.4.2 Sodium (Na)

Sodium (in compound form) is sometimes applied to grazed swards in an attempt to improve palatability and increase dry matter intake. To improve pasture palatability, 10 kg/ha Na_2O could be applied with nitrogen throughout the season. There is some limited evidence that sodium application might also reduce the risk from grass staggers in some situations by influencing K:N ratios in herbage.

20.5.4.3 Sulphur (S)

Sulphur is an essential nutrient for maximising dry matter yield and protein levels in both grazed and conserved grass. Sulphur deficiency is now common in grassland, especially in later cuts or where high rates of nitrogen are used. ADAS trials in the 1990s showed yield increases from cut grass of 20%–30% where sulphur deficiency has occurred, mainly as a result of a marked decline in the level of industrial pollution. Grass growing on free draining soils is vulnerable and situations where winter rainfall

Table 20.5 Phosphorus and potassium fertiliser recommendations for grassland (kg/ha P_2O_5 or K_2O).

Management system	**Phosphorus soil index level**				
	0	**1**	**2**	**3**	**3+**
Ley establishment (seedbed)	120	80	50	30	0
Grazing (annual application)	80	50	20	0	0
Cutting for silage 1st cut	100	70	40	20	0
2nd cut	25	25	25	0	0
3rd cut	15	15	15	0	0
4th cut	10	10	10	0	0
Management system	**Potassium soil index level**				
	0	**1**	**2**	**3**	**3+**
Ley establishment (seedbed)	120	80	50	0	0
Grazing (annual application)	60	30	0	0	0
Cutting for silage 1st cut	60 (autumn) 80 (spring)	30 (autumn) 80 (spring)	70	30	0
2nd cut	120	100	75	40	0
3rd cut	80	80	60	20	0
4th cut	70	70	55	20	0

Notes:
1. N, P, and K can be applied as "straights" or as compound or blended fertilisers of a suitable analysis.
2 Where there is a known risk of hypomagnesaemia, application of potash on grazing fields in spring should be avoided. Except in the case of serious potash deficiency, applications to grazing areas should be made in the June/July period.
3. To minimise luxury uptake of potash, no more than 80 kg/ha should be applied in the spring for the first cut. The balance of the recommended rate should be applied in the previous autumn.
4. At soil K Indices 2+ or below, extra potash is needed after cutting. In one- or two-cut systems, an extra 60 kg potash per ha should be applied after the last cut or by the autumn whereas in a three-cut system and an extra 30 kg should be applied after cutting. Where grazing follows cutting, this may be applied as an extra 30 kg potash per ha per grazing on up to two occasions.
5. Where applications of FYM or slurry have taken place account must always be taken of the nutrient contribution from these sources.
6. On soils at risk of sulphur deficiency, apply 40 kg SO_3/ha before each cut of silage and 20–30 kg SO_3/ha for grazed fields when up to 100 kg N/ha is applied with an additional 20–30 kg SO_3/ha for each additional 100 kg N/ha applied.
Based on AHDB Nutrient Management Guide RB209 (2021).

is unusually low have also shown this deficiency. Farms where little or no FYM or slurry is applied to grass may also be susceptible. Sulphur, at up to 40 kg/ha of SO_3, should be applied in deficient situations before each cut. For grazing, 20–30 kg SO_3/ha should be supplied when up to 100 kg N/ha is applied with an additional 20–30 kg SO_3/ha for each additional 100 kg N/ha applied. A deficiency can best be assessed by an analysis of herbage (the critical level is 0.25% total sulphur or if the N:S ratio is greater than 13:1).

20.6 Irrigation of grassland

In dry areas with low summer rainfall (i.e., generally deemed to be a low grass growth class) it is the summer months of June to August which normally give rise to the poorest rates of grass growth. Soil moisture deficits (SMDs) during this period can frequently reach 100 mm or more in the south and east of England. Grass growth is usually severely impaired when the SMD reaches 50 mm. A 20 year study carried out in the south-east of England and reported in the 1980s, referred to an increase in grass dry matter yields of 25% as the result of the sustained use of irrigation. Irrigation can be a way of overcoming such severe summer droughts and (in effect) transforming the farm from grass growth class low to medium. Only a relatively few farms have installed irrigation equipment sufficient to make a worthwhile impact on grass growth and the cost of purchasing and operating irrigation equipment as well as the lack of available water generally make it very difficult to justify. Most frequently, when grass is irrigated, it is through the marginal use of equipment which is already being justified by its use on high value arable crops such as potatoes, salads or vegetables.

20.7 Key points

- Newly established leys constitute the most productive and nutritious form of grassland and reseeding is often a means of greatly increasing grassland production.
- The decision to reseed is normally based on the proportion of desirable plant species falling below 50% of ground cover.
- Reseeding can be a total reseed, with seeds sown either direct or under sown to a cover crop, or a partial reseed, by oversowing or slot-seeding into the existing pasture.
- Once the decision has been made to reseed, the timing, technique and early management need to be planned and implemented effectively.
- In a sward which contains more than 25% of desirable plant species, it should be possible to improve the composition without the introduction of seeds as long as soil health, plant nutrients and grazing or cutting are well managed.
- A healthy, well drained and aerated soil is a fundamental key to productive grassland and the starting point for pasture improvement.
- Nitrogen is the most important nutrient required by grasses and fertiliser and manure recommendations are based on the requirement of the crop, adjusted for the amount of nitrogen available from the soil (Soil Nitrogen Supply—SNS).
- Legumes, such as clovers and lucerne, will make a significant contribution to the nitrogen requirements of the crop, reducing or eliminating the need for nitrogen application.
- Other nutrients must be supplied in the correct quantities and balance. On cut swards, in particular, phosphorus and, especially, potassium are required in large amounts.
- Organic manures or slurry can complement, or even replace, the use of inorganic fertilisers for grass and forage crops when stored and applied effectively.
- Effective grazing and cutting management play a key role in maintaining productive pastures.

Sources of further information and advice

AHDB, 2018. Healthy Grassland Soils (SRUC, Newcastle University and ADAS).
AHDB, 2019. Grassland Reseeding Guide.
AHDB, 2021. Nutrient Management Guide—Section 1: Principles of Nutrient Management and Fertiliser Use.
AHDB, 2021. Nutrient Management Guide—Section 2: Organic Materials.
AHDB, 2021. Nutrient Management Guide—Section 3: Grass and Forage Crops.
AHDB Beef and Lamb, 2013. Improving pasture for better returns beef and sheep. BRP Manual 1.
AHDB Beef and Lamb, 2016. Improving soils for better returns beef and sheep. BRP Manual 3.
AHDB Beef and Lamb, 2016. Managing clover for better returns beef and sheep. BRP Manual 4.
AHDB Beef and Lamb, 2016. Managing nutrients for better returns beef and sheep. BRP Manual 7.
British Grassland Society, 2021. In: Multi-species swards Proceedings of the 13th BGS Research Conference, March 2nd–4th 2021.
Defra, 2008. *Manure Planning in Nitrate Vulnerable Zones* (PB3577).
Defra, 2009. *Protecting Our Water, Soil and Air*: A Code of Good Agricultural Practice for Farmers, Growers and Land Managers.
Defra, 2018. Using Nitrogen Fertilisers in a NVZ Nitrate Vulnerable Zones—GOV.UK. www.gov.uk.
Frame, J., Laidlaw, A.S., 2013. Improved Grassland Management. Crowood Press.
Hopkins, A. (Ed.), 1999. Grass: Its Production and Utilization, third ed. Wiley-Blackwell.
Institute of Arable Crops Research, 1991. Guide to the Classical Field Experiments. Rothamsted Experimental Station, AFRC.
Younie, D. (Ed.), 2012. Grassland Management for Organic Farmers. The British Grassland Society.

Grazing management

21

21.1 Introduction

The utilisation of grass by grazing is undoubtedly the most economical means of production for ruminant livestock as well as horses. Reference has already been made to the low cost of energy as grazing compared with conserved grass. Research has repeatedly shown that grazed forage costs between 30% and 50% of those of conserved feeds and even less when compared with the costs of energy as bought-in feed. There is extensive evidence available that well managed grazed grass can provide adequate energy and protein for production, although this will vary substantially from farm to farm. There is general agreement, however, that maximising the contribution from grazing, whatever the system of production, is a sensible approach to reducing costs and maintaining an acceptable degree of profitability, even in (or perhaps especially in) a time of depressed prices and/or increasing input costs.

The approaches to grassland and grazing for equine enterprises vary even more than those for agricultural livestock, encompassing the whole range from exercise areas to full nutrient provision.

21.2 Stocking rate

By convention, stocking rates are usually expressed as "Grazing Livestock Units" (LU or GLU) per hectare; a unit being a 500 kg Friesian/Holstein cow. Stocking rates should be accurately calculated from the information contained in Table 21.1, and based on a monthly reconciliation of actual stock numbers in the various categories on a farm and the total forage hectares. Historically, there has often been a strong correlation between stocking rate and farm profitability. High stocking rates (e.g., >3 GLU/ha) would normally only be associated with the very best grass growing conditions and with near optimum applications of nitrogen fertiliser. Lower rates (1.5−2.0 GLU/ha) might be associated with organic farms, lower input systems or simply with poorer grass growing conditions.

Stocking rate calculations are also usually an integral part of meeting the requirements of various Agri-Environment Schemes, details of which are supplied by Defra, Natural England or the devolved Governments. It needs to be borne in mind that individual schemes might not, however, use the values quoted in Table 21.1, occasionally adopting their own interpretation of livestock unit coefficients or only including specific categories of livestock.

Lockhart and Wiseman's Crop Husbandry Including Grassland. https://doi.org/10.1016/B978-0-323-85702-4.00006-6

Table 21.1 Grazing Livestock Unit (GLU) figures currently in use for farm management.

Cattle	GLU	Sheep	GLU
Dairy cows (Friesian/Holstein)	1.00	Lowland ewes	0.11
Beef cows (excluding calves)	0.75	Upland ewes	0.08
In—calf heifers	0.8	Hill ewes	0.06
Bulls	0.65	Breeding ewe hogs	0.06
Other cattle 0–1 years old	0.34	Other sheep over 1 year old	0.08
Other cattle 1–2 years old	0.65	Store lambs under 1 year old	0.04
Other cattle over 2 years old	0.8	Rams	0.08

Source: Nix, 2021, as advised by Defra.

Horses may be typically considered to be roughly equivalent to cattle of the same weight.

21.3 Principles of grazing management

Grazing management necessitates balancing the feed requirements of stock against the potential availability and nutritive value of grass or other forages and requires a good knowledge of both. It can entail manipulating forage growth through forage selection and management, modifying annual and seasonal animal forage requirements by choice of livestock species and species balance, stocking rate and seasonality of production in addition to controlling forage availability by means of grazing management.

It is affected by the location and nature of the farm as well as the livestock production system practiced. In fact, the choice of production system (especially the date of calving or lambing) is often strongly influenced by the grass and forage production characteristics of a particular farm. Such considerations would include "earliness" of grass growth and the potential date of turnout, the potential for extended grazing into the autumn and early winter as well as factors such as the likelihood of a mid-season drought. Consequently, the synchronisation of lambing or calving with the expected growth of grass so as to turn out stock to grass ready to maximise utilisation from grazing is often seen as a good approach to minimising production costs and optimising profitability. There are, of course, other considerations that need to be borne in mind, such as price seasonality and continuity of product supply such as milk and finished livestock throughout the year that would also have a bearing on the system adopted. An additional consideration is likely to be the requirements of any environmental land management schemes where both grazing severity, as determined by annual and seasonal stocking rate, and grazing system could be prescribed.

21.3.1 Animal potential

So far as milk production is concerned, calving in late winter/early spring should ensure that cows can be turned out to grass immediately after calving in order to maximise the contribution of grazed grass to their production. Peak lactation will then coincide with peak grass growth and nutritive value in late spring and early summer and the cows will normally be dried off as grass supply wanes in late autumn or early winter. Autumn calvers, too, can achieve good levels of production from grazing, occasionally from autumn grass prior to housing, and, particularly, in the spring towards the latter half of the lactation when no feed supplementation should be necessary but cows might still be yielding up to 25–30 litres per day. The majority of the milk they produce, however, will be based on winter rations and the production of sufficient quantities of high quality silage will often feature at the forefront of grassland management decisions. The susceptibility of a farm to summer drought should also influence the decisions about calving date.

There is potential for grazing to provide sufficient energy and protein for a daily milk yield of upwards of 30 litres from animals capable of producing it. This figure could fall to below 20 litres in late summer as grass nutritive value declines, particularly if not well managed over the summer months, although animals might also struggle to consume sufficient grass to achieve high yields from fresh grass alone as its moisture content is also liable to increase.

In the case of growing stock (young cattle or lambs) the potential is for grazing to provide sufficient nutrients for individual daily live weight gains of up to 1.5 kg for beef cattle and 300 g for lambs to be achieved from high quality grassland. Again, the need to maintain high pasture nutritive value as the grazing season advances and balancing increasing animal demand as animals grow with decreasing grass productivity requires careful management. In all breeding/rearing enterprises the combined grazing of adults and juveniles prior to weaning can also lead to the almost inevitable build-up of parasitic worm populations. Grazing management must address the need to integrate parasite control measures together with the provision of sufficient quantities of nutritious grass at the correct stage of growth throughout the grazing season.

21.3.2 Grass potential

The factors which affect grass yields such as sward species and varieties, grass growing conditions and fertiliser (especially nitrogen) application, have been dealt with at length in Chapters 19 and 20. Seeds mixtures suitable for grazing in various situations have also been described in detail in Section 19.8.2 and in Table 21.2 and Tables 19.6–19.8. Other site factors such as altitude and aspect, and soil factors such as texture and drainage, affect the suitability for both grass growth and grazing. This is particularly important when considering the suitability of a field or farm for extended early spring and autumn grazing, as exemplified by systems often referred to as involving the New Zealand approach and being increasingly adopted in Ireland and the UK. On some farms specific crops such as autumn sown rye, rye with Italian or hybrid ryegrass or triticale are sown specifically for early grazing.

Winter wheat and oats, sown as grain crops, may also be grazed in early spring so long as soil conditions are suitable and the crop has not reached the "stem extension" growth stage. Although there may be some suppression of grain yield, the strategic gain in terms of cost saving to the livestock enterprise may well outweigh this.

Table 21.2 Suitability of seed mixtures.

Purpose	Types of forage plant that should be used
For early grazing, i.e., early bite	Mainly Italian ryegrass and hybrid ryegrass
For optimum production throughout the season	Ryegrass (supported by either non-ryegrass or permanent pasture)
For winter-grazing cattle (provided conditions allow)—deferred grazing	Non-ryegrass species
For the grazing block	Herbage plants such as perennial ryegrass that produce a closely knit sward

Note: As far as possible, the heading dates of individual varieties should correspond because, with only a few exceptions, the digestibility of a plant starts to decline when the stem extends and the ear shows signs of emerging.

Increasingly, however, the adoption of tightly controlled grazing systems that enable grass to be gradually accumulated over the very late autumn and winter months have allowed many producers to provide much earlier grazing than hitherto from the farm's existing grassland.

Summer drought may require special provision to be made to feed stock during times of shortage of grazing. This might include sowing an area of quick-growing turnips in spring for summer grazing or the inclusion of a field of drought tolerant forages such as lucerne, with its potential for zero grazing ("cutting and carting") during the summer months. Many dairy enterprises ensure that additional silage is available specifically for feeding during times of grass shortage, and silage or haylage bales are commonly fed to beef and sheep to supplement grazing both at spring turnout and late in the grazing season as well as during periods of drought that might be encountered during the summer months.

21.3.3 Sward measurement

Good grass utilisation is the key to successful grassland farms, reducing wastage by grazing at the right time with grass at the correct stage of growth. Historically, this was a largely subjective process with decisions usually being made somewhat intuitively. More recently, a far more objective and measured approach has been possible as a result of research.

Management decisions can be based on the morphology of the plants being grazed. If plants are prevented from getting to the fourth leaf stage then maximum grass utilisation occurs without allowing plants to start producing less nutritious stemmy growth or leafy material to die and decay so the aim of many producers is to ensure that grass is always grazed at the three leaf stage. At peak grass growth in late spring and early summer, three new leaves can be produced within as little as 14–21 days since the last grazing while in winter this might take as many as 100 days or more and grazing management decisions need to be made and adjusted accordingly.

Estimates of undisturbed sward height were the first real attempt at providing an objective measure of performance at grass from all classes of stock, however, providing both a measure of grass productivity and nutritive value as well as grazing animal performance.

As a sward is grazed increasingly closer to the ground herbage intake and animal performance as well as grass growth will be progressively reduced. On the other hand, as grass is allowed to become progressively taller it will develop less nutritious stemmy growth and the uneaten herbage will decay resulting in reduced utilisation. There is, therefore, a compromise grass height for different categories of livestock that will result in optimal pasture utilisation and animal performance while also leading to good grass growth and persistence. The ideals are represented in Table 21.3 which gives information about continuously and rotationally grazed pastures. For continuously grazed animals, which tend to remain on the same pasture for prolonged periods, the aim is to achieve a "steady state" with the ideal grass height being maintained consistently. On the other hand, for rotationally grazed animals, which are routinely moved from one grazing area to another in some form of rotational pattern and allowing grazed grass to recover and regrow, the height of the "residue" indicates when the animals should be moved on to the next grazing in the interest of livestock performance and grass regrowth.

Estimates of sward height are best made with a sward stick such as the one shown in Fig. 21.1, but in the absence of anything so sophisticated, a ruler or even a mark on the side of a boot are more than adequate substitutes. To get a reliable estimate of average sward height, at least 20 measurements should be taken in a field at random from both grazed and rejected areas. Sward height estimates are likely to be of most use to farmers grazing animals on a continuous basis.

If grass height exceeds target, especially in the period of rapid grass growth in late spring, the inevitable stem production and accumulation of dead foliage leads to

Table 21.3 Suggested target range of sward heights for continuously grazed animals or postgrazing (residual) sward heights for animals on rotational grazing systems.

	Target sward height
Sheep	
In spring and summer	
Ewes and lambs (medium growth rates)	4–5 cm
Ewes and lambs (high growth rate)	5–6 cm
Dry ewes	3–4 cm
In autumn	
Flushing ewes	6–8 cm
Finishing lambs	6–8 cm
Store lambs	4–6 cm
Cattle	
Dairy cows	7–10 cm
Finishing cattle	7–9 cm
Beef cows and calves	7–9 cm
Store cattle	6–8 cm
Dairy replacements	6–8 cm
Dry cows	6–8 cm

Figure 21.1 A sward stick designed by the Hill Farming Research Organisation.

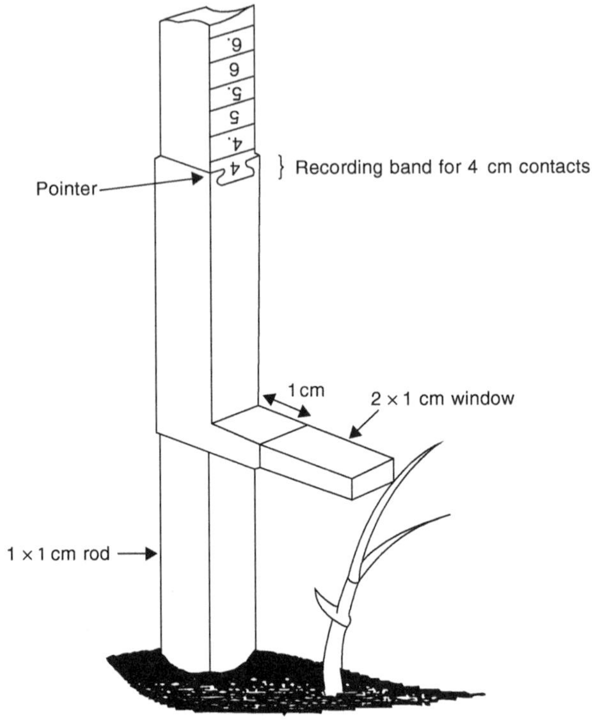

reduced animal production and subsequent pasture growth and structure will be compromised. In order to prevent this from happening grass could be topped to the appropriate height but, in order to avoid wastage of grass, grazing pressure should be increased by either increasing livestock numbers or tightening up the area available for grazing with the surplus grass on the ungrazed area being harvested for hay or silage. On the other hand, if grass height falls below target then livestock numbers should be reduced or the area of land available should be increased. Alternatively, additional nitrogen fertiliser should be applied if conditions are suitable for grass growth or supplementary feeding with forage or concentrates will need to be introduced.

Horses have been found to be very selective in species choice and sward height, usually grazing the shortest forage. The best way to maintain appropriate grass height is to mix graze horses with alternative livestock species, either jointly or sequentially, in order that the grass rejected by horses is grazed.

21.3.4 Dry matter yield estimates

The use of grass height, while still useful for continuous grazing systems, has been largely superseded by the measurement of pasture yield or mass, representing density as well as height, expressed in kg DM/ha.

Figure 21.2 A rising plate metre.

The most practical method is the use of a rising plate metre (RPM), which can be mechanical or electronic (Fig 21.2). Electronic plate metres measure the yield or "cover" in kg DM/ha automatically while the measurements from the former have to be converted manually or digitally using either prescribed conversion equations supplied by the manufacturer or those derived on farm following calibration of the metre.

The use of plate metres has become increasingly popular in recent years, initially almost exclusively on dairy farms but, latterly, more frequently on beef and sheep farms, providing reliable estimates of sward dry matter yields. Again, it is important to walk the fields or paddocks in a random pattern and make 30 readings in each. The data from an electronic plate metre can be directly downloaded into software programmes which are increasingly being used as decision support tools. The technology is developing rapidly with sensors attached to farm vehicles and unmanned aerial vehicles as well as satellite technology likely to play increasingly important roles in future.

AHDB provide excellent resources on grassland management, and suggest that, for cattle, grass should not be allowed to grow beyond 2000–2800 kg DM/ha before grazing to a residue of 1500 kg DM/ha, with a corresponding figure for sheep of 1800–2000 kg and 1500 kg DM/ha.

To be successful, this approach to grassland management involves devising a plan and setting targets, in essence a grass "budget," monitoring grass weekly by taking measurements and responding by making appropriate management decisions either with or without the use of decision support software.

This technique is likely to be mainly of use to farmers grazing on a rotational system and the key concepts are setting seasonal targets for overall farm cover i.e., the average cover over the entire grazing area and, most importantly, acquiring a weekly value for the cover in every individual field or paddock. The daily feed requirements of the herd or flock will also be calculated to determine the demand. The aim is to create a "grazing wedge" with the target cover for introducing animals in the next field or paddock to be grazed and progressively smaller covers in each field or paddock to be grazed subsequently. Following anticipated grass growth and based on the feed demand of the herd or flock, the ideal cover will then be present in each field or paddock as it is required for grazing.

Initially, calculations would have been carried out manually but enterprising farmers and consultants created their own spreadsheets and a number of commercial packages e.g., AgriNet have also been available for several years. These packages enable the data gathered weekly on farm to be processed in order that grass budgets can be drawn up and grazing rotations planned with the yield or availability of grass plotted against anticipated demand. This is particularly valuable during early spring and late autumn when attempting to extend the grazing season at times of the year when grass growth can be slow and ground conditions challenging. Seasonal and annual grass yields and fertiliser use are also calculated in order that grass production can be determined and evaluated, less productive fields or paddocks might be identified for soil or pasture improvement and fertiliser use can be targeted more effectively. Crucially, the software also provides a means of "benchmarking" a farm's data and its subsequent use by agricultural consultants or as a means of comparison within a farm discussion group.

21.3.5 Stocking rates and density at grass

The output from any grazing system is not only strongly correlated with the overall stocking rate carried but also with the stocking densities maintained at different stages throughout the grazing season. The rate of grass growth in May can be up to 100 kg DM/ha/day or more but declines to 20–30 kg DM/ha/day as the season progresses which means that grazing intensity, as measured by stocking density, needs to be adjusted over the grazing season in order to balance supply and demand.

The area devoted to grazing is usually expanded as the grazing season advances as a result of the progressive introduction of silage or hay aftermaths into the grazing area. This means that the integration of grazing and forage conservation is one of the key management tools at the farmer's disposal in order to manage grassland effectively. On intensively managed paddock systems, dairy cows might be allocated about 10 kg DM/head at spring turnout. With an initial grazing cover of 2400 kg DM/ha and target residue or residual of 1500 kg DM/ha, 900 kg DM/ha would be available for grazing and such a herd would require 1.1 ha/100 cows/day.

Ewes with lambs would normally start the grazing season at about 15–25 ewes with lambs/ha. However, if spring grass growth is poor, it may be necessary to allocate increased grazing to ewes with lambs at the expense of the forage conservation area in the expectation that a hay or silage cut may become available later in the season when grass growth improves.

With young cattle or dairy replacements at grass, the stocking rate is often expressed in terms of the liveweight of animals per hectare since individual weights can vary so much. On well-fertilised grassland, about 2.5 t of liveweight/ha would be a good target in the early season falling subsequently to 1–2 t/ha later in the season. Fig 21.3 describes the way grass areas could be allocated throughout the season.

21.4 Grazing systems

In essence, the choice of grazing system adopted rests between some form of rotational grazing, with animals being allocated new grazing strips, paddocks, fields or blocks in sequence and the previously grazed areas allowed to recover and regrow, or continuous grazing where livestock (not necessarily the same number of animals or area of land to allow for fluctuation in grass growth) are not routinely moved.

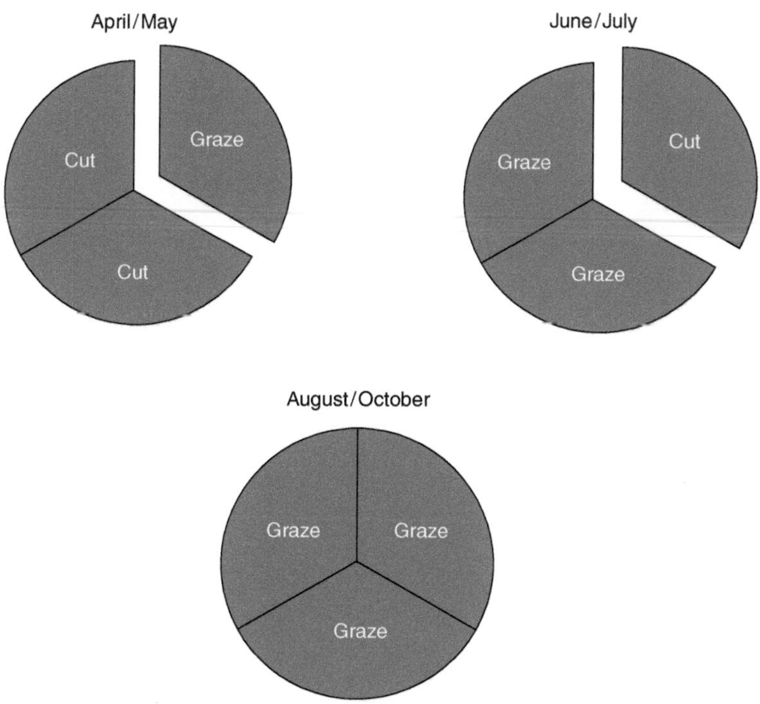

Figure 21.3 The 1, 2, 3 or 1/3:2/3 system for grazing young cattle.

Much has been made of the impact of choice of grazing system on grassland and animal production but this can be, and often is, overstated. As long as the system is managed in accordance with the principles already outlined with effective planning, monitoring and a degree of flexibility, production at moderately high stocking rates will not be greatly affected.

On the other hand, at higher stocking rates and with highly variable grass growing conditions, in particular, the degree of grazing control associated with rotational grazing will result in improved grassland and animal performance.

Some forage species either respond better to being grazed rotationally or have to be grazed rotationally in order to survive - examples include white clover for sheep and red clover or lucerne for all animal species.

21.4.1 Strip grazing

Strip grazing (Fig. 21.4) represents the ultimate in terms of controlled grazing. The system allows the animals access to a limited, flexible area of fresh grass, typically either once or twice daily for dairy cows and usually at slightly longer intervals for other animals and is controlled by portable electric fencing placed in front of the animals. A back fence can also be used in order that the area most recently grazed is almost immediately fenced off and protected from further grazing. This is to protect the recovering

Figure 21.4 Strip grazing.

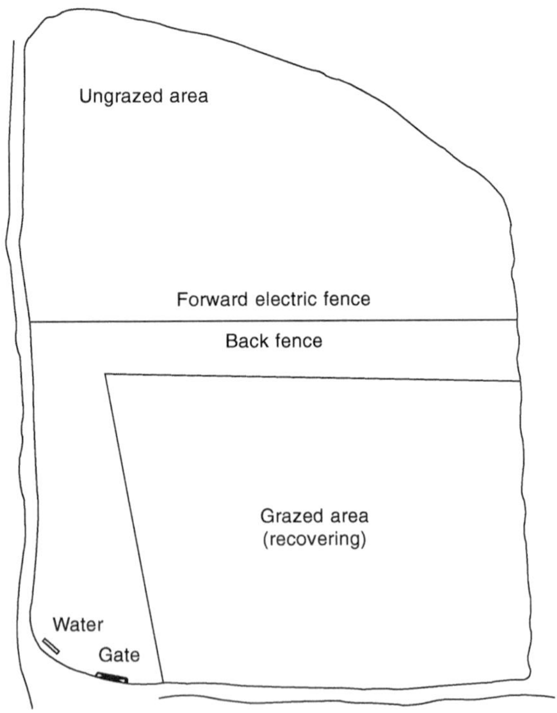

sward from constantly being regrazed as well as being trampled or poached, slowing its recovery and regrowth (Fig. 21.4).

Strip grazing is time-consuming and requires constant decision making about how much grass is needed for the livestock as well as a feeding face long enough to allow all animals simultaneous access. Regular grass measurement now makes this far simpler by removing much of the decision making and any guesswork involved. It is always important to maintain good access to water. Excellent grass or crop utilisation can be achieved but wet soil conditions can lead to serious poaching along the line of the fence.

Strip grazing is also a suitable system for utilising crops such as forage rye for early bite, or turnips, rape and kale later in the season. Beef cattle can be strip grazed on direct drilled turnips after barley with baled straw or silage, usually placed at the edge of the fields, allocated alongside the grazed crop in order to add dietary roughage or a more balanced diet. Ewes or store lambs grazing rape and turnips during the autumn and early winter are also usually rationed in this way, with blocks of forage being introduced for periods of 3—4 days with two, three or four strands of electric fencing. Welfare regulations require a dry (grass) run-back to allow animals lying areas off wet soil.

21.4.2 Rotational or paddock grazing

The principle of this system is a more formalised or rigid approach to rotational grazing (Fig. 21.5). The grazing area is divided into equal-sized paddocks by means of permanent or semi-permanent fencing with each paddock usually designed to provide one day's grazing before the animals are moved on. Good fencing, access tracks and a reliable water supply to each paddock are essential features of the system. On a dairy farm the grazing paddocks will normally occupy land close to the milking parlour.

There is significant variation in the number of paddocks considered to be "ideal," but the common feature is a system which allows animals to graze for a limited period (typically half a day or a full day for dairy cows but often several days for beef cattle and sheep) followed by a period of recovery. A typical one day paddock grazing system will be based on anywhere between 21 and 40 paddocks with paddocks being regularly taken out of the grazing rotation and harvested as silage as grass growth and availability dictate. At the peak of spring growth it would be expected that animals would rotate around a paddock system on a 21-day cycle or less, dependent on levels of grass growth. As the season progresses the rate at which grass recovers after grazing will decline. At this time the grazing obtained from part of the area which has already been cut can be introduced. Fertilising with nitrogen after each grazing (Table 20.4) and field operations such as "topping" of rejected material and weeds will maintain the quality of the grazing throughout the season. Many producers have introduced pre-grazing mowing of any paddocks where grass is threatening to get out of control and stemmy rather than topping after grazing but there is conflicting evidence of the utilisation of premown material by grazing animals.

"Leader/follower" grazing is practiced on some farms with higher priority stock leading, and being given access to almost unlimited grass before being moved on,

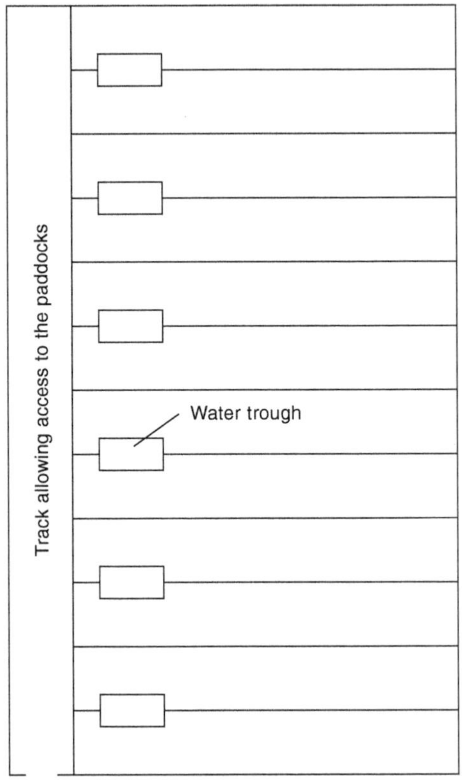

Figure 21.5 Paddock grazing.

with lower priority animals following in order to remove vegetation to the desired height or cover e.g., high yielding cows grazing ahead of a low yielding group. Such a system can lead to very high levels of pasture utilisation.

The outcome is a very predictable and easily managed system once established as long as the number of paddocks and the length of the grazing rotation are adjusted in response to seasonal grass growth in order to ensure that each paddock contains the appropriate amount of grass to meet the needs of the animals at each grazing. Establishing and maintaining the grazing wedge outlined previously is an essential component of the system.

Young cattle and sheep, when managed on rotational grazing systems, usually have access to fewer paddocks i.e., 4–6 would be normal, but the length of the complete rotation and the recovery period allowed would be roughly the same as for dairy cows. AHDB recommends efficient grazing strategies such as well managed paddock systems for both beef and lamb production and claims that the financial returns justify the effort and establishment costs. "Forward creep" grazing, where lambs graze ahead of ewes through a creep gate, can provide better nutrition for the growing lambs and also potentially play a role in reducing their intake of parasites.

The use of a paddock system for horses, with rotational use and rest phases, mimics the habits observed in semi-feral nomadic equine populations. Groups tend to return to areas in a cycle of 10–21 days which allows for regrowth, and in suitable (warm, dry) weather conditions can conceivably reduce the parasite burden by dehydration of the infective larval stage. Permanent wooden or electric tape fencing are the most suitable divisions for equine paddocks.

21.4.3 Continuous grazing (not to be confused with set stocking)

The majority of beef cattle, sheep and horses in the UK are managed on continuous grazing systems where animals remain in the same grazing area for extended periods. The obvious appeal of continuous grazing is the simplicity of management, but it is extremely important to ensure that the swards are of suitable quality (short, dense and leafy) and, particularly, that the grazing pressure, as determined by the stocking density at grass, is at the correct level to maintain the optimum sward height or mass. Continuous grazing should not be confused with "set stocking" where livestock numbers remain constant and often lead to alternating periods of over and undergrazing as grass growth fluctuates. Optimum sward height recommendations are set out in Table 21.3 and these should be measured regularly (i.e., weekly) with adjustments made to the stocking density by adding or removing animals, reducing the size of the grazing area (e.g., closing additional land up for silage or hay) or introducing extra grazing (e.g., silage or hay aftermaths) as the need arises. Dairy cows are occasionally continuously grazed, but it is generally agreed that better output from grass is obtained from rotational or strip grazing largely due to better grass utilisation. Continuous grazing is often used for dry cows and heifers, however.

Fertiliser application to continuously grazed areas, where applicable, can follow one of two patterns, either by application to the whole area at set intervals, or to a part of the area (e.g., a quarter) on a more regular (e.g., weekly or fortnightly) basis. It has been observed, with sheep in particular, that animals tend to avoid the most recently fertilised areas for several days and this can lead to a kind of rotational grazing without fences.

The biggest problem of single species equine grazing areas is the development of lawns and roughs due to horses using latrine areas to defaecate and subsequently avoiding grazing that area. Mixed grazing with ruminants is recommended to balance equine species and area selectivity, improving the consistency of the sward, and potentially reducing the levels of species-specific parasite larvae.

21.4.4 Mixed grazing

Grazing more than one livestock species on the same pasture can be highly beneficial to both the grassland and the animals when the opportunity to do so arises. Different animal species exhibit different grazing behaviour. Sheep and horses tend to nibble plants and demonstrate a high degree of selection in their diet while cattle tend to be more indiscriminate, wrenching plant material away. Cattle and horses also avoid grazing immediately around their droppings or dung pats, often rejecting large quantities of grass which sheep would quite readily consume. One animal species can therefore consume grass that another would avoid or reject, leading to better pasture utilisation and more productive pastures. The animals can graze together or in sequence if preferred.

Another benefit of mixed grazing is the fact that, due to different animal species acting as hosts to different internal parasites, this can also serve as an approach to parasite control (Section 21.5).

21.4.5 Zero grazing

"Cutting and carting" green forage to animals which are housed or stocked on a loafing area of grassland involves substantial extra costs as well as wear and tear on machinery. On the majority of UK farms it would primarily be regarded as a technique for green feeding of forage crops such as lucerne, kale or forage maize where either the nature of the crop or conditions underfoot precluded the use of strip or block grazing or where there was a need to cut green forage from outlying fields in times of shortage of grazing. Intensive dairy units sometimes use direct cut grass in the Total Mixed Ration (TMR), instead of allowing cows to walk to fields to graze. The utilisation of grass is at its highest when zero grazed, as there is no wastage due to faecal contamination or any other factors that lead to rejection.

21.4.6 Storage feeding

Some livestock production systems entail very limited turnout of animals to graze with the bulk of the grass and forage grown being preserved as silage. This approach is normally associated primarily, if not solely, with high yielding dairy cows fed forages as part of a TMR and allowing for the feeding of a consistently nutritious ration and the avoidance of either changes in the nutritive value of grass or the impact of adverse weather on the time spent grazing and the milk yields produced. Utilisation of grass and forage can be exceptionally high and there is no need for the daily harvesting necessary when zero grazing but this approach is labour intensive and costly in terms of machinery and also buildings to handle and store conserved forage and manure.

21.5 Strategies to minimise parasitism at grass

The provision of "clean" (i.e., relatively parasite free) grazing for young animals is a requirement of good health and welfare management. There is evidence of resistance in ruminant and equine gastrointestinal (GIT) nematodes to most of the common anthelmintic (wormer chemical) groups. This has developed from the continual use of chemical control of infestation of youngstock exposed to high challenge before developing a degree of immunity. The SCOPS group was formed to develop sustainable strategies for parasite control in sheep as well as to facilitate and oversee the delivery of their recommendations to the industry. The challenge is to manage both the parasites and their resistance to anthelmintics and grazing management is a potentially key component.

"Leader/follower" and "forward creep" systems have already been mentioned and a reduction in parasitism can be brought about in this way by reducing the extent to which the young cattle or lambs, as leaders, are forced to graze heavily contaminated material at the base of the sward. Pastures previously or concurrently grazed with older animals of the same species present the highest risk.

The fact that parasites are usually highly host-specific provides an additional opportunity for their control. One animal species is largely immune to another species' parasites and two contrasting grazing animal species can play a key role in reducing each other's parasite burden on grassland. Where the grazing pressure of both species is roughly equivalent, mixed species grazing (mixed grazing) where the two species graze together or in sequence, can greatly reduce the parasite problem.

The variable and highly unpredictable life cycles of many parasites, however, mean that a successful approach to their control needs to take a strategic season-long or multi-season approach with complementary use of anthelmintics where effective and allowed. The choice of anthelmintic and timing of its administration is often integral to successful control. The basic premise is that the longer the break from grazing by a particular animal species then the lower the parasite population on that pasture will be but the duration of this break has to be a minimum of several months or even one or more years to be effective and for the pasture to be considered "clean." Strategies for minimising this problem include alternating sheep and cattle on different fields for grazing, year on year.

The regular introduction of newly sown leys into the system provides clean grazing, and moving young cattle or weaned lambs onto silage or hay regrowth or aftermaths, usually after worming, in mid-season is also extremely effective. The "1, 2, 3" or "1/3: 2/3" system for young cattle, illustrated in Fig. 21.3, incorporates this latter principle. It must be remembered that exposure to infective nematode larvae is necessary for the development of immunity when adult, therefore a low level of challenge should be allowed.

There is evidence that the inclusion in a sward of herbaceous plants high in tannins, such as legumes, chicory and plantains gives some anthelminthic activity, and improves the growth rates and health of grazing young cattle and lambs.

Recent wet seasons have also increased the incidence of liver fluke for which mud snails act as an intermediate host. Improving the drainage of liver fluke-infected pastures can reduce the risk, although consideration needs to be given to the cost and to the environment.

21.6 Measuring and setting targets for grassland production and output from grassland

Assessment of the real value of the output from grass and forage on livestock farms is difficult for practical reasons. The measurement of the yield of a crop of wheat or potatoes is easy as there is a quantifiable, physical and marketable yield. With grass and forage crops, although the yield of dry matter is obviously present and is increasingly being measured directly (historically largely through counting loads of grass or numbers of bales but currently by weekly measurement of "cover" in kg DM/ha), it has to be utilised by farm animals before any financial or marketable output is created. Where a high proportion of the animal production is obtained from grazing these estimates are particularly difficult.

Most approaches to target setting involve little more than the quantity of milk or animal liveweight gain produced per hectare. Some producers have recorded milk yields in excess of 20,000 kg/ha and cattle liveweight gains of 1000 kg/ha and, in theory, this seems a logical approach. For livestock systems that are largely, if not entirely, reliant on grass e.g., a spring calving dairy herd on an extended grazing system, this approach is frequently adopted and appears to work fairly well as a means of both setting targets and comparing the performance of different units. On the other hand, unless the animals are fed exclusively on grass, allowance needs to be made for concentrates and forages bought or brought in. Another possible anomaly is that the level of production is also often inflated if it is based only on the grazing area, or "grazing platform", with additional or off-lying land used for silage production being overlooked.

Variability of inputs, including fertiliser, reseeding, etc., and their associated costs, are a potentially further complication and an alternative school of thought is to base any target setting, monitoring and benchmarking on financial rather than physical criteria with a measure of profitability as the ultimate aim. Changes to both input costs and commodity prices can have a significant impact on financial performance and, possibly, lead to frequent short-term changes in focus.

Utilised Metabolisable Energy (UME), measured in GJ/ha, was one of the earliest attempts to devise a measuring and monitoring system on dairy farms and appeared to offer promise. This approach took into consideration the dietary energy requirements of the milk produced on the farm, made allowances for the energy supplied by all purchased or imported feeds and forages and attributed the remainder to the energy supplied by home produced grass and forage.

21.7 Future developments

It is highly likely that the reliance on grazing will continue to increase within livestock production systems but that the aims of and drivers for the approaches taken might well differ.

Where agricultural production is key, there appears to be further scope for fine tuning grassland measurement techniques and the decision making process. Remote data gathering and real time processing are likely to become increasingly important across all livestock sectors, rather than being largely confined to milk production as has been the case up until now.

The adoption of precision technology has also been limited to date when compared to other sectors of the industry, possibly due to technical issues such as the heterogeneity of pasture, but there appears to be considerable potential for the uptake of further technologies such as the use of a grass algorithm to adjust fertiliser application in real time.

Knowledge transfer will also play a significant role with AHDB already making significant headway and providing a great deal of technical information and support. A grass monitoring project provides up to date grass growth and quality data from across the UK already and Grazing Mentors can help farmers to unlock the potential of their pasture with one-to-one guidance on a grazing strategy specifically tailored to their farm.

New approaches to grazing may well be an integral part of this approach. Novel techniques such as Cell Grazing, where areas of grassland are subdivided into "lanes" by bespoke temporary electric fencing and further subdivided for grazing as cells, has been successfully adopted by beef cattle producers with claims of excellent grass production, utilisation and livestock performance, could well become much more commonly adopted. Grazing of large blocks of grassland, in either a conservation or production setting, can also be controlled by virtual fencing where the area to be grazed is defined and audio signals are sent wirelessly to the grazing animals. Monitoring of grazing animals by means of the technology is another potential advantage.

Climate change and the uncertainty of grass growth associated with dry spells as well as the challenge of grazing in wet and stormy conditions could also force producers to consider making changes to their current grazing systems. There could be a need to adopt a more flexible or versatile approach to deferred grazing i.e., creating a reserve of grass to graze during drought or conserving grass in order to ensure the availability of reserves of silage and hay.

It is likely that grazing will also increasingly be used to manage land primarily viewed as conservation grassland to create and maintain wildlife habitats with the likelihood of prescriptions for stocking rates, grazing season and severity, grazing methods and even choice of breed. Such approaches have been prescriptive to date but plans are afoot to attempt to base management, and any changes to management, on outcomes i.e., the habitats or biodiversity created or maintained.

A regenerative approach to grassland management is another potentially significant development with the adoption of a slightly less conventional grazing management technique. This is based on a rotational system with short duration, high density grazing and longer grass recovery periods than usual. The technique is frequently referred to as "mob grazing", but "holistic", "tall grass" or "regenerative grazing" are often used as alternative terms for the same general approach. The general principle is to graze much larger grass covers or masses than would normally be employed. More importantly, much more vegetation would normally be left behind as an ungrazed residual or residue (conceivably up to a two-thirds of the original cover). This has been shown to result in increased soil carbon levels and appears to be a particularly valuable approach where soil carbon or organic matter levels are low or have become depleted. The outcome is often much lower production costs because of a longer grazing season with increased biodiversity as well as improved soil biology and health being reported. This could result in the return of more grass-based livestock systems on arable farms with the main aim being to build up soil organic matter and carbon and improve soil structure and health.

21.8 Key points

- The use of grass and other forages for grazing often proves to be the most economical means of production for agricultural livestock.
- Grazing has the potential to provide sufficient energy and protein for a daily milk yield of upwards of 30 litres from dairy cows and for individual daily live weight gains of 1.5 kg in beef cattle and 300 g in lambs.
- Grazing management involves balancing the feed requirements of stock against the potential availability and nutritive value of grass or other forages.
- Good grass utilisation is the key to successful grazing management, reducing wastage by grazing fields at the right time with grass at the correct stage of growth.
- Grazing intensity, as measured by stocking density, needs to be adjusted over the grazing season in order to balance supply and demand
- Regular estimates of sward height can provide a measure of grass productivity and nutritive value as well as grazing animal performance. The use of devices such as a rising plate metre enable direct estimation of pasture mass or yield with data being downloaded into software programs which are increasingly being used as management decision support tools.
- The choice of grazing system is between rotational grazing, with animals being allocated new grazing areas in sequence and the previously grazed grass allowed to recover and regrow, or continuous grazing where livestock (not necessarily the same number) are not routinely moved.
- Although more expensive to establish and involving more decision making and labour, rotational grazing offers more control over grazing and is essential for some forages.
- Grazing strategies can be devised to play a key role in the prevention and control of internal parasites in livestock.
- Assessing the output from grass and forage on livestock farms is difficult since they have to be utilised by farm animals before any financial or marketable output is created. Measures of milk or liveweight output per hectare offer scope for target setting and benchmarking.
- It is likely that the reliance on grazing will continue to increase within livestock production systems but the approaches taken will differ depending on whether livestock production, habitat management, climate change or the need to regenerate soils are the drivers.

Sources of further information and advice

AHDB Beef and Lamb, 2010. Managing clover for better returns beef and sheep. BRP Manual 4.

AHDB Beef and Lamb, 2016. Planning grazing strategies for better returns beef and sheep. BRP Manual 8.

AHDB Beef and Lamb, 2017. TechnoGrazing - Investigating the Potential.

AHDB Beef and Lamb, 2022. Better Returns Programme. http://www.ahdb.org.uk.

AHDB Grass, 2021. AHDB Grass Campaign. http://www.ahdb.org.uk/knowledge-library/ahdb-grass.

British Grassland Society, 2013. In: Proceedings of the BGS 11th Research Conference – Science and Practice for Grass-Based Systems, September 2013.

Crofts, A., Jefferson, R.G. (Eds.), 2007. The *Lowland Grassland Management Handbook*, second ed. English Nature/The Wildlife Trusts.

Dairy Development Centre, 2014. Grass Value. Improving the Dairy Supply Chain.
Defra and Natural England, 2020. Grazing and Pasture: Sustainable Management Schemes. https://www.gov.uk/grazing–and-pasture–sustainable-management–schemes.
Frame, J., Laidlaw, A.S., 2013. Improved Grassland Management, Crowood.
Hopkins, A. (Ed.), 2000. Grass: Its Production and Utilization, third ed. Blackwell Science.
Hybu Cig Cymru, 2014. Grassland Management. www.hccmpw.org.uk.
Ledgard, S., Schils, R., Eriksen, J., Luo, J., 2009. Environmental impacts of grazed clover/grass pastures. Irish Journal of Agricultural & Food Research 48, 209−226.
Newton, J., 1999. Organic Grassland and Sheep Husbandry: 15 Questions Answered. Newton Publishing.
Nix, J., 2021. Farm Management Pocketbook 2022, 52nd Edition. Agro Business Consultants Ltd (annual).
O'Beirne-Ranelagh, 2005. Managing Grass for Horses (J.A. Allen).
Pearson, C.J., Ison, R.L., 1997. Agronomy of Grassland Systems, second ed. Cambridge University Press.
Stubbings, L., Bartley, D., Busin, V., Lovatt, F., Page, P., Rose Vineer, H., Skuce, P., 2020. SCOPS Technical Manual. https://doi.org/10.17605/OSF.IO/SQA4E.
Teagasc, 2010. Grassland for Horses. Agriculture Food Development Authority.
Teagasc and Irish Farmers Journal, 2016. Grazing Guide 2. Grazing Guide for dairy, beef and sheep. Irish Farmers Journal.
Thomas, C., Reeve, A., Fisher, G.E.J., 1991. Milk from Grass. The British Grassland Society.
Younie, D., 2012. Grassland Management for Organic Farmers, Crowood.

Websites

www.ahdb.org.uk.
www.ahdb.org.uk/knowledge-library/forage-for-knowledge.
www.agrinet.ie.
www.britishgrassland.com.
www.europeangrassland.org.
www.gov.uk.
www.grasscheckgb.co.uk.
www.grasslanddev.co.uk.
www.hccmpw.org.uk.
www.naturalengland.org.uk.
www.scops.org.uk.
www.soilassociation.org.
www.teagasc.ie/crops/grassland/grassland-management/.

Conservation of grass and forage crops

22

22.1 Introduction

22.1.1 Reasons for the conservation of forage and forage conservation techniques

Grasses, as well as a range of other forages, are usually conserved during periods of peak growth in late spring, summer and early autumn in order to create a stored product that can be fed during periods of reduced grass growth. Traditionally, this has almost exclusively been during the cold and wet late autumn, winter and early spring months but, increasingly, conserved forage is also being used to supplement grazed grass during prolonged cold, wet or, particularly, dry spells at other times of the year. Alternatively, some of the more intensive dairy and beef production systems offer housed or grazed livestock conserved forage on a year-round basis, often as part of a total mixed ration formulated for optimal liveweight gain or milk yields.

A key additional role of conserving grass is as a flexible management tool to manipulate the availability of grass for grazing with any surplus grass being conserved when it threatens to 'get out of control'.

It must be borne in mind that conserving forage can be costly, particularly when the costs of feeding and the possible need for handling livestock manure are added to the costs of harvesting and storage. On the other hand, the value of having a quantity of conserved forage in reserve to accommodate for the variability of grass growth e.g., during a late spring, summer drought or early autumn and the flexibility this offers, cannot be overestimated.

22.1.2 Forage conservation techniques

Two main principles underpin the most commonly adopted conservation techniques:

Dehydration—the removal of sufficient moisture so that the forage can be stored aerobically e.g., hay, where the drying is normally carried out entirely in the field, or artificially dried grass, where the drying is carried out artificially at high temperatures.

Anaerobic fermentation—the anaerobic storage and acid fermentation of moist forage e.g., silage, made in horizontal (clamps) or vertical (tower) silos and bales, or 'haylage', high dry matter silage usually made in bales.

Lockhart and Wiseman's Crop Husbandry Including Grassland. https://doi.org/10.1016/B978-0-323-85702-4.00018-2

22.1.3 The quality of conserved forage

The quality of any conserved forage is a combination of two factors, the nutritive value i.e., energy, protein and mineral content largely determined by the type of crop harvested and its maturity, and the preservation quality—the effectiveness of the preservation and storage of the crop.

Both nutritive value and preservation quality will be clearly reflected in a laboratory analysis of the forage but the appearance and smell of a crop can also give a useful estimate of the latter in particular.

Although the nutritive value of a forage will indicate its potential feed value to the animal, the quality of preservation will usually have as large a bearing on its actual value in practice as it will influence its palatability i.e., the quantity that will be eaten. It will also indicate the likelihood of any possible health risks to humans and animals from the presence of undesirable organisms (e.g., botulism, listeria, fungi and moulds) or their metabolic products (mycotoxins).

22.1.4 Choice of forage conservation technique

The choice of forage conservation technique employed on farm is largely determined by practicalities such as location (especially rainfall), the scale of operation, the availability of machinery or contractors and the storage facilities available on farm as well as the feeding system employed and the needs of the animals.

The aims are to produce a conserved forage that is as similar in nutritional value to the original crop as possible, preserved with minimal loss of 'yield', manageable throughout the harvesting and storage as well as the feeding stages, cost effective and which poses little or no threat to the environment.

Silage is either harvested by a forage harvester (trailed or self-propelled) or forage wagon and stored in a horizontal silo, bunker or clamp or, occasionally, a vertical or upright tower silo. Alternatively, silage may be baled as large cylindrical or rectangular bales normally weighing 500–1000 kg. The harvesting process usually incorporates a period of field wilting before the forage is placed in store.

Hay, on the other hand, is normally field dried for a minimum of 3 or 4 days before being baled and stored under cover.

22.1.5 Choice of forage conservation strategy

The aim for any farm or holding is to achieve the best balance between forage quality and yield. This is primarily determined by the timing and date of harvesting, particularly of the first cut of the season.

A young leafy crop will have a high nutritive value but will produce a relatively low yield. If the crop is allowed to grow and develop it will become more stemmy and its yield will be considerably higher but its nutritive value will fall (Fig. 22.1).

Typically, during the course of a week during May or June, a productive sward could produce an additional 700 kg DM/ha or more in good growing conditions but

Figure 22.1 Changes in the nutritive value and yield of perennial ryegrass (AHDB).

its D value (digestibility) would fall by approximately 1.5 units during this time. The decision about when to harvest will need to be based on the most appropriate compromise between yield and quality for the farming system and farm concerned while still responding to prevailing weather and ground conditions.

It is possible to take more than one harvest from most grasses and legumes over the course of the season and many farms take anywhere between two and five cuts in total between late April and late September depending on locality and the farming system practiced. Although frequent harvesting will reduce the yield at each harvest, and can reduce total annual crop yield, taking more than one cut annually leads to better 'quality' forage with younger material harvested. There will also be quicker recovery or regrowth of the crop after harvest as the residue or stubble contains more green leaf as well as better quality pastures because plant density is better maintained due to the shorter interval between harvests. An added advantage is a greater flexibility of crop management with the timing of harvests and the area of land harvested being adjusted as the growing season progresses in response to prevailing crop and weather conditions.

22.1.6 Reasons for the dominance of silage

Hay was the traditional method for conserving grass in Western Europe. Hay, 'cured' in the field, needs a prolonged drying period for the reduction of its moisture content to a safe level for storage and feeding, which is not something that can be relied upon in the climate of the UK in particular. The quantity of hay conserved virtually halved during the latter part of the twentieth century, while the popularity of silage, both clamped and baled, and haylage rose. The most recent figures available indicate that at least 2.5 million hectares of land are used for silage production, yielding more than 15 million tonnes of conserved forage annually, making it by far and away the dominant form of forage conservation in the UK. This is in keeping with much of the remainder of the world.

The changes in British farming systems, increased herd and flock sizes and the reliance on mechanised feeding systems rather than labour-intensive handling of bulk feeds, has also driven the change to fermented, rather than dehydrated, winter forage. Grass and legume varieties have been bred that produce very high yielding, highly digestible crops, making silage making almost the inevitable choice. Other influences include higher stocking rates and the need for a consistently highly nutritious forage which, coupled with improvements in ensiling techniques and reliable systems for mechanically handling and feeding silage, have allowed increased output from forage-based systems. Field machinery has also improved, and the advent of the widespread use of contractors for silage making has meant that farmers now have access to sophisticated and efficient machinery for making silage. Speed of operation has become very important since declines in the D value of grass (see Fig. 22.1) and the need for timeliness of harvesting are well appreciated and a self-propelled forage harvester can achieve an output of as much as 400 t/h. It is therefore possible for a farmer, typically, to expect an efficient team of contractors to make an entire first cut of silage within 1 or 2 days, a job which would at one time have taken weeks.

22.1.7 Environmental threat of silage effluent

Unfortunately, there is also a high potential for losses if silage making is not carried out well and the greatest potential for losses, with the resultant threat of environmental pollution, comes from silage effluent. It is possible to produce in excess of 200 litres of effluent per tonne of silage from very wet direct cut or lightly wilted material.

Silage effluent has a very high biochemical oxygen demand (BOD) which makes it over 200 times as polluting as raw domestic sewage. Pollution of groundwater or watercourses by silage effluent has resulted in many prosecutions and some very heavy fines. The relevant legislation in England is laid out in the Water Resources (Control of Pollution) (Silage, Slurry and Agricultural Fuel Oil) (England) Regulations 2010 and as amended 2013 (SSAFO). It should be borne in mind that Scotland, Wales and Northern Ireland also have their own legislation.

All silage stores must be constructed to an approved standard with nonporous floors and side walls of an appropriate load bearing capacity, perimeter drains to intercept all possible seepage and a sealed effluent collection and storage pit. This needs to be emptied regularly, especially during the period immediately after ensiling when effluent production is greatest. Silage should not be made or stored within 10 m of inland or coastal waters. Care must also be taken to avoid pollution and to adhere to NVZ and any other relevant regulations when spreading effluent on land. If it is to be applied to grassland or a growing crop, the recommendation is to dilute effluent at a 1:1 ratio to avoid leaf scorch.

Silage effluent can be stored and fed to livestock as a nutritious liquid feedstuff, 10 litres of undiluted effluent having the nutritional value of roughly 0.55 kg of barley, but the short period over which it is produced as well as the storage implications limit its attraction.

In practice, bearing in mind the threat effluent poses to the environment, its production should be minimised. One approach has been to add an absorbent material (e.g., rolled cereals, confectionary by-products or dried sugar beet pulp) as an additive at ensiling. However, the simplest expedient is to wilt silage crops in the field to 25%–30% dry matter, in which case the production of effluent is minimised. In the case of forage maize and other cereal crops, harvesting of the standing crop is delayed until the dry matter reaches in excess of 30%.

22.2 The silage-making process

22.2.1 Silage fermentation

Most silage is made from grass. Therefore, most of the information in this section concerns the making of grass silage.

Silage is the product formed when a crop (grass or other materials e.g., legumes, cereals, brassicas) containing sufficient moisture to spoil as a result of the activity of aerobic organisms is stored anaerobically. The forage normally undergoes an acid fermentation in which bacteria produce lactic, acetic and butyric acids from soluble carbohydrates or sugars present in the material itself. The fermentation is conducted by bacteria and enzymes occurring naturally on the crop itself and the net result is a reduction in pH (increased acidity).

The combination of acidity (which prevents the growth of the vast majority of anaerobic spoilage organisms) and a lack of oxygen (which prevents the growth of aerobic organisms) ensures preservation of the crop. The efficiency of the ensilage process is judged according to the relative proportions of the fermentation acids produced:

- Good—if lactic acid is predominant
- Poor—if butyric acid is predominant (a weaker acid which does not prevent the growth of spoilage organisms as effectively and which is also associated with proteolysis—protein breakdown to ammonia)

The two main aspects of successful fermentation are the rapid establishment of anaerobic conditions and a rapid fall in pH to inhibit anaerobic spoilage organisms.

On the growing crop, aerobic organisms dominate (lactic acid producing bacteria are found in small numbers on dead or dying parts). Once cut, aerobic organisms continue to dominate and their activity, together with that of plant enzymes, results in a loss of soluble plant contents (initially negligible but much more rapid after 24–48 h if the crop remains wet). When being placed in store, and until an airtight seal is created, respiration and enzyme activity continue and can be accompanied by an increase in temperature within the material (which should be to avoided) which continues until all the oxygen that is trapped within the store is used up. Any heat detected indicates that air has not been excluded and soluble nutrients in the crop are being lost by respiration. This is why it is important to minimise the presence of oxygen by packing or consolidating the material thoroughly and sealing it effectively and quickly.

Once any trapped oxygen is consumed, fermentation commences. The pH of the crop ($\approx$6.0) should fall rapidly as lactic acid bacteria such as *lactobacilli* convert sugars to lactic acid. Most acid is produced within the first few days and fermentation is virtually complete within the first 7 days, a stable pH of 4.0–4.2 being achieved in typical grass silage with a dry matter of 25%–30%. Such a silage will have a light brown colour as well as a distinctive sharp smell and plant components will retain their structure on being rubbed.

The greatest threat is posed by *clostridia*, hydrophilic (water loving) bacteria usually associated with wet crops and conditions as well as soil and manure contamination of the crop or storage area. *Clostridia* convert sugars into the weaker (less acidic) butyric acid. The conversion process itself is less efficient than that involved in the formation of lactic acid and leads to greater losses of both nutritive value and yield. Butyric acid is not such an effective crop preservative, meaning the crop is also less well protected against other potentially damaging organisms. The process also gives rise to proteolysis (protein damage and loss when amino acids and proteins in the forage are converted into ammonia).

Clostridia can also convert any lactic acid already formed to butyric acid, this process is known as secondary fermentation (a term that is often used incorrectly to describe the changes and losses seen in silage at feeding) and will be accompanied by a rise in pH. The pH of butyric silage is often in the region of 5, and it will have a somewhat slimy and easily fragmented texture, usually accompanied by a darker colour and a foul smell. The cause of the pungent, unpleasant smell will often be the ammonia formed as a result of proteolysis.

A clear relationship exists between the dry matter content and pH of well fermented silage, the lower the dry matter content i.e., the wetter the silage, the lower is the pH required to ensure that the crop has been well preserved (Table 22.1).

As long as the silage store is airtight and anaerobic conditions remain, the silage remains stable. But, when a silage store is opened, oxygen has free access to the surface of the material and promotes the development of aerobic spoilage microorganisms.

Table 22.1 Fermentation quality in relation to silage pH and dry matter content.

Silage dry matter content (%)	Target pH	pH reflecting very poor fermentation
15	3.7	4.7+
20	4.0	5.0+
25	4.2	5.3+
30	4.4	5.5+
35	4.7	5.5+
40	5.0	5.5+
45+	5.5	5.5+

22.3 Crops for silage making

22.3.1 Grasses

All types of grass can be ensiled. Nutritive value will be very dependent on the stage of growth and digestibility (D value) or energy content of the crops (Fig. 22.1). The fermentation quality, or ease with which it can be fermented, will depend on a great many other factors, but the dry matter content and the amount of water soluble carbohydrates (sugars) contained are two of the most important. It must be remembered that immature crops, although of high nutritive value, are generally low in soluble carbohydrates and this, allied to their higher moisture content, makes such crops more difficult to ensile and greater thought needs to be given to field wilting and/or additives. Mature crops, despite being 'nutrient poor', are generally more easily fermented.

Fermented forage for equine use tends to be predominantly grass based.

22.3.2 Legumes

Most legumes are more difficult than grass to make into well fermented silage. This is because, despite their high nutritive value, they usually contain relatively low levels of sugar as well as a lower dry matter content. Legumes are often grown in mixtures with grasses or cereals to improve the yield produced but this also improves the fermentation characteristics of the crop since grasses and cereals contain much higher levels of water-soluble carbohydrates.

Although red and white clovers are usually made into silage in mixtures with various grasses, occasionally red clover may be grown and harvested as a pure stand. Lucerne is another legume often grown on farms in the drier parts of the country, most frequently as a pure stand but also mixed with red clover or with grass. D values of red clover and lucerne when cut for silage are often slightly lower than for grasses such as ryegrass, but their protein content is much higher (up to 20%–22% crude protein at early flowering). To make well fermented silage, harvested crops containing substantial proportions of either lucerne or clover should be field wilted to at least 30% dry matter and an effective silage additive should probably be used.

Annual legumes such as peas (forage or grain varieties are both suitable) and vetches can also be made into high protein silage. These crops are also often mixed with spring barley or oats as 'arable silage mixtures' in order to improve the yield as well as fermentation characteristics of the crop. Such mixtures can be harvested at any stage from when the cereal grain is at the 'milky' up to the 'cheesy' stage and when the peas or vetches have pods largely filled. Forage varieties of lupin have also been used as an arable silage crop, and trials have also been undertaken with forage varieties of soya bean. Both crops have potential in their own right or could be grown as 'bicrops' with cereals including forage maize.

22.3.3 Forage maize

Maize as a silage crop continues to increase substantially in popularity and detailed information about its agronomy is given in Section 18.1.1. According to the region where it is being grown, a variety of maize should be selected on the basis of its maturity group, which determines its ability to yield ripe cobs which contain starch. At harvest the cobs should be about 50% of the total yield and starch should comprise 25%–30% of the dry matter. This usually occurs when the dry matter percentage of the whole crop reaches 30%–35%, typically when 65% or more of the grain on the cob will be yellow. If it is just possible to make a dent with a finger or thumbnail in the grains on a number of cobs then the crop is judged ready to harvest.

Fine (1.5–2 cm) chopping and grain cracking are essential for making good maize silage, with recent work suggesting that the longer chop length is better for digestive health and for milk solids. The crop must be well consolidated, which can be difficult with a longer chop length, in order to exclude air to avoid overheating and the growth of moulds, particularly when the silage store is opened for feeding. Care must be taken at this time to keep a uniform clamp face and to work quickly across the clamp. It is recommended to have narrow clamps for maize silage in particular (or to divide wider ones with temporary barriers) to avoid aerobic deterioration and mould growth at the face. Since there is so much starch and some sugar still in maize when it is harvested, additives should not normally be needed to ensure a satisfactory fermentation but appropriate additives should be considered for application at harvest to reduce losses due to aerobic deterioration at feeding.

'Ground ear maize' (GEM—whole cob and sheath—45% DM) and 'crimped maize' (CM—kernels only—25 to 35% DM) are the products of cob-only harvesting options using a maize snapper header on a forage harvester (for GEM) or a combine harvester (for CM) with the crops stored in a sealed anaerobic environment to make very high energy silages. Crimped maize is occasionally fed to pigs as well as ruminants in other parts of Europe.

22.3.4 Other cereals for 'wholecrop' silage

Any type of cereal can be taken for silage, but the best yields and quality are usually obtained from winter wheat. Advantages are that it is an excellent source of cereal starch (although maize is generally favoured by livestock farmers) and it can substitute for forage maize in the north of England, Scotland and Northern Ireland where maize cob and starch yields may not be so good because of a less suitable growing climate. Any surplus can be left to ripen for combining at a later date and harvest takes place in July leaving many more postharvest options open for subsequent cropping on the fields in question.

There are two basic systems for making wholecrop cereal silage. One is to cut the crop when it is still moderately green at about 35%–40% dry matter in July for acid fermented wholecrop silage. The other is to leave the crop until it is nearly ripe for combining (higher than 50% dry matter) and to use feed-grade urea as an additive at the clamp. When the clamp is sealed (good sealing is again absolutely essential for this technique) the heat of the initial fermentation releases ammonia gas from the urea. This has the effect of preserving the grain and straw as a moist relatively nonfermented high pH product. Alternatively, a product known as 'alkalage' is made in the same way, but from a crop at nearly 80% dry matter which undergoes no fermentation, and requires the use of a forage harvester with special adaptations. Where clamp space is not available, these crops can be stored in large stretchable bags or tubes on suitable hard standing. Rodent control is essential for most conserved cereal crops.

Other systems exist for ensiling moist grain after crimping (partial grain cracking through fluted rollers) and with the use of preserving agents such as propionic acid. The use of by-products such as brewers' grains for ensiling has also become commonplace, either layered within grass clamps on its own or, occasionally, mixed with an absorbent such as dried sugar beet pulp (grainbeet).

22.3.5 Alternative crops

A number of alternative crops have been successfully ensiled and, although such crops frequently fall out of favour for various reasons, highlight the possibilities that different crops offer.

'Kaleage' was the name given to leafy kale varieties ensiled after mowing and wilting for 24–48 hours. Sowing at a high seed rate gives a thick crop which could be wilted on a fairly high stubble and the kale could then be harvested for storage in a clamp or, more frequently, for round baling. Other options investigated in the past have involved growing kale as a bicrop with spring cereals for making into clamp silage but such crops are seldom grown.

Fodder beet was also popular for ensiling in some European countries such as Denmark, but has been eclipsed by the rise in popularity of forage maize. In the UK some pioneering work was done at ADAS Rosemaund with a system involving varieties which grew with a large proportion of the root above the ground which could be lifted by its tops in reasonably dry field conditions in September to give a moderate yield of fairly clean beet. A special harvester then chopped the whole crop and ensiling took place between alternate layers of chopped straw (sometimes ammonia treated) and an absorbent such as dried sugar beet pulp. A more commonly adopted alternative is to harvest and clamp the roots in the normal way after harvesting with a harvester with a top saver and to make silage from the tops or simply to graze them in the field following wilting.

In both kale and beet crops, excessive soil contamination is a potentially enormous problem that will have a negative impact on the quality of fermentation. Currently these crops have almost completely disappeared from UK agricultural use as forages conserved in this way but these examples at least highlight some alternative possibilities.

22.4 Factors affecting silage fermentation

22.4.1 Type of crop

Crops with high levels of fermentable carbohydrates (sugars or starch) are usually the easiest from which to make well preserved silage. Ryegrasses, especially Italian and tetraploid ryegrasses with their high sugar content, as well as the large number of high sugar ryegrasses (HSGs) available, or maize and other cereals with high starch content, usually ferment well with no need for additive application to aid fermentation. Low sugar crops such as short leafy grass and legumes at most stages of growth are more difficult to ferment and need some extra treatment such as wilting, the use of an effective additive, or both.

22.4.2 Fertiliser and manure treatment

This can affect the sugar content of grasses in particular. It is important to stress that the nitrogen recommendations and timings set out in Table 20.4 should be adhered to. Late or excessive applications of fertiliser can lead to the ensiling of grass containing large quantities of nitrogen in an ammoniacal or nitrate form and the presence of excessive levels of ammonia in the final silage. Heavy nitrogen applications can also lead to the depression of grass sugar content. Where fertiliser and manure applications to silage crops are delayed because of poor weather or soil conditions, rates should be reduced appropriately or the harvest delayed.

Crops can be analysed to determine both ammoniacal or nitrate nitrogen as well as sugars prior to harvest and, provided a rapid turnaround time from taking a sample in the field to the return of the results can be ensured or an instantaneous field technique developed, this is potentially a decision-making tool that offers much promise and it is somewhat surprising that it appears to have been so little used in the UK to date.

Despite being valuable sources of potash, in particular, slurry and manure are also potentially sources of *Clostridia* which interfere with effective silage fermentation.

22.4.3 Weather prior to and during harvesting

Dry, sunny weather is to be preferred for silage making as this increases the dry matter content of the crop and makes conditions less suitable for *Clostridia*. Plant sugar levels are also likely to be highest in sunny weather. Grass mown in the afternoon of a sunny day has been shown to have maximum levels of sugars or 'water soluble carbohydrates' (WSCs). Wilting conditions will also be improved and the possibility of soil contamination is minimised.

22.4.4 Minimising soil and slurry contamination

The control of moles and the rolling of silage fields are important preparatory tasks in the spring in order to avoid contamination of the crop at harvest. On most farms they are standard practice. Slurry application, although common on silage fields, should be avoided if possible unless it can be injected into the soil or placed at ground level by means of a trailing shoe applicator. Where there is no alternative, it should be applied as early as possible relative to the proposed cutting date. The heavy contamination of silage grass with slurry can lead to poor fermentation and subsequent losses in feed value. Other fairly obvious ways in which contamination can be minimised include power washing the clamp prior to ensiling, tipping the crop on a clean concreted area and ensuring that the wheels of the vehicles driving on the material while spreading and consolidating the crop (usually with a buckrake) are as clean as possible. Levels of contamination can most easily be assessed by reference to the percentage of 'ash' in the silage analysis report. A figure of less than 10% is acceptable whereas 15% or more would indicate that significant levels of contamination have taken place.

Mowing height is also important as cutting too low will increase the risk of soil contamination. In dense swards the residual height could be as low as 5 cm, but 7.5–10 cm is advisable in more open crops, particularly bearing in mind the uneven nature of many fields and the risk of 'scalping' turf. Cutting lower than this can also mean that the recovery or regrowth of the crop will be slower.

22.4.5 *Field wilting*

In the last 30 years the average dry matter content of clamp silage in the UK has increased from 20% to 25% to close to 30%. Wilting in the field after cutting has now become a standard treatment when making grass or legume crops into silage.

A crop with a relatively high dry matter content contains a higher concentration of sugars, is generally easier to ferment as it is less likely to undergo undesirable butyric fermentation, will be less acidic (see Table 22.1) and usually results in better feed intake by livestock as well as being less bulky to store and transport. The biggest impact, however, is on the volume of harmful silage effluent produced and which needs to be collected, stored and disposed of (Table 22.2).

Simply leaving the mown swath in the field, however, can be counterproductive and lead to a degree of respiration and enzyme activity within the swath, which is undesirable. Cut swaths should be subsequently spread to full ground cover in order to obtain optimal wilting. High performance mower/conditioners now mow, condition and spread the crop in one operation, speeding up the rate of drying and minimising wilting time in the field as well as reducing the risk of soil contamination since all that is required subsequently is the rowing up of the crop prior to forage harvesting. Conditioning the crop involves physically damaging the waxy cuticle at the perimeter of leaves and stems with nylon brushes, tines or rollers and increases the rate of wilting substantially so that, in good drying conditions, it is easily possible to achieve dry

Table 22.2 Typical silage effluent production (litres/tonne of forage ensiled) at different crop dry matter contents.

Dry matter content (%)	Typical effluent production (l/tonne of forage)
15	Up to 200
20	50–75
25	10–30
30	0–10
35	0

matter contents of 30% or more from a 24 hour wilt. Most farms use a rotary tedder and may ted the crop prior to rowing up for harvesting with a precision chop harvester. This process will lead to an increase in dry matter content, but it can also lead to higher levels of field losses and increase the risk of contamination with soil and needs to be managed with care.

Wilting to higher dry matter levels of 30%–40% is required for ensiling in towers so that the equipment will work well. This normally requires a 36–48 hours wilt and some additional tedding. A similar, or even higher, DM content is usual for big bale silage especially for that known as 'haylage' and increasingly used as equine forage.

A dry matter content of at least 30% is highly desirable for ensiling all legumes and lucerne and red clover can benefit from the action of roller/crimpers which crack the thick stems of these plants and improve the wilting rate, although this must be weighed up against the risk of leaf damage and shatter as the majority of the protein is found in the leaves.

Maize and wholecrop cereals are 'self wilting' crops since harvesting does not normally take place until they have started the natural dehydration process associated with ripening. Crops like kale for ensiling should also be wilted for about 48 hours. The best technique is to mow the crop with a mower-conditioner and leave it on a fairly high stubble (about 10 cm) prior to baling direct from the swath.

22.4.6 Forage harvesting

On the majority of farms grass silage is now harvested using high output metred chop machines. Wilted grass is usually chopped to between 25 and 75 mm, whereas maize benefits from even shorter chopping to between 15 and 20 mm. When harvesting maize, the grains should be cracked (to break the hard seed coat) to aid fermentation and subsequent digestion by farm animals fed the silage. Finely chopped material usually consolidates better and it is possible to achieve anaerobic conditions in the clamp more quickly. Fine chopping also results in the more rapid release of sugars from the plant for fermentation.

22.4.7 Baled silage

Approximately 25% of the silage made on UK farms currently is harvested by big balers. This is a system of production which has substituted for hay on many of the smaller stock farms where a silo or clamp would be impractical or unviable. The avoidance of the large capital expense of a clamp silo has obvious attractions for many farmers and baled silage also provides an additional option if existing silage clamp capacity has been filled or if individual silages are to be stored separately for strategic feeding. Baled silage also offers flexibility for buffer feeding grazing livestock, particularly during the warmer summer months when aerobic deterioration at the face of an opened silage clamp would be substantial.

The management of the crop at harvest is very similar to that for clamp or silo storage but with a marginally greater emphasis on field wilting and the achievement of a higher dry matter content as effluent production needs to be avoided. Earlier balers were undeniably better equipped to handle longer, more fibrous crops but modern machines can handle all lengths and forms of grasses and legumes effectively. Crops are normally wilted up to about 40% dry matter prior to baling in high density round or square bales. As soon as the grass has been baled it should be wrapped by means of a bale wrapping machine, using polythene stretch film. Four overlapping layers of wrap is the minimum, with six being preferred to increase protection of the crop and to ensure anaerobic conditions are maintained in the bale.

The quicker wrapping takes place after baling, the more efficient will be the exclusion of air and the better the quality of the final product and many balers are now combined baler-wrappers that will wrap a bale as soon as it has been formed. It is extremely important to avoid puncturing the wrap during handling and stacking, although prompt 'patching' with suitable tape will minimise the risk of poor fermentation if any damage occurs. Stacks should be located on a hard standing, fenced so that stock cannot damage them, netted to avoid bird damage, and protected against rodents. It is easier to stack square profile bales, which are available in many lengths and widths, but round bales can often be easier to feed in self-feed ruminant systems.

22.4.8 Baled 'haylage'

Haylage is the name given to high dry matter content silage wilted to a minimum dry matter content of 40%+, often harvested specifically for horses but also increasingly being fed to a wide range of farm livestock. High dry matter silage or haylage relies less on fermentation and the lowering of crop pH than more typical clamp silage at 25%–30% dry matter, due to the reduced threat of *Clostridia*. The pH of well preserved haylage will often be higher than 5.0 and it will have a far less pronounced distinctive silage smell. In such high dry matter crops, excluding air and maintaining anaerobic conditions becomes even more important as high dry matter material will deteriorate far more rapidly on exposure to air.

Crops of grass for very high dry matter baled 'haylage' for horses are normally made from reasonably mature grass (D value 60–65) which is tedded and wilted for up to 3 days to achieve a dry matter content of 60% or above. For competition horses, tetraploid Italian or hybrid ryegrasses ensure a high sugar content and a satisfactory preservation to provide a very nutritious forage that will reduce the need for high concentrate feeding, improve equine gut health and cut costs. On many equestrian enterprises haylage is also preferred to hay as a result of its dust free characteristics, thus promoting good respiratory health. Wrapping with at least six layers of high quality stretch film is essential, as horses are particularly susceptible to organisms such as botulism, which can be present in soil-contaminated and poorly sealed bales. Commercial firms produce several other types of haylage, including high fibre (late cut ryegrass) or species specific haylage (such as timothy) for a lower sugar content and overall nutritional value suited to leisure horses and native ponies and even lucerne haylage for horses in hard work.

22.4.9 Filling a horizontal (clamp) silo

In the first instance, it should be ensured that the clamp is as airtight as possible and side sheets should be used unless the side walls are sealed. Filling a clamp silo should then take place as rapidly as possible. Buckraking and subsequent consolidation should aim at creating a wedge-shaped clamp in the initial stages. Rolling and consolidation should be a continuous process during clamp filling and particular attention should be paid to consolidating the edges ('shoulders') of the clamp, a process made somewhat easier by having a clamp with sloping side walls. If possible, the 'shoulders' should be built a little higher than the centre of the clamp during filling. A final layer of heavier, unwilted or less heavily wilted grass will help complete consolidation. If the process of filling the clamp is likely to take several days, or is delayed by breakdowns or unfavourable weather conditions, it is advisable to sheet the clamp each night as this will help to minimise the intake of air into the clamp.

22.4.10 Final sheeting and sealing

Sealing should take place as soon as filling and consolidation have finished. The majority of clamps will be sealed with two thicknesses of plastic sheet and the 'shoulders' of the clamp may well be covered by a third layer, from the overlapped side sheets. Weight should be placed evenly on top of the sheet over the entire clamp. Historically, this was done with used car tyres or, in roofed stores, straw bales placed tightly together.

The recent development of cling film and oxygen barrier films have revolutionised the sealing of the top of clamps. The usual black plastic silage sheets are permeable to air, as are the original cling films, but oxygen barrier films used with close-weave netting and sand-bag type weights, or even straps ratcheted into place, are significantly better at maintaining anaerobic fermentation conditions. These techniques can reduce DM losses by 10%–20% due to uniform air exclusion across the clamp top and sides, with less risk of contamination from mould growth and the associated mycotoxins which can seriously affect animal and human health.

Ideally, clamps should not be opened after ensiling until feeding starts.

22.4.11 The use of silage additives

Silage additives are natural or manufactured products that can be added to or applied onto the crop at harvesting in liquid or solid form to improve the efficiency of silage preservation at different stages from fermentation to feeding. They are most frequently applied as liquids via applicators on the forage harvester or baler, although granule or powder applicators have also been developed.

Silage additives take many chemical and physical forms and can either stimulate fermentation, encouraging lactic fermentation by either providing or releasing carbohydrate or adding lactic acid bacteria, or inhibit fermentation, discouraging or bypassing fermentation by direct acidification or by partial or complete sterilisation of the crop. Some of the earliest additives used included strong acids such as hydrochloric and sulphuric acids or the carbohydrate-rich molasses. In spite of quite different modes of action, both of these approaches gave rise to improved fermentation, either by direct acidification or by increasing the supply of sugar and leading to an increased rate of lactic acid production in the store. Organic acids such as formic acid were subsequently extensively used with the acid having both an acidifying and an antimicrobial effect with formalin often being added for its inhibitory or sterilising effect. However, the extremely unpleasant and corrosive nature of acids or acid/formalin mixtures (particularly unwelcome on expensive contractor-operated machinery), as well as significant health and safety concerns, have led to a wide range of alternatives being investigated and developed.

Molasses is still in use and can be extremely effective in the ensiling of low sugar crops but the inconvenience of applying large quantities of liquid molasses makes its viability questionable.

The most widely used additives today are inoculant additives which work by the application of very large numbers of live, lactic acid producing bacteria (at least 100,000, but often in excess of one million, per gramme of grass), at the time of ensiling. They are based on *homofermentative lactic acid bacteria* such as *Lactobacillus plantarum*, converting sugar to lactic acid and giving rise to a rapid fall in pH and an efficient fermentation, preserving both energy and protein levels in the crop. These additives occasionally include enzymes to release extra sugar for the bacteria to work on or chemical preservatives such as potassium sorbate that have been shown to improve the aerobic stability of the silage at feeding in particular. Recent attention has also been paid to some of the seemingly less efficient *heterofermentative* lactic acid bacteria, such as *Lactobacillus buchneri*, that produce both lactic and other acids such as propionic and acetic, resulting in a slower and less efficient fermentation but inhibiting yeasts and moulds, especially during the feed-out phase. Inoculants have achieved a high degree of popularity over the years because of their greatly improved efficacy as well as their total safety. Organic standards limit the use of additives to the inoculant type.

There is an array of commercially available additives of different types, but all should have met the EU 1831/2003 Feed Additive Regulations. They can cost between 50 pence and £5 per tonne of grass and therefore become quite significant items of expenditure on livestock farms. Although extensive trials with beef and dairy cattle have shown that both silage intake and daily liveweight gain or milk output were often significantly improved with additive treated silage, deciding whether to apply an additive and, if so, which one is very difficult. Unfortunately, there is no good way to predict the effectiveness of microbial inoculants or other forms of additives and there are products that have little or no independent research to support claims of improved fermentation or animal performance.

It can be concluded that additives are not necessarily essential for good silage fermentation but that they can assist by reducing dry matter and feed value losses e.g., by better fermentation, reduced effluent production or increased aerobic stability at feeding leading to improved animal performance (the true indicator of success) e.g., greater intake or more efficient utilisation. Additives were traditionally used as a ‘tool’ when poor weather or other factors, such as low crop sugar were likely to result in a poorly fermented product. However, it is now recognised that, even when harvesting and crop conditions appear to be ideal, an appropriate additive can further improve silage quality as well as reduce the losses and wastage often associated with silage making and feeding.

22.4.12 Respiration and aerobic spoilage at feeding

Despite successfully excluding air throughout the storage period, 'aerobic deterioration' or spoilage can occur when the silage is being fed. This is particularly true if the weather is warm, if the rate at which a silage is being used is fairly slow or if the surface area of the clamp or silo face exposed is large. Aerobic bacteria, yeasts and moulds can affect the silage and, if the face is left exposed for a long period of time, there can be significant deterioration. Bacteria of the *Listeria* family can be particularly damaging in the case of baled silage where the wrapping has been punctured and the contents exposed and, in some cases, this can lead to a serious brain infection (listeriosis) in stock being fed on the silage. Maize silage is particularly susceptible to aerobic deterioration at feed out because of its high carbohydrate and dry matter content contents which provide a readily available food source and ideal conditions for spoilage organisms. Really good consolidation at ensiling coupled with ensiling in a long narrow clamp are ways in which such deterioration can be minimised. An appropriate silage additive can also greatly improve the stability of silage at feed-out. The use of shear grabs or mechanical block cutters which leave a clean, uniform surface when feeding all types of silage also minimises the extent to which the face of the clamp is disturbed.

Before any conserved forage is fed, a sufficiently representative sample should be analysed for feed value.

22.5 Hay

With the intensification of grassland management has come a marked reduction in the popularity of hay. On many smaller or more traditional farms, more predictable i.e., less weather dependent big bale silage systems have replaced hay. For many small equestrian enterprises, however, hay remains the most manageable winter forage in terms of storage, handling and feeding, and there is a ready market for well-made small bale hay in the UK and elsewhere.

Hay cropping can be composed of a species monoculture, usually ryegrass 'seeds hay' in the UK (often lucerne in N. America), a mixture of selected species such as ryegrass and clover or, more traditionally, an indigenous hay meadow that has as many as 20 or more species of grasses and herbaceous plants. The high dry matter content of hay can result in a higher nutrient intake as well as the use of less bedding material.

22.5.1 Traditional hay production

Making hay traditionally i.e., 'field cured hay' is entirely weather-dependent and often a risky method of forage conservation in UK conditions. The target is to reduce the moisture content of mown grass from approximately 80%−85% to 12%−15% for safe barn storage. This can be achieved in 3 to 4 days of hot sunny weather but can frequently mean that the hay is in the field for a week or more when the weather is unsettled. The crop is subjected to respiratory, physical, and leaching losses that result from plant respiration while still wet and the impact of frequent tedding and rowing up as the crop becomes drier and more brittle as well as the action of dews or rain showers. Such losses, from material which may be already of fairly low feed value (e.g., often a D value as low as 60 when cut), frequently result in hay with a feeding value which is little better than straw. Further hazards can arise as a result of baling and storing the hay at too high a moisture content. The heating of bales with the consequent growth of bacteria and moulds, at best, results in a further deterioration of feeding value and, at worst, in the spontaneous combustion of the stack. Handling and feeding mouldy hay can be hazardous to stock as well as causing substantial hazards to humans. The human medical condition known as 'farmers lung' can be caused in this way.

There are a few 'dos and don'ts', however, which may have a beneficial effect on hay quality. As with silage, if the sward is cut before full heading or flowering then the feed value is higher, but this is not the traditional time to harvest a 'hay meadow'. Grass species such as timothy, cocksfoot and meadow fescue are often favoured as they dry relatively quickly. The more sappy ryegrasses, particularly the tetraploid varieties with their higher moisture content, although ideal for making silage and haylage, are more difficult to dry sufficiently well to make good quality field hay. Another important point is not to condition too severely, and certainly not flail, any crop which is to be made into field-cured hay. Leaching losses in the event of rain showers from such crops could be excessive. Additives to restrict microbial activity in hay bales based on propionic or benzoic acid or their salts are available but have not been widely taken up. For most farmers, field-cured hay will remain merely an opportunist method of conservation when the right crop and weather conditions occur.

Increases in the designations of traditional species-rich hay meadows for conservation purposes may well increase the quantity of this type of material conserved in the future. The management agreements for many of these meadows usually stipulate very late (e.g., from July onwards) harvesting to facilitate the setting of seeds by the desired species. By this time the feeding value of the hay can be very low indeed.

Other than for those dedicated hay producers largely supplying hay for sale to the equine sector, hay has become something of an opportunistic crop with many farmers taking full advantage of favourable drying weather to make hay but with the insurance of baled silage or haylage to rely on if the weather changes. A 'staggered' multi-cut silage system with individual fields or groups of fields being harvested routinely, largely as bales, often mean that harvesting will be taking place regularly throughout the summer and that high quality hay can be secured as and when weather conditions are favourable.

22.5.2 Barn-drying or conditioning

The obvious value of some form of artificial drying of hay in store was recognised in the 1950s and 1960s when barn-drying installations involving electric or diesel driven fans were popular to a limited extent. Hay was dried continuously with drying taking up to 21 days. The obvious disadvantages of high fuel costs, and the very high demand for labour when baled hay often had to be double handled, have subsequently almost seen the total demise of these installations. Those very few which remain now tend to produce specialist high value products for the equine market. It is possible, by producing low density bales (about 100 kg/m^3), to bale and carry hay with up to 40% moisture content and to dry to less than 20% moisture in store but these bales can be very difficult to handle subsequently. This can also be achieved with loose or even chopped hay, and systems exist for handling and drying relatively small tonnages of such material. Although most legume hay is imported into the UK, home-produced lucerne or sainfoin hay, destined for the equine market, can also be successfully barn-conditioned i.e., the final stages of drying completed in store in order to avoid shattering and loss of the quicker drying leaves in the field and the stemmy nature of both of these enables air movement through the bales and facilitates drying.

22.6 Green-crop drying

Green-crop drying can no longer be considered as a farm enterprise but as an industrial operation using agricultural crops. Most of the output is milled and cubed as grass and lucerne nuts, in effect as a concentrate feed, but in some cases the material is sold long as dried grass or lucerne or chopped and mixed with molasses and chopped straw (e.g., Dengie equine products such as 'Alfa–A', Hi-Fi).

Crops for drying may be drawn from a single farm source or contracted out to a number of farms and are often grown as a break from arable cropping. Grass and lucerne constitute the main output. Tall fescue has been favoured in some cases because of its resilience in dry conditions and its long growing season as well as its rapid drying rate which reduces the fuel or energy requirement of the crop drying process. Occasionally, other crops such as wholecrop maize or cereals have been dried and sold as blends with the main output of grass or lucerne nuts. In most cases the nuts are sold on to feed compounders, but some farmers and horse keepers feed them directly to stock. One of the valuable constituents of artificially dried grass 'meal' is the range of pigments in it; this is one of the main reasons for its inclusion in poultry rations, for example, to improve the colour of egg yolks.

The system of production is extremely simple and grass is cut and transported to be dried immediately. Efforts to partially wilt crops in the field in order to reduce the fuel consumption associated with drying have highlighted a lack of uniformity of field drying as a potential problem. The driers themselves are usually large rotating drums and the fresh crop is introduced into a stream of very hot air (over 1000°C) fuelled by oil, gas or electricity. The crop spends about 3 minutes in the drier, emerging at a temperature of about 120°C before being milled and cubed at a moisture content of about 10%. Obviously, the logistics of such an operation need to be very efficiently organised, and during the main production season the drying plants are run 24 h a day and 7 days a week.

Crop agronomy and fertiliser inputs will be much the same as for a livestock-based unit, except that the absence of any return of organic manures necessitates greater reliance on both nitrogen and potash fertilisers on grass and mainly potash fertiliser on lucerne. In the latter case total potash (K_2O) application may need to be in excess of 300 kg/ha in order to replace that removed by the crop.

Green-crop drying has never developed in the UK to the same extent as in the rest of Europe where co-operative ventures have been set up or in North America capitalised on the availability of financial aid to establish drying facilities.

22.7 Key points

- Grasses and other forages can be preserved effectively during periods of peak growth in late spring, summer and early autumn to create a conserved product that can be fed strategically or during periods of reduced grass growth.
- The quality of any conserved forage is a combination of its nutritive value i.e., the crop harvested and its maturity, and the preservation quality i.e., the effectiveness of the preservation process and storage.
- The aim for any farm or holding is to achieve the best balance between forage quality and yield which is primarily determined by the timing and number of harvests.
- The selection of a suitable forage conservation technique and strategy for a farm depends on location, the intensity and level of crop and animal production required as well as the machinery and storage facilities available.
- The main preservation technique adopted involves fermentation as silage which involves anaerobic storage and a rapid fall in pH to inhibit spoilage organisms.
- Silage quality is influenced by the choice of crop and harvesting date, processes such as field wilting and chopping, the use of additives as well as the harvesting process and subsequent crop storage.
- The alternative to anaerobic fermentation is to rely on drying the crop as hay but making hay traditionally i.e., 'field cured hay' is entirely weather-dependent and often a risky method of forage conservation in UK conditions.

Sources of further information and advice

AHDB, 2018. Growing and Making Maize Silage for Better Returns. AHDB.
AHDB, 2019. Making Grass Silage for Better Returns. AHDB.
British Grassland Society, Forage Conservation Fact Sheets.
British Grassland Society, Big Bale Silage: The Technology of Making Silage in Big Bales and its Place in the General Forage Conservation Scene. Proceedings: BGS Conference, 1989.
Frame, J., Laidlaw, A.S., 2013. Improved Grassland Management. Crowood Press.
Hopkins, A. (Ed.), 2000. Grass: Its Production and Utilization, third ed. Blackwell Science.

Park, R.S., Stronge, M.D., 2005. Silage Production and Utilisation. Wageningen Academic Publishing.
Raymond, W.F., Waltham, R., 1996. Forage Conservation and Feeding, fifth ed. Farming Press.
Teagasc, 2016. Quality Grass Silage for Dairy and Beef Production Systems—A Best Practice Guide.
The Environment Agency, Fact Sheets and Forms Relating to the SSAFO Regulations.
Wilkinson, J.M., 2005. Silage. Chalcombe Publishing.

Websites

www.ahdb.org,.uk/dairy.
www.ahdb.org.uk/beef-lamb.
www.britishgrassland.com.
www.environment-agency.gov.uk.
www.gov.uk.
ahdb/forage-for-knowledge.

Appendices

Appendix 1

Soil texture assessment in the field

The texture of a soil (i.e., the amount of sand, silt and clay present) can be measured by mechanical analysis of a representative sample of the soil in the laboratory. But it is important for the farmer and adviser to be able to assess the texture of the soil in the field, not only as a guide for cultivations and general management, but also because it affects the recommended application rate of soil-acting pesticides (mainly herbicides) and lime. Many of the soil-acting chemicals are adsorbed by the clay and/or organic matter in the soil and so higher dose rates may be required on soils which are rich in these materials. On sandy soils, surface-applied residual herbicides may be washed into the root zone of the crop too easily and so cause damage.

With practice, it is possible to become reasonably skilled at assessing soil texture by feeling the soil in the following way:

Carefully moisten a handful of stone-free soil until the particles cling together (avoid excess water). Work it well in the hand until the structure breaks down; rub a small amount between the thumb and fingers to assess the texture according to how gritty, silky or sticky the sample feels. The handful of moist soil can also be assessed by the amount of polish it will take, and the ease or difficulty of moulding it into a ball and other shapes.

Sands feel gritty but are not sticky when wet (loose when dry) and do not stain the fingers.

Clays (at the other extreme of particle size) take a high polish when rubbed, are very sticky, bind together very firmly and need some pressure to mould into shapes.

Silty soils have a smooth silky feel and the more obvious this is, the greater is the amount of silt present. The amount of polish the sample takes and its grittiness are guides to the amount of clay and sand present.

Loams have a fairly even mixture of sand, clay and silt and, because these tend to balance each other, loam soils are not obviously gritty, silky or sticky and they take only a slight polish. A ball of moist loam soil is easily formed, and the particles bind together well.

These are the main texture grades, but a wide range of intermediate grades exist, each having different amounts of sand (of various sizes), silt and clay particles. All this can be complicated by the amount of organic matter present which has a soft silky feel and is usually dark brown or black in colour.

Once the particular qualities described are recognised, it should be possible to use the Texture Key to arrive at a single texture (Fig. 1).

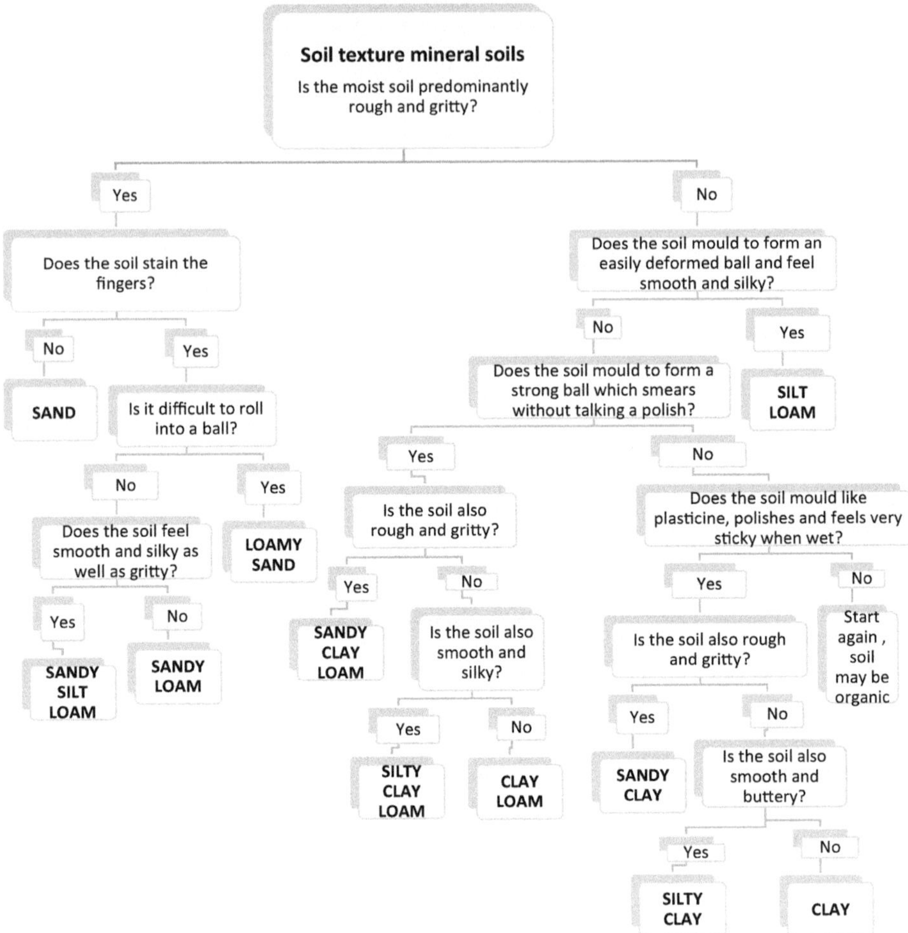

Figure 1 Soil texture assessment of mineral soils.
Adapted from RB209 Agriculture and Horticulture Development Board.

The prefix *organic* can be applied to the above classes if organic matter levels are relatively high, i.e. an organic mineral soil having between 6 and 20% organic matter (depending on clay content), *Peaty soils* between 20 and 35% organic matter and peats >35% organic matter. The prefix *calcareous* (Calc) can be applied if more than 5% calcium carbonate is present.

The above textural groups are used in advisory work for making recommendations on dose rates for soil-acting herbicides, nitrogen application rates for some crops, lime recommended rates and for assessing available-water capacity, suitability for mole drainage, workability, stability of soils and risk of soil erosion. Soil texture is particularly important when assessing ability of a soil to retain nitrogen and hence soil nitrogen supply. In nitrate vulnerable zones, the dates for the closed periods when manufactured nitrogen fertiliser or manures can be applied vary with some soil textures.

Appendix 2

Nomenclature of common plants including crop plants

(A) annual, (B) biennial, (P) perennial

Common names	Botanical names
Barley (A)	*Hordeum vulgare*
– Meadow (P)	*Hordeum secalinum*
– Wall (A)	*Hordeum murinum*
Beans—field and broad (A)	*Vicia faba*
– Green (A)	*Phaseolus vulgaris*
Beet—sugar and fodder (B)	*Beta vulgaris*
Bent—black (P)	*Agrostis gigantea*
– Common (P)	*Artemisia capillaris*
– Creeping (watergrass) (P)	*Artemisia stolonifera*
Bindweed—black (A)	*Fallopia convolvulus*
– Field (P)	*Convolvulus arvensis*
Bistort amphibious (P)	*Persicaria amphibia*
Bitter cress—hairy (A)	*Cardamine hirsuta*
Black-grass (A)	*Alopecurus myosuroides*
Borage (A)	*Borage officinalis*
Bracken (P)	*Pteridium aquilinum*
Bristly oxtongue (A) or (B)	*Picris echioides*
Broccoli (A)	*Brassica oleracea* ssp. *italica*
Brussels sprouts (A)	*B. oleracea* ssp. *gemmifera*
Brome—soft (A) or (B)	*Bromus hordeaceous*
– Rye (A) or (B)	*Bromus secalinus*
– Meadow (A) or (B)	*Bromus commutatus*
– Barren (sterile) (A) or (B)	*Anisantha sterilis*
– Great (A)	*Anisantha diandra*
Broomrape—common (A or P)	*Orobanche minor*
Buckwhcat (A)	*Fagopyrum esculentum*
Burdock—greater (P)	*Arctium lappa*
Burnet—salad (P)	*Sanguisorba minor* ssp. *minor*
– Fodder (P)	*S. minor* ssp. *muricata*
Buttercup—bulbous (P)	*Ranunculus bulbosus*
– Corn (A)	*Ranunculus arvensis*
– Creeping (P)	*Ranunculus repens*
– Meadow (crowfoot) (P)	*Ranunculus acris*
Cabbage (A)	*Brassica oleracea* ssp. *capitata*
– Spring green (A)	*B. oleracea* ssp. *viridis*
Cauliflower (A)	*B. oleracea* ssp. *botrytis*
Campion—red (B)	*Silene dioica*
– Bladder (P)	*Silene vulgaris*
– White (B) or (P)	*Silene latifolia*
Canary-grass—awned (bristle-spiked) (A)	*Phalaris paradoxa*

Continued

Common names	**Botanical names**
Carrot—cultivated and wild (B)	*Daucus carota*
Cat's-ear (P)	*Hypochaeris radicata*
Chamomile—corn (A) or (B)	*Anthemis arvensis*
Charlock (yellow) (A)	*Sinapis arvensis*
Chervil—rough (B)	*Chaerophyllum temulum*
Chickweed—common (A)	*Stellaria media*
– Common mouse-ear (A)	*Cerastium fontanum*
Chicory (P)	*Cichorium intybus*
Cleavers (A)	*Galium aparine*
Clover—red (P)	*Trifolium pratense*
– White (P)	*Trifolium repens*
– Alsike (P)	*Trifolium hybridum*
– Crimson (P)	*Trifolium incarnatum*
Cocksfoot (cock's-foot) (P)	*Dactylis glomerata*
Colt's-foot (P)	*Tussilago farfara*
Corncockle (A)	*Agrostemma githago*
Cornflower (A)	*Centaurea cyanus*
Corn mint (P)	*Mentha arvensis*
Corn salad—common (A)	*Valerianella locusta*
Couch—common (P)	*Elytrigia repens*
– Tall or false (onion) (P)	*Arrhenatherum elatius (var. bulbosum)*
Cowbane (P)	*Cicuta virosa*
Cow-wheat—field (A)	*Melampyrum arvense*
Crane's-bill (A)	*Geranium* spp.
Cress—swine (A)	*Coronopus squamatus*
Crested dog's tail (P)	*Cynosurus cristatus*
Cuckoo flower (P)	*Cardamine pratensis*
Daisy (P)	*Bellis perennis*
Dandelions (P)	*Taraxacum agg.*
Darnel (A)	*Lolium temulentum*
Dead nettle—red (A)	*Lamium purpureum*
– Henbit (A)	*Lamium amplexicaule*
Docks—broadleaved (P)	*Rumex obtusifolius*
– Curled (P)	*Rumex crispus*
Evening primrose (B)	*Oenothera biennis*
Fat hen (A)	*Chenopodium album*
Fescue—red (P)	*Festuca rubra agg.*
– Sheep's (P)	*Festuca ovina*
– Meadow (P)	*Festuca pratensis*
– Tall (P)	*Festuca arundinacea*
– Rat's-tail (A)	*Vulpia myuros*
– Squirreltail (A)	*Vulpia bromoides*
Fiddleneck common (A)	*Amsinckia micrantha*
Fleabane—Canadian (A)	*Conyza canadensis*
Flixweed (A)	*Descurainia sophia*

Common names	Botanical names
Fool's parsley (A)	*Aethusa cynapium*
Forget-me-not (field) (A)	*Myosotis arvensis*
Foxglove (B) or (P)	*Digitalis purpurea*
Foxtail—meadow (P)	*Alopecurus pratensis*
Fritillary (P)	*Fritillaria meleagris*
Fumitory—common (A)	*Fumaria officinalis*
Gallant-soldier (A)	*Galinsoga parviflora*
Garlic—field (P)	*Allium oleraceum*
Gorse (whin, furze) (P)	*Ulex europaeus*
Goosefoot—many seeded (A)	*Chenopodium polyspermum*
Gromwell—field (A)	*Lithospermum arvense*
Ground-elder (P)	*Aegopodium podagraria*
Ground-ivy (P)	*Glechoma hederacea*
Groundsel (A)	*Senecio vulgaris*
Hair-grass—tufted (P)	*Deschampsia cespitosa*
– Wavy (P)	*Deschampsia flexuosa*
Hawkbit—autumn (P)	*Leontodon autumnalis*
– Rough (P)	*Leontodon hispidus*
Hawk's-beard—rough (B)	*Crepis biennis*
– Smooth (A)	*Crepis capillaris*
Heather (P)	*Calluna vulgaris*
– Bell (P)	*Erica cinerea*
Hedge mustard (A)	*Sisymbrium officinale*
Hedge-parsley—spreading (A)	*Torilis arvensis*
Hemlock (B)	*Conium maculatum*
Hemp-nettle—common (A)	*Galeopsis tetrahit*
Henbane (A)	*Hyoscyamus niger*
Hogweed (P)	*Heracleum sphondylium*
– Giant (P)	*Heracleum mantegazzianum*
Horsetail—field (P)	*Equisetum arvense*
– Marsh (P)	*Equisetum palustre*
Kale (A)	*B. oleracea* ssp. *viridis*
Knapweed—common (P)	*Centaurea nigra*
Knawel—annual (A) or (B)	*Scleranthus annuus*
Knotgrass (A)	*Polygonum aviculare*
Knotweed—Japanese (P)	*Fallopia japonica*
Leek (B)	*Allium ampeloprasum*
Lettuce (A)	*Lactuca sativa*
Linseed and flax (A)	*Linum usitatissimum*
Loose silky-bent (A)	*Apera spica-venti*
Lucerne (P)	*Medicago sativa* ssp. *sativa*
Lupins (A)	*Lupinus angustifolius*
Maize (A)	*Zea mays*
Marigold—corn (A)	*Chrysanthemum segetum*
Mat grass (P)	*Nardus stricta*

Continued

Common names	**Botanical names**
Mayweed—scented (A)	*Matricaria recutita*
– Scentless (A)	*Tripleurospermum inodorum*
– Stinking chamomile (A)	*Anthemis cotula*
Meadow grass—annual (A)	*Poa annua*
– Rough (P)	*Poa trivialis*
– Smooth (P)	*Poa pratensis*
Medick—black (A) or (P) (yellow clover)	*Medicago lupulina*
Mercury—annual (A)	*Mercurialis annua*
– Dog's (P)	*Mercurialis perennis*
Mignonette—wild (B) or (P)	*Reseda lutea*
Mugwort (P)	*Artemisia vulgaris*
Mustard—black (A)	*Brassica nigra*
– White (A)	*Sinapis alba* subsp. *dissecta*
– Treacle (A)	*Erysimum cheiranthoides*
Nettle—common (P)	*Urtica dioica*
– Small, annual (A)	*Urtica urens*
Nightshade—black (A) or (B)	*Solanum nigrum*
– Deadly (P)	*Atropa belladonna*
Nipplewort (A) or (P)	*Lapsana communis*
Oat (A)	*Avena sativa*
– Bristle (A)	*Avena strigosa*
– Spring, wild (A)	*Avena fatua*
– Winter, wild (A)	*Avena sterilis* subsp. *ludoviciana*
Oat-grass—downy (P)	*Helictotrichon pubescens*
– False or tall (onion couch) (P)	*Arrhenatherum elatius (var. bulbosum)*
Oilseed rape (A)	*Brassica napus* ssp. *napus*
Onion—wild (P)	*Allium vineale*
– Bulb (B)	*Allium cepa*
Orache—common (A)	*Atriplex patula*
Pansy—field (A)	*Viola arvensis*
– Wild (A) or (P)	*Viola tricolor*
Parsley—cow (P)	*Anthriscus sylvestris*
– Fool's (A)	*Aethusa cynapium*
- Sheep's (P)	*Petroselinium crispum*
Parsley-piert (A)	*Aphanes arvensis*
Pearlwort (P)—procumbent	*Sagina procumbens*
Peas (A)	*Pisum sativum*
Penny-cress—field (A)	*Thlaspi arvense*
Persicaria—pale (A)	*Persicaria lapathifolia*
Pineappleweed (rayless mayweed) (A)	*Matricaria discoidea*
Plantain—greater (P)	*Plantago major*
– Ribwort (narrow-leaved) (P)	*Plantago lanceolata*
Poppy—corn or common (A)	*Papaver rhoeas*
Potatoes (A)	*Solanum tuberosum*
Primrose (P)	*Primula vulgaris*
Purple moor grass (P)	*Molinia caerula*
Quinoa (A)	*Chenopodium quinoa*

Common names	Botanical names
Radish—wild (A) or (B)	*Raphanus raphanistrum* subsp. *raphanistrum*
Ragwort—common (B) or (P)	*Senecio jacobaea*
– Oxford (P)	*Senecio squalidus*
Ramsons (P)	*Allium ursinum*
Red bartsia (A)	*Odontites vernus*
Redshank (A)	*Persicaria maculosa*
Reed—common (P)	*Phragmites australis*
Restharrow—common (P)	*Ononis repens*
– Spiny (P)	*Ononis spinosa*
Rush—jointed (P)	*Juncus articulatus*
– Soft (common) (P)	*Juncus effusus*
– Hard (P)	*Juncus inflexus*
– Heath (P)	*Juncus squarrosus*
Rye (A)	*Secale cereale*
Ryegrass—perennial (P)	*Lolium perenne*
– Italian (P)	*Lolium multiflorum*
– hybrid (P)	*Lolium multiflorum x perenne*
Sainfoin (P)	*Onobrychis viciifolia*
Saffron—meadow (P)	*Colchicum autumnale*
St John's wort (P)—perforate	*Hypericum perforatum*
Scabious—field (P)	*Knautia arvensis*
Scarlet pimpernel (A)	*Anagallis arvensis*
Sedges (P)	*Carex* spp.
Selfheal (P)	*Prunella vulgaris*
Shepherd's-needle (A)	*Scandix pecten-veneris*
Shepherd's-purse (A)	*Capsella bursa-pastoris*
Silverweed (P)	*Potentilla anserina*
Soft-grass—creeping (P)	*Holcus mollis*
Sorrel—common (P)	*Rumex acetosa*
– Sheep's (P)	*Rumex acetosella*
Sow-thistle—corn or perennial (P)	*Sonchus arvensis*
– Smooth, milk (A)	*Sonchus oleraceus*
Soya (A)	*Glycine mas*
Speedwell—common, field (A)	*Veronica persica*
– Germander (P)	*Veronica chamaedrys*
– Green field (A)	*Veronica agrestis*
– Ivy-leaved (A)	*Veronica hederifolia*
– Wall	*Veronica arvensis*
Spinach (B)	*Spinacia oleracea*
Spurge—sun (A)	*Euphorbia helioscopia*
– Dwarf (A)	*Euphorbia exigua*
Spurrey—corn (A)	*Spergula arvensis*
Strawberry (P)	*Fragaria x ananassa*
Stork's-bill—common (A) or (B)	*Erodium cicutarium*
Sunflower (A)	*Helianthus annuus*
Sweet vernal-grass (P)	*Anthoxanthum odoratum*

Continued

Common names	Botanical names
Swedes (B)	*Brassica napus* ssp.*napobrassica*
Thistle—creeping (field) (P)	*Cirsium arvense*
– Spear (Scotch) (B) or (P)	*Cirsium vulgare*
– Marsh (P)	*Cirsium palustre*
Timothy (P)	*Phleum pratense*
Toadflax—common (P)	*Linaria vulgaris*
Trefoil—hop (A)	*Trifolium campestre*
– Lesser (A)	*Trifolium dubium*
– Common bird's-foot (P)	*Lotus corniculatus*
Triticale (A)	*X Triticosecale*
Turnips (B)	*Brassica rapa* ssp.*rapa*
– Stubble (A)	*B. rapa* ssp. *oleifera*
Venus's looking glass (A)	*Legousia hybrida*
Vetch—common (tares) (A) or (B)	*Vicia sativa*
– Hairy (P)	*Vicia villosa*
– Yellow (P)	*Vicia lutea*
– Kidney (P)	*Anthyllis vulneraria*
Viper's bugloss (B)	*Echium vulgare*
Water-dropwort (hemlock) (P)	*Oenanthe crocata*
Wheat (A)	*Triticum aestivum*
– Durum (A)	*Triticum durum*
– Spelt (A)	*Triticum spelta*
Willowherb (rosebay) (P)	*Chamerion angustifolium*
Wood-rush—field (P)	*Luzula campestris*
Yarrow (P)	*Achillea millefolium*
Yellow iris or flag (P)	*Iris pseudacorus*
Yellow-rattle (A)	*Rhinanthus minor*
Yorkshire-fog (P)	*Holcus lanatus*

For plant distribution in the UK, see www.brc.ac.uk/plantatlas/.

Appendix 3

Nomenclature of crop diseases

Crop names	Common names	Botanical names
Cereals		
All cereals	Powdery mildew*	*Blumeria graminis*
Wheat, barley, triticale	Yellow (stripe) rust	*Puccinia striiformis*
Barley	Brown leaf rust	*Puccinia hordei*
Wheat	Brown leaf rust	*Puccinia triticina*

Crop names	Common names	Botanical names
Rye, triticale	Brown leaf rust	*Puccinia recondita*
Oats	Crown rust	*Puccinia coronata*
Wheat	Stem rust	*Puccinia graminis*
Barley, rye, triticale	Rhynchosporium (leaf scald)	*Rhynchosporium commune*
Wheat, barley, rye	'Septoria nodorum' leaf + glume blotch	*Parastagonospora nodorum*
Wheat	'Septoria tritici' leaf blotch	*Zymoseptoria tritici*
Oats	Leaf spot, seedling blight	*Pyrenophora avenae, 'Drechslera avenae'*
Barley	Net blotch*	*Pyrenophora teres, 'Drechslera teres'*
Wheat, barley, rye	Tan spot	*Pyrenophora tritici-repentis, 'Drechslera tritici-repentis'*
All cereals	Black point, sooty mould	*Alternaria* spp. and *Cladosporium* spp.
Barley	Halo spot	*Selenophoma donacis*
Barley	Ramularia leaf spot	*Ramularia collo-cygni*
All cereals	Ascochyta leaf scorch (spot)	*Didymella exitialis*
All cereals	Pink snow mould	*Monographella nivalis, 'Microdochium nivale'*
All cereals	Ergot	*Claviceps purpurea*
Wheat,	Bunt, stinking smut	*Tilletia tritici*
Wheat, barley, rye, triticale	Dwarf bunt	*Tilletia controversa*
Barley	Bunt, covered, smut	*Ustilago hordei*
Oats	Bunt, covered, smut	*Ustilago segetum*
Oats	Loose smut	*Ustilago avenae*
Wheat	Loose smut	*Ustilago tritici*
Barley	Loose smut	*Ustilago nuda*
Rye	Stripe smut	*Urocystis occulta*
Wheat	Flag smut	*Urocystis agropyri*
Barley	Leaf stripe	*Pyrenophora graminea 'Drechslera graminea'*
All cereals	Cephalosporium leaf stripe	*Hymenula cerealis*
All cereals	Eyespot−R type	*Oculimacula acuformis, 'Helgardia acuformis'*
All cereals	Eyespot−W type	*Oculimacula yallundae, 'Helgardia herpotrichoides'*
All cereals	Sharp eyespot	*Ceratobasidium cereale, 'Rhizoctonia cerealis'*
All cereals	*Rhizoctonia* stunt	*Rhizoctonia solani*
Wheat, barley, rye	Take-all	*Gaeumannomyces tritici*
Oats, wheat, barley, rye	Take-all	*Gaeumannomyces avenae*

Continued

Crop names	Common names	Botanical names
All cereals	Brown foot rots and ear blight	*Fusarium* spp. and *Microdochium nivale*
All cereals	Foot rot	*Cochliobolus sativus*, '*Drechslera sorokiniana*'
All cereals	Black (sooty) mould	*Cladosporium* spp. and *Alternaria* spp.
Winter wheat, barley	Snow rot	*Typhula incarnata*
Wheat, barley	Omphalina patch	*Omphalina pyxidata*
Maize	Damping-off	*Pythium* spp.
Maize	Bunt, covered, smut	*Ustilago maydis*
Maize	Stalk rot	*Fusarium* spp.
Maize	Eyespot	*Kabatiella zeae*
Maize	Northern corn leaf blight	*Setosphaeria turcicum*
Potatoes		
	Foliage blight and tuber blight	*Phytophthora infestans*
	Pink rot	*Phytophthora erythroseptica*
	Early blight	*Alternaria alternata* or *Alternaria solani*
	Common scab	*Streptomyces scabiei*
	Powdery scab	*Spongospora subterranea*
	Gangrene	*Phoma exigua* var. *foveata*
	Watery wound rot	*Pythium* spp.
	Wart disease	*Synchytrium endobioticum*
	Sclerotinia white mould	*Sclerotinia sclerotiorum*
	Blackleg and soft rot	*Pectobacterium* spp. and *Dickeya* spp.
	Skin spot	*Polyscylalum pustulans*
	Black scurf and stem canker	*Thanatephorus cucumeris* '*Rhizoctonia solani*'
	Dry rot	*Fusarium* spp.
	Brown rot	*Ralstonia solanacearum*
	Black dot	*Colletotrichum coccodes*
	Silver scurf	*Helminthosporium solani*
	Ring rot	*Clavibacter michiganensis* subsp. *sepedonicus*
	Verticillium wilt	*Verticillium albo-atrum*
Sugar beet		
	Blackleg	*Pleospora betae*
	Downy mildew	*Peronospora farinosa*
	Powdery mildew	*Erysiphe betae*
	Ramularia leaf spot	*Ramularia beticola*
	Beet rust	*Uromyces betae*

Crop names	Common names	Botanical names
	Violet root rot	*Helicobasidium brebissonii* 'Rhizoctonia crocorum'
	Vector of rhizomania	*Polymyxa betae*
	Cercospora leaf spot	*Cercospora beticola*
	Yellow leaf spot	*Stemphylium* sp.
Legumes		
Peas, beans	Downy mildew	*Peronospora viciae*
Peas	Leaf and pod spot Ascochyta blight	*Mycosphaerella pinodes*, 'Ascochyta pisi'
Beans	Leaf and pod spot	*Didymella fabae*, 'Ascochyta fabae'
Beans	Bean rust	*Uromyces viciae-fabae*
Beans, peas	Chocolate spot and grey mould	*Botryotinia fuckeliana* '*Botrytis* spp.'
Peas, beans	White mould, stem rot	*Sclerotinia* spp.
Red clover, lucerne	Clover rot	*Sclerotinia trifoliorum*
All legumes	Damping–off of seedlings	*Pythium* spp.
Peas	Pea wilt	*Fusarium oxysporum f.* sp. *pisi*
Lucerne	Verticillium wilt	*Verticillium albo-atrum*
Peas	Powdery mildew	*Erysiphe pisi*
Peas	Bacterial blight	*Pseudomonas syringae pv. pisi*
Peas, beans	Foot rot	*Fusarium solani f.* sp. *pisi*
Brassicae		
	Club root (finger and toe)	*Plasmodiophora brassicae*
	Powdery mildew	*Erysiphe cruciferarum*
	Downy mildew	*Peronospora parasitica*
	White blister	*Albugo candida*
	Light leaf spot	*Pyrenopeziza brassicae*
	Dark leaf and pod spot	*Alternaria* spp.
	Stem canker and phoma leaf spot	*Leptosphaeria maculans* and *Leptosphaeri biglobosa*, 'Phoma lingam'
	Ring spot	*Mycosphaerella brassicicola*, 'Astromelia brassicae'
	Stem rot	*Sclerotinia sclerotiorum*
	White leaf spot	*Mycosphaerella capsellae*
	Verticillium wilt	*Verticillium* spp.
	Grey mould	*Botryotinia fuckeliana*, 'Botrytis cinerea'
	Black rot	*Xanthomonas campestris pv. campestris*

Continued

Crop names	Common names	Botanical names
Linseed		
	Alternaria blight	*Alternaria linicola*
	Pasmo	*Mycosphaerella linicola*, '*Septoria linicola*'
	Powdery mildew	*Podosphaera lini*
	Grey mould	*Botryotinia fuckeliana*
Onions		
	White rot	*Sclerotium cepivorum*
	Downy mildew	*Peronospora destructor*
	Neck rot	*Botrytis allii*
Grasses		
	Drechslera leaf spot	*Pyrenophora* spp., '*Drechslera* spp.'
	Ergot	*Claviceps purpurea*
	Powdery mildew	*Blumeria graminis*
	Rhynchosporium leaf scald	*Rhynchosporium* spp.
	Crown rust	*Puccinia coronata*
	Brown rust	*Puccinia loliina*
	Stem rust	*Puccinia graminis*

*Different races or forms found on different cereals (*forma specialis*)

Note: The sexual (teleomorph) stages for some diseases have only just been identified. These are given first. The 'anamorph', imperfect or asexual classification, is still used and is also given. Where an old botanical name is commonly used such as Septoria, it is usually not written in italics.

Appendix 4

Land and soil classification UK

Land classification maps and reports

Agricultural lands in England and Wales have been classified into five grades by the Department for Environment Food and Rural Affairs (Defra Agricultural land classification—ALC). In Scotland, the Land Capability for Agriculture Classification was developed by the Macauley Institute, Aberdeen. Here there are seven main grades or classes.

The information available in these classification schemes is of importance to farmers, advisers, conservationists and even to planners.

The classifications are based mainly on the following:

Climate—rainfall, accumulated temperature.
Relief—slope, surface irregularities, flood risk.
Soil—wetness, depth, texture, structure, stoniness and moisture balance.

These characteristics can affect the range of crops which can be grown, the potential crop yield and the consistency of and the cost of obtaining the yield.

Agricultural land classification in England and Wales

Agricultural land classification in England and Wales is graded into five grades; under this classification, the third grade is divided into two sub-grades.

Grade 1—Excellent. Land with no, or very minor, limitations to agricultural use. A very wide range of agricultural and horticultural crops can be grown. Yields are high and consistent. This grade occupies 3% of agricultural land in England and Wales and is coloured dark blue on the maps.

Grade 2—Very good. Land with minor limitations which affect crop yield, cultivations or harvesting. A wide range of crops can be grown, but there may be difficulties with very demanding crops. The level of yield is generally high, but it can be lower or more variable than Grade 1. This grade occupies 14% of agricultural land and is coloured light blue.

Grade 3—Good to moderate. Land with moderate limitations which affect choice of crop, timing and type of cultivations, harvesting or level of yield. Yields are generally lower or more variable than on land in Grades 1 and 2. Grade 3 is the largest group of soils, occupying approximately 50% of agricultural land and is coloured green.

Sub-grade 3a—Moderate/high yields of a narrow range of crops (especially cereals) or moderate yields of a wide range of crops. Some good grassland.

Sub-grade 3b—Moderate yields of a narrow range of crops and lower yields of a wide range of crops.

Grade 4—Poor. Land with severe limitations restricting range and/or yield levels. Suited to grass plus some low and variable yields of cereals. It also includes droughty arable land. It occupies approximately 20% of agricultural land and is coloured yellow.

Grade 5—Very poor. Land with very severe limitations. Permanent pasture or rough grazing. It occupies 14% of agricultural land and is coloured light brown.

Urban areas are coloured red and non-agricultural areas are coloured orange on the maps.

Land capability for agriculture—Scotland

Land capability classification for Scotland is divided into 7 classes with 13 divisions. Classes 1–4 correspond with the grades 1–4 in England and Wales.

Class 5—Land capable of supporting improved grassland.

Class 6—Land capable only of use as rough grazing.

Class 7—Land of very limited agricultural value.

Lands capable of arable farming (classes 1–3.1) make up only 8% of the Scottish land area compared with 20% for land capable of mixed farming (classes 3.2–4.2), 18% of land capable of supporting improved grassland (classes 5.1–5.3) and 51% of land only capable of supporting rough grazing (classes 6.1–7).

Index

'*Note:* Page numbers followed by "f" indicate figures and "t" indicate tables.'

H

I

Kolibri 360 GmbH
Ferdinand-Jühlke-Straße 7
99095 Erfurt, Deutschland
produktsicherheit@kolibri360.de